IET HEALTHCARE TECHNOLOGIES SERIES 11

Wearable Technologies and Wireless Body Sensor Networks for Healthcare

The IET International Book Series on Sensors

IET Book Series on e-Health Technologies – Call for Authors

Book Series Editor: Professor Joel P.C. Rodrigues, the National Institute of Telecommunications (Inatel), Brazil, and Instituto de Telecomunicações, Portugal.

While the demographic shifts in populations display significant socio-economic challenges, they trigger opportunities for innovations in e-Health, m-Health, precision and personalized medicine, robotics, sensing, the Internet of Things, cloud computing, Big Data, Software Defined Networks and network function virtualization. Their integration is, however, associated with many technological, ethical, legal, social and security issues. This new Book Series aims to disseminate recent advances for e-Health Technologies to improve healthcare and people's wellbeing.

Topics considered include Intelligent e-Health systems, electronic health records, ICT-enabled personal health systems, mobile and cloud computing for eHealth, health monitoring, precision and personalized health, robotics for e-Health, security and privacy in e-Health, ambient assisted living, telemedicine, Big Data and IoT for e-Health, and more.

Proposals for coherently integrated international multi-authored, edited or co-authored handbooks and research monographs will be considered for this Book Series. Each proposal will be reviewed by the Book Series Editor with additional external reviews from independent reviewers. Please email your book proposal for the IET Book Series on e-Health Technologies to: Professor Joel Rodrigues at joeljr@ieee.org or joeljr@inatel.br, or alternatively contact author_support@theiet.org.

Wearable Technologies and Wireless Body Sensor Networks for Healthcare

Edited by
Fernando José Velez and Fardin Derogarian Miyandoab

The Institution of Engineering and Technology

Published by The Institution of Engineering and Technology, London, United Kingdom

The Institution of Engineering and Technology is registered as a Charity in England & Wales (no. 211014) and Scotland (no. SC038698).

The Institution of Engineering and Technology
Michael Faraday House
Six Hills Way, Stevenage
Herts SG1 2AY, United Kingdom

www.theiet.org

British Library Cataloguing in Publication Data
A catalogue record for this product is available from the British Library

ISBN 978-1-78561-217-6 (hardback)
ISBN 978-1-78561-218-3 (PDF)

Typeset in India by MPS Limited
Printed in the UK by CPI Group (UK) Ltd, Croydon

Contents

8 Two innovative energy efficient IEEE 802.15.4 MAC sub-layer protocols with packet concatenation: employing RTS/CTS and multichannel scheduled channel polling **241**
Fernando J. Velez, Luís M. Borges, Norberto Barroca, and Periklis Chatzimisios

11 Integration of sensing devices and the cloud for innovative e-Health applications 361

Albena Mihovska, Aristodemos Pnevmatikakis, Sofoklis Kyriazakos, Krasimir Tonchev, Razvan Craciunescu, Vladimir Poulkov, Harm op den Akker, and Hermie Hermens

12 VitalResponder®: wearable wireless platform for vitals and body-area environment monitoring of first response teams 387

João Paulo Silva Cunha, Susana Rodrigues, Duarte Dias, Pedro Brandão, Ana Aguiar, Ilídio Oliveira, José Maria Fernandes, Catarina Maia, Ana Rita Tedim, Ana Barros, Orangel Azuaje, Eduardo Soares, and Fernando de La Torre

Foreword

Since 1975, the Polytechnic Institute of Covilhã, and then, in 1979, the University Institute of Beira Interior taught bachelor (3 years) and *licenciatura* (5 years) degrees in Textile Engineering, in its different areas. The University of Beira Interior (UBI) was created on 30 April 30, 1986 and an even richer blend of courses and subjects in the areas of production and textile manufacturing started to be offered in the Textile Engineering five-year degree.

In the late 1980s or the beginning of the 1990s, the Center for Textile and Paper Materials was among the first research units that were created in fledging UBI. In 1992/93, the Ph.D. degree in Textile Engineering was created, one among several doctoral programmes (e.g., Electrical Engineering) that were accredited. In 2000, following the trend form industry on the creation of strong competences in the product domain, the *licenciatura* (5 year) degree in Fashion Design was created. Also in 1999/2000, the 5-year degree in Electrical Engineering (later Electrical and Computer Engineering) was created in UBI.

In 29 May, 2002, a group of young professors who had recently concluded their doctoral studies, created the Covilhã site of Instituto de Telecomunicações (IT), as an external laboratory (that in 2008 was promoted to delegation from IT-Lisbon while UBI joined the IT partnership of Universities and Telecommunication Industries by becoming an institutional associated).

In the first decade of the twenty-first century, research on smart and functional textiles, embedded electronic sensors in smart fabrics and clothing and wireless body sensor networks (WBSNs), in the broad field of wearable technologies, attracted the attention of researchers from UBI and its research centres (FibEnTech, CICS and IT). This group of enthusiastic researchers, together with colleagues from the Department of Physics, Health Sciences Faculty and Hospital Pero da Covilhã, embraced basic and applied collaborative research with industry, also aiming at innovation, e.g., in the framework of Smart-clothing, SIVIC, PROENERGY-WSN and CREaTION research projects while creating an international network of collaborations.

The book on 'Wearable Technologies and Wireless Body Sensor Networks for Healthcare' to be included in the Healthcare Technologies series from IET puts together the research and development efforts of the editors and co-authors of more than one decade of research on wearable technologies, IoT and WBSNs for healthcare within their own groups. Recent research on wearable technologies and WBSNs for healthcare comes here to light.

The book topics span from scenarios and WSBN communication-applications to sensor devices and systems, activity recognition, smart textiles and their applications to smart sensing, radio frequency (RF) propagation aspects, modelling of this very complex communication environments, measurements and cognitive radio, link layer, medium access and control sub-layer protocols and synchronisation aspects. It also covers the rich blend of medical applications of WBSNs as well as the underlying wearable solutions and the need for standardization.

The book, on the one hand, illustrates conceptual aspects and applications, and it provides a new vision in characterising wearable technologies and the need for interoperability, on the other hand. Energy harvesting within wearable solutions is a key issue since it allows for improving energy efficiency and reliability in wearable antenna and sensor devices, algorithms, protocols and networks.

It will certainly motivate new generations of students, researchers and practitioners to start their own path in this very promising field of research while seeking innovation.

Covilhã, February 2019
Manuel José dos Santos Silva
Full Professor from UBI

Preface

About the book: Instrumenting the human body and building wireless sensor networks constitutes a challenge beyond instrumenting the personal area and its networks, a goal that was achieved in the early 2000s. In 2018 and coming years, not only investigating low-power solutions and technologies that are harmless to human being are challenging for the research community but also to address ways to give intelligence to the way sensing and processing are performed, as well as applying cognition to radio and network communications together with energy efficiency and harvesting. The quick growth of smart textiles, high-performance low-power multi-hop networks, efficient processing techniques for smart antennas and ultra wideband represent a stimulus to provide new applications for on-, in- and body-to-body wearable communications for the human body well-being.

Hence, continuous advances in wearable technologies, sensors and smart Wireless Body Area Networks for different applications, e.g., sport activity and health monitoring, radio communication aspects, including radio channel characterisation, and especially energy efficient solutions with energy harvesting and storage have been sufficient reasons that have encouraged the necessity to collect a variety of experience in this field. Thanks to that, a new book has been prepared whose title is "Wearable Technologies and Wireless Body Sensor Networks for Healthcare" to be included in the Healthcare Technologies series from IET. The book's topics span from scenarios and WBSN communication-applications to sensor devices and systems, activity recognition, smart textiles and their applications to smart sensing, RF propagation aspects, modelling of this very complex communication environments, measurements and cognitive radio, link layer, Medium Access and Control (MAC) sub-layer protocols and synchronisation aspects, the rich blend of medical applications of WBSNs as well as the underlying wearable solutions, and the need for standardization.

The book, on the one hand, illustrates conceptual aspects and applications, and it provides a new vision in characterising wearable technologies and the need for interoperability, on the other hand. Energy harvesting within wearable solutions is a key issue since it allows for improving energy efficiency and reliability in wearable antenna and sensor devices, algorithms, protocols and networks.

Contents: Apart from "Introduction" (Chapter 1) and "Conclusion" (Chapter 15), the book is organized into six main parts:

1. *WBSN Communication-applications and Scenarios (Chapters 2 and 3)*
2. *Devices and Systems (Chapter 4)*
3. *Textile Materials for Wearable Applications (Chapter 5)*
4. *Propagation Aspects and CR (Chapters 6 and 7)*

5. *Link Layer, MAC Sub-layer and Synchronization Aspects (Chapters 8 and 9)*
6. *Applications of Wearable Technologies and WBSNs (Chapters 10–14)*

In more detail, the book is organized as follows: Chapter 2 presents a comprehensive overview from deployments scenarios and the underlying sets of applications as well as the characterization of wearable technologies and solutions that support them. A wearable BAN system to use for both medical and non-medical applications, including large number of sensors (<250) and high data rate (<35 MHz) communication demands, is presented in Chapter 3. A development platform that extends the principles of physical computing to the physiological domain is addressed in Chapter 4. Chapter 5 covers the characterization of textile materials and is mainly focused on the analysis of the dielectric properties of normal fabrics. Chapter 6 discusses the development of the human movement identification using the radio signal strength in WBAN and includes aspects of data collection, data pre-processing, and classifier training. A viable architecture of an MBAN with practical CR features based on ultra-wideband radio technology is proposed in Chapter 7. Chapter 8 presents an IEEE 802.15.4 MAC layer performance enhancement by employing RTS/CTS combined with packet concatenation. Performance evaluation of these protocols includes the analysis of minimum delay, maximum throughput and the energy trade-off. Chapter 9 presents a one-way method for synchronization at the MAC sub-layer of nodes and a circuit based on that in a wearable sensor network. The proposed approach minimizes the time skew with an accuracy of half of clock cycle on average.

A wearable data capture system for gait analysis is presented in Chapter 10. It consists of pantyhose with embedded conductive yarns interconnecting sensor nodes forming a mesh like network. The sensor nodes are customized sensing electronic modules that capture inertial and electromyographic signals and sends aggregated information to an external processing device through a Bluetooth link. Chapter 11 describes the eWALL e-Health environment and the required technological innovations to design an affordable, easy-to-implement smart, "caring home" cognitive environment that "senses" intuitively the wishes and "learns" the needs of the person that lives in this house, providing unobtrusive daily support, notifying informal and formal caregivers when necessary and serving as a bridge to supportive services offered by the outside world. Chapter 12 presents a new wearable technology, including ECG device, capable to acquire medical quality ECG using only two electrodes with a small form factor, also acquiring body temperature and actigraphy. A wearable environmental sensing system capable to monitor temperature, humidity, pressure, luminosity, altitude, toxic gases, and, if needed, GPS has also been developed. In Chapter 13, an easy-to-wear belt with a telemedicine system for continuous monitoring of the foetal health is proposed. Chapter 14 presents the state-of-the-art in RF EH for wearable biomedical sensors specifically targeting the global system of mobile 900/1800 cellular and 700 MHz digital terrestrial television networks as ambient RF energy sources. Chapter 15 not only gives a final overview of the main topics covered in the book but also presents a taxonomy for the classification of wearable devices, gives insights on the primary areas of innovation in wearable healthcare, and identifies the need for standardization, the path towards the interoperability with other wireless systems.

Acknowledgments

The editors would like to acknowledge Valerie Moliere, Senior Commissioning Book Editor at the IET, responsible for the field of healthcare technologies, who rescued the project of writing the book in the critical phase of its completion. Special thanks go to Anderson Ramos for his hard effort in helping to revise the final manuscript.

This work is funded by FCT/MEC through national funds and when applicable co-funded by FEDER – PT2020 partnership agreement under the project UID/EEA/ 50008/2019. The work from Chapters 1, 2, 7, 8 and 15 has been partially supported and funded by the Ph.D. FCT grant FRH/BD/66803/2009, by CREaTION, UID/EEA/ 50008/2013, COST IC0905, COST CA15104, SFRH/BSAB/113798/2015, 3221/BMOB/16 CMU Portugal Faculty Exchange Programme grant, CONQUEST (CMU/ECE/030/2017) and ORCIP.

Authors from Chapter 2 would like to acknowledge the IEEE and John Wiley and Sons for permissions (License Numbers 4555931121700, 25 March 2019, and 4555611077624, 24 March 2019, respectively) to reuse text extracts from the papers 'Design and evaluation of multi-band RF energy harvesting circuits and antennas for WSNs' (authors: Luís M. Borges, Norberto Barroca, Henrique M. Saraiva, Jorge Tavares, Paulo Gouveia, Fernando J. Velez, Caroline Loss, Luisa Salvado, Pedro Pinho, Ricardo Gonçalves, Nuno Borges Carvalho, Raúl Chávez-Santiago and Ilangko Balasingham) and 'A small fully digital open-loop clock and data recovery circuit for wired BANs' (authors: Fardin Derogarian, João Canas Ferreira and Vítor Grade Tavares), respectively.

Authors from Chapter 2 would also like to acknowledge the IEEE (two papers) and Elsevier for permissions (License Numbers 4599480561755, 31 May 2019; 4599500401079, 31 May 2019; and 4599491188482, 31 May, 2019) to reuse tables and figures from the papers 'Wireless body area networks: A survey' (authors: Samaneh Movassaghi, Mehran Abolhasan, Justin Lipman, David Smith and Abbas Jamalipour), 'Cognitive radio for medical body area networks using ultra wideband' (authors: Raul Chavez-Santiago, Keith E. Nolan, Oliver Holland, Luca De Nardis, João Ferro, Norberto Barroca, Luís M. Borges, Fernando J. Velez, Vânia Gonçalves and Ilangko Balasingham) and 'Application specific study, analysis and classification of body area wireless sensor network applications' (authors: Adnan Nadeem, Muhammad Azhar Hussain, Obaidullah Owais, Abdul Salam, Sarwat Iqbal and Kamran Ahsan), respectively.

The work from Chapter 5 is funded by FCT/MEC through national funds and co-funded by FEDER-PT2020, when needed, in partnership agreement under the projects UID/EEA/50008/2013, UID/EEA/50008/2019 and UID/Multi/00195/2013.

The authors wish to thank the European COST Action IC1301WiPE for the support given to this research, CAPES Foundation for the PhD grant, process no. 9371-13/3 and LMA Leandro Manuel Araújo Lda for supplying the 3D textile samples.

Authors from Chapter 6 would like to acknowledge the IEICE for permission (Permission Number 16RA0032, 13 July 2016), to reuse a figure from the paper 'Human motion classification using radio signal strength in WBAN' (authors: Sukhumarn Archasantisuk, Takahiro Aoyagi, Tero Uusitupa, Minseok Kim and Jun-ichi Takada).

Authors from Chapter 7 would like to acknowledge the Springer Nature for the permission (License Number 4444311033149, 8 October 2018) to reuse text extracts from the book chapter 'Case Studies for Advancing CR Deployment,' Chapter 6 in the book *Cognitive Radio Policy and Regulation: Techno-Economic Studies to Facilitate Dynamic Spectrum Access*, edited by Arturas Medeisis and Oliver Holland, Springer, London, UK, 2014, pp. 309–348 (authors: Dariusz Więcek and Fernando José Velez).

Authors from Chapter 9 would like to acknowledge the IEEE for the permission (License Number 4441541270677, 3 October 2018) to reuse text extract from the paper 'A Precise and Hardware-Efficient Time Synchronization Method for Wearable Wired Networks', published in the IEEE Sensors Journal, vol. 16, no. 5, 1 March 2016 (authors: Fardin Derogarian and João Canas Ferreira).

The work from Chapter 12 has been financed by National Funds through the FCT – Fundação para a Ciência e Tecnologia (Portuguese Foundation for Science and Technology) within the project 'VR2Market' Grant CMUP-ERI/FIA/0031/2013 and project NanoSTIMA funded through the North Portugal Regional Operational Program (NORTE 2020), under the PORTUGAL 2020 Partnership Agreement, and through the European Regional Development Fund (ERDF). The authors would like to thank all the previous and current members of the project team that involves also end-users from firefighters from Albergaria-a-Velha and Vila Real-Cruz Verde, Polícia de Segurança Pública (PSP) from Porto Metropolitan Command, consultants and researchers. Regarding contributions and suggestions to the work developed, authors would like to thank Emanuel Lima and Pedro Santos.

The research work from Chapter 13 was supported by UDR (Unidade de Detecção Remota), Department of Physics from University of Beira Interior, by IST-UNITE, by the PhD FCT (Fundação para a Ciência e Tecnologia) grant SFRH/BD/38356/2007, by Fundação Calouste Gulbenkian, by 'Projecto de Re-equipamento Científico' REEQ/1201/EEI/2005 (a Portuguese Foundation for Science and Technology project), and by the 'Smart-Clothing' project. The authors would like to acknowledge the fruitful discussions and contributions from all the colleagues from the 'Smart-Clothing' project. In special, we would like to thank Mrs. Andreia Rente and Prof. Rita Salvado for helping us in the production of the Flex Sensor and on-off belts.

The research from Chapter 14 was supported by UID/EEA/50008/2013, UID/EEA/50008/2019, the European Social Fund, PROENERGY-WSN, INSYSM, CREaTION, HANCAD, ECOOP, ORCIP, EFAtraS and COST IC1004. Special thanks go to Henrique Saraiva. Authors from Chapter 14 would like to acknowledge the Elsevier for the permission (License Number 4555591360585, 24 March 2019) to reuse text extract from the paper 'Strategies for high-performance supercapacitors for HEV', published in Journal of Power Sources, vol. 174, no. 1, 22 November 2007, pp. 89–93 (authors: Marina Mastragostino and Francesca Soavi).

Authors from Chapter 15 would like to acknowledge the Springer Nature for the permission (License Number 4555571279421, 24 March 2019) to reuse text extract from the paper 'Toward a Taxonomy of Wearable Technologies in Healthcare', published in New Horizons in Design Science: Broadening the Research Agenda: 10th International Conference, DESRIST 2015, Dublin, Ireland, 20–22 May 2015, Proceedings, vol. 9073, p. 496 (authors: Mayda Alrige and Samir Chatterjee).

Editors also gratefully acknowledged the team from IT-Covilhã (Radio Systems Group) by the continuous collaboration, incentive and joy during the hard work to write the Chapters.

Finally, we would like to thank our families by the support, love, patience and understanding by our delays and some absences, in the period of writing the book.

List of reviewers

António S. Lebres, Departamento de Física, Universidade da Beira Interior, Covilhã, Portugal

Caroline Loss, FibEnTech Research Unit – Universidade da Beira Interior, Covilhã, Portugal

Emanuel M. Popovici, School of Engineering, University College Cork, Ireland

Fardin Derogarian, Instituto das Telecomunicações and DEM, Universidade da Beira Interior, Faculdade de Engenharia, Covilhã, Portugal

Fernando J. Velez, Instituto das Telecomunicações and DEM, Universidade da Beira Interior, Faculdade de Engenharia, Covilhã, Portugal

João Canas Ferreira, Departamento de Engenharia Eletrotécnica e de Computadores, Faculdade de Engenharia da Universidade do Porto and INESC TEC, Porto, Portugal

José Machado da Silva, Departamento de Engenharia Eletrotécnica e de Computadores, Faculdade de Engenharia da Universidade do Porto and INESC TEC, Porto, Portugal

Luis Bernardo, Instituto de Telecomunicações and Universidade Nova de Lisboa, Monte da Caparica, Portugal

Luís M. Borges, Nokia, Alfragide, Portugal

Michele Magno, Integration System Laboratory, ETH Zürich, Zürich, Switzerland

Minseok Kim, Graduate School of Science and Technology, Niigata University, Japan

Mohsen Koohestani, ESEO-TECH, RF-EMC Research Group, Angers, France, and Institute of Electronics and Telecommunications of Rennes, Rennes, France

Mona Ghassemian, Faculty of Computer Science and Engineering, Shahid Beheshti University, Tehran, Iran, and School of Informatics, King's College London, London, UK

Ricardo Gonçalves, Departamento de Eletrónica, Telecomunicações e Informática, Universidade de Aveiro and Instituto de Telecomunicações, Aveiro, Portugal

Sukhumarn Archasantisuk, Department of Human System Science, Graduate School of Decision Science and Technology, Tokyo Institute of Technology, Japan

Takahiro Aoyagi, Department of Electrical and Electronic Engineering, School of Engineering, Tokyo Institute of Technology, Japan

Vitor Grade Tavares, Departamento de Engenharia Eletrotécnica e de Computadores, Faculdade de Engenharia da Universidade do Porto and INESC TEC, Porto, Portugal

Vladimir K. Poulkov, Faculty of Telecommunications, Technical University of Sofia, Sofia, Bulgaria

Author biographies

Fernando J. Velez (M'93–SM'05) received the Licenciado, MSc and PhD degrees in Electrical and Computer Engineering from Instituto Superior Técnico, Technical University of Lisbon in 1993, 1996 and 2001, respectively. Since 1995, he has been with the Department of Electromechanical Engineering of Universidade da Beira Interior, Covilhã, Portugal, where he is an Assistant Professor and also a researcher at Instituto de Telecomunicações. Fernando was an IEF Marie Curie Research Fellow in King's College London in 2008/2009 (OPTIMOBILE IEF) and a Marie Curie ERG fellow at Universidade da Beira Interior from 2010 until March 2013 (PLANOPTI ERG). He made or makes part of the teams of several European and Portuguese research projects on mobile communications, and he was or is the coordinator of six Portuguese projects. In the time span 2006–15, Fernando has got four national projects approved as coordinator (CROSSNET, MobileMAN, OPPORTUNISTIC-CR and PROENERGY-WSN) and almost 20 projects approved as participant, one European project (UNITE) and CREaTION (Cognitive Radio Transceiver Design for Energy Efficient Data Transmission), a prestigious project in research lines of excellence approved by FCT in 2013. In 2018/2019, he is leading a CMU Portugal Exploratory project, and he got a Marie Skłodowska-Curie ITN Action (TeamUp5G) that has started in 2019. He has authored 2 books, 14 book chapters, around 150 papers and communications in international journals and conferences, plus 39 in national conferences. Prof. Velez was the coordinator of the WG2 (on cognitive radio/software defined radio coexistence studies) of COST IC0905 "TERRA." His main research areas are cellular planning tools, traffic from mobility, cross-layer design, radio resource management, spectrum sharing/ coexistence/aggregation, hybrid analogue–digital solutions for massive multiple input multiple output (MIMO), radio frequency (RF) energy harvesting, wearable sensors and wireless body area networks (WBANs), and cost/revenue performance of advanced mobile-communication systems.

Fardin Derogarian Miyandoab received BS in communication engineering from Tabriz University in 1997 and MS in electronic engineering from Urmia University in 2003 and PhD in telecommunication from Porto University in 2015. His thesis focused on wearable systems and body area networks. He was a lecturer at Telecommunication Company of West Azerbaijan from 2003 to 2009 and at Urmia Azad University from 2007 to 2009. Recently, he has joined the Radio Systems group in Instituto de Telecomunicações as a postdoctoral fellow under supervision by Prof. Fernando J. Velez. His current research interests include telecommunication, integrated circuits, wireless sensor networks, body area network, wearable health systems, cognitive radios, energy harvesting and mobile networks.

Raúl Chávez-Santiago obtained his PhD degree from Ben-Gurion University of the Negev, Israel, in 2007. He held postdoctoral positions at the Informatics Research Laboratory (LRI) of the University Paris-Sud XI, France, and Bar-Ilan University, Israel. He joined the Intervention Center, Oslo University Hospital, Norway, in 2009, where he investigated ultra-wideband (UWB) radio technology for medical applications. He was a Management Committee member in the COST Actions IC0902 and IC0905 on cognitive radio.

Luís M. Borges is currently a senior NPO engineer at NOKIA (EPDC). He worked as a RAN Engineer at NetOne (Angola) within the WiMAX and LTE project as well as worked in Ericsson within the Vodafone PT SRAN modernization project. Luís M. Borges received the Licenciatura and PhD degrees in Electrical Engineering from Universidade da Beira Interior, Covilhã, Portugal, in 2006 and 2013, respectively. Luís M. Borges was a Post-Doc researcher in CREaTION project. He was also research assistant at Instituto de Telecomunicações, Covilhã. He made or makes part of the team of COST 2100 and COST IC1004 European projects and participated in SMART-CLOTHING and PROENERGY-WSN Portuguese projects. Luís M. Borges is a member of the Ordem dos Engenheiros (EUREL). His main research areas are or were Mobile networks and optimization, wireless sensor network (WSN), medium access protocols, cross-layer design, hardware development, network modeling, and application development, cognitive radio and energy harvesting.

Norberto Barroca received his Licenciado degree in electronics and telecommunications engineering from Polytechnic Institute of Castelo Branco in 2007 and his MSc and PhD degrees in electrical and computer engineering from the University of Beira Interior in 2009 and 2014, respectively. He was a research assistant at Instituto de Telecomunicações, while pursuing research for the PhD degree. His research interests included wireless sensor networks and cross-layer design.

Ilangko Balasingham received the MSc and PhD degrees from the Department of Electronic Systems, Norwegian University of Science and Technology (NTNU). He performed his Master degree thesis at the Department of Electrical and Computer Engineering, University of California Santa Barbara, USA. He is the Head of Section for Medical ICT R&D at Oslo University Hospital and Professor of Medical Signal Processing and Communications at NTNU. He was the professor by courtesy at Nagoya Institute of Technology in Japan in 2016/2017. His research interests include body area sensor network, microwave short range sensing of vital signs and nanoscale communication networks. He has authored or coauthored over 250 journal and full conference papers, 7 book chapters, 6 patents and 20 articles in popular press.

João Canas Ferreira is an associate professor (with tenure) in the Department of Electrical and Computer Engineering, Faculty of Engineering, University of Porto, Portugal, and a senior researcher with INESC TEC, Porto. His current research interests include FPGA-based reconfigurable computing, hardware acceleration of embedded systems, flexible energy-efficient platforms for IoT and embedded machine learning. He is handling editor for the journal microprocessor and microsystems. He is a senior member of the IEEE and a member of ACM and Euromicro.

Vítor Grade Tavares received the Licenciatura and MS degrees in electrical engineering from the University of Aveiro, Aveiro, Portugal, in 1991 and 1994, respectively, and the PhD degree in electrical engineering from the Computational NeuroEngineering Laboratory, University of Florida, Gainesville, FL, USA, in 2001. In 1999, he joined the University of Porto, Porto, Portugal, as an Invited Assistant, where he has been an Assistant Professor since 2002. In 2010, he was a visiting professor with Carnegie Mellon University, Pittsburgh, PA, USA. He is also a Senior Researcher with the Instituto de Engenharia de Sistemas e Computadores, Tecnologia e Ciência, Porto. His current research interests include low-power, mixed-signal and neuromorphic integrated-chip design and biomimetic computing, CMOS RF integrated circuit design for wireless sensor networks and transparent electronics.

Hamed Rezaie received his PhD in software engineering at the school of Computer Science and Engineering, Shahid Beheshti University in 2017. He currently works as IT director at Hamedan water and waste water company. His research interests are wearable activity recognition and WBAN MAC layer design and analysis, and he has published a number of IEEE journals and conferences.

Mona Ghassemian currently works as a senior researcher at British Telecom with a focus on 5G and IoT systems. She is a senior IEEE member and serves as IEEE, United Kingdom and Ireland vice-chair. She is a member of Tactile Internet working group (IEEE 1918.1) since 2016, served as technical programme committee (TPC), track cochair and invited speaker of a number of IEEE conferences since 2006, and tutorial presenter at IEEE PIMRC 2016 on Internet of Medical Things, and the area editor of Elsevier Ad hoc networks (2016–17). She received the IEEE Region 8 SB counselor award in 2012, University of Greenwich Early Career Research Excellence Awards in 2011 and PhD grant at King's college London awarded by NTT DoCoMo, 2001–04.

Caroline Loss received in 2009 the Licenciado degree in Fashion and Style from University of Caxias do Sul, Brazil, with honor. In 2010, she moved to Portugal, where she received the MSc (2012) and the PhD in Textile Engineering (2017), both from the University of Beira Interior. Currently, she is postdoctoral fellow at FibEnTech Research Unit, working in the TexBoost Project – Less Commodities, More Specialities. The focus of her research is the wearable antennas and bio-radar, including the electromagnetic characterization of the textile materials, development of the textile antennas for on-body communication and their integration in clothing. Dr. Loss has authored or coauthored three book chapters and more than 20 papers for international journals and conferences. She has awarded four distinctions for her work on textile antennas, including an Honorable Mention in the Student Paper Award at the International Microwave Workshop Series on Advanced Materials and Processes Conference (IMWS-AMP), and third place at The Fiber Society Graduate Student Paper Competition, both in 2017.

Marco Rossi was born in Italy in 1988. He studied Optoelectronic Engineering at the University of Pavia, Italy, where he graduated in 2012. From 2012 to 2016, he was with the Electromagnetics Group, INTEC (Information Technology) Department at Ghent University, Belgium, where in 2016 he received his PhD with a thesis titled

"Stochastic full-wave modeling for the variability analysis of textile antennas." In 2016, he was also awarded the URSI Commission B Young Scientist Award at the 2016 URSI Commission B International Symposium on Electromagnetic Theory (EMTS). As of 2017, he is with the R3S (RF & Smart Sensor Systems) Department at the Fraunhofer Institute for Reliability and Microintegration (IZM), Berlin, where he is currently Group Manager of the RMC (RF Materials & Components) Group. His current research interests are design of antennas, interconnects and passives and material characterization for System-in-Package solutions for mmWave applications.

Sam Agneessens was born in Belgium, in 1986. He received the MS degrees in electrical engineering from the University of Ghent, Ghent, Belgium, in 2011. He is currently with the Electromagnetics Group, Department of Information Technology (INTEC) where he is working toward his PhD degree. His research focuses on the design and development of textile antennas for wearable applications and implementation of substrate integrated waveguide techniques on textile materials. Sam Agneessens was awarded the URSI Young Scientist Award 2014.

Ricardo Gonçalves was born in Lisbon, Portugal, in 1988, and is a Leader Researcher at Evoleo Technologies, in Maia, Portugal (Wireless Circuits and Systems). He received the BSc and the MSc (Magna Cum Laude) on Electronics and Telecommunications Engineering from Instituto Superior de Engenharia de Lisboa in 2010 and 2012, respectively. In 2010/2011 he was awarded with merit scholarship for exceptional academic performances. He recently got the PhD degree in Electrical Engineering at the University of Aveiro, where he was a researcher at Instituto de Telecomunicações. His main research interests include wireless power transfer systems, radio frequency identification (RFID) and wireless passive sensor networks, with focus on the development of printed antennas and microwave circuits in nonconventional materials such as textiles, ceramics, plastic, paper and cork, for the former applications.

Hendrik Rogier received the Electrical Engineering and the PhD degrees from Ghent University, Gent, Belgium, in 1994 and in 1999, respectively. He is a currently a Full Professor with the Department of Information Technology, Ghent University, a Guest Professor with the Interuniversity Microelectronics Centre, Ghent, Belgium, and a Visiting Professor with the University of Buckingham, Buckingham, United Kingdom. From October 2003 to April 2004, he was a visiting scientist at the Mobile Communications Group of Vienna University of Technology. He authored and coauthored about 155 papers in international journals and about 180 contributions in conference proceedings. His current research interests are antenna systems, radiowave propagation, body-centric communication, numerical electromagnetics, electromagnetic compatibility and power/signal integrity. He was twice awarded the URSI Young Scientist Award. Moreover, he received the 2014 Premium Award for Best Paper in IET Electronics Letters and several awards at conferences. He is an Associate Editor of IET Electronics Letters, IET Microwaves, Antennas and Propagation, and the IEEE Transactions on Microwave Theory and Techniques. He acts as the URSI Commission B representative for Belgium and is a Senior Member of the IEEE.

Pedro Pinho was born in Vale de Cambra, Portugal in 1974. He received the Licenciado and Master degrees in Electrical and Telecommunications Engineering, and the PhD degree in Electrical Engineering from the University of Aveiro, Portugal, in 1997, 2000 and 2004, respectively. He is currently an Assistant Professor in Electronics, Telecommunications and Computers Engineering Department with Instituto Superior de Engenharia de Lisboa (ISEL/IPL) and a Senior Member of the research staff with Instituto de Telecomunicações (IT), Aveiro, Portugal. Dr. Pinho is also a senior member of the IEEE, and associate editor of the IET Microwaves Antennas and Propagation Journal. Dr. Pinho serves as Technical Program Committee in different conferences and reviewer of several IEEE journals. He has authored or coauthored more than 200 papers for conferences and international journals.

Rita Salvado received the European PhD degree in Textile Engineering, from the University of Beira Interior (UBI) and Université de Haute Alsace in 2002. Currently, he is the auxiliary professor at the Department of Textile Science and Technology in UBI and director of the Master degree (second cycle) in textile engineering and vice-director from the Wool Museum of the UBI. She was the director of the PhD degree in Textile Engineering from UBI and director of the bachelor degree in fashion design of UBI (Nov. 2006—Nov. 2013) as well as ERASMUS coordinator (Sep. 2004—Jan. 2014). Rita has authored more than 50 publications in scientific journals and conferences. In 2003, she received the "The Fiber Society Student Award" and, in 2013, the Pedagogical Merit Award from UBI.

Sukhumarn Archasantisuk was born in Thailand, in 1990. She received BEng. degree from Chulalongkorn University, Thailand in 2012, and MEng and DEng degrees from Tokyo Institute of Technology in 2014 and in 2017, respectively. Her research interests include body area network and human motion classification. She has been working as an associate visionary architect at Kasikorn Business Technology Group in Thailand since 2017.

Takahiro Aoyagi was born in Yokohama, Japan, on November 1970. He received B. Eng., MEng., and DEng degrees from Tokyo Institute of Technology, Tokyo, Japan, in 1993, 1995, 1998, respectively. He has been an Associate Professor of the Department of Electrical and Electronic Engineering, School of Engineering, Tokyo Institute of Technology. His field of research is electromagnetic compatibility and electromagnetic wave engineering. He has received the young scientist award of the 27th symposium on ultrasonic electronics, 2007. Dr. Aoyagi is a member of the Acoustical Society of Japan, the Japan Society of Applied Physics and the Institute of Electronics, Information and Communication Engineers.

Tero Uusitupa, MSc and DSc in electrical engineering, worked as a researcher in Aalto University, Espoo, Finland, where his last research project focused on WBANs. After Aalto University, he has had various tasks involving electromagnetic-field simulation, RF/antenna design, measurements, programming and consulting. Currently, he is a freelancer and interested in RFID, IoT and technology in general.

Minseok Kim is with the Graduate School of Science and Technology, Niigata University, Niigata, Japan, as an Associate Professor. His current research interests include radio propagation channel measurement and modeling, millimeter waves, antenna array signal processing, body area network and ITC healthcare. He is a senior member of IEEE.

Jun-ichi Takada received Doctor of Engineering in electrical and electronic engineering from Tokyo Institute of Technology (Tokyo Tech) in 1992. After serving as a Research Associate at Chiba University in 1992–94 and as an associate professor at Tokyo Tech in 1994–2006, he has been a Professor at Tokyo Tech since 2006. Currently, he is a Professor in the Department of Transdisciplinary Science and Engineering, School of Environment and Society, Tokyo Tech. From 2003 to 2007, he was also a researcher at the National Institute of Information and Communication Technology, Japan. He served as the chair of measurement working group (WG) in ITU-R TG 1/8 on compatibility between UWB devices and radiocommunication services in 2005. He served as the cochair of special interest group (SIG) in body communications in European COST action 2100 "Pervasive Mobile & Ambient Wireless Communications" by 2010. His current research interests include radio-wave propagation and channel modeling for mobile and short-range wireless systems, radio spectrum sharing among heterogenius radio systems and information and communication technology (ICT) applications for international development. He is a fellow of Information and Communication Engineering (IEICE), Japan, a senior member of IEEE, and a member of Japan Society for International Development.

Jorge Tavares received the Licenciatura and MSc degrees in Electromechanical Engineering from Universidade da Beira Interior, Covilhã, Portugal, in 2006 and 2009, respectively. Luís M. Borges was an MSc researcher in the PROENERGY-WSN project. He was also a research assistant at Instituto de Telecomunicações, Covilhã. He made or makes part of the team of *COST Action IC0905 TERRA* and participated in PROENERGY-WSN Portuguese projects. His main research areas are or were WSN, medium access control protocols, cross-layer design, hardware and application development, cognitive radio and radio-frequency energy harvesting.

Periklis Chatzimisios serves as an Associate Professor and the Director of the Computing Systems, Security and Networks Research Lab, Department of Informatics at Alexander TEI of Thessaloniki. He is also a visiting fellow in the Faculty of Science & Technology, at Bournemouth University, United Kingdom. Dr. Chatzimisios is/has been involved in several standardization and IEEE activities serving such as the IEEE 5G Initiative, the Standards Development Board for IEEE Communication Society (ComSoc), the IEEE ComSoc Standards Program Development Board and the IEEE ComSoc Education Services Board. Dr. Chatzimisios is the editor/author of 8 books and more than 130 papers and book chapters on the topics of performance evaluation and standardization of mobile/wireless communications, Internet of Things, Vehicular Networking and Big Data.

José Machado da Silva received the Licenciatura and PhD degrees, both in Electrical and Computer Engineering from the Faculdade de Engenharia da Universidade do

Porto (FEUP), Portugal, in 1984 and 1998, respectively. He is currently an Associate Professor at FEUP and a Senior Researcher at INESC TEC, with teaching and research responsibilities in the domains of design and test of electronic circuits, namely, for wearable and biomedical applications. He coedited one book, six book chapters, and more than ninety journal and conference papers. His research interests include analog, mixed-signal, and RF test and design for testability, systems dependability, embedded instrumentation, analog and digital signal processing, VLSI design, and electronic bioimplants.

Albena Mihovska obtained her PhD from Aalborg University (2008) and is an Associate Professor at Aarhus University, Department of Business Development and Technologies, where she is with the CTIF Global Capsule research group. Her main current activities relate to research in the area of smart dense connectivity and related applications; 5G ultradense access networks, Internet of Things technologies for healthcare and smart grid, and most recently, holographic communications. In addition, she focuses on aspects of digital innovations and their impact on business. She has been an active contributor to ITU-T and European Telecommunications Standardization Institute (ETSI) standardization activities in the areas of Internet of Things. She is a member of IEEE, and INFORMS (the leading international association for Operations Research and Analytics professionals). She has authored and coauthored more than 170 publications, including peer-reviewed international books, journal and conferences publications.

Aristodemos Pnevmatikakis is a Professor at Athens Information Technology, with more than 20 years' experience in signal processing. His prime research interest is in systems for multimodal detection, tracking and identification, aiming at activity recognition in intelligent spaces. To this extend, he utilizes signals from cameras, microphones and wearable, environmental and domotics sensors. He is the coauthor of the books *Audio-Visual Person Tracking: A Practical Approach* and *Delta-Sigma Modulators, Modeling, Design and Applications* (Imperial College Press, 2011 and 2003). His research has resulted to multiple scientific publications and has been featured on media and events, while the resulting systems have been successfully evaluated at international evaluation campaigns. He has been involved in numerous EU and national research and industrial projects.

Sofoklis Kyriazakos is an associate professor in Aarhus University in the Business Development and Technology Development. He has a Master degree in Electrical Engineering and Telecommunications in RWTH Aachen and has obtained his PhD and MBA from the National Technical University of Athens. Sofoklis has published more than 100 publications in international conferences, journals, books and standardization bodies and has more than 400 citations. At the same time, he has an entrepreneurial carrier, having established three startup companies. He has member of Board of Directors in several companies and Athens Information Technology, a Center of Excellence for Research and Education.

Krasimir Tonchev is a leading researcher at the Faculty of Telecommunications, Technical University of Sofia, Bulgaria. His main interests are kernel-based support vector machines, age and gender recognition from faces, 3D facial analysis, affective computing, artificial intelligence. He has participated in several scientific projects both national and international for the development of reliable face and emotion recognition methods and ambient assisting living systems. Collaborating with other authors, he has published more than 40 scientific papers in conferences and journals. He is an IEEE member.

Razvan Craciunescu is an assistant professor at the University Politehnica of Bucharest, and his main interests are in the field of wireless communications, sensor networks and Internet of Things. He is coauthor to more than 30 articles in international journals and conferences a participated and managed several international research and development project both in universities and in the private sector in the field of IoT and e-health.

Vladimir Poulkov, PhD, received his MSc and PhD degrees at the Technical University of Sofia, Bulgaria. He has more than 35 years of teaching, research and industrial experience in the field of Telecommunications. His expertise is related to information transmission theory, power control and resource management for next-generation telecommunications networks. He is the author of more than 100 scientific publications and is teaching BSc, MSc and PhD courses in Information Transmission Theory and Next Generation Access Networks. Currently, he is head of the "Teleinfrastructure R&D" Laboratory at the Technical University of Sofia, Chairman of the Bulgarian Cluster of Telecommunications and vice-chairman of the European Telecommunications Standardization Institute (ETSI) General Assembly. He is a Senior IEEE Member.

Harm op den Akker is a senior researcher at Roessingh Research and Development, The Netherlands, responsible for the research track on "Human-Computer Interaction for Personal Care." Harm holds a degree in technical computer science (BSc) and human media interaction (MSc) obtained from the University of Twente in Enschede, The Netherlands. In 2014, Harm obtained his PhD degree with his thesis on "Smart Tailoring of Real-Time Physical Activity Coaching Systems." Harm has worked in various European and national research projects (FP6: AMIDA, FP7: eWALL, AAL: IS-ACTIVE). His current research focus lies on developing and evaluating innovative ICT services that improves the human–computer interaction element of telemedicine/eHealth systems through state-of-the-art AI and personalization.

Hermie J. Hermens studied biomedical engineering at the University of Twente; then he became the head of the research group at Roessingh, Centre for Rehabilitation. In 1990, he was the cofounder and the first director of RRD. He did his PhD on surface EMG, became professor in Neuromuscular Control (2001) and later (2010) professor in Telemedicine. Currently, he supervises 15 PhD students. Twenty six PhD students finished their PhD under his (co)supervision. He is the (co)-author of over 280 peer reviewed publications, and his work was cited over 13,000 times (H-index 55). He coordinated three successfully completed European projects (Seniam, Crest, and Impulse), is coordinator of the recently acknowledged European project COUCH (ranked fourth of 188 proposals), and co-acquired and participated in over 25 other European projects in the area of surface EMG, functional electrical stimulation and ICT. He is a fellow and past president of the Int. Society of Electrophysiology and Kinesiology (ISEK), editor in chief of the JBMR and coordinator of the Seniam group leading to the first surface EMG recommendations. His present research is focused on combining Biomedical Engineering with ICT to create new end-to-end Telemedicine applications for people with chronic disorders.

João P. Cunha is an associate professor (with "Agregação") of Electrical and Computer Engineering and Bioengineering at the Faculty of Engineering of the University of Porto, coordinator of the U. Porto Centre of Competence in Future Cities and senior researcher at the INESC-TEC where he cofounded the *Center for Biomedical Engineering Research* (C-BER) that aggregates ~30 researchers. He earned a degree in Electronics and Telecommunications Engineering (1989), a PhD (1996) and an "Agregação" degree (2009) in Electrical Engineering at the U. Aveiro. He is a *Senior Member* of IEEE (2004) where he joined the Engineering in Medicine and Biology Society (EMBS) in 1986 as a student member. He presently serves as Scientific Director of the *Carnegie-Mellon|Portugal* program where he is a faculty since 2007. He coauthored +250 publications and 6 patents, holding an h-index 21 (GoogleScholar).

Susana Rodrigues is a post-doctoral researcher at C-BER, INESC TEC Porto, Portugal. She has been involved in several projects related with human sensing and health care. She completed her PhD in Psychology at University of Porto in 2016. Her R&D activities are mainly focused in occupational health and psychophysiology. She has gained substantial experience working in the psychophysiological analysis of stress and fatigue among different occupational groups (police officers, firefighters, air traffic controllers, neurosurgeons, students) using multi-methods and novel wearable devices. Her work has been presented in national and international conferences (17) and published in several peer-reviewed journals and book chapters (10).

Duarte Dias is a biomedical engineer with a transversal expertise in wearable health devices, human physiology, hardware and firmware development, signal processing and data analysis. Since his Master degree, he has been working as a R&D technician C-BER, INESC TEC, being responsible for wearable prototypes development in several international projects with companies and researches institutes, master thesis coordination, field trials coordination and project management. His transversal scientific knowledge enables him to easily interact with different technology and scientific areas. He is the coauthor in more than ten scientific publications, including a first-author review in "Sensors." His interest for entrepreneurship and transfer lead him to support and be involved in the creation of two spin-offs.

Pedro Brandão is an assistant professor from University of Porto, School of Science, current director of the Master in Medical Informatics and a researcher from Instituto de Telecomunicações. Pedro did his PhD on middleware for body sensor networks in the University of Cambridge. His current research dwells in medical informatics, namely, mHealth for diabetes and a health kiosk project for autonomic use. He also works on ad hoc networks for first responders. This led to the current involvement in the VR2Market, Health Kiosk, MyDiabetes and MobiWise projects and several conference publications. Pedro has also reviewed or been TPC member of Computer Communication (Elsevier), International Journal of Medical Informatics (Elsevier), MDPI—Sensors, IEEE Comm. Letters, Med-Hoc-Net, etc.

Ana Aguiar is an assistant professor at the University of Porto (UP) and researcher at Instituto de Telecomunicações (IT). Her research focuses on wireless and mobile systems, concretely vehicular networks, Wi-Fi networks with mobility, mobile IoT, and spatiotemporal data science for smart mobility. She coordinates the Networked Systems Group at IT/UP. She is adjunct coordinator of the Center of Competences for Future Cities at UP, operating the UrbanSense environmental sensors and the crowdsensing tool SenseMyCity. She leads the VOCE project on voice stress detection and participated in more than 15 projects, in 9 of them as co-PI or WP leader. She published more than 60 journals and conference articles, is reviewer for several IEEE transactions and expert with the EC.

Ilídio C. Oliveira, PhD, is a faculty member with the Department of Electronics, Telecommunication and Informatics, at the University of Aveiro, where he teaches software engineering and software quality subjects. His research interests include health informatics, mobile health, pervasive systems and mobile computing.

José Maria Amaral Fernandes is the assistant professor at the Dept. of Electronics, Telecommunications and Informatics of the Universidade de Aveiro, Portugal. He is also a researcher at IEETA, a R&D institute at University of Aveiro (http://www.ieeta.pt/). He has been working on IT solutions for online multimodal monitoring and data analysis. He has been focused mainly in two scenarios: human physiological and behavioral response in real environments (gait analysis, stress scenarios) and characterization of city based (buses telemetry and distributed sensors in the city of Porto). His main focus is to achieve better spatiotemporal characterization of domain relevant events to support decision/action in useful timeframe.

Catarina Maia is the head of the Technology Transfer Office at INESC TEC, in charge of Intellectual Property management and licensing, and the European IPR Helpdesk Ambassador for Portugal, helping promoting the strategic usage of Intellectual Property Rights by SMEs. She is also an invited lecturer at the Faculty of Engineering and the Faculty of Sciences at University of Porto and Porto Business School. She holds a Master in Management—EGP University of Porto Business School and has graduated in Microbiology from Escola Superior de Biotecnologia, Universidade Católica Portuguesa. Her main interests are related with entrepreneurship, business strategy, innovation management and intellectual property strategy.

Ana Rita Tedim is an expert in web and mobile applications security in e-commerce, banking, finance, healthcare, defense, broadcasting and advertisement markets. She holds a Master degree in bioengineering from the Faculty of Engineering, U.Porto.

Ana Cristina Barros is a senior researcher at INESC TEC in the area of operations and technology management. She is a Member of EurOMA, the European Operations

Management Association. Ana Barros has been visiting researcher at Carnegie Mellon University (2014/2015; 2012), Massachusetts Institute of Technology (2008–11), and Cornell University (2002). She obtained a PhD in Engineering and Management (Technical University of Lisbon, 2011), an MBA in Logistics and Entrepreneurship (Technical University of Munich, 2004), and an MS and BS in Chemical Engineering (University of Porto, 2000). Before joining INESC TEC, Ana Barros worked several years in the procurement and production planning departments of German and Portuguese companies.

Orangel Azuaje received his BEng degree from Universidad Simón Bolívar (USB), Venezuela in 2012, MEng degree from Faculdade de Engenharia da Universidade do Porto (FEUP), Portugal in 2015 and currently, he is pursuing the PhD degree on the MAP-tele Doctoral Programme in Telecommunications, a joint venture of Universidade do Minho, Universidade de Aveiro and Universidade do Porto (MAP). His research interests are in the areas of wireless communications with a focus lately on provisioning of quality of service and channel modeling in wireless network.

Eduardo Soares is a PhD student in University of Porto and Instituto de Telecomunicações, researching multicast for MANET, WSN and Mesh networks. With interest in communication, systems and services for decentralized networks such as MANET, WSN, VANET and UAV networks.

Fernando De la Torre received his BSc degree in Telecommunications as well as his MSc and Ph. D degrees in Electronic Engineering from La Salle School of Engineering at Ramon Llull University, Barcelona, Spain in 1994, 1996, and 2002, respectively. In 1997 and 2000, he became an Assistant and Associate Professor in the Department of Communications and Signal Theory in Enginyeria La Salle. In 2003, he joined the Robotics Institute at Carnegie Mellon University, and currently he is a Research Associate Professor. His research interests are in the fields of Computer Vision and Machine Learning.

José Martinez-de-Oliveira was born in February 12, 1949 in Portugal and obtained the MD degree in the Porto School of Medicine in 1974. He became specialist in Gynecology & Obstetrics in the Hospital de São Joäo, Porto, Portugal, in 1973. He concluded his PhD degree (GynOb) from the Porto School of Medicine in 1988. He has been Full Professor of Obstetrics & Gynecology in the School of Health Sciences, University of Beira Interior, Covilhã, Portugal, since 2005, and became an Emeritus Professor in 2019. From 2005 until 2019, he was the Director, Head of Child & Mother Health Department in Centro Hospitalar Cova da Beira (Covilhã). He retired in 2019. Presently, José is the President of University of Beira Interior Ethics Committee, CICS-UBI researcher (Health Sciences Research Center—UBI), Member of ISSVD Presidents' Council (International Society for the Study of Vulvovaginal Disease), ISIDOG (International Society of Infectious Disease in Obstetrics & Genecology) 2019 International Congress Organizer (Porto, Portugal). His main scientific areas of interest are vulvovaginal infections and POC tests development.

António S. Lebres received the Licenciado degree in Electrical Engineering from Instituto Superior Técnico (IST), Technical University of Lisbon in 1974 and PhD degree in Electrotechnical Engineering from University of Beira Interior (UBI) in 1998. From 1974 to 1985, he worked with National Laboratory Nuclear Reactor Research team (LFEN) designing nuclear electronics, and in 1989, he received the *Researcher* degree from Laboratório Nacional de Engenharia e Tecnologia Industrial (LNETI), Lisbon. In 1991, he received an MBA degree in General Management from Universidade Nova de Lisboa, and in 1994, he got an MBA degree in Finances Internacionales from Haute École Commerciale—HEC Management, Paris. Since 1989, he has been with the Physics Department of University of Beira Interior, Covilhã, Portugal, where he is an assistant professor running electronics matters. António was a researcher at Unidade de Detecção Remota (UDR)—Applied Physics and Telecommunications group until 2016. He participated in SAMURAI and SMART-CLOTHING Portuguese projects, being coordinator of last one. Prof. Lebres has coauthored two book chapters, several papers and communications in international journals and conferences. His main research areas are small signal acquisition and processing systems, low noise and fast signal amplifiers and instrumentation design.

Chapter 1

Introduction

Fernando J. Velez[1] and Fardin Derogarian[1]

1.1 Motivation

Instrumenting the human body and building human-centred wireless sensor networks (WSNs) constitute a challenge that goes beyond simply installing and networking electronics in the personal area, a goal that is being pursued since the early 2000s. In 2018 and coming years, investigating low-power solutions and technologies that are harmless to the human being remains a challenge for the research community, which is required to address also the means to give intelligence to the way how sensing and processing are performed, as well as applying cognition to radio and network communications, paring with energy efficiency and harvesting. The fast growth of smart textiles, high performance low-power multi-hop networks, efficient processing methods for smart antennas and ultra-wideband (UWB) represents a stimulus to provide new applications for on-, in- and body-to-body wearable communications for the human body well-being.

Hence, continuous advances in wearable technologies, sensors and secure wireless body area networks (WBANs), radio communication aspects, including radio channel characterization, and especially energy efficient solutions with energy harvesting (EH) and storage, have been sufficient reasons that have encouraged the necessity to collect a variety of knowledge and experience in this field.

Wearable systems with embedded sensors in textile for healthcare are of one the growing application fields of wireless body sensor networks (WBSNs) and underlying applications. According to market research reports [1], the global wearable technology market is predicted to grow from 29.92$ billion in 2016 to 71.23$ billion by 2021, growing at a compound annual growth rate (CAGR) of 18.9% in 6 years. However, there are still many challenges ahead. Many promising technological advances in electronics, information technology and communication systems have paved the way to new concepts in healthcare and other personal applications. For example, telemedicine, e-hospital and ubiquitous healthcare are enabled by emerging new electronic devices and wireless broadband communication technologies. While initially

[1] Instituto de Telecomunicações and Departamento de Engenharia Electromecânica – Universidade da Beira Interior; Faculdade de Engenharia, Portugal

becoming mainstream for portable devices such as tablets and smart phones, wireless communications are evolving towards wearable solutions, which may even include implantable solutions. These healthcare solutions relying on sensor networks, and WBSNs are becoming critical devices for the emerging concept of connected health.

Wearable systems and technologies, a subcategory of personal health system (PHS), are a solution for continuous health monitoring through non-invasive biomedical, biochemical and physical (PHY) measurements. They have attracted a lot of attention from the research community and industry during the last decade, as evidenced by the numerous and yearly increasing research and development efforts. These systems can be characterized as integrated platforms that are worn on the body. Typical examples include wrist-worn devices and biomedical clothes. As the benefits of wearable technologies and WBSNs become obvious, and the attitude of users towards the application of new technologies becomes more positive, it is expected that wearable systems will expand significantly in the near future. For these reasons, research on wearable technologies has been attracting increased interest from various technological fields. Portable monitoring devices that use micro-electromechanical integrated circuits (ICs) and systems that realize miniaturized accelerometers and gyroscopes and other types of sensors and actuators have been introduced. These have been successfully applied in areas that go from entertainment to sports and rehabilitation, in both research and clinical environments, as well as in daily routine and leisure activities.

Usually, communication between on-body parts and an external computer or personal digital assistant (PDA) is done over a wireless link, but communication among the nodes in the wearable part may be established in a different way. In fact, the wearable part of the system can be categorized according to the communication medium. In this context, the available technologies are wired, wireless and human body communication (HBC), a communication technique in which the body is used as signal propagation medium [2].

The main advantage of utilizing wireless devices is their easy deployment, both for adding new devices or removing them. In turn, high energy consumption, low data rate, interference, low reliability, low security and health issues due to radio frequency (RF) signal effects on the body are some of drawbacks of wireless wearable systems.

By considering the possibility of having available conductive yarns or copper wires, the aforementioned issues related with wireless devices lead to an alternative use of wired communication in body area networks (BANs). Wired networks, as a second type of communication infrastructure for BAN applications, provide high-speed, reliable and low-power solutions. However, they suffer from fixed node placement, lack of standards and are cumbersome.

HBC is a communication technique in which the body is used as signal propagation medium. Electrical signals can easily propagate through the human body. Recently, several HBC schemes have been suggested for BAN where the data is transferred through the person's skin. Since communication is restricted to the human body, interception is more difficult than in other wireless methods. HBC consumes less power than wireless but more than wired communication implementations.

There are however challenges regarding interfacing with the body (probe sizes), dealing with skin irritation, lack of communication standard, low available data rates and reliability.

Sensing electronic devices that capture inertial and electromyographic signals facilitate measuring vital signs, recording and transmission of biometric and performance parameters for athlete monitoring. Together with signal processing methodologies, these devices allow for advanced monitoring of human gait in sport activity or for medical purposes. The capture of electromyographic signals from several muscles may be performed with the use of several electrodes embedded in the clothes and connected to different sensor nodes. For the sake of coordination of data from different origins, resource-efficient synchronization mechanisms must be applied. As flexible antennas and different electronic circuits and microwave devices are embedded onto the garment, the conductive and dielectric properties of textile materials used for their manufacturing need to be characterized in detail. Together with wearable sensors placed around the body and machine-learning techniques, radio signals can be used for human movement identification, e.g. for activity recognition purposes.

BANs are a subcategory wireless sensor network (WSN). However, in comparison with WSN, BANs only consist of few nodes, usually less than 10. Besides, reliability, service quality and high data rate are also very relevant aspects, even with higher importance than in WSNs. In the majority of recent applications, single hop communications are considered because of the reduced number of nodes and due to the direct connection to a sink node. Nevertheless, with the increasing demand for higher number of nodes, multi-hop communication is inevitable. In the latter case, routing and media access control (MAC) sub-layer protocols have to be carefully selected or designed to satisfy the reliability requirements of the system, while providing high throughput with reasonable amount of energy. Medical BANs are mostly introduced in unlicensed frequency bands, where the risk of mutual interference with other electronic devices can be relevant. Hence, techniques that consider aspects of frame capture and the application of cognitive radio (CR) techniques can potentially alleviate these problems in medical communication environments. Together with energy efficiency, aspects arising from energy saving considerations addressed in the context of MAC sub-layer and network protocols, EH is a promising technique. In particular, RF-EH techniques paring with energy storage with super capacitors are promising to facilitate the provision of autonomous wearable technologies and WBSNs. Furthermore, wake-up radio-based power-minimization methodologies provide energy-saving mechanisms while increasing lifetime within wearable systems.

Different applications include specific products and devices such as mobile wearable surveillance for first response and hazardous professions, or wearable sensors for foetal monitoring in the last 5 weeks of pregnancy. The description of their functionality and performance provides a set of examples of how wearable technologies will be very useful to the population or target groups in different scenarios, including the e-health environment. In particular, recent research and development projects, like eWALL or VitalResponder®, have been identifying required technological innovations to conceive easy-to-implement smart 'caring home'.

1.2 Brief history of wearable technologies and WBSNs

Wearable technology is commonly understood as the broad group of smart electronic devices or accessories that can be worn on the body or implanted under the skin. Their design often incorporates practical functions and features. The history of wearable technologies comprises a pioneering phase along the past centuries, and a more recent phase, where ubiquitous computing and wearable computers that make technology pervasive by interweaving it into daily life are in the centre of their development.

Examples of wearable devices that have been proposed along the centuries, with innovations that addressed different types of applications [3], are as follows:

- Eyeglasses (1286) – Before the production of the convex lens, ingenious ways were found to overcome visual impairments; squinty Emperor Nero looked through an emerald in order to see at gladiator fights.
- Nuremberg egg (1510) – The Nuremberg egg was one of the first portable mechanical timekeeping devices. Cumbersome, inaccurate and worn around the neck (like Flava Flav), these watches were a key status symbol in the sixteenth century in Europe.
- Abacus ring (1600) – Dating back to the early days of China's Qing dynasty, before the calculator watch, there was the abacus ring.
- Air conditioned top hat (1800s) – The Victorians were technology pioneers, and also obviously quite sweaty, being thus natural that an air conditioning top hat was patented.
- Illuminated ballet girls (1884) – The introduction of illuminated ballet girls has added to the attractions of the grand stage. Girls with electric lights on the foreheads and batteries concealed in the recess of their clothing [4].
- Hearing aid (1898) – The hearing aid is another interesting device. The first hearing aid, also called an ear trumpet, was invented in the seventeenth century, but the first electronic hearing aid was developed in the 1890s after the invention of the telephone. Digital hearing aids hit the market in 1987 with the Nicolet 'Phoenix'. They continue to be redesigned and to shrink in size with each new version [5].
- Electric Girl Lighting Company (1890) – In 1890, the *New York Times* wrote that in a very short time private houses would be lighted by girls, instead of stationary lights. These girls came as a courtesy of the Electric Girl Lighting Company and wore dresses fitted with batteries and bulbs.
- First wearable camera (1907) – Although at first a novelty, the Pigeon Camera found real use during the First World War, being used to capture aerial photographs behind enemy lines.
- First virtual reality 'headset' – In 1960, cinematographer Morton Heilig created a new way of enjoying movies through what he called the 'immersive arcade experience'. It is the first recorded attempt at blending cinema with virtual reality. Morton called his invention the 'Stereophonic Television Head-Mounted Display' and patented the invention. Two years later, in 1962, he patented another invention the "Sensorama Simulator" (US Patent #3.050.870). An upgraded virtual reality

simulator with handlebars, binocular display, vibrating seat, stereophonic speakers, cold air blower and a device close to the nose that would generate odours, all synchronized with the action in the film [4].

- Roulette Shoe (1961) – MIT (Massachusetts Institute of Technology) mathematicians Edward O. Thorp and Claude Shannon built a computer small enough to fit inside a shoe. The wearer would observe the rotations of the roulette ball, tap the shoe accordingly and then receive a vibration telling them which number to bet on. Thorp's and Shannon's invention consists of two parts. One part is hidden in a shoe, and another part is fitted inside of a cigarette pack. The data-taker indicates the speed of the roulette wheel, and the computer sends the data to a hearing aid. A simple but efficient system that let the mathematicians predict the outcome of many roulette games, in times when the computers were the size of rooms. The wearable computer was invented in 1961, but it was mentioned for the first time only in 1966, in the revised edition of *Beat the Dealer* book by Edward Thorp. Later on, Thorp disclosed another wearable computer in the *LIFE Magazine*, on 27 March 1964 edition, pp. 80–91. A model for winning at the *Wheel of Fortune* gambling game was also published in *LIFE Magazine*.
- TV glasses (1963) – A pioneer of modern science fiction, Hugo Gernsback, thought that in the future, traditional television sets would be abandoned in favour of TV glasses, essentially a small portable television screen strapped to the face.

More recent advances include the following ones:

- The first wristwatch computer (1975) – The first ever 'wristwatch calculator' was introduced by Pulsar in 1975, just before Christmas time. The first "Limited Edition" of 100 pieces were available in 18 kt solid gold for 3,950$. Nevertheless, just like the gold Apple Watch, it was a huge success. A few months later, a more affordable 'stainless steel' version was offered for just 550$. As mentioned in [4], 'The Calculator' was promoted as the watch 'For the man who had everything until now'.
- Calculator watch (1977) – The 1980s and the 1990s were the times of the so-called commercial pioneering in wearable computing. In fact, the first wearable devices with mass market impact arrived in the late 1970s. One of the first wearables with real commercial success was the calculator watch, launched by Hewlett-Packard in 1977. Named the HP-01, the watch was sold at a top price of 850$. That is the equivalent to 3.500 US$ in 2015. Apart from the calculator functions, the HP-01 was able, just like any modern smartwatches, to store names, addresses, phone numbers and even appointments. The history of wearable technology can also not ignore the Japanese company Casio and its 'statement', at that time, the 'Databank' watch. It was so influential that the Police lead singer Sting wore one in the making of the song 'Wrapped Around Your Finger' and Michael J. Fox showcased his watch too, in the movie 'Back to the Future'.
- The first wearable music player (1979) – According to [4], the first low cost stereo, the Sony Walkman, changed how the world listens to music, and arguably has avoided having to listen to unwanted public transport conversations.

- The first portable computer (1981) – While still in high-school, Steve Mann wired a 6,502 computer (as used in the Apple-II) into a steel-frame backpack to control flash-bulbs, cameras and other photographic systems. The display was a camera viewfinder cathode-ray tube (CRT) attached to a helmet, giving 40 column text. Input was from seven microswitches built into the handle of a flash-lamp, and the entire system (including flash-lamps) was powered by lead-acid batteries.
- SEIKO UC 2000 Wrist PC (1981) – Marketed as a portable PC, the UC 2000 lets users tell the time, add up and input up to 2 kB of data via a keyboard strapped to the arm.
- Nelsonic Space Attacker Watch (1984) – Nelsonic made gaming portable with the Space Attacker Watch, with two front buttons, allowing wearers to play a basic space style arcade game.
- Private eye (1989) – A head mounted screen, hand held input device, 85 MB hard drive and motorbike battery made up this device, one of the earlier attempts at a portable Google Glass style computer.
- Sneaker phone (1990's) – Sports illustrated came up with the sneaker phone as a free promo in the early 1990s. A shoe and a corded phone all-in-one!
- The mBracelet (1999) – Developed in 1999 by Studio 5050 in New York, the mBracelet was 'innovation on the wrist' and included contactless payment. According to [4], with regret, the mBracelet remained just at the prototyping stage. Seen as a far too futuristic idea at that time, the project was let to vanish into obsolescence, and it died before it made to the market. Just like many other amazing ideas, the mBracelet was ahead of its time.
- Levi's ICD+ Jacket (2000) – It could be read in an advertisement in late 1990s on Philips Wearable Technology Centre 'Philips technology in every shirt and skirt'. It is in the year 2000 when together Levi's and Philips launched the world's first fashion tech garment, the ICD+ anorak. A conductible fabric harness allowed for a mobile phone, MP3 player and headphones to be integrated within the jacket, with a built-in button allowing wearers to switch between these. This project was led by Massimo Osti, a designer who also created the 'Ice Jacket', rubber flax and rubber wool, and colour variable with temperature material.
- Bluetooth headset (2002) – Using the new Bluetooth technology, Nokia introduced the Bluetooth headset, allowing users to take call hands free.
- Nike+ (2006) – A collaboration with Nike and Apple, Nike+ has been a fitness tracking kit. With a shoe embedded tracker, they could view the time, distance, pace and calories covered during workout on an iPod Nano screen.
- Fitbit Classic (2008) – In 2008, Fitbit, a wearable technology start-up, launched its Fitbit Classic wristband. A fitness tracker that allowed the wearer to monitor the taken steps, the travelled distance, the kilocalories consumed, intensity levels and sleep patterns.
- Google Glass (2013) – A pair of smart glasses also called a head-mounted Android smartphone. On paper, Google Glass project was ground breaking. In reality, the trial release of the Glass did not go quite as expected. There were major privacy concerns, endless headache complaints and even reports of potential addiction caused by the use of the glasses for a prolonged period of time. It was the start of

Figure 1.1 Garmin activity tracker and wearable antennas for RF energy harvesting

a new era, the use of wearable tech in fitness, healthcare and even fashion thanks to the new generations of smartwatches, smart jewellery and luxurious fitness trackers. Google glass was an evolution of the Lizzy research project, developed at the MIT Media Lab by Thad Starner and Doug Plat, since 1993. It began as a research project to develop a wearable platform that could be used for general-purpose computing applications. It was created as an open-source project with an assembly guide, hoping that many other people would adapt and repurpose it for their needs. Starner has continued to wear a customized computer since he created Lizzy, updating the parts with new technologies and features over the years, and then lead Google's Project Glass.

- Solar Powered Jacket (2014) – In late 2014, Tommy Hilfiger released a jacket with embedded solar panels, allowing users to charge their phone on the go.
- Activity Trackers (2014) – Fitness and activity trackers have been some of the most accessible wearable technology items of recent years, with models allowing users to view steps taken, walking speed, heart rate, sleeping patterns and even monitor UV ray exposure as shown in Figure 1.1.
- Apple watch (2015) – As well as texting, fitness tracking, TV control and ticket storage, the Apple clock can also tell the time.
- Ringly (2015) – Ringly offers escape from ones' phone, alerting wearers of notifi-cations via a discrete series of vibrations and light displays. It assumes that users will only check their phone when alerted of something really important.
- Quell (2015) – When placed near the body, Quell recognizes the signs of oncoming chronic pain and acts to stimulate nerves and block pain signals to the brain.
- bPay (2015) – Wearers of Barclay's bPay can make contactless payments via a personalized wristband, removing both the hassle of remembering pin code and carrying a bank card.

- Oculus rift (2015–16) – Via a motion sensing headset, headphones and built-in display, this virtually reality tool lets users step inside games and other virtual worlds to experience these more vividly than before.

According to [6], in 1998, the Tampere University of Technology, the University of Lapland and Reima Ltd decided to explore research on wearable technology. In the shared project they jointly created, the different kinds of prototypes, concepts and the few commercial products, needed to be organized into groups for inspection, and this identified need motivated them to establish a set of definitions for 'wearable computer', 'wearable electronics', and 'intelligent clothing', as explained in [6]. In particular, they identified that clothing is intelligent when it adds something traditionally unclothing-like to the garment, without taking away or compromising any traditional characteristics such as washability or wearability. Ideally, an intelligent garment offers a nontraditional garment function, such as health or sport activity monitoring, in addition to its traditional function of protecting the body. For example, it could collect data and either transfer it wirelessly and automatically to an external computing unit or process the data itself and respond to the computed conclusions without any user interfacing. When wearable computers evolved beyond PC hardware built in a backpack, and the number of people involved in wearable technologies increased in the late 1990s, the issue of textiles became relevant. Also, according to [6], initial collaboration with the "design" community was started when a wearable had to get an ergonomics update to enable long-term use, and a textile or clothing design student was asked to assist. They were the 'softer value' trespassing in the tech geek domain. The authors from [6] give a comprehensive description on how one team from Carnegie Mellon University in Pittsburgh, PA, showed scientifically why design is critical to the development process and provided the tools with which wearable technologies could be built more comfortably.

Arnault, in *The History of Wearable Technology – Past, Present and Future* [4], foresees a future of wearable technology, with embedded sensors and smart fabrics, with the beginning of new verticals and industries, digital health, fashion 'tech' and augmented reality. He foresees humanity immersed in an ocean of big data, interacting with smart clothes and objects, autonomous cars, smart cities, i.e. the 'digital persona', in a better world, connected, synchronized and safe.

The development of personal area networks (PANs) and WBSNs grew out of the research done by a number of different groups working at MIT in the 1990s [7]. Their idea was to interconnect information appliances that were carried on the body. They could send data through the body by modulating electric fields and determine relative positioning using electric field sensing. At that time, Zimmerman developed technology that had the effect of allowing the body to act like a copper cable [8]. The Physics and Media Group in MIT led by Gershenfeld was applying a method known as 'near-field coupling' to particular problems [9]. This method allows an accurate determination of the position of one part of the body in relation to another. They placed pairs of antennas on parts of the body (e.g. on the hand and elbow), and it was noted that by running an electric current between them, the circuit's capacitance changed as the parts moved (e.g. by flexing the elbow). So, the determination of the positions of the antenna was possible by measuring the capacitance change. However, such a measurement method returns wrong results when a hand is placed between the antennas. Zimmerman solved the problem and showed that some of the electric current was passing through the body, therefore affecting the measurement [9].

At the same time, a group working at the Media Lab asked these two researchers to develop a network to connect together all the electric gadgets that a person might be carrying on them. Many people carried around a number of digital devices such as a mobile phone, a PDA, and a digital watch but none of them could communicate with each other. Zimmerman and Gershenfeld were involved in the project and noticed that if they modulated the electric field flowing through a person's body, they could make it represent a 1 or 0 symbols, thus allowing the body to act as a communication medium to carry digital information. They also observed that if the frequency and power used were both kept very low, then the signal would not propagate far beyond the body. This would mean that only devices on the body or in direct contact with it were able to detect the signal. The current and voltage level used were very small, totally unnoticed by the person. The aforementioned ideas and efforts to developed a PAN were the earliest attempts to create a BAN.

1.3   State-of-the-art and recent advances

It is worthwhile to overview recent publications in the field of wearable technologies and WBSNs for healthcare, since several books and surveys have been published on these topics, in the last years. A review of research and development on wearable biosensor systems for health monitoring has been presented in [10]. This survey is focused on multi-parameter physiological sensing system designs, providing reliable vital signs measurements and incorporating real-time decision support for early detection of symptoms or context awareness. Authors have addressed the challenges and issues that need to be resolved for wearable systems to become more applicable to real-life situations.

Authors from [11] cover topics on wearable and autonomous systems, including devices. It renders some aspects in wearable system, such as, advanced wearable sensors for enabling applications, solutions for arthritic patients in their limited and conditioned movements, wearable gate analysis, EH, physiological parameter monitoring, communication, pathology detection, and also illustrates theoretical aspects and applications.

Authors from [12] review the main components of wearable systems, such as sensors, energy generation, signal processing and communications systems. Then they present some key applications of wearable systems in a variety of fields. Finally, the book covers aspects of commercial, social and environmental factors affecting development and use of wearables.

Microelectronics and complementary metal oxide semiconductor (CMOS) ICs provide very small size and ultra-low-power systems for biomedical applications, which is the subject of [13]. This book provides an overview of how to design and apply CMOS integrated circuits (ICs) for biomedical applications and also partially addresses wearable applications. The main topics include vital signal sensing and processing, biomedical wireless communications and examples of biomedical ICs.

Authors from [14] overview various types of wearable sensors, their medical applications like electrocardiography (ECG), electroencephalography (EEG), blood pressure, detection of blood glucose level, pulse rate, respiration rate and non-medical applications like daily exercise monitoring and motion detection of different body parts. They also address different types of noise removing filters that are helpful to remove noise from ECG signals.

Human activity recognition (HAR) is an important task in pervasive computing with many applications in medical, security, entertainment and tactical scenarios which is addressed by the authors of [15]. This paper presents the state-of-the-art in HAR based on wearable sensors. The main issues and challenges of HAR systems are presented, which is also followed with the main solutions to each one of them.

Authors of [16] present an overview of non-invasive physiological monitoring instrumentation with a focus on electrode and optrode interfaces to the body, and micropower-IC design for unobtrusive wearable applications. The paper addresses to technologies for electrode-tissue impedance measurement, photoplethysmography, functional near-infrared spectroscopy and signal coding. Examples of wearable and unobtrusive systems for physiological monitoring in medical applications are also presented.

Measuring of human vital signs should be as unobtrusive as possible. This kind of measuring sensors is directly placed on the human body as a patch or integrated into garments, which requires the devices to be very thin, flexible and sometimes even stretchable. The authors of [17] present an overview of recent technology developments in this wearable systems segment and also introduces application examples.

The relevant developments in the field of wearable sensors and systems to the field of rehabilitation are reviewed in [18]. The described applications have been utilized in health and wellness, safety, home rehabilitation, assessment of treatment efficacy and early detection of disorders. The integration of wearable and ambient sensors in the context of achieving home monitoring of older adults and subjects with chronic conditions has also been discussed.

An overview of the bluetooth low energy (BLE) technology for wearable sensor-based healthcare systems is presented in [19]. In comparison to traditional wired or other wireless communication modules, BLE is beneficial for short-range, low-power applications such as wearable sport bends, smart toothbrushes and intelligent watches. This paper addresses to challenges and hardware and software developments of wireless products equipped with Bluetooth modules.

Sensing of vital health signs is a challenging issue which is addressed in [20]. Being non-intrusive to people's comfort and safety, while providing good accuracy, noise effects on weak signals, and utilizing different kind of sensing elements are the main research and development challenges. The authors have presented a survey on the health sensing technologies using BSNs and mobile phones and classified related works by their application goals.

The book *Wearable Sensors* [21,22] provides a wide variety of topics associated with the development and application of wearable sensors. It provides an overview and coherent summary of many aspects of current wearable sensor technologies and includes the most current knowledge on the advancement of light-weight hardware, EH, signal processing, and wireless communications and networks. Authors have also addressed practical problems with smart fabrics, biomonitoring and health informatics, end user centric design, ethical and safety issues.

Research, latest developments and future directions of BSNs are the main topics of [23]. Authors present researches in the context of BANs including system integration, sensor miniaturization, low-power sensor interface, wireless telemetry, wearable and implantable *in vivo* monitoring systems and signal processing.

The authors of [24] have provided a structured literature survey of research in wearable technology for upper extremity rehabilitation, e.g. after stroke, spinal cord injury, for multiple sclerosis patients or even children with cerebral palsy. They have classified the main topics into three categories, depending on their functionality: movement and posture monitoring, monitoring and feedback systems that support rehabilitation exercises and serious games for rehabilitation training.

The work in [25] presents an overview of commonly available mobile and wearable technology (MWT) and examines how it can be used in health and wellness systems. Authors have provided case studies from two research projects, and the issues and challenges that arise in the use of MWT are also discussed.

The book *Wearable Electronics Sensors* [26] presents research and advancements in the area of wearable sensors, wireless sensors and sensor networks, protocols, topologies, instrumentation architectures, measurement techniques, EH and scavenging, signal processing, design and prototyping.

Research and developments carried out on wearable computing, WBSNs, wearable systems integrated with mobile computing, wireless networking and cloud computing are presented in [27]. It provides a specific focus on advanced methods for programming body sensor networks (BSNs) and features of online website to support readers in developing their own BSN application/systems.

Wearable sensor systems are able to transmit the information obtained from infants' body or even they can perceive external threats, such as falling or drowning. Some available wearable sensor systems for infants have been reviewed in [28]. Furthermore, authors have introduced the different modules of the framework in the sensor systems, and summarized and discussed the methods and techniques applied in wearable sensor systems.

The security in wearable communications is addressed [29], where an overview of security concerns for typical wearable applications and state-of-the-art research and state-of-the-practice development for security in wearable computing in both industry and academia is provided. A layered adaptive security architecture is proposed and discussed possible mechanisms to enhance the security of each layer.

A survey on current wearable mobile medical monitoring system (WMMMS) with emphasis on devices based on textile and wireless sensing networks is provided in [30]. Authors review several system implementations and compare them according to different classifications.

A broad spectrum of wearable sensing techniques and their applications in m-health have been provided in [31]. This book includes design methodologies of wearable sensors and systems that are end-to-end integrated and connected into worn items such as intelligent textile cloth and provides review materials on sensing technologies, signal processing, power harvest and signal transmission.

The authors of [32] have discussed the differences between wearable and traditional biometrics for computer systems, such as fingerprints, eye features or voice. They also have reviewed and provided a categorization of wearable sensors which is useful for capturing biometric signals, analysed the computational cost of the different signal processing techniques and reviewed and classified the most recent proposals in the field of wearable biometrics.

The threats and opportunities in the development and the acceptance of immersible and wearable technologies are addressed in [33]. Authors identify the hardware and software challenges and the bottlenecks of the current technologies and the limitations raised to the development of these technologies and discuss the social and commercial challenges related to innovation and acceptability.

Some characteristics of ultrawideband (UWB) radio signals make them highly suitable for less invasive systems, e.g. medical applications. UWB is a radio technology that can use a very weak signal over very wide spectrum for short-range, high-bandwidth communications [34]. Authors from [35] survey on UWB technology for medical sensing and communications. Authors have addressed challenges and suggested some research perspectives.

The state-of-the-art in radio frequency (RF)–energy harvesting (EH) for wearable biomedical sensors specifically for the global system for mobile (GSM) 900/1,800 MHz cellular and 700 MHz digital terrestrial television (DTTV or DTT) networks, as ambient RF energy sources has been presented in [36]. Authors provide guidelines for the choice of the number of stages for the RF energy harvester, which is dependent on the requirements from the embedded system to power supply.

In [37] an overview is provided on a number of activity recognition methods for a wearable sensor system. To study the accuracy against energy efficiency of transmission and processing power, authors have applied three methods for data transmission, namely 'stream-based', 'feature-based' and 'threshold-based' scenarios. For each method, the impact of variation of sampling frequency and data-transmission rate on energy consumption of wireless motes are also addressed.

Wearable antennas, made of flexible construction materials such as textiles are planar structures. A survey on the key factors for the design and development of textile antennas, from the choice of the textile materials to the framing of the antenna, has been presented in [38]. Authors describe also the most widely used textile materials.

1.4 Wearable medical technologies and devices, networks and frequency bands

Demands and benefits of the wearable medical devices have accelerated the growth of this technology in recent years. They have expanded widely in many applications from

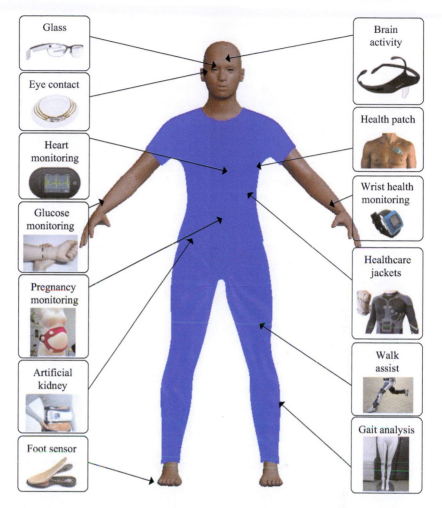

Figure 1.2   Illustration of recently proposed wearable healthcare technologies

blood pressure measuring to brain activity monitoring. There are four most preva-
lent chronic illnesses: congestive heart failure, diabetes, hypertension and chronic
obstructive pulmonary disease which have motivated the development of medical
devices specifically targeting each one of them [39]. Smart eye contact for measuring
blood glucose, heart vital signal monitoring, blood glucose monitoring patch, pregnancy monitoring, artificial kidney, foot sensor, brain activity monitor, health patch,
wrist health monitoring, healthcare jackets, walk assist and gait analysis are just a
few examples of available technologies as shown in Figure 1.2.

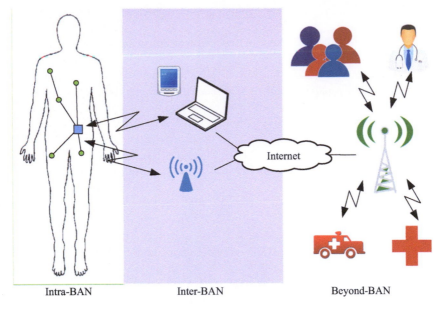

Intra-BAN Inter-BAN Beyond-BAN

Figure 1.3 Communication architecture of BAN

1.4.1 Communication architecture

The communication architecture for wearable systems and BANs applications can be categorized into three main classes as shown in Figure 1.3: Intra-BAN, Inter-BAN and Beyond-BAN. Intra-BAN includes in-body or on-body sensor nodes, network between them, which can be wired, wireless (<2 m coverage) or HBC. Inter-BAN refers to the communication between the Intra-BAN via a link (usually a wireless interface) and a computer, PDA and or access points (AP). Inter-BAN communication aims to interconnect BANs with various networks, such as including cellular networks and the internet. Beyond-BAN communication is for use in metropolitan areas. A gateway, such as a PDA, is needed to bridge the connection between the inter-BAN and this network. Beyond-BAN communication allows accessing all necessary information of a patient to help on diagnosis or can be notified of an emergency status.

1.4.2 Physical layer

The PHY layer consists of the networking hardware and transmission technologies of communication network. This layer provides electrical, mechanical and procedural interface to the transmission medium. The choice of the PHY layer depends on the application. IEEE 802.15.6 specifies three different PHY layers: HBC, narrow band (NB) and UWB [40]. NB physical (PHY) is responsible for data transmission/reception, activation or deactivation of the radio transceiver and Clear Channel Association (CCA) in the current channel. HBC PHY defines the electrostatic field communication requirements covering modulation, preamble/start frame delimiter and packet structure.

The UWB PHY layer is used for communication between on-body devices and for communication between on-body and off-body devices [41]. Transceivers in a UWB PHY generate similar signal power levels to those used in the Medical Implant Communication Service (MICS) band and also allow low implementation complexity. Based on the specifications for UWB PHY layer, the Physical Protocol Data Unit (PPDU) bits are converted into RF signals for wireless transmission. The UWB PPDU consists of a synchronization header, a PHY header (PHR) and PSDU. The PHR carries information about the scrambler seed, length of the payload and the data rate of the PHY Payload (PSDU). The receiver uses the information in the PHR to decode the PSDU [40].

Issues related to wireless devices such as high energy consumption, low data rate, interference, security and health problems due to signal propagation on body, and the possibility of having available conductive yarns or copper wires, lead to an alternative use of wired communication in some WBSNs applications. The use of fabrics with embedded conductors such as copper wires and conductive yarns can provide the required electrical connections and would make the product more user friendly and comfortable [42]. The design of such a wearable system must take into account the fact that electrical characteristics of conductive yarns are not exactly those of normal wires, since their electrical properties may exhibit significant variations in relaxed and stretched modes. Besides, conductive yarns are prone to tearing and wearing out; thus, to achieve high reliability, the network must be tolerant to catastrophic and parametric faults.

1.4.3   MAC sub-layer protocols

In WSNs and WBSNs, the communication block of sensor nodes consumes most of the energy, which needs to be considered in energy-efficient applications [43,44]. The MAC protocol plays significantly important role in controlling the communication module while reducing the energy consumption, increasing throughput and reducing end-to-end delay. Regardless of the network type, MAC sub-layer protocols are grouped into contention-based and schedule-based protocols [45]. In contention-based MAC sub-layer protocols, such as carrier sense multiple access with collision avoidance CSMA/CA, nodes contend for the shared channel to transmit data. If the channel is occupied, the node defers its transmission until it becomes idle. Although CSMA/CA protocols are scalable and have no strict time synchronization constraints, they impose a significant protocol overhead. In contrast, schedule-based protocols, such as the time division multiple access (TDMA) protocol, divide the channel in time slots of fixed or variable duration. These slots are assigned to the existing nodes and each node transmits only in its own slot. Schedule-based protocols consume less energy than contention-based protocols, but for appropriate operation, they require time synchronization.

One of the most important attributes of a good MAC protocol for BAN is the energy efficiency [46]. Among existing MAC protocols, IEEE 802.15.4 and IEEE 802.15.6 can be mentioned as standard energy efficient methods available for BANs application. In some applications, the device supplies with harvested power from a RF energy harvesting (RF-EH) system or should support a battery lifetime of months or years, without replacing or charging, while other applications may require a battery life of a few hours. For example, cardiac defibrillators and pacemakers have a lifetime of more than 5 years, while swallowable camera pills have a lifetime of 12 h [40]. One energy-consumption reducing method for the MAC sub-layer is the low-power listening (LPL) [47] (also known as channel polling). In this approach, nodes wake up for a short duration to check the channel activity. If the channel is idle, the nodes go into sleep mode; otherwise, they continue to be active on the channel to receive data.

Quality of service (QoS) is also an important factor to be considered in MAC protocols for BAN applications [48]. QoS also includes point-to-point delay and delay variation. In some cases, such as fitness and medical surgery monitoring applications, real-time communication is required. The latency of data transmission depends on the application. For example, for multimedia applications, latency should be less than 250 ms, while the jitter should be kept below 50 ms [43].

Rapidly growing the number of wireless devices lead to increased traffic on radio spectrum, especially on industrial scientific medical (ISM) bands. CR technology has emerged as a key technique in wireless networks to optimally use licensed or white space bands. It enables unlicensed users (or secondary users (SUs)) to opportunistically access underutilized licensed spectra whenever the licensed users (or primary users (PUs)) are idle. The MAC layer plays a key role in several CR functions, such as channel sensing, resource allocation, spectrum mobility and spectrum sharing [49].

1.4.4   Routing protocols

In a multi-hop communication, a routing protocol is needed to establish end-to-end data delivery. A lot of research is being done towards energy efficient routing in *ad-hoc* networks and WSNs [50]. Routing protocols for WBSN borrow many concepts from WSN. However, some of the proposed solutions are inadequate for BANs. For example, in WSNs having high throughput and low routing overhead is more important than minimal energy consumption. Energy efficient *ad-hoc* network protocols attempt to find routes in the network that minimize energy consumption in nodes with small energy resources, thereby ignoring many parameters such as the amount of operations and energy required to transmit and receive. Unlike BANs, in WSNs, node failure is not problematic. In BANs, the number of devices worn by a patient is smaller than the typical number of nodes in a WSN. Besides, some WSNs protocols only consider networks with homogeneous sensors [51], which does not apply when considering different devices in BAN. In brief, although challenges faced by BANs are in many ways similar to sensor networks and WSNs, there are intrinsic differences between the two, which require special attention in each context.

1.4.5   IEEE 802.15.6 and 11073 standards

All medical devices and wearable systems must as well meet standards and obey regulations. An overview of international, USA and European standards is available in

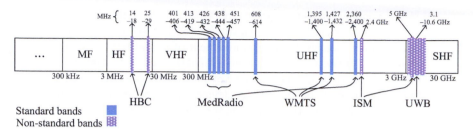

Figure 1.4   Radio frequency bands for medical wireless devices

[52–54]. This section presents two IEEE standards 802.15.6 and 11073 for low-power wireless communication and interoperability.

IEEE 802.15.6-2012 [55] is a standard for medical or non-medical applications and determines low-power, short-range, highly reliable wireless communications in the vicinity of, or inside, a body (but not limited to humans) including wearable systems. This standard determines the data rate up to 10 MHz which is needed to satisfy all existing medical applications. The utilized radio channels are ISM or other bands compliance with applicable medical and communication regulatory authorities. To minimize specific absorption rate (SAR) into the body and increase battery life time, transmitters operate on very low-power. In most cases, medical systems have to support QoS for emergency messages and provide communication security for sensitive signals.

IEEE 11073 standard for health informatics, medical, and health device communication enables communication between medical, healthcare and wellness devices with external computer systems in a manner that enables plug-and-play interoperability [56]. Real-time plug-and-play interoperability for medical devices and efficient exchange of care device data, acquired at the point-of-care, in all care environments, are the main goals of IEEE 11073.

1.4.6   Frequency bands

Figure 1.4 depicts standards and non-standard radio bands that are used for medical wireless devices. The Medical Device Radiocommunications Service (MedRadio) is a specification introduced by Federal Communications Commissions (FCC) for communication spectrum needs for diagnostic and therapeutic purposes of medical implants and body-worn medical devices [57]. A previous specification from FCC, called the MICS, established in 1999 that MedRadio is in the radio spectrum of 401–406, 413–419, 426–432, 438–444 and 451–457 MHz range. In 2012, the band of 2,360–2,400 MHz was also assigned for medical body area network (MBAN). Due to being out of ISM band, interferences can significantly be reduced. In Europe, ETSI has considered the 401–402 and 405–406 MHz ranges as a standard called MEDS (Medical Data Service Devices) [58]. This set of frequencies is also being considered in other parts of the world.

Wireless medical telemetry service (WMTS) is another standard from FCC in United States for wireless medical applications [59]. WMTS was created in 2000 to avoid radio interferences due to establishment of digital television. Three radio bands have been defined for WMTS; 608–614 MHz (UHF TV channel 37), 1,395–1,400 MHz and 1,427–1,432 MHz (all used by the Federal Government). Due to lack of international agreement on these bands, devices operating on WMTS bands cannot be marketed or used in countries other than the United States.

1.4.7   Wireless devices

In comparison with sensing and processing blocks, wireless transceivers consume a significant higher amount of energy. Hence, selection of appropriate communication devices and protocols for wearable systems running on small batteries for long time is very important. Table 1.1 shows a list of available low-power wireless technologies for wearable systems.

Table 1.1   Wireless technologies and standards for WBSNs

Mote	Frequency band	Range	Data rate	Applications
NFC [60]	13.56 MHz ISM	20 cm	424 kbps	RFID, very short range communication
Zarlink [61]	402–405 MHz 433–434 MHz	2 m	200, 400, 800 kbps	Implantable devices, BAN
Sensium [62]	862–870 MHz 902–928 MHz	3 m	50 kbps	Medical devices
Bluetooth LE [63]	2.4 GHz ISM	10 m	2 Mbps	Healthcare, PAN, BAN
ZigBee [64]	868 MHz, 2.4 GHz ISM	20 m	20~250 kbps	Multipurpose, healthcare, PAN, BAN
RuBee [65]	131 kHz	30 m	1.2 kbps	Harsh environment, high security
ANT [66]	2.4 GHz ISM	30 m	60 kbps	Sports and fitness, multipurpose
BodyLAN [67]	2.4 GHz ISM	100 m	250 kbps 1 Mbps	WSN, PAN, medical

1.5   European- and global-funded research projects

Many research projects in the field of wearable technologies have been or are being carried out. A list of current and ongoing international and European projects in the context of wearable systems is shown in Table 1.2. These projects are mainly proposed to monitor health parameters, e.g. human gait monitoring, remote health monitoring, gesture detection, wound monitoring and therapy, monitoring of PHY activity, health status monitoring in harsh environments, heart monitoring and telecardiology. They

Table 1.2 List of projects

Project	Application	Network	Sensors
ProLimb [68]	Human gait monitoring	Mesh, conductive yarns, bluetooth	EMG, ACC, GYRO
ASNET [69]	Remote health monitoring	WiFi, internet, GSM	Temperature, blood pressure
WiMoCA [70]	Sport, gesture detection	Star, bluetooth, internet	ACC
MIMOSA [71]	Ambient intelligence	Bluetooth, wibree, mobile, internet	RFID sensors
WM [72]	Monitoring of vital signs during spaceflights	wire	Temperature, ECG
CROW^2 [73]	Life-critical and rescue operation	Wireless, body-to-body	Any sensor
SWAN-iCare [74]	Wound monitoring and therapy	WLAN, GPRS, internet	MMPs, CRP, TNF-alpha, PH, temperature
SVELTE [75]	Monitoring of physical activity	Bluetooth	ACC, MGN
ProeTEX [76]	Health status monitoring in harsh environments	RS485 bus, bluetooth, WiFi	ECG, SpO$_2$, temperature
WE-CARE [77]	Heart monitoring, telecardiology	WiFi, GPRS, 3G	ECG
INTERACTION [78,79]	Remote monitoring of physical interaction with the environment	Wire, wireless	EMG, goniometer, strain sensor
Nephron+ [80,81]	Wearable artificial kidney	RS485 bus	Physiological sensors
Smart-Clothing [82]	Health monitoring and sport	WSN, GSM, internet	ECG, temperature, movement

utilize a variety of sensors such as electromyography (EMG), ECG, accelerometer (ACC), gyroscope (GYRO), magnetometer (MGN), temperature, radio frequency identification (RFID), PH meter, SpO$_2$ and goniometer.

1.6 Challenges

The recent advances of wearable technologies from their advent till commercial usage has only taken few decades. There are still many challenges ahead that need to be addressed as follows:

- **Design and development issues**: A typical wearable system is composed of textiles, hardware, protocols and software. Because of limited amount of resources in these kind of systems, efficiency is a key factor. Different types of metrics include the energy efficiency, channel utilization and lifetime. Besides, the cost of wearable products is important for their popularity. Wearable devices can be categorized into three subclasses, based on their price [33,83]: (a) less expensive devices such as Fitbit, which cost as low as 50–100$, (b) moderate price devices, Sony and Samsung smartwatches, which cost around 300–400$ and (c) expensive devices, like Google Glasses that cost around 1,500$. Medical devices used to monitor vital signals, such as ECG and or EMG, involve very low level signals in the range of few milivolts, in a noisy environment especially 50/60 Hz signals propagated from electricity grid. Using appropriate electrodes, protecting against noise, filtering and amplification of such weak signals to provide clear output, must be taken into the account [84]. Because of wearable systems portability requirements, they must be small and light. Miniaturized devices are not only more attractive for the most of people but also they are very light, comfortable and usually consume low energy. Besides, the shape of the these devices is to be curved or strapped to fit for wearing purposes [33,85]. Nevertheless, continuous usage of wearable devices may cause skin allergies and/or rashes [86]. To avoid such health issues, the materials utilized for wearable products must be carefully selected.

- **Social issues**: Many wearable health systems require internet connection for e-health monitoring and ambulatory messages. However, internet is not yet available everywhere, particularly in developing countries. In the absence of a wide internet coverage, many of wearable technologies can neither evolved nor succeed. Access to personal data generated or captured by wearable systems must obey some rules and policies. In fact, considerable legislation is needed to deal with the advancement in the context of wearable technologies. The European Union, in its directive 95/46/EC, has asserted that during all the transition, processing or storage of data, the integrity, privacy and confidentiality of the personal data must be secured at any cost [33]. The performance of the systems is a trade-off between technology limits and demands. Integration techniques provide adequate computational power for wearable devices. However, demands for new applications, reduction of weight, power, size and acceptability of wearable devices, impose difficulties for the device designers, while including such capabilities that allow the heterogeneous workability of a tiny wearable device [87].
- **Security**: There are some privacy and security issues that are emerging with the development of wearable systems [88]. Privacy issues are increasing, perhaps because of the fact that developments at both ends are not consistent with each other and the users demand for more functionality has been traded-off for compromised privacy [89]. Privacy in wearable computing is among the major challenges of all time, especially the mechanisms utilized to transfer confidential information. Wearable devices work in close integration with the human body and closely monitor the human behaviour, likes, movements and activities. Without adequate security controls in place, personal data can be stolen and used for committing fraud and crime. Healthcare applications store user data on a centralized system to

provide the user with access to the data later for reference usage [88]. However, the privacy terms of the service provider are generally vague and mention statements such as 'your data may be viewed by a third party'. Such situations are considered neither by the user nor by the group of their friends, and the access to data is given to the device-maker to further make it available to third parties. Giving the third party access to sensitive health data makes them vulnerable to diverse privacy threats [90]. It is of immense importance to determine how secure the information is with the device-maker and what encryption standards are being followed, and what other measures the device maker may take to further strengthen the security issues with the increasing threats to data [33].

1.7   Main objectives and structure of the book

The new book *Wearable Technologies and Wireless Body Sensor Networks for Healthcare* is included in the *Healthcare Technologies* series from IET. This specific book is dedicated to topics that span from scenarios and WSBN communication-applications to sensor devices and systems, activity recognition, monitoring of the human gait through the application of dedicated IC, smart textiles and their applications to smart sensing, RF propagation aspects, detailed modelling of this very complex communication environments, measurements and CR, link layer, MAC sub-layer protocols and synchronization aspects, aspects of network layer, the medical applications of WBSNs as well as the underlying wearable solutions.

The book not only illustrates conceptual aspects and applications but also provides new vision in characterising devices and systems. EH within wearable solutions is a key issue since it allows for improving energy efficiency and reliability in wearable antenna and sensor devices, algorithms, protocols and networks.

Apart from Chapter 1, the book is organized into six main parts, as follows:

1. Overview of WBSN Communication-applications and Scenarios (Chapters 2 and 3) – a motivation for the wearable technologies by covering the applications involved in WBSNs scenarios is addressed through the proposal of a classification taxonomy and discussion of their characterization parameters, followed by the discussion of some laboratory trials including large number of sensors. In these trials, a dedicated circuit is applied in the implementation of a wearable BAN system to be used for both medical and non-medical applications.

   Chapter 2 presents a comprehensive overview from deployments scenarios and the underlying sets of applications for wearable technologies and WBSNs, as well as their classification and main characterization parameters, while discussing the advantages, disadvantages and opportunities for RF-EH to power WBSN nodes. A reliable wearable BAN system to use for both medical and non-medical applications, including large number of sensors (<250) embedded in textile and high data rate (<18 Mbps) communication demands is presented in Chapter 3.

2. Devices and systems (Chapter 4) – Automatic recognition of PHY activities, which is known HAR, is addressed for a wearable sensor system. The HARs are used in rehabilitation centres to monitor the activity of daily living while assessing daily functional status of elderly. Besides providing a comprehensive background to the HAR procedures, aspects of design, implementation and evaluation are also addressed.

    Chapter 4 presents a survey on HAR systems with an application-oriented approach.

3. Textile materials for wearable applications (Chapter 5) – The characterization of the conductive and dielectric properties of textile materials used for the manu-facturing of flexible antennas and different devices that are embedded onto the garment need to be characterized in detail. As little information can be found in literature on the electromagnetic properties of regular textiles, it is worthwhile to present an overview of the different methods considered to characterize the permittivity of textile materials and underlying experimental techniques.

    Chapter 5 covers the characterization of textile materials and is mainly focused on the analysis of the dielectric properties of normal fabrics.

4. Propagation aspects and CR (Chapters 6 and 7) – The application of radio signal strength captured by sensors placed around the human body in human movement identification is behind a method proposed to identify the human movement in WBANs. Different types of human motion are considered and machine-learning techniques are applied, including neural networks and decision trees. To avoid mutual interference with other electronic devices, a viable architecture of an MBAN with practical CR features based can be applied while enhancing the coexistence with other collocated wireless systems.

    Chapter 6 investigates the feasibility of using radio signal strength of sensors placed around the human body for human movement identification purposes. A viable architecture of a WBAN with practical CR features based on UWB radio technology is proposed in Chapter 7 while addressing MAC protocols that facilitate the optimization of channel use while considering aspects of RF-EH.

5. Link layer, MAC sub-layer and synchronization aspects (Chapters 8 and 9) – The available power determines the maximum lifetime of the WBSN nodes and wearable systems. The proposal of an IEEE 802.15.4 MAC layer performance enhancement that employs RTS/CTS combined with packet concatenation (e.g. block acknowledgement) facilitates the improvement of channel efficiency by decreasing the deferral time before transmitting a data packet. The so-called multi-channel scheduled channel polling (MC-SCP-MAC) protocol explores the influential range concept, denial channel list, extra resolution phase algo-rithm and frame capture effect to achieve the maximum performance in terms of delivery ratio and reduced energy consumption. It is therefore important to compare MC-SCP-MAC with traditional protocols, such as SCP-MAC or MC-LMAC, while verifying how the influential range concept allows for reducing the redundancy level and the energy consumption in the network. As a com-plementary aspect, synchronization at the MAC sub-layer can be applied in the

simultaneous acquisition of surface electromyographic signals of several muscles. These synchronization protocols are used in the router IC designed for wearable systems.

Chapter 8 presents an IEEE 802.15.4 MAC layer performance enhancement by employing RTS/CTS combined with packet concatenation. Chapter 9 presents a one-way method for synchronization at the MAC sub-layer of nodes and a circuit based on that in a wearable sensor network. The proposed approach minimizes the time skew with an accuracy of half of clock cycle in average. Performance evaluation of these protocols includes the analysis of minimum delay, maximum throughput and the energy trade-off.

6. Applications of wearable technologies and WBSN (Chapters 10–14) – Wearable sensor networks for human gait analysis or cardiac and coronary monitoring, mobile wearable surveillance product for first response and hazardous professions or wearable sensors for foetal monitoring in the last 5 weeks of pregnancy are examples of recent projects that developed comprehensive applications of wearable technologies. The description of their functionality and performance provides a set of examples of how wearable technologies will be very useful to the population or target groups in different scenarios, including the e-health environment. In particular, recent research and development projects, as eWALL, have been identifying required technological innovations to conceive easy-to-implement smart 'caring home'. Finally, a very recent disruptive technology is EH (e.g. RF-EH) that can pave the road towards the massive utilization of wireless wearable sensors for patient self-monitoring and daily healthcare.

A wearable data-capture system for gait analysis is presented in Chapter 10. It consists of a pantyhose with embedded conductive yarns interconnecting sensor nodes forming a mesh-like network. The sensor nodes are customized sensing electronic modules that capture inertial and electromyographic signals and sends aggregated information to an external-processing device through a bluetooth link.

Chapter 11 describes the eWALL e-health environment and the required technological innovations to design an affordable, easy-to-implement smart, 'caring home' cognitive environment. Chapter 12 presents a new wearable technology, including ECG device, capable to acquire medical quality ECG using only two electrodes with a small form factor, also acquiring body temperature and actigraphy. A wearable environmental sensing system capable to monitor temperature, humidity, pressure, luminosity, altitude, toxic gases and, if needed, Global Positioning System (GPS) has also been developed. In chapter 13, an easy-to-wear belt with a telemedicine system for continuous monitoring of the foetal health is proposed. Chapter 14 presents the state-of-the-art in RF-EH for wearable biomedical sensors specifically targeting the global system of mobile 900/1,800 MHz cellular and 700 MHz DTTV or DTT networks as ambient RF energy sources.

Chapter 15 not only gives a final overview of the main topics covered in the book but also presents a taxonomy for the classification of wearable devices, gives insights on the primary areas of innovation in wearable healthcare and identifies the need for standardization, the path towards the interoperability with other wireless systems.

References

[1]   Wearable Technology Market, *Scalar Market Research*; 2016. Available from: https://www.scalarmarketresearch.com/market-reports/wearable-technology-market.

[2]   Yoo H.J., Hoof C.V. *Bio Medical CMOS ICs*. 1st ed. Boston, MA, USA: Springer; 2011.

[3]   Desjardins J. *The History of Wearable Technology*; 2015. Available from: http://www.visualcapitalist.com/the-history-of-wearable-technology/.

[4]   Arnault L. *The History of Wearable Technology – Past, Present and Future*; 2015. Available from: https://wtvox.com/featured-news/history-of-wearable-technology-2/.

[5]   Guler S.D., Gannon M., Sicchio K. Chapter: A Brief History of Wearables. In: Crafting Wearables. Springer; 2016. pp. 3–10.

[6]   McCann J., Bryson D. *Smart Clothes and Wearable Technology*. Boston, MA, USA: CRC Press; 2009.

[7]   Reddy G.P., Reddy P.B., Reddy R. 'Body Area Networks'. *Journal of Telematics and Informatics*. 2013 12;1(1):36–42.

[8]   Zimmerman T.G. 'Personal Area Networks: Near-field Intrabody Communication'. *IBM Systems Journal*. 1996;35(3.4):609–617.

[9]   Zimmerman T.G., Smith J.R., Paradiso J.A., Allport D., Gershenfeld N. 'Applying electric field sensing to human-computer interfaces'. *Intl. Conf. on Human Factors in Computing Systems (SIGCHI)*; New York, NY, USA: ACM Press/Addison-Wesley Publishing Co.; 1995. pp. 280–287. Available from: http://dx.doi.org/10.1145/223904.223940.

[10]   Pantelopoulos A., Bourbakis N.G. 'A Survey on Wearable Sensor-Based Systems for Health Monitoring and Prognosis'. *IEEE Transactions on Systems, Man, and Cybernetics, Part C (Applications and Reviews)*. 2010;40(1): 1–12.

[11]   Lay-Ekuakille A., Mukhopadhyay S.C. *Wearable and Autonomous Biomedical Devices and Systems for Smart Environment*. Berlin: Springer; 2010.

[12]   Bonfiglio A., Rossi D.D. *Wearable Monitoring Systems*. Boston, MA: Springer; 2011.

[13]   Yoo H.J., Hoof C.V. *Bio-Medical CMOS ICs*. Boston, MA: Springer; 2011.

[14]   Rehman A., Mustafa M., Javaid N., Qasim U., Khan Z.A. 'Analytical survey of wearable sensors'. *2012 Seventh International Conference on Broadband, Wireless Computing, Communication and Applications*; 2012. pp. 408–413.

[15]   Lara O.D., Labrador M.A. 'A Survey on Human Activity Recognition Using Wearable Sensors'. *IEEE Communications Surveys Tutorials*. 2013;15(3): 1192–1209.

[16]   Ha S., Kim C., Chi Y.M., *et al.* 'Integrated Circuits and Electrode Interfaces for Noninvasive Physiological Monitoring'. *IEEE Transactions on Biomedical Engineering*. 2014;61(5):1522–1537.

[17]   van den Brand J., de Kok M., Sridhar A., *et al.* 'Flexible and stretchable electronics for wearable healthcare'. *2014 44th European Solid State Device Research Conference (ESSDERC)*; 2014. pp. 206–209.

[18]   Patel S., Park H., Bonato P., Chan L., Rodgers M. 'A Review of Wearable Sensors and Systems with Application in Rehabilitation'. *Journal of NeuroEngineering and Rehabilitation.* 2012;9(21):2–17.

[19]   Zhang T., Lu J., Hu F., Hao Q. 'Bluetooth low energy for wearable sensor-based healthcare systems'. *2014 IEEE Healthcare Innovation Conference (HIC)*; 2014. pp. 251–254.

[20]   Song L., Wang Y., Yang J.J., Li J. 'Health sensing by wearable sensors and mobile phones: A survey'. *2014 IEEE 16th International Conference on e-Health Networking, Applications and Services (HealthCom)*; 2014. pp. 453–459.

[21]   Sazonov E., Neuman M. *Wearable Sensors.* Waltham, MA: Academic Press; 2014.

[22]   Sazonov E. *Wearable Sensors: Fundamentals, Implementation and Applications.* Boston, MA: Academic Press; 2014.

[23]   Yang G.Z. *Body Sensor Networks.* 2nd ed. London: Springer; 2014.

[24]   Wang Q., Chen W., Markopoulos P. 'Literature review on wearable systems in upper extremity rehabilitation'. *IEEE-EMBS International Conference on Biomedical and Health Informatics (BHI)*; 2014. pp. 551–555.

[25]   Amor J.D., James C.J. 'Setting the scene: Mobile and wearable technology for managing healthcare and wellbeing'. *2015 37th Annual International Conference of the IEEE Engineering in Medicine and Biology Society (EMBC)*; 2015. pp. 7752–7755.

[26]   Mukhopadhyay S.C. *Wearable Electronics Sensors.* Palmerston North: Springer; 2015.

[27]   Fortino G., Gravina R., Galzarano S. *Wearable Systems and Body Sensor Networks: From Modelling to Implementation.* Hoboken, NJ: John Wiley & Sons Inc; 2015.

[28]   Zhu Z., Liu T., Li G., Li T., Inoue Y. 'Wearable Sensor Systems for Infants'. *Sensors.* 2015;15(2):3721–3749.

[29]   Wang S., Bie R., Zhao F., Zhang N., Cheng X., Choi H.A. 'Security in Wearable Communications'. *IEEE Network.* 2016;30(5):61–67.

[30]   Liang T., Yuan Y.J. 'Wearable Medical Monitoring Systems Based on Wireless Networks: A Review'. *IEEE Sensors Journal.* 2016;16(23):8186–8199.

[31]   Zhang Y.T. *Wearable Medical Sensors and Systems.* Springer; 2020.

[32]   Blasco J., Chen T.M., Tapiador J., Peris-Lopez P. 'A Survey of Wearable Biometric Recognition Systems'. *ACM Computing Surveys.* 2016;49(3):43: 1–43:35. Available from: http://doi.acm.org/10.1145/2968215.

[33]   Saleem K., Shahzad B., Orgun M.A., Al-Muhtadi J., Rodrigues J.J.P.C., Zakariah M. 'Design and Deployment Challenges in Immersive and Wearable Technologies'. *Journal of Behaviour & Information Technology.* 2017;36(7):687–698.

[34]   Ultra-wideband (UWB); 2014. Available from: https://en.wikipedia.org/wiki/Ultra-wideband#cite_note-1.

[35]   Chávez-Santiago R., Balasingham I., Bergsland J. 'Ultrawideband Technology in Medicine: A Survey'. *Journal of Electrical and Computer Engineering.* 2012;2012:3:1–3:9. Available from: http://dx.doi.org/10.1155/2012/716973.

[36] Borges L.M., Chávez-Santiago R., Barroca N., Velez F.J., Balasingham I. 'Radio-frequency Energy Harvesting for Wearable Sensors'. *Healthcare Technology Letters*. 2015;2(1):22–27.

[37] Rezaie H., Ghassemian M. 'Implementation Study of Wearable Sensors for Activity Recognition Systems'. *Healthcare Technology Letters*. 2015;2(4): 95–100.

[38] Salvado R., Loss C., Goncalves R., Pinho P. 'Textile Materials for the Design of Wearable Antennas: A Survey'. *Sensors*. 2012;12(11):15841–15857. Available from: http://www.mdpi.com/1424-8220/12/11/15841.

[39] Brassard P., Ferland A., Marquis K., Maltais F., Jobin J., Poirier P. 'Impact of Diabetes, Chronic Heart Failure, Congenital Heart Disease and Chronic Obstructive Pulmonary Disease on Acute and Chronic Exercise Responses'. *The Canadian Journal of Cardiology*. 2007;25(B):89B–96B.

[40] Kwak K.S., Ullah S., Ullah N. 'An overview of IEEE 802.15.6 standard'. *3rd Intl. Symp. on Applied Sciences in Biomedical and Communication Technologies (ISABEL)*; 2010. pp. 1–6.

[41] Movassaghi S., Abolhasan M., Lipman J., Smith D., Jamalipour A. 'Wireless Body Area Networks: A Survey'. *IEEE Communications Surveys Tutorials*. 2014;16(3):1658–1686.

[42] Gatzoulis L., Iakovidis I. 'Wearable and Portable eHealth Systems'. *IEEE Engineering in Medicine and Biology Magazine*. 2007;26(5):51–56.

[43] Ullah S., Higgins H., Braem B., *et al.* 'A Comprehensive Survey of Wireless Body Area Networks: On PHY, MAC, and Network Layers Solutions'. *Journal of Medical System*. 2012;36:1065–1094.

[44] Rahim A., Javaid N., Aslam M., Rahman Z., Qasim U., Khan Z.A. 'A comprehensive survey of MAC protocols for wireless body area networks'. *17th Intl. Conf. on Broadband, Wireless Computing, Communication and Applications (BWCCA)*; 2012. pp. 434–439.

[45] Gopalan S., Park J.T. 'Energy-efficient MAC protocols for wireless body area networks: Survey'. *Intl. Congress on Ultra Modern Telecommunications and Control Systems and Workshops (ICUMT)*; 2010. pp. 739–744.

[46] Ullah S., Shen B., Riazul Islam S.M., *et al.* 'A Study of MAC Protocols for WBANs'. *Journal of Sensors*. 2009;10(1):128–145.

[47] Cano C., Bellalta B., Sfairopoulou A., Oliver M., Barceló J. 'Taking Advantage of Overhearing in Low Power Listening WSNs: A Performance Analysis of the LWT-MAC Protocol'. *Mobile Networks and Applications*. 2011;16(5): 613–628. Available from: http://dx.doi.org/10.1007/s11036-010-0280-4.

[48] Karapistoli E., Stratogiannis D., Tsiropoulos G., Pavlidou F. 'MAC protocols for ultra-wideband ad hoc and sensor networking: A survey'. *4th Intl. Congress on Ultra Modern Telecommunications and Control Systems and Workshops (ICUMT)*; 2012. pp. 834–841.

[49] Bhandari S., Moh S. 'A Survey of MAC Protocols for Cognitive Radio Body Area Networks'. *Journal of Sensors*. 2015;15(4):9189–9209.

[50] Akkaya K., Younis M. 'A survey on routing protocols for wireless sensor networks'. *Journal of Ad Hoc Networks*. 2005;3:325–349.

[51] Dargie W., Poellabauer C. *Fundamentals of Wireless Sensor Networks: Theory and Practice*. 1st ed. Wiley; Chichester: 2010.

[52] List of International Standards for Medical Devices; 2015. Available from: http://www.mdb.gov.my/mdb/documents/standards/international%20standard.pdf.

[53] U.S. FDA Medical Device Regulations; 2016. Available from: http://www.registrarcorp.com/fda-medical-device/?lang=en&s_kwcid=TC|9242|fda%20regulatory%20medical%20device||S|b|175693784841&gclid=CMWXq-m3l9ICFcMy0wod8hcGaQ.

[54] European Union Updates List of Standards for Medical Device Directives; 2015. Available from: https://www.emergogroup.com/blog/2015/07/european-union-updates-list-standards-medica-device-directives.

[55] IEEE. *IEEE Standard for Local and metropolitan area networks, Part 15.6: Wireless Body Area Networks*. IEEE Std 802-2014 (Revision to IEEE Std 802-2001); Feb 2012.

[56] International Standard ISO/IEEE 11073-10418:2014; 2016. Available from: https://www.iso.org/obp/ui/#iso:std:iso-ieee:11073:-10418:ed-1:v2:cor:1:v1:en.

[57] Medical Device Radiocommunications Service (MedRadio), Federal Communications Commissions (FCC); 2017. Available from: https://www.fcc.gov/general/medical-device-radiocommunications-service-medradio.

[58] Specific Conditions for Medical Data Service Devices (MEDS) Operating in the 401 MHz to 402 MHz and 405 MHz to 406 MHz Bands, European Telecommunications Standards Institute (ETSI); 2009. Available from: http://www.etsi.org/deliver/etsi_en/301400_301499/30148929/01.01.01_60/en_30148929v010101p.pdf.

[59] Wireless Medical Telemetry Service (WMTS), FCC; 2017. Available from: http://wireless.fcc.gov/services/index.htm?job=about&id=wireless_medical_telemetry.

[60] Near Field Communication (NFC); 2017. Available from: http://nearfieldcommunication.org/.

[61] Zarlink ZL70101, Medical Implantable RF Transceiver; 2009. Available from: http://ulp.zarlink.com/zarlink/zweb-zl70101-datasheet-dec09.pdf.

[62] Sensium TZ1030; 2015. Available from: http://www.all-electronics.de/wp-content/uploads/migrated/document/37566/412-document.pdf.

[63] Bluetooth; 2019. Available from: https://www.bluetooth.com/.

[64] ZigBee; 2018. Available from: http://www.zigbee.org/.

[65] RuBee; 2019. Available from: http://standards.ieee.org/develop/wg/RuBee.html.

[66] ANT; 2019. Available from: https://www.thisisant.com/.

[67] BodyLAN: Ultra Low Power Wireless; 2019. Available from: http://www.sonicboomwellness.com/static/wims/512e86ef761fd08b3d000000/bodylan-wireless-protocol.pdf.

[68] ProLimb project, FEUP, INESC TEC; 2013. https://www.inesctec.pt/ctm-en/projects/projects/prolimb/; 2013.

[69]   Sheltami T., Mahmoud A., Abu-amara M. 'Warning and monitoring medical system using sensor networks'. *18th National Computer Conf. (NCC18)*; 2006. pp. 63–68.

[70]   Farella E., Pieracci A., Benini L., Rocchi L., Acquaviva A. 'Interfacing Human and Computer with Wireless Body Area Sensor Networks: The WiMoCA Solution'. *Multimedia Tools and Applications*. 2008;38:337–363.

[71]   Jantunen I., Laine H., Huuskonen P., Trossen D., Ermolov V. 'Smart Sensor Architecture for Mobile-Terminal-Centric Ambient Intelligence'. *Sensors and Actuators A Physical*. 2008;142(1):352–360.

[72]   Rienzo M.D., Vaini E., Lombardi P. 'Wearable monitoring: A project for the unobtrusive investigation of sleep physiology aboard the International Space Station'. *2015 Computing in Cardiology Conference (CinC)*; 2015. pp. 125–128.

[73]   Hamida E.B., Alam M.M., Maman M., Denis B., D'Errico R. 'Wearable body-to-body networks for critical and rescue operations 2014; The CROW project'. *2014 IEEE 25th Annual International Symposium on Personal, Indoor, and Mobile Radio Communication (PIMRC)*; 2014. pp. 2145–2149.

[74]   Texier I., Xydis S., Soudris D., *et al.* 'SWAN-iCare project: Towards smart wearable and autonomous negative pressure device for wound monitoring and therapy'. *2014 4th International Conference on Wireless Mobile Communication and Healthcare – Transforming Healthcare Through Innovations in Mobile and Wireless Technologies (MOBIHEALTH)*; 2014. pp. 357–360.

[75]   Doron M., Bastian T., Maire A., *et al.* 'Estimation of physical activity monitored during the day-to-day life by an autonomous wearable device (SVELTE project)'. *2013 35th Annual International Conference of the IEEE Engineering in Medicine and Biology Society (EMBC)*; 2013. pp. 4629–4632.

[76]   Magenes G., Curone D., Caldani L., Secco E.L. 'Fire fighters and rescuers monitoring through wearable sensors: The ProeTEX project'. *2010 Annual International Conference of the IEEE Engineering in Medicine and Biology*; 2010. pp. 3594–3597.

[77]   Chen C., Bian K., Huang A., *et al.* 'WE-CARE: A wearable efficient telecardiology system using mobile 7-lead ECG devices'. *2013 IEEE International Conference on Communications (ICC)*; 2013. pp. 4363–4367.

[78]   INTERACTION project: Training and monitoring of daily-life physical INTERACTION with the environment after stroke; 2013. Available from: https://joinup.ec.europa.eu/community/epractice/case/interaction-training-and-monitoring-daily-life-physical-interaction-environ.

[79]   Paradiso R., Mancuso C., Toma G.M.D., Caldani L. 'Textile sensing platforms for remote monitoring of physical interaction with the environment'. *2014 8th International Symposium on Medical Information and Communication Technology (ISMICT)*; 2014. pp. 1–5.

[80]   Markovic M., Rapin M., Correvon M., Perriard Y. 'Design and Optimization of a Blood Pump for a Wearable Artificial Kidney Device'. *IEEE Transactions on Industry Applications*. 2013;49(5):2053–2060.

[81] European Project: ICT Enabled Wearable Artificial Kidney and Personal Renal Care System; 2014. Available from: http://www.nephronplus.eu/.

[82] Borges L.M., Rente A., Velez F.J., *et al.* 'Overview of progress in Smart-Clothing project for health monitoring and sport applications'. *2008 First International Symposium on Applied Sciences on Biomedical and Communication Technologies*; 2008. pp. 1–6.

[83] Gereffi G., Wyman D.L. *Manufacturing Miracles: Paths of Industrialization in Latin America and East Asia*. Princeton, NJ: Princeton University Press; 2016.

[84] Ianov A.I., Kawamoto H., Sankai Y. 'Development of noise resistant hybrid capacitive-resistive electrodes for wearable robotics, computing and welfare'. *2013 IEEE/RSJ International Conference on Intelligent Robots and Systems*; 2013. pp. 4249–4254.

[85] Mathkour H.I., Shahzad B., Al-Wakeel S. 'Software risk management and avoidance strategy'. *International Conference on Machine Learning and Computing, IPCSIT*; 2011. pp. 477–481.

[86] Pandian P., Mohanavelu K., Safeer K., *et al.* 'Smart Vest: Wearable Multi-Parameter Remote Physiological Monitoring System'. *Medical Engineering & Physics*. 2008;30(4):466–477.

[87] Garcia-Gomez J.M., de la Torre-Diez I., Vicente J., Robles M., Lopez-Coronado M., Rodrigues J.J. 'Analysis of Mobile Health Applications for a Broad Spectrum of Consumers: A User Experience Approach'. *Health Informatics Journal*. 2014;20(1):74–84.

[88] Zhou J., Cao Z., Dong X., Lin X. 'Security and Privacy in Cloud-assisted Wireless Wearable Communications: Challenges, Solutions, and Future Directions'. *IEEE Wireless Communications*. 2015;22(2):136–144.

[89] Saleem K., Derhab A., Al-Muhtadi J., Shahzad B. 'Human-Oriented Design of Secure Machine-to-Machine Communication System for e-Healthcare Society'. *Computers in Human Behavior*. 2015;51(B):977–985.

[90] Helbing D. *Thinking Ahead – Essays on Big Data, Digital Revolution, and Participatory Market Society*. Switzerland: Springer; 2015.

Chapter 2

Scenarios and applications for wearable technologies and WBSNs with energy harvesting

Fernando J. Velez[1], Raúl Chávez-Santiago[2],
Luís M. Borges[1], Norberto Barroca[1],
Ilangko Balasingham[2], and Fardin Derogarian[1]

The tremendous advances in radio communications and ultra-low power (ULP) electronics have enabled the development of communicating biomedical sensors for the continuous monitoring of patients' physiological signals. In the last decade, much research has been done towards the advancement of wireless body area network (WBAN) technology, resulting in the release of the IEEE 802.15.6-2012 standard for the interconnection of wearable and implantable biomedical sensors. This chapter provides an overview of the sets of wireless body sensor network (WBSN) applications, as well as of their characterization parameters. One of the key requirements for widespread adoption of the WBSN technology for daily healthcare is, however, the ease of use of the sensing devices. From the patient's perspective, this means that besides being small, unobtrusive and ergonomic, WBAN nodes have to be capable of long-term operation without the need to frequently charge, recharge or even use batteries. Although the recent advances in ULP electronics have reduced the power consumption of major WBSN node components to the sub-milliwatt (sub-mW) level, the vision for uninterrupted self-powered WBANs has yet to be realized. Energy harvesting (EH), i.e. taking energy from ambient sources to power autonomous wireless networked systems, is a developing technology with a tremendous potential to complement ULP electronics towards the realization of this vision, thereby enabling the perpetual remote monitoring of a patient's vital signs. An integrated circuit (IC) that integrates most of network functionalities onto low power wearable systems has been introduced as well.

[1]Instituto de Telecomunicações and Departamento de Engenharia Electromecânica, Universidade da Beira Interior; Faculdade de Engenharia, Portugal
[2]Wireless Sensor Network Research Group, Intervention Centre, Oslo University Hospital, Norwegian University of Science and Technology, Norway

2.1   Introduction

As the population suffering from chronic diseases increase worldwide, the real-time monitoring of various physiological signals becomes indispensable for the treatment and management of medical conditions like diabetes and cardiovascular illnesses. *Wearable* and *implantable* biomedical sensors can facilitate timely medication and early prehospital care by transmitting wirelessly these vital signals to portable units for display and analysis. The interconnection of these biomedical sensors, referred to as a WBAN, has already been standardized in IEEE 802.15.6-2012. One of the key requirements for widespread adoption of the WBAN technology for daily healthcare is, however, the ease of use of the sensing devices [1]. From the patient's perspective, this means that besides being small, unobtrusive, and ergonomic, WBAN nodes have to be capable of long-term operation without the need to frequently charge, recharge, or even use batteries. Although the recent advances in ULP electronics have reduced the power consumption of major WBAN node components to the sub-mW level [2], the vision for uninterrupted self-powered WBANs has yet to be realized. EH, i.e. taking energy from ambient sources to power autonomous wireless networked systems [3], is a developing technology with a tremendous potential to complement ULP electronics towards the realization of this vision, thereby enabling the perpetual remote monitoring of a patient's vital signs.

Four main groups of EH techniques for wearable and implantable WBAN nodes can be identified depending on the ambient energy sources: *mechanical, thermal, biochemical* and *electromagnetic* [2]. In WBAN environments, the human body can be the source of some types of ambient energy, e.g. mechanical energy from body and muscle movements, whereas other types of energy like electromagnetic radiation require external sources. EH from external sources to power implantable biomedical sensors is generally problematic as it imposes the need for additional wireless power transmission (WPT) solutions [4]. Thus, hereinafter we consider EH for wearable sensors only.

Efficient EH for WBANs poses major challenges because of the variable placement of wireless nodes on the human body, resulting in uncertainty about the proper exposure to ambient energy sources. For this reason, WBAN nodes with EH capabilities (EH-WBAN nodes) have to be designed in a different way from battery-operated nodes in order to intelligently manage the harvested energy [3], thereby ensuring uninterrupted operation even when the energy source is temporarily unavailable. Ambient electromagnetic sources, specifically radio frequency (RF) transmissions from commercial networks, represent a steady source of energy to power EH-WBAN nodes due to the ubiquitous deployment of telecommunication infrastructure in urban and suburban areas. RF energy harvesting (RF-EH) techniques aim to convert the received RF energy into direct current (DC) power with the use of a rectifying antenna (rectenna), which consists of a receiving antenna followed by a nonlinear RF-DC conversion circuit. The main components of an RF-DC converter are a high-frequency (HF) filter, a rectifier, a DC filter and a load [5].

Wearable and implantable biomedical sensors can facilitate timely medication and early prehospital care by transmitting wirelessly these vital signals to portable units for display and analysis. The interconnection of these biomedical sensors is referred to as a WBSN. Wearable technologies and WBSNs provide the means to support on- and in-body sensing, processing, actuation, communication, and EH and storage abilities as a solution to the challenges of ubiquitous personal monitoring in applications such as healthcare, daily life monitoring, protection and safety. Accordingly, the new generation of clothing is expected to be able to sense, communicate data and harvest energy in an almost non-intrusive way.

Power consumption and size are two main bottlenecks of wearable technologies. Reduction of power consumption minimizes the circuit size as we need smaller battery or EH. Integrating most of the system functionalities in a single IC not only reduces the size but also significantly deduces power consumption. In this chapter, we introduce a communication IC which is able to handover acquired data to end user while consuming very small amount of energy.

Thus, in the remainder of this chapter, we identify the applications for wearable technologies and WBSNs, as well as their classification and main characterization parameters, while discussing the advantages, disadvantages and opportunities for RF-EH to power WBAN nodes. We survey some of the RF-EH solutions for WBAN that have been recently reported in the literature. We also provide a detailed description of the current status of our own research and, as a step forward, we sketch some characteristics for opportunistic RF-EH based on the principles of spectrum sensing for cognitive radio (CR). Finally, we address to an IC which is designed to establish data communication for low power wearable systems.

2.2 Classification of applications and characterization parameters

The WBAN standardization working group (IEEE 802.15.6) has produced the first draft of a document specifying the physical (PHY) and medium access control (MAC) layer characteristics of the radio interfaces for WBAN applications [6]. The standard introduces the framework for WBSNs, which includes the network topology and the reference model explaining the supporting functionality of nodes and centric device. According to [7], a WBSN enables novel uses of this networking technology, especially in healthcare, fitness, entertainment, sports, etc. A WBSN allows users to interconnect wireless devices they carry on or with them.

A medical BAN (MBAN, i.e. a WBAN for medical applications) comprises multiple sensor nodes, each capable of sampling, processing and communicating one or more vital signals. This biomedical information is transmitted to a body network controller [8]. The MBAN constitutes the first tier of a WSBN environment, as shown in Figure 2.1, which is referred to as intra-WBSN communications. Besides, IEEE 802.15.6-compliant WBAN nodes will facilitate a more active involvement of patients in the daily management of chronic diseases like diabetes and hypertension, thereby allowing for reducing the overall expenditure on medical treatment and healthcare.

The exponential growth that is currently witnessed in wearable technologies for healthcare is creating opportunities for using WBSNs in various application domains.

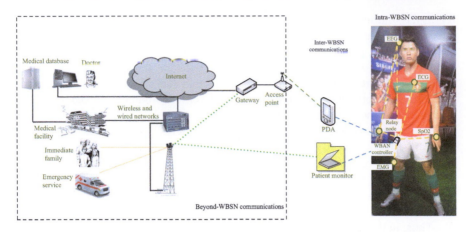

Figure 2.1   Three-tier architecture for a WSBN environment. © 2012. Adapted, with permission, from Reference [9]

A set of surveys and overview papers have been published on the characterization of applications for WSBNs. The authors from [10] have presented a survey of challenges and applications of Body area networks (BANs) in healthcare. They have also proposed the role of BANs as a virtual doctor by defining its architecture. In [7], the authors present an analysis and classification of body area wireless sensor network applications and highlight that WBSNs can play a vital role in the provision of efficient healthcare services while reducing the burden on the clinical system. The authors from [7] performed an in depth review and analysis of WBSN applications in various walks of life. On the basis of this analysis, WBSN applications have been classified in four major domains as shown in Figure 2.2: healthcare, disability assistance, sports and human-activity monitoring. These main areas are further divided into subdomains, each playing an important role in human life. Any other application of WBSNs which could not fit into the first four categories was included in the others category. In particular, the healthcare applications are divided into WBSNs for general healthcare, neonatal healthcare monitoring and animal healthcare.

In the disability assistance group, the following WBSNs application-specific subcategories are identified: rehabilitation, activity monitoring (real-time), posture detection, way-finding for blind/deaf-blind person and support for elderly person. In sports, WBSNs can be applied to training, monitoring, self-assessment and performance enhancement. Human-activity monitoring is another relevant group, whose usage includes the following situations: emergency situation (e.g. firefighters, volunteers and rescuers), remote monitoring (e.g. staff, kids and elderly), training (interactive dance, stage performance, etc.), security (soldiers) and safety (e.g. personal safety and elderly). Novel applications, such as WBSNs, to identify frostbites or to assist blind swimmers, as well as WBSNs to assist diabetic patients (with immediate alerts to a person, its caregiver, or clinical staff when the sugar level is above or below the thresholds) were also identified in [7]. In [11], the authors present

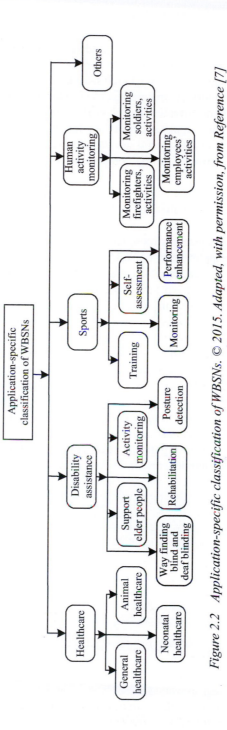

Figure 2.2   Application-specific classification of WBSNs. © 2015. Adapted, with permission, from Reference [7]

a review on telemedicine-based WBAN framework for patient monitoring. In particular, a detailed characterization of BAN applications is performed by presenting their functional requirements in detail (data rate, bandwidth, latency, accuracy and reliability). The considered applications are electrocardiogram (ECG), electromyography (EMG), electroencephalography (EEG), blood saturation (peripheral oxygen attenuation, SpO2), glucose monitoring, temperature, motion sensor, among others. In [12], a broader survey on the characterization and classification of WSN applications is proposed which includes several applications in the telemedicine and healthcare but also a wider range of applications. Concepts are introduced, a taxonomy for the classification of applications is proposed, a very complete list of characterization parameters is identified and their range of variation is analyzed for a multitude of applications. This comprehensive survey presents a roadmap and future trends in the classification and characterization of these applications, identifying different generations (1G, 2G, 3G and beyond).

The authors of [13] present a more specific taxonomy by proposing a classification of the various applications of WBANs in different sectors of medical and non-medical fields, in a categorization according to IEEE 802.15.6, as shown in Figure 2.3. The IEEE 802.15.6 working group established the first draft of the communication standard of WBANs in 2009/2010, and ratified in 2012, optimized for low-power on-body/in-body nodes for various applications. Its aim is to develop a communication standard for low power devices and operation on, in or around the human body (but not limited to humans) to serve a variety of applications including medical, consumer electronics, personal entertainment and others. The main requirements of WBANs in IEEE 802.15.6 are discussed in the Section IV of [13]. According to [13], using WBANs in medical applications allows for permanent monitoring of the patient physiological attributes such as blood pressure (BP), heart beat and body temperature. In cases where abnormal conditions are detected, data being collected by the sensors can be sent to a gateway such as a cell phone. The gateway then delivers its data via a cellular network or the Internet to a remote location such as an

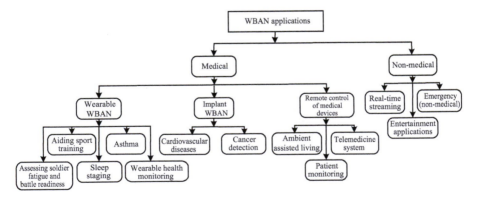

Figure 2.3   Classification of WBAN applications. © 2014. Adapted, with permission, from Reference [13]

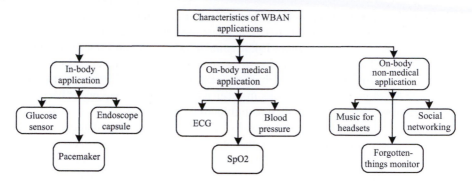

Figure 2.4   Main characteristics of in/on-body applications. © 2014. Adapted, with permission, from Reference [13]

emergency centre or a doctor's room based on which an action can be taken. Additionally, WBANs will be a key solution in early diagnosis, monitoring and treatment of patients with possibly fatal diseases of many types, including diabetes, hypertension and cardiovascular-related diseases.

Medical applications of WBANs can be further classified into (i) wearable WBAN (either disability assistance or human performance management ones), (ii) implantable WBAN and (iii) remote control of medical devices subcategories. Non-medical applications of WBANs include the (i) real-time streaming, (ii) entertainment applications, (iii) emergency (non-medical), (iv) emotion detection and (v) secure authentication subcategories.

As the technological requirements of WBANs are application-specific, the characterization parameters for the in-body and on-body applications from Figure 2.4 are identified in Table II from [13].

Characteristics of WBANs include not only the type and number of nodes, the type of topology and the communication architecture – intra-, inter- (infrastructure based or *ad-hoc* based architectures) and beyond-WBAN communications but also the range of data rates, duty cycle, power consumption, Quality of Service (QoS) (sensitivity to latency) and privacy.

2.3   Spectrum opportunities for RF-EH

Although RF-EH can dramatically extend the operating lifetime of WBAN nodes and potentially render them battery-free, the main drawback of this technique is the low power density that can be scavenged from RF ambient sources in the order of $0.01–0.1\,\mu W/cm^2$. Contrastingly, power densities within $10\,\mu W/cm^2–10\,mW/cm^2$ can be obtained with the use of photovoltaic EH techniques and $20\,\mu W/cm^2–10\,mW/cm^2$ with thermal ones [5]. Moreover, the amount of RF energy that can be harvested from operating telecommunication networks, e.g. cellular networks, depends very much on the varying spatial-temporal traffic patterns; in most cases, however, these traffic fluctuations can be predicted to facilitate the optimization of the energy management policy for the EH systems [3]. In spite of these disadvantages, the ubiquitous deployment and continuous operation of cellular communication networks and other systems like Wi-Fi in urban and suburban areas make RF a reliable ambient energy source, even though transient drops in received power level may occur because of different propagation phenomena like multipath fading.

The spectrum opportunities for RF-EH in ultra HF have been identified through extensive measurement campaigns within 300–3,000 MHz in the urban and suburban areas of Covilhã, Portugal and London, United Kingdom [14,15]. In both cases, the most promising RF-EH opportunities in terms of highest power density came from the global system for mobile communications (GSM) frequency bands in 900 MHz (GSM 900) and 1,800 MHz (GSM 1800). For instance, the measurement campaign in London resulted in average input RF power density of 36 and 84 nW/cm^2 for base station transmit (BTx) in the GSM 900 and GSM 1800 bands, respectively [15]. Digital TV and 3G cellular network bands were also identified as possible RF energy contributors.

Although the GSM 900/1800 bands appear to be the most obvious choices for RF-EH, it is worthwhile taking into account the trade-off among antenna size, antenna efficiency, channel propagation losses and absorption, to design a RF energy harvester for WBANs [16]. While the use of frequencies above 1 GHz reduces the size of the antenna for RF-EH and provides higher antenna efficiency, the propagation losses also increase, resulting in less power density available at the input of the RF energy harvester. Additionally, because of the frequency-dependent dielectric properties of biological tissues, human body absorption increases with frequency; high body absorption may deteriorate the antenna efficiency, especially when it is placed directly on the patient's skin.

2.4   RF-EH solutions for WBAN

A number of RF-EH solutions for WBANs have been proposed in the literature. Early solutions were designed to harvest energy within a single narrow frequency band. Generally, these solutions utilize a dedicated base station that radiates the necessary energy to be scavenged by the EH-WBAN nodes. We argue, however, that the use of a dedicated RF source does not strictly constitute EH from ambient energy sources, but it is rather a form of WPT. On the other hand, since the availability of multiple RF sources can increase the amount of harvested energy [5], more recent research aims to develop RF-EH systems with the capability of operating over multiple narrow frequency bands or a single wide frequency band. These multi-band solutions do not require a dedicated RF source as they are designed to harvest energy from cellular networks, wireless local area networks, and other operating radio communication systems. Below we survey the characteristics of some of the *single-band* and *multi-band* RF-EH solutions for WBANs, highlighting the aspects of more interest from a radio communication perspective.

2.4.1 Single-band RF-EH solutions

An RF-EH solution for a wearable tag that can monitor phonocardiogram, ECG, blood oxygen saturation, respiratory sounds, BP and body temperature is described in [17]. The tag was equipped with a low-power IC that harvests RF energy from base stations transmitting in 900 MHz and 2.4 GHz. Nevertheless, the necessary distance between the radiation source and the tag to ensure battery-free operation in indoor environments is approximately 12 m for 900 MHz and 3 m for 2.4 GHz. Thus, while a single base station in 900 MHz could provide enough RF energy to power the tag in a room, the use of 2.4 GHz would require a high density of RF sources, e.g. in the high density deployment scenario of wearable devices, such as public transportation, proposed in [18].

A system-on-a-chip (SoC) solution for multi-node EH-WBANs was demonstrated in [16]. In this solution, each EH-WBAN node consists of three main blocks, namely (i) sensor data capture and RF transmission, (ii) multi-node clock synchronization, and (iii) far-field RF-EH. The multi-node EH-WBAN system utilizes two RF channels, one for sensor data communication and the other for energy transmission. A network coordinator (NC) transmits a large RF signal in the 915 MHz industrial-scientific-medical (ISM) frequency band (specifically in a channel centred at 904.5 MHz) during a predetermined period of time known as the *EH phase*. After this phase, the NC instructs the EH-WBAN nodes to switch into the *data transmission phase*, during which sensor data are transmitted to the NC in the medical implant communication system frequency band (specifically in a channel centred at 402 MHz) using an on-off keying modulation scheme at 250 kb/s. Time division multiple access supports the multi-node operation.

An energy-efficient application-specific IC (ASIC) for WBANs nodes was implemented and tested in [19]. In this work, two types of WBAN nodes were identified, namely *sensing nodes* and *stimulating nodes*. The sensing nodes measure and transmit the patient's vital signs, and thus their functions are performed periodically, whereas the stimulating nodes can either be periodical or event-driven as they actuate based on feedback from the sensing nodes, e.g. drug delivery and nerve stimulating. Based on this distinction, a network protocol that meets the requirements of both types of nodes was devised. Basically, the ASIC has two standby modes, *active* and *passive*, for the sensing and stimulating nodes, respectively. In the former, only an ULP low-frequency timer is active, which periodically powers up a sensing node to transmit physiological information in a primary communication channel. In the latter, the node is power silent while a passive RF receiver is used for RF-EH in a secondary channel. Hence, the ASIC consists of a microcontroller unit (MCU), a power management unit, reconfigurable sensor interfaces, communication ports controlling a wireless transceiver and a passive RF receiver offering the EH-WBAN stimulating nodes the capability of work-on-demand with zero standby power. The MCU accomplishes the MAC and network protocols. An EH-WBAN prototype was implemented to test the performance of the ASIC. The work-on-demand protocol used a primary channel in the 433 MHz ISM band and a secondary channel in 915 MHz ± 10 MHz. PHY experiments demonstrated efficient operation of the EH-WBAN prototype for typical biomedical duty cycles, e.g. 1 ms/5 min.

2.4.2   *Multi-band RF-EH solutions in the GSM 900/1800 bands*

An architecture for a textile-based tri-band wearable rectenna is described in [20], which can potentially harvest RF energy from GSM 900/1800 and Wi-Fi in 2.4 GHz upon user request. The major advantage of this device is its purported ability to start autonomously from a fully discharge state, in principle rendering it battery free. For this purpose, an external inductor and an input capacitor are needed, which reportedly do not affect the overall dimensions. However, to the best of our knowledge, the operation of the proposed IC has been assessed via simulation only and the PHY implementation is currently under way.

A wideband RF-EH rectenna operating within 900–2,450 MHz for outdoor WBANs was introduced in [5]. Different rectifier topologies were discussed therein, concluding that the single series diode structure is a good compromise between output voltage, V_{outDC}, and conversion efficiency, η_0. In this case, the conversion efficiency is defined as the ratio of the useful output power that can be provided to the WBAN node, P_{outDC}, to the received power at the input of the RF-EH circuit, P_{inRF} [21,22]. An RF-DC converter circuit was designed and simulated considering a two-tone signal (1.8 and 2.4 GHz) at the RF input, which demonstrated that the DC output was 20% higher in average in comparison to the case when a single-tone input was used. Then, a test circuit was fabricated with commercial off-the-shelf (COTS) components and measurements of the single-tone and two-tone cases were undertaken, which confirmed the conclusion derived from simulations.

In [22], we presented an RF-EH system specifically developed to operate in the GSM 900/1800 bands and a wearable dual-band antenna that can be embedded in textiles. We designed and evaluated via computer simulations a five-stage Dickson voltage multiplier with an impedance-matching circuit; a voltage multiplier converts alternating current (AC) power from a lower voltage to a higher DC voltage using a network of capacitors and diodes. During the fabrication of a prototype with COTS components, we noticed that different printed circuit board fabrication techniques resulted in different values of conversion efficiency. Hence, we built three different prototypes referred to as Prototypes 1, 2 and 3 and measured their characteristics. Figure 2.5 shows the output voltage and Figure 2.6 the conversion efficiency of the prototypes with a load impedance of 100 kΩ in all the cases. The superimposed simulation results in Figures 2.5 and 2.6 were obtained through a harmonic balanced analysis (i.e. a frequency domain method) that evaluates the steady state solution of a non-linear circuit. Two levels for V_{outDC} are particularly highlighted by solid lines in Figure 2.5, namely the minimum voltage required to power a WBAN node (1.8 V) and the recommended voltage (3 V). These voltage values allow for assessing the sensitivity, which is defined as the minimum value of P_{inRF} necessary to power a WBAN node. Obviously, this definition depends on the specific device and application, as different IC technologies and protocols can cause the sensitivity to change. The measured sensitivity values for $V_{outDC} = 1.8$ V of our different prototypes are summarized in Table 2.1. The maximum conversion efficiency, η_{0max}, and maximum harvested power (i.e. the power dissipated in the load), $P_{EH_{max}}$, for a load impedance of 100 kΩ are also listed in Table 2.1. It is observable that Prototype 2 allows for achieving the highest

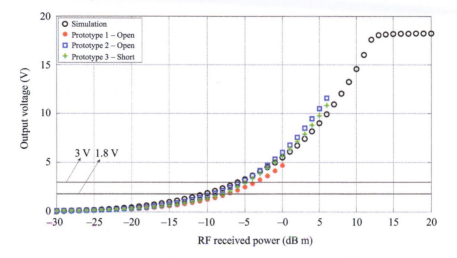

Figure 2.5   Output voltage as a function of the RF received power

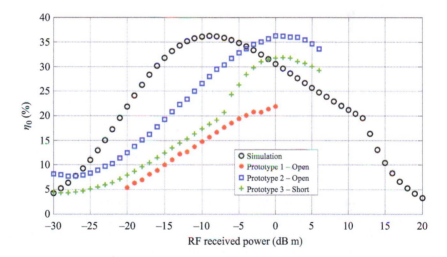

Figure 2.6   Conversion efficiency as a function of the RF received power

value for the conversion efficiency. It also enables the highest value of maximum power collected but at higher values of the RF received power $P_{inRF} = 6$ dB m. For the sake of comparison, the RF harvester in [5] exhibited a measured sensitivity of more than 10 dB m for $V_{outDC} = 1.8$ V in the two-tone operation mode with a resistive load $R_L = 10$ kΩ. Thus, our prototypes represent significant improvement due to the use of a voltage multiplier. However, considering that $P_{EHmax} = 1.33$ mW for Prototype 2 was obtained with $P_{inRF} = 6$ dB m, we conclude that our RF energy harvester

Table 2.1   Characteristics of the RF-EH prototypes developed in [22]

RF-EH circuit	Sensitivity	Maximum conversion efficiency, $\eta_{0_{max}}$	Maximum harvested power, $P_{EH_{max}}$
Prototype 1	-7 dB m for $V_{outDC} = 1.8$ V	22 % at $P_{inRF} = 0$ dB m	0.22 mW at $P_{inRF} = 0$ dB m
Prototype 2	-9 dB m for $V_{outDC} = 1.8$ V	36 % at $P_{inRF} = 0$ dB m	1.33 mW at $P_{inRF} = 6$ dB m
Prototype 3	-8 dB m for $V_{outDC} = 1.8$ V	32 % at $P_{inRF} = 1$ dB m	1.16 mW at $P_{inRF} = 6$ dB m

may not power up a WBAN node continuously since the power from ambient sources rarely exceeds $P_{inRF} = 0$ dB m. Hence, an effective energy storing system (ESS) has to be devised.

Multi-band RF-EH solutions in the digital terrestrial television band are addressed in Chapter 14.

2.4.3   Supercapacitor-based energy storing system

The addition of an ESS that accumulates the harvested energy and allows its use when it is sufficient to power an EH-WBAN node is a viable way to improve our dual-band GSM 900/1800 MHz RF-EH circuit. Although the most common storing element is a rechargeable battery, this solution presents some disadvantages like imprecise estimation of the remaining energy, limited number of recharge/discharge cycles and negative environmental impact. A more convenient alternative is the use of supercapacitors, which exhibit a much longer lifetime than batteries and allow for determining precisely the remaining amount of stored energy. Commercially available lithium-ion supercapacitors offer the highest gravimetric energy density to date, reaching 15 W h/kg. However, ongoing research in this field focuses on improving energy density with the use of nanomaterials for the fabrication of supercapacitors. For instance, the use of curved graphene sheets has led to supercapacitors with energy density of 85.6 W h/kg, whereas the use of manganese dioxide intercalated nano-flakes has resulted in energy density of 110 W h/kg. Other issues that are being addressed by researchers are reducing internal resistance, expanding temperature range, increasing lifetimes and minimizing production cost.

We propose the use of a supercapacitor-based ESS to be coupled with our five-stage Dickson voltage multiplier as shown in Figure 2.7. We took inspiration from [23], in which a supercapacitor block was used as the storing element for the energy harvested by solar panels. Our proposed solution consists of four main blocks: (i) the front end to manage, control and transfer the energy from the RF energy harvester into the supercapacitor bank; (ii) a front-end control unit that runs the control algorithm for the front-end block; (iii) a back-end circuit that converts the energy stored in the supercapacitor bank into a constant voltage and (iv) the embedded system block, which is the WBAN node in this case.

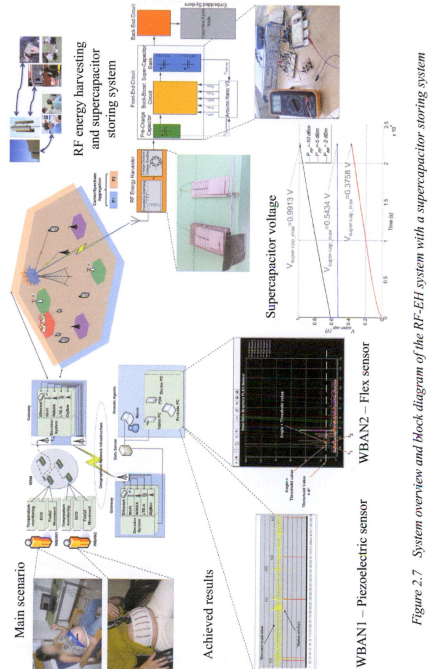

Figure 2.7 System overview and block diagram of the RF-EH system with a supercapacitor storing system

The front-end circuit is controlled by an ArduinoTM Nano V3 microcontroller. As the microcontroller itself consumes energy to operate, a starting battery-free solution has to be devised in order to make the entire system autonomous and self-sustainable. We consider two approaches to power up the microcontroller. The first approach consists of using an OLIMEXTM board with a solar panel and a rechargeable battery that delivers an output voltage of 3.3 V. The ArduinoTM microcontroller can be supplied by this photovoltaic system, while the supercapacitor powers the embedded system, i.e. the WBAN node. The second approach considers the use of a dedicated pre-charged 1,000 μF capacitor attached to the microcontroller as a starting power supply, so the microcontroller can control the flow of energy from the RF energy harvester to the supercapacitor. When the supercapacitor is charged with a stable voltage and the starting 1,000 μF capacitor drains all its energy, then the microcontroller can switch to using the stored energy for self-sustained management activity. Currently we are working towards the implementation of these two approaches using COTS components.

2.5   MAC for opportunistic RF-EH for WBAN

Even with the addition of an ESS, the amount of harvested energy from ambient RF sources may still be insufficient to continuously power a WBAN node. Hence, MAC protocols that take into account the operation characteristics of EH-WBAN nodes have to be devised to assist the intelligent management of the harvested energy. A good example of this approach is the MAC protocol proposed in [19] for a single-band system. As the continuation of this idea, we sketch some of the features that a MAC protocol for opportunistic multi-band RF-EH should possess. We also take inspiration from CR, a technology that aims to exploit unoccupied parts of the radio spectrum in an opportunistic manner to transmit information from unlicensed secondary users (SUs) without interfering with licensed primary users (PUs).

2.5.1   *Double stage MAC for radio cognitive networks*

In a cognitive radio network (CRN), the SUs have the ability to sense the spectrum occupancy in a given frequency band in order to identify unused space-time-frequency slots (spectrum holes) that can be utilized to transmit information opportunistically without causing interference. This characteristic ranks CRNs as effective solutions to alleviate the increasing demand for radio spectrum. A CR system in which the SUs synchronously and periodically sense the channel occupancy in order to determine the level of PU transmission activity was presented in [24]. In this approach, the SUs sense the PU transmission activity by using an energy-based sensing scheme, which relies on a classical energy detector; however, other spectrum sensing techniques can be used too. The time frame organization for SUs consists of two time periods; spectrum sensing is performed in the first time period in order to evaluate whether a node is allowed to access the channel and then transmit during the second time period. As depicted in Figure 2.8, the total time frame duration is T_F^{SU}. The first N_S slots of a time frame with total time duration equal to T_S^{SU} are used for spectrum sensing,

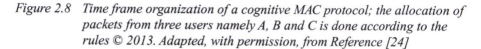

Figure 2.8   Time frame organization of a cognitive MAC protocol; the allocation of packets from three users namely A, B and C is done according to the rules © 2013. Adapted, with permission, from Reference [24]

whereas Θ is the duration of the synchronization tone sent by the access point (i.e. network controller) in each idle frame. The main idea behind this MAC protocol is the adoption of two different stages for scheduling the medium access during the transmission period. In the first stage and part of the second one, the SUs express their willingness to send data. The second stage also includes the time period during which the SUs access the channel.

2.5.2   Opportunistic RF-EH for WBAN

Analogous to the CRN protocol described above, we think of a scenario in which ULP EH-WBAN nodes in a secondary network harvest ambient RF energy from nearby transmitters in a primary network, while opportunistically access the licensed spectrum. In such a scenario, the double stage MAC protocol in [24] can be of interest since it is simple and easy-to-implement and exhibits good results in terms of aggregated throughput. Additionally, the information from the supercapacitor ESS about available energy can be used to reach a balance between the harvested, stored and consumed energy of the EH-WBAN node to avoid compromising its normal operation.

Figure 2.9 shows a sketch of an envisaged scenario for opportunistic RF-EH based on the analogy with CRNs. The EH-WBAN nodes detect the spectrum holes to send their data while taking advantage of the spectrum sensing period, T_S^{SU}, to opportunistically harvest RF energy from nearby PU transmissions. The harvested energy is stored in the supercapacitor ESS, which can be employed when it is fully charged. The envisaged scenario considers that the PUs and SUs are spread over an area where two different zone types can be identified, namely Safe Zones to protect the PU transmissions where SUs are not allowed to transmit, and Harvesting Zones, located within the safe zones where EH is particularly favourable because of the large ambient RF power density available. Additionally, three different operation modes for the EH-WBAN node should be considered:

- Harvesting, when the EH-WBAN node is located inside the harvesting zone of an active PU and its supercapacitor ESS is not fully charged.
- Transmitting/Receiving, when the EH-WBAN is not located inside any safe zone of active PUs and its supercapacitor ESS is fully charged.
- Idle, when the EH-WBAN node is located inside any safe zone of active PUs and its supercapacitor ESS is fully charged, or when the EH-WBAN node is neither fully charged nor inside any of the harvesting zones.

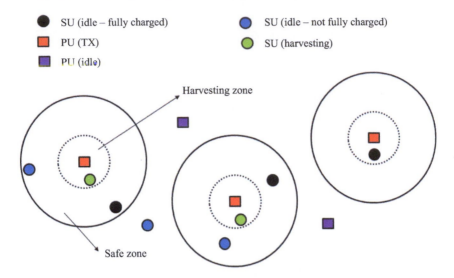

*Figure 2.9   Envisaged scenario for opportunistic RF-EH based on the analogy
with CRNs*

2.6   An integrated circuit for low power wearable system

In this section, an IC for low power wearable system for capturing human locomotion
is introduced. Such a system is able to be power supplied from small coin batteries
or RF-EH. It is composed of a set of sensors embedded in textile connecting each
other in a mesh topology with conductive yarns and an on body wireless module to
collect data from wearable part and send them to a computer or PDA. Figure 2.10
plots E-legging for capturing human locomotion with a wired network composed
of conductive yarns. Research shows that the most comfortable and easiest way to
monitor physiologic signals consists in using garments with conductive yarns [25,26].
Using conductive yarn also prevents wireless interference problem, reduce power
consumption and increase throughput of the system.

As is mentioned, a mesh topology is utilized for communication and a central
node collects data from sensors. End-to-end communication between central node
and sensors is based on multi-hop packet delivery. So each sensor must be able to
act as router and handover received packets to neighbour nodes. For that, a communication IC has been implemented using a four-metal, 0.35 μm CMOS technology.
Each IC is a four port router, including all sub-module to receive, buffer, process
and send packets. Some of low power characteristics and sub-modules of the IC are
as follow:

- **An ULP clock synchronization:** From the standpoint of the network, each sensor
  is a four-port router with bidirectional links to other sensors. Sensors are equipped
  with a local crystal oscillator. When sensors communicate, the receiver node

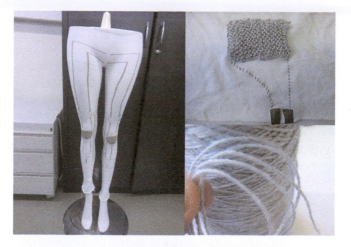

Figure 2.10   E-legging for capturing human locomotion with a wired network composed of conductive yarns

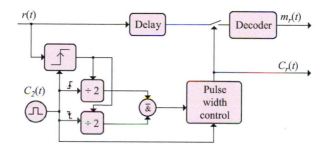

Figure 2.11   Block diagram of the clock synchronization circuit. © 2016. Adapted with permission, from Reference [27]

must be able to synchronize its clock to the incoming data. Because there is no global synchronization, a local synchronization method is necessary. For this IC, a very small, fully digital open-loop CDR method has been designed and evaluated in [27]. Figure 2.11 shows the block diagram of the circuit. At the receiver, a stable crystal quartz oscillator generates the system clock $C_2(t)$, whose frequency is twice the data rate and nominally equal to the clock frequency of the transmitter.

The synchronization circuit consists of eight logic gates in total and occupies 0.0022 mm^2 in a 0.35 μm CMOS process. Experimental results shows that circuit consumes 15 μW@35 Mbps which is much lower than most of exist method. Such a ULP operation makes this circuit a good choice for RF-EH systems.

- **A low power and precise time-synchronization method:** Because the information about the relative time occurrence of events is generally important to be known, so time synchronization, with appropriate accuracy, is required in many distributed data acquisition systems. Because of clock drift, sensors must periodically synchronize their time with reference. Keeping sensors synchronized consumes a significant amount of energy. The time synchronization protocol for establishing time synchronization between sensor nodes of the wearable part of the system has been implemented in IC. The synchronization is based on one-way master-to-slave message exchange implemented in the MAC sub-layer. Experimental evaluation with IC implementation shown an average one-hop clock skew of 4.6 ns. The sub-microsecond skew values provided by this approach satisfy the requirements of many BAN applications. The proposed circuit achieves the maximum synchronization performance that could be achieved by precision time protocol (PTP) (a standard time synchronization protocol), but with fewer timing messages and calculations, less complexity and better energy efficiency. The time synchronization method is explained in detail in Chapter 12.
- **Packet, circuit and hybrid switching:** In networking, a switching algorithm, such as packet switching, specifies the end-to-end packet delivery method. Collisions happen in shared environments when two or more nodes try to use a common channel or any other resources. In packet switching, each node stores and forwards the packets whenever the links are free. Packet switching in a wearable system may adversely affect power consumption, throughput and end-to-end delay.

   One of the methods that can overcome the aforementioned drawbacks is to use circuit switching at MAC sub-layer, because it is a bufferless method. In bufferless networks, such as optical networks or some network-on-chip architectures, a packet arrives at the destination node over intermediate nodes without any buffering. This approach usually reduces the total network power consumption and delay under low network workloads, but routing performance for randomly generated packets is significantly worse than for traditional buffered routing (store-and-forward method) for high network workloads. In fact, a bufferless algorithm without communication scheduling at the nodes may be significantly ineffective.

   Each packet and circuit switching has their own advantages and disadvantages. An end-to-end packet delivery may establish on a hybrid switch of both packet and circuit switching. Such a hybrid switching scheme significantly improves system performance in terms of end-to-end delay, throughput and power consumption.

   A sub-module has been implemented in IC which is able to perform each of packet, circuit or hybrid switching. Such a hybrid architecture provides less power consumption, less buffering in intermediate nodes and less end-to-end delay.

2.7   Discussion and concluding remarks

This chapter started by presenting taxonomies to the classification of WBSN applications while introducing the main characterization parameters and highlighting aspects of their standardization within IEEE 802.16. In particular, WBANs will allow for continuous monitoring of patients in medical applications, capable of early detection of abnormal conditions resulting in major improvements in the quality of life. Importantly, even basic vital signs monitoring (e.g. heart rate) can enable patients to engage in normal activities as opposed to being home bound or nearby specialized medical services.

RF-EH is becoming a mature technology with the enormous potential to realize the vision for uninterrupted self-powered networks of biomedical sensors, capable of perpetual monitoring of patients' vital signs. During the last 5 years, we have witnessed the evolution of RF-EH for WBAN from single-band solutions with a dedicated RF energy source to multi-band solutions with the ability to scavenge RF energy from various commercial telecommunication systems. Along this way, several important observations have been reported, which we summarize below as we consider them the departing points for future research:

- The use of multiple RF sources can increase the amount of harvested energy, at least 20% higher in average in comparison to the case when a single source is used. Therefore, the development of multi-band RF-EH solutions should be prioritized.
- The GSM 900 and GSM 1800 bands provide the highest power density for RF-EH in urban and suburban environments. Improvement in the conversion efficiency and sensitivity of rectennas for these frequency bands is one of the biggest challenges.
- Two types of WBAN nodes can be distinguished: sensing nodes and stimulating nodes. Their different operating characteristics, periodical and event-driven, should be taken into account for the design of MAC protocols that can help to manage intelligently the harvested energy.

The successful development of battery-free WBANs calls for the participation of experts in various fields of engineering in a cross-disciplinary joint effort. Although most of the work on RF-EH for WBANs has been undertaken by the microwaves and electronics communities, communication engineers can contribute significantly to solve major challenges like the design of optimal MAC and routing protocols. For instance, the concept of energy cooperative network, i.e. a network in which the nodes share energy by wirelessly transmitting data and energy at the same time, can bring major improvements to the use of RF-EH in the human body environment and widen the research opportunities. Much of the ongoing research in radio communication engineering can be directly applied to RF-EH solutions, e.g. the development of efficient wideband textile antennas in the context of WBAN communications.

Advances in other disciplines will certainly foster new ideas for the improvement of RF-EH solutions for WBANs, like the creation of much better supercapacitors with the use of nanomaterials like graphene. Chip designers will play an important role too, as miniaturized SoC will have to replace the relatively bulky RF-EH prototype circuits built with COTS components. An IC has been introduced which is designed to use in low power wearable systems. The aforementioned IC integrates most of network functionalities and reduces significantly the power consumption and circuit size. Many possibilities for research and innovation lie ahead in this field, and with this overview, we have tried to bridge the different aspects involved in the design of RF-EH solutions for networks of wireless biomedical sensors operating in the human body environment.

References

[1]   Hanson M.A., Powell H.C. Jr., Barth A.T., *et al.* 'Body area sensor networks: Challenges and opportunities'. *Journal of Computer*. 2009;42(1):58–65.

[2]   Lo B., Thiemjarus S., Panousopoulou A., Yang G.Z. 'Bioinspired design for body sensor networks [Life Sciences]'. *IEEE Signal Processing Magazine*. 2013;30(1):165–170.

[3]   Gunduz D., Stamatiou K., Michelusi N., Zorzi M. 'Designing intelligent energy harvesting communication systems'. *IEEE Communications Magazine*. 2014;52(1):210–216.

[4]   Olivo J., Carrara S., Micheli G.D. 'Energy harvesting and remote powering for implantable biosensors'. *IEEE Journal of Sensors*. 2011;11(7):1573–1586.

[5]   Kuhn V., Seguin F., Lahuec C., Person C. 'A multi-tone RF energy harvester in body sensor area network context'. *in Proc. Loughborough Antennas and Propagation Conference (LAPC), Loughborough, UK*; 2013. pp. 238–241.

[6]   IEEE. *IEEE 802.15 IEEE 802.15 WPAN Task Group 6 (TG6) Body Area Networks [online]*. Available from http://www.ieee802.org/15/pub/TG6.html [Accessed 19 Aug 2016]; 2009.

[7]   Nadeem A., Hussain M.A., Owais O., Salam A., Iqbal S., Ahsan K. 'Application specific study, analysis and classification of body area wireless sensor network applications'. *Computer Networks*. 2015;83:363–380.

[8]   Chen M., Gonzalez S., Vasilakos A., Cao H., Leung V.C. 'Body area networks: A survey'. *Mobile Networks and Applications*. 2011;16(2):171–193.

[9]   Chávez-Santiago R., Nola K.E., Holland O., *et al.* 'Cognitive radio for medical body area networks using ultra widebands'. *IEEE Wireless Communications – Special Issue on Cognitive Radio – A Practical Perspective*. 2012;19(4):74–81.

[10] Barakah D.M., Ammad-uddin M. 'A survey of challenges and applications of wireless body area network (WBAN) and role of a virtual doctor server in existing architecture'. *in Proc. of Third IEEE International Conference on Intelligent Systems, Modelling and Simulation (ISMS), Kota Kinabalu, Sabah, Malaysia*; 2012. pp. 214–215.

[11] Chakraborty C., Gupta B., Ghosh S.K. 'A review on telemedicine-based WBAN framework for patient monitoring'. *Telemedicine and e-Health*. 2013;19(8):619–626.

[12] Borges L.M., Velez F.J., Lebres A.S. 'Survey on the characterization and classification of wireless sensor networks Applications'. *IEEE Communications Surveys and Tutorials*. 2014;16(4):1860–1890.

[13] Movassaghi S., Abolhasan M., Lipman J., Smith D., Jamalipour A. 'Wireless body area networks: A survey'. *IEEE Communications Surveys and Tutorials*. 2014;16(3):1658–1686.

[14]   Barroca N., Saraiva H.M., Gouveia P.T., *et al.* 'Antennas and circuits for ambient RF energy harvesting in wireless body area networks'. *in Proc. of the 24th Annual IEEE International Symposium on Personal, Indoor and Mobile Radio Communications (PIMRC), London, UK*; 2013. pp. 532–537.

[15]   Piñuela M., Mitcheson P.D., Lucyszyn S. 'Ambient RF energy harvesting in urban and semi-urban environments'. *IEEE Transactions on Microwave Theory and Techniques*. 2013;61(7):2715–2726.

[16]   Xia L., Cheng J., Glover N.E., Chiang P. '0.56 V, −20 dBm RF-powered, multi-node wireless body area network system-on-a-chip'. *IEEE Pervasive Computing*. 2010;9(1):71–77.

[17]   Mandal S., Turicchia L., Sarpeshkar R. 'A low-power, battery-free tag for body sensor networks'. *IEEE Pervasive Computing*. 2010;9(1):71–77.

[18]   Pyattaev A., Johnsson K., Andreev S., Koucheryavy Y. 'Communication challenges in high-density deployments of wearable wireless devices'. *IEEE Wireless Communications*. 2015;22(1):12–18.

[19]   Zhang X., Jiang H., Zhang L., Zhang C., Wang Z., Chen X. 'An energy-efficient ASIC for wireless body sensor networks in medical applications'. *IEEE Transactions on Biomedical Circuits and Systems*. 2010;4(1):11–18.

[20]   Dini M., Filippi M., Costanzo A., *et al.* 'A fully-autonomous integrated RF energy harvesting system for wearable applications'. *in Proc. 43rd European Microwave Conference (EuMC), Nuremberg, Germany*; 2013. pp. 987–990.

[21]   Valenta C.R., Durgin G.D. 'Harvesting wireless power: Survey on energy-harvester conversion efficiency in far-field, wireless power transfer systems'. *IEEE Microwave Magazine*. 2014;15(4):108–120.

[22]   Borges L.M., Barroca N., Saraiva H.M., *et al.* 'Design and evaluation of multi-band RF energy harvesting circuits and antennas for WSNs'. *in Proc. 21st International Conference on Telecommunications (ICT), Lisbon, Portugal*; 2014. pp. 308–312.

[23]   Fahad A., Soyata T., Wang T., Sharma G., Heinzelman W., Shen K. 'SOLAR-CAP: Super capacitor buffering of solar energy for self-sustainable field systems'. *in Proc. IEEE International SOC Conference (SOCC), Niagara Falls, NY*; 2012. pp. 236–241.

[24]   Oliveira R., Borges L.M., Velez F.J. 'A double stage random access scheme for decentralized single radio cognitive networks'. *in Proc. 10th International Symposium on Wireless Communication Systems (ISWCS), Ilmenau, Germany*; 2013. pp. 1–5.

[25]   Lymberis A., Gatzoulis L. 'Wearable health systems: From smart technologies to real applications'. *in 28th Annual International Conference of the IEEE Engineering in Medicine and Biology Society, 2006. EMBS'06*; vol. Supplement. IEEE; 2006. pp. 6789–6792.

[26]   Gatzoulis L., Iakovidis I. 'Wearable and portable eHealth systems'. *IEEE Engineering in Medicine and Biology Magazine*. 2007;26(5):51–56.

[27]   Derogarian F., Ferreira J.C., Tavares V.G. 'A small fully digital open-loop clock and data recovery circuit for wired BANs'. *International Journal of Circuit Theory and Applications (IJCTA)*. 2016;44(3):503–548.

Chapter 3

A reliable wearable system for BAN applications with a high number of sensors and high data rate

Fardin Derogarian[1], João Canas Ferreira[2],
Vítor Grade Tavares[2], José Machado da Silva[2],
and Fernando J. Velez[1]

This chapter addresses a wearable body area network (BAN) system for both medical and nonmedical applications, especially those including a large number of sensors at BAN scale (<250), embedded in textile and with high data rate (<9+9 MHz) communication demands. The overall system includes an on-body central processing module (CPM) connected to a computer via a wireless link and a wearable sensor network. Due to the fixed location of the sensors and the possibility of using conductive yarns in textiles, a wired network has been considered for the wearable components. Employing conductive yarns instead of using wireless links provides a more reliable communication, higher data rates and throughput, and less power consumption. The wearable unit is composed of two types of circuits, the sensor nodes (SNs) and a base station (BS), all connected to each other with conductive yarns forming a mesh topology with the base node at the center. The reliability analysis shows that communication in a multi-hop connection of sensors in mesh topology is more reliable than in the conventional star topology. From the standpoint of the network, each SN is a four port router capable of handling packets from destination nodes to the BS. The end-to-end communication uses packet switching for packet delivery from SNs to the BS or in the reverse direction, or between SNs. The communication module has been implemented in a low power field programmable gate arrays (FPGA) and a microcontroller. The maximum data rate of the system is 9+9 Mbps while supporting tens of sensors, which is much more than current BAN applications need. The suitability of the proposed system for utilization in real applications has been demonstrated experimentally.

[1]Instituto de Telecomunicações and Departamento de Engenharia Electromecânica, Faculdade de Engenharia, Universidade da Beira Interior, Portugal
[2]Department of Electrical and Computer Engineering, Faculty of Engineering, University of Porto and INESC TEC, Portugal

3.1   Introduction

In recent years many promising technological advances in electronics, information technology and communication systems have paved the way to new concepts in health-care. For example, telemedicine, eHospital, and ubiquitous healthcare are enabled by emerging new electronic devices and wireless broadband communication technologies. While initially becoming mainstream for portable devices such as tablets and smart phones, wireless communications are evolving toward wearable solutions; even implantable solutions are being introduced. These health-care devices rely on sensor networks and BANs.

BAN refers to a subcategory of sensor networks mainly used for measuring or monitoring vital body parameters without affecting the lifestyle of human beings. These systems can be used at care centers or at home by patients (or even by healthy people) who want to improve or monitor their health conditions. It is expected that in the coming years, BANs will help bring about revolutionary changes in health-care systems and applications.

A BAN consists of wearable or implantable sensors (possibly associated with local computing devices) and a communication network for data collection [1–6]. Usually, an on-body system is responsible for collecting data from sensors via an intra-network and send them to a computer or personal digital assistant (PDA) via a wireless link. According to the application and design parameters, the intra-network can be wired, wireless or even use the human body as communication medium. Wired networks, as a second type of communication infrastructure for BAN applications, provide high-speed, reliable and low-power solutions [7–9]. This chapter describes the design and evaluation of a new wearable BAN to support the measurement of human locomotion parameters in a practical, comfortable and noninvasive way. The entire communication infrastructure, including hardware, software and protocols, has been designed for this purpose. Although the design of the signal acquisition modules, driver software and data-processing programs are beyond the scope of the present work, the designed communication system was embedded together with those facilities and tested in real data-acquisition experiments. This chapter then focuses on the networking component, in its different levels. The proposed system can also be used in many other BAN applications, especially those including a large number of sensors (compared to the usual BAN scale), embedded in textile and with high data-rate communication demands.

Employing conductive yarns instead of using wireless links in the wearable unit provides communication that is more reliable, has higher data rates and throughput, and consumes less power. The wearable unit is composed of two types of circuits, SNs and a BS, all connected to each other with conductive yarns forming a mesh topology with the base node at the center. In comparison with traditional serial or star topologies, a mesh connection is more reliable, provides higher data-rate communication and supports significantly more sensors.

End-to-end data packet delivery is based on packet switching by using source routing for minimum cost forwarding (SRMCF) routing protocol (cf. Section 3.5). This protocol has been designed to establish a spanning tree of all minimum cost forwarding paths with the base node at the center. The physical and media access control (MAC) layers have been implemented on a low-power FPGA (Actel IGLOO AGLN125). The maximum data rate of the system is 9+9 Mbps while supporting tens of sensors, which is much more than current BAN applications need. The main contributions of this chapter are as follows:

- Hardware prototypes of SNs (excluding acquisition circuitry), BS and CPM with physical and MAC layers implemented on FPGA. The prototypes are multitask devices that support asynchronous communication and concurrent send/receive operations.
- The SRMCF reactive routing protocol for both wired and wireless networks. The protocol has been implemented on the designed wearable system.

All designs have been analyzed theoretically and validated experimentally. The rest of the chapter is organized as follows: Section 3.2 describes related work and Section 3.3 presents wearable system architecture. Experimental results obtained from prototype are presented in Section 3.4. In Section 3.5, SRMCF routing protocol is described. Finally, Section 3.6 concludes the chapter.

3.2   Related work

In the last years, a number of wired and wireless networks for wearable applications have been proposed. Several surveys on BANs and wearable systems can be found in the recent literature [10–15]. As for surveys on applications, a special emphasis is given to medical and health-care areas in [16–20].

One of the early attempts to manufacture a wearable system was reported by Post and Orth [21]. They built an electronic system by utilizing conductive yarns as a data bus and described some techniques for building circuits from commercially available fabrics, yarns, fasteners and components. They also have shown that all of the input devices can be made by seamstresses or clothing factories, entirely from fabric.

Based on theoretical analysis and experimental results, the ability of conductive yarns to carry both data and power supply has been shown by Wade and Asada [22]. They introduced a wearable direct current (DC) power-line communication network for health sensing and rehabilitation. Their system provides a network through which multiple sensors and actuators can send and receive both information and power. Their method applies to wires or conductive yarns with high conductivity.

A wearable personal network (WPN) system based on a fabric serial bus with conductive yarns was introduced by Lee *et al.* [8]. Their system is a four-layered wearable personal area network (PAN) with a bus topology controlled by a host node, and their protocol works with a variety of microcontrollers. The system was implemented on an FPGA running at 50 MHz and works at up to 10 Mbps. The profile layer is provided to make the application development process easy. The data link layer exchanges frames in a master–slave manner in either reliable or best effort mode. The lower part of the data link layer and the physical layer of the WPN is built on a fabric/serial-bus interface that is capable of measuring bus signal properties and adapting to medium variation. They have implemented WPN communication modules (WCMs) on small flexible printed circuit boards (PCBs) and they have also designed a WPN shirt prototype using implemented WCMs and conductive yarns.

An inductive transceiver with a fault-tolerant network switch for multilayer wearable BAN applications has been introduced by Yoo and Lee [23]. The inductive coupling transceiver is made from conductive yarns and employs a resonance compensator (RC) with a digitally controlled on-chip capacitor bank and a variable hysteresis Schmitt trigger to compensate dynamic and static variances of the woven inductor. It communicates at a 10 Mbps data rate. Each node consists of two chips: one with transceiver and switches, and one with the RC. Both were fabricated in a 0.25 μm complementary metal oxide semiconductor (CMOS) technology and occupy 2 and 0.8 mm², respectively.

Cho *et al.* [24] presented an interference resilient 60 kbps-to-10 Mbps body channel transceiver for multimedia and medical data transactions in BANs. The transceiver uses the human body as a signal transmission medium in the 30–120 MHz frequency range for energy-efficient and scalable data communication around the body. In the frequency range of body channel communication (BCC), the human body may operate as a receiving antenna and pick-up large interferences, degrading its signal-to-interference ratio (SIR) to 22 dB. In order to avoid body-induced interferences, a 4-channel adaptive frequency hopping (AFH) scheme that monitors channel status continuously and selects only the clean channels was added to the BCC module. To support 10 MHz wideband frequency shift keying (FSK) signaling with the fast AFH, the direct-switching modulator and the delay-locked loop (DLL)-based demodulator are incorporated in the transceiver design. The dual frequency synthesizers in the modulator reduce the frequency hopping overhead to 4.2 μs without degradation of the phase noise. The transceiver, fabricated in 0.18 μm CMOS technology, withstands a −28 dB SIR and operates at 1.8 m with −25 dB SIR.

An impulse radio (IR) type transceiver for human body communication (HBC) has been presented by Shikada and Wang [25]. In their system, the transmitter employs an IR scheme in which the information data is converted to pulse trains. First, information with a date rate of 1.2 Mbps is encoded with eight pulses for bit "1" and nothing for bit "0." The pulses are produced by a 4.9 MHz oscillator and an exclusive OR (XOR) gate, so they have a duration of 10 ns. The modulated pulses are then spectrum-shaped by a band-pass filter to restrain their main spectrum components to the 30–50 MHz range. The output signal drives an electrode which is placed on the body. The receiver employs the envelope detection scheme. The received signal is filtered, amplified and then adjusted to an adequate level by an automatic gain controller (AGC). The envelope detection is realized by a diode and a low-pass filter. The detected signal is then assigned a logical 1 or 0 by a comparator after the envelope detector.

Magnetic human body communication (MHBC) employing a quasi-magnetostatic field was proposed by Ogasawara *et al.* [26]. The system is designed to mitigate the influence of the surroundings on the communication channel. It uses the human body as one component of a loop that creates magnetic coupling between two transmitters. A circuit model was devised, and its validity was experimentally demonstrated. They have reported that the results show the validity of the MHBC channel model and its robustness with regard to the surrounding environment. They also show that HBC based on magnetic coupling is a feasible technology. The available data rate of the system is not reported.

A wearable wireless electrocardiography (ECG) sensor was presented by Nemati *et al.* [27]. This system combined an appropriate wireless protocol for data communication with capacitive ECG signal sensing and processing. The ANT protocol was used as a low data-rate wireless module to reduce the power consumption and size of the sensor. Small capacitive electrodes were integrated in a cotton T-shirt together with a signal processing and transmitting board on a two-layer standard PCB. Signal conditioning and processing were implemented to remove motion artifacts.

The wearable system presented in the remaining part of this chapter consists of embedded sensors in textiles, all connected to each other with conductive yarns in a mesh topology. The main advantages of the proposed system, compared to previous works, are low energy consumption and the support for large number of sensors, while keeping the data rate high.

3.3 Wearable system architecture

This section describes the prototype of the proposed wearable system including network and hardware. The hardware of the system consists of SN, BS, CPM and connections between the SNs, which are established by conductive yarns in a mesh network fashion. The system includes a wearable infrastructure and an on-body CPM, which communicates with a computer as shown in Figure 3.1. The functionality of each part is as follows:

1. Wearable infrastructure: This part includes a set of SNs equipped with electromyography (EMG) electrodes for capturing electrical signals from skin tissue, and also accelerometers and gyroscopes for measuring kinetic parameters. To ensure a high data-rate and reliable communications, a mesh topology was selected. For that, SNs that contain all necessary communication parts are connected to each other with conductive yarns. A BS node is responsible for collecting all the data captured by the SNs. In fact, each SN acts as a router device to handover packets from the source SN to the BS. The routing operation of the SN is based on the SRMCF routing protocol (cf. Section 3.5).

2. CPM: This part is a low-power microprocessor-based system. The CPM, as an on-body device, gathers the information from the wearable network via a communication link to the BS module. The CPM accesses the different SNs dispersed throughout the pantyhose using the wire mesh network embedded in the technical fabric. To transfer data to a computer or saving it locally, the CPM has been equipped with a wireless Bluetooth module, a USB port and a micro SD card.

3. Computer or PDA: The information collected by the CPM needs to be post-processed, analyzed and stored in a computer. A software program may concurrently plot the real-time data transferred by CPM and shows the node status and connections.

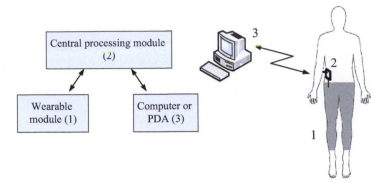

Figure 3.1   General architecture of the system

Table 3.1   Measured values per unit [30] (at 10 MHz)

Yarn	Relaxed (natural length)		Stretched (1.5 × natural length)	
	Ω/cm	nH/cm	Ω/cm	nH/cm
1	0.65	16.4	0.47	6.68
2	6.76	35	4.43	33

3.3.1   Characterization of the conductive yarns

The most comfortable and easiest way to monitor physiologic signals consists in using garments with conductive yarns as wire links [28,29]. These kind of conductors can be modeled as an R–L series impedance [30]. Because of the dependency of resistance and inductance on signal frequency and wire length, the impedance of a line of length l can be expressed as

$$Z(f,l) = l \times (R(f) + j2\pi f L(f))$$
(3.1)

Table 3.1 shows the measured values of these parameters at 10 MHz for two different typical conductive yarns. One of the inherent properties of most conductive yarns is their ability to stretch, and their parameter values are different for relaxed and stretched modes. Both resistivity and inductivity decrease with stretching. In comparison with normal copper wires, the equivalent inductive and resistive parameters are reasonably higher in yarns. For this reason, in practice, several yarns must be put together in parallel to reduce the resistivity and inductivity to levels closer to those of copper wires. The number of parallel yarns needed is application dependent. Although some capacitive coupling between parallel yarns exist, it is lower than that found in a twisted pair of copper wires (for the frequencies under consideration), so it was not taken into account in (3.1).

Supplying power to all parts of a wearable system is one important design consideration. Single battery operation, in alternative to having one battery per node, simplifies the apparatus. However, in order to enable power-supplying to sensors from one single central battery, the resistivity of the conductive yarns needs to be small to prevent losses.

3.3.2   Intra-network

The intra-network of SNs in wearable systems includes a set of SNs all connected to a collector node. The topology of the network depends on the number of SNs and data rate. The star and serial bus topologies are useful for a small number of the SNs. With increasing number, the performance of the aforementioned networks degrades: the number of ports increases in a star network; the same happens to contention and electrical load on the bus in a serial network.

In networks with a high number of SNs, the mesh topology is a good way to overcome these limitations while keeping high data rates. To evaluate the channel behavior, consider a network with n SNs connected in series to a collector node as shown in Figure 3.2. C1 to Cn stand for the capacitive impedance of SNs, and R1 to Rn and L1 to Ln for the conductive yarn-equivalent circuits. The combination of the electrical properties of the conductive yarn with the capacitance of the ports results in a nth-order RLC low-pass filter for a bus network and a second-order filter in a mesh network. Considering that the parameters of the conductive yarn depend on its length, the behavior of such a low-pass filter will also be a function of the yarn length. When the number of nodes in the wearable network increases, the network performance of a serial bus connection will degrade faster than the performance of a mesh network. Therefore, a mesh topology is a logical choice for systems with a high number of nodes and high data rate.

3.3.3   Network layers

According to the open system interconnection (OSI) architecture, a network consists of a stack of several layers, each with its own specific tasks. The network functions to establish end-to-end communication between SNs and BS are grouped in five layers:

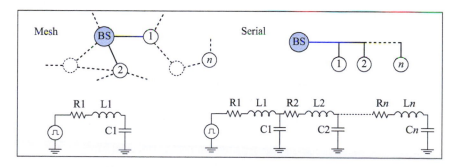

Figure 3.2   Serial bus and mesh interconnection modes

physical, MAC, network, middleware and application layers. They are described in the following subsections.

3.3.3.1   Physical layer

The communication medium, transmitter, receiver, coding and modulation define the first layer of the OSI model, the physical layer. To avoid the use of separated lines for transmitting and receiving and to reduce the complexity of the yarn network that embedded in the fabric, the communication between SNs is bidirectional, as shown in Figure 3.3. Because the node-to-node communication is established over dedicated lines, a baseband communication method has been selected. Data is mapped to signal levels by the non-return to zero inverted (NRZI) encoding method. NRZI can be operated with either DC or AC coupling because the information is uniquely defined by signal transitions. Such a property makes the design of the communication module more flexible. For evaluation of the proposed circuit, DC coupling has been utilized. In the idle state, the line is free and both nodes monitor the line for data. In fact, in this case, the line is in high-impedance mode and both nodes are able to seize the line to start transmission. To keep the signals in zero level in the idle state, lines are pulled down.

3.3.3.2   MAC layer

In BAN applications, reliability, sharing of the line and energy efficiency are important features to take into account when devising a MAC protocol. To achieve the aforementioned features in the present system, communication at the MAC layer is controlled by the RTS/CTS handshaking mechanism. In fact, in mesh networks it is very useful that each node is able to handle communication requests from different neighbors simultaneously. In order to efficiently manage the requests, the sender and receiver nodes should be aware of the status of each other. Otherwise, a significant number of data packets may be lost, requiring retransmission. The use of RTS/CTS handshaking helps to ensure reliable communication with low packet loss and good resource management. The receiver has to recognize the beginning of the data in the data frame. For that, frame-sync bits (3-bit Barker code) are placed just before the data stream.

Each SN is a packet generator independent from the other nodes. Independent packet-generation results, in the long run, in packets arriving at the intermediate nodes in a random sequence. Such a network operation can be considered as a Poisson

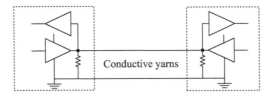

Figure 3.3   Connection between the nodes in physical layer

process, with an exponential time interval between consecutive packets received at each node [31]. The throughput at the BS node is

$$S_{BS} = \frac{Pe^{-\lambda_r \tau}}{(1/\lambda) + T_r + 2\tau + e^{-\lambda_r \tau}(T_c + P + 3\tau)} \tag{3.2}$$

where λ denotes the total packet rate, T_r the RTS time, τ the propagation time including the signal propagation and detection time, λ_r the packet rate in the direct from BS to nodes, T_c the CTS time and P the total packet time. For any other SN, the same conditions apply.

The arrival rate λ is the time a packet takes while waiting in queue plus the packet serving time. Encapsulating the header of the packet at the receiver node for processing and routing purposes also takes some time. When BS is connected to only one node, the idle time cannot be smaller than the packet serving time. Considering this time, the maximum channel utilization can be calculated as

$$S_{BS} = \frac{Pe^{-\lambda_r \tau}}{T_{idle} + T_r + 2\tau + e^{-\lambda_r \tau}(T_c + P + 3\tau)} \tag{3.3}$$

where T_{idle} is the idle time.

3.3.3.3 Network layer

End-to-end data communication between SNs and BS is established over intermediate nodes. So a routing protocol has to be used. For that, SRMCF, an energy efficient routing protocol, is adopted. The protocol will be explained in detail in Section 3.5.

3.3.3.4 Middleware and application layers

The middleware layer, placed between the network and application layers, is responsible for making the appropriate interface between the hardware, network and applications, so that they operate as a whole. It provides services for driver applications and provides a runtime environment that can support and coordinate multiple applications.

In the application layer, all acquired data is saved to the microcontroller RAM. The middleware controls the length of data in the RAM and builds the payload. The first byte of data payload defines the data type that it is used for addressing the payload contents.

Depending on the application, a sensor can be used to monitor a phenomenon, sensing, event detection and identification, etc. [32]. The application layer uses the underlying network layers to establish process-to-process communication. In the present system, each SN is able to capture EMG signals and kinetic information by using three sensors: EMG electrode, accelerometer and gyroscope.

EMG sensor includes electrodes for capturing electrical signals from skin tissue. Raw EMG signals can be in the range ± 5 mV (for athletes!) and typically the frequency ranges between 6 and 500 Hz, showing most signal power between 20 and 150 Hz [33]. The microcontroller used in the SN contains a multichannel analog to digital (ADC); one of them, with 10-bit resolution, is used for EMG. The internal

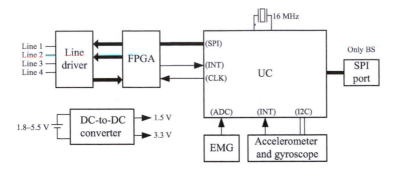

Figure 3.4   Overall organization of the SNs and BS

RAM is used to buffer all data acquired by the sensors. Both accelerometer and gyro-scope sensors are three-axes MEMS devices, which means that each sample includes three 16-bit values sampled at 50 Hz. The EMG samples are 16-bit values and the corresponding sampling rate is set to 1 kHz.

3.3.4   SNs, BS and CPM circuits

Figure 3.4 shows a block diagram of the SN. The circuit contains a 16-bit micro-controller and one FPGA (low-power Actel AGLN125 [34]) on the main PCB; the FPGA is used to implement the physical and MAC layers of the network. Although the FPGA is able to drive the lines directly, a separated board is used for protection and flexibility. Line drivers and sensors are on separate PCB boards.

   Each SN has four bidirectional ports for connecting to other nodes in a mesh network. The power supply is a coin-type battery (LIR2450), which, together with a DC-to-DC converter, generates a 3.3 V power supply voltage for microcontroller and sensors, and a 1.5 V supply voltage for FPGA core and line driver circuits.

   The internal oscillator of the microcontroller (16-bit, PIC24FJ64GA104 with 8 kB of RAM) is the clock source for both microcontroller and FPGA. The micro-controller implements the network, middleware and application layers; acquires the signals from EMG sensors and kinematic data from the accelerometer and gyroscope. The code was written in the C language with MPLAB X IDE, and the microcontroller was programmed with the PICkit 3 programmer. The circuit implemented in the FPGA contains a number of control registers: all of them can be read or written by the microcontroller via an SPI port.

   The BS circuit is similar to that developed for the SN, with the exception of an extra SPI port used to connect with the CPM board. The CPM board shown in Figure 3.5 uses the same microcontroller and connects to the BS.

3.3.5   FPGA-based implementation of the physical and MAC layers

The hardware implementation of the physical and MAC layers was described first in Verilog, synthesized with the Libero IDE Project Manager V9.1 from Microsemi [34]

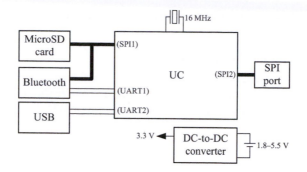

Figure 3.5   Block diagram of CPM

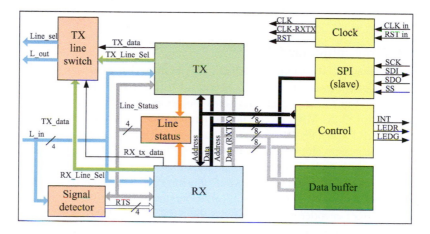

Figure 3.6   Communication module implemented on an FPGA

and used to program the FPGA with the FlashPro4 device programmer. Figure 3.6 depicts the block diagram of the communication module implemented in the IGLOO nano FPGA [35]. The Control module provides access to the settings of the circuit and enables reading or writing data from/to internal RAM via an internal bus and an SPI port. The SPI port provides a high-speed communication channel to the microcontroller. The acronym TX represents the transmitter module that includes all the submodules needed for communication: packet handover, encoding, generating and detecting MAC information, and also for frame generation. Conversely, the label RX indicates the receiver. This module also includes all submodules needed for decoding, generating and detecting MAC frames, buffering packets carried in data frames and for error checking. Both RX and TX modules are independent in order to enable communication with two nodes simultaneously. The TX Line Switch module is responsible for seizing the lines for data transmission, either coming from the TX or RX modules. The Signal Detector module, as its name implies, detects the incoming preamble signals and RTS messages. The Clock module generates the internal clock, resets the circuit and sets the bit rate.

3.3.5.1   Internal bus architecture

The communication module includes two 8-bit internal buses; one for memory and buffer access by the microcontroller and the other for buffer sharing among RX, TX and microcontroller. Such a separation ensures the required high-speed operation.

3.3.5.2   Signal detector

This module includes four RTS Detector submodules. Each RTS Detector is connected to one of the incoming lines and is able to detect RTS messages and inform the RX module to proceed with packet reception. In fact, the RTS message is a string of 1s with at least four consecutive 1s. Each module counts the number of consecutive 1s and activates the output. Any string of consecutive 1s that is smaller than four is considered as an error detection when the port is in idle mode.

3.3.5.3   Transmitter module

Figure 3.7 shows all the submodules that implement packet processing at the transmitter. *TX Control* contains all the registers for transmission. It is responsible for controlling all transmission steps and holds the necessary submodules to realize the communication at the MAC level. The TX Timing circuit module generates the timing clock CLKTX, which determines the transmission data rate. Its frequency is one quarter of the system clock CLK. The encoder unit includes all submodules for data encoding and insertion of the synchronization bits in the data frame. It generates the RTS messages and determines the end of transmission. The input of the encoder is serial, supplied by a PISO (parallel-in serial-out) module that serializes the buffer output. An NRZI encoder module converts the NRZ output of the encoder.

Transmission starts when the packet stored in the buffer is ready to be sent. Buffer control controls the status of the packets (first data set of each segment in the buffer). If the status indicates that a packet is ready for transmission, then buffer controls informs the TX by activating PKT2SEND2TX. First, TX starts the transmission process by reading the packet information, including the length and the port number (1–4). Then

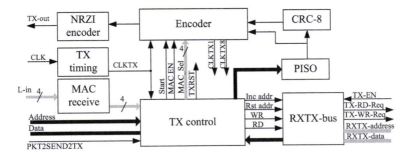

Figure 3.7   Block diagram of the transmitter module

TX checks if the line is available by using the line-status signals generated by the line status module. If the line is free, then TX control activates MAC-EN for generating the RTS message through the encoder module. After sending RTS, TX control waits for a CTS reply from the receiver node. By utilizing the MAC receive submodule, the TX module has the ability to receive MAC messages directly and independently from the RX.

The absence of a CTS reply means that for some reason, the receiver is not able to receive a packet at that moment. So, the transmitter waits for a while which is configurable from byte to a full packet length time and attempts another transmission. The transmitter starts to send the packet after receiving CTS. A counter in TX control counts the number of outgoing data bytes and compares it with the packet length. While sending the packet, the cyclic redundancy check (CRC)-8 module generates a checksum of all the bits in the packet for error detection; the generated CRC value is added by the encoder to the end of the packet before sending. If the destination node receives the packet correctly (no CRC error), it replies to the sender with an ACK message. Upon completion of this cycle, the TX releases the packet in the buffer by changing its status to free. The absence of an ACK response or the reception of an ERR message means that an unsuccessful transmission took place and the TX needs to resend the packet. The number of packet transmission retries is determined by the repeat RTS register in TX control.

3.3.5.4   Receiver module

All submodules of the RX module work together to implement the packet reception process according to the block diagram of Figure 3.8. RX control is responsible for controlling all steps in packet reception. Like TX, RX is an independent module capable of generating and sending MAC messages and communicating directly with the sender. RX timing generates a clock signal synchronized with the incoming stream for data recovery and also detects the frame starting bits for alignment. The decoder decodes the received data and generates the CLKRX1 and CLKRX8 signals. These are used by the serial-in parallel-out (SIPO) and CRC-8 check modules for byte formation and separation of synchronization bits from data.

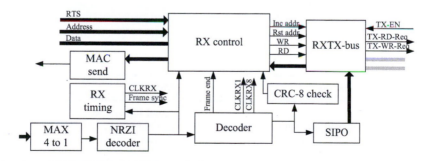

Figure 3.8   Block diagram of the receiver module

Receiving starts if the signal detector module detects an RTS message and if RX is free. If clear, the RX replies with a CTS message generated by the MAC send module and waits for frame sync bits to be detected by the RX timing circuit. The decoder is responsible for separating the synchronization bits. Regardless of the quality of the information conveyed, all received data is stored in the buffer through the RXTX-bus. In fact, it is at the end of reception that the CRC-8 check determines the validity of the received data. If this module confirms reception without error, then the RX changes the status of the corresponding buffer segment notifying a successful reception and then releases the buffer. Otherwise, the RX module just releases the buffer without further action.

3.4　Experimental results

This section presents and discusses the experimental results obtained with the actual implementations of the SNs, BS and CPM shown in Figure 3.9. The supply voltage, clock frequency and FPGA resource usage are summarized in Table 3.2. For the

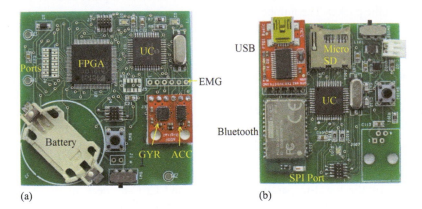

(a)　　　　　　　　　　　　　　　　　　　　　(b)

Figure 3.9　Picture of the prototype; (a) SN and BS ($6 \times 6\,cm^2$), (b) CPM ($5 \times 5.6\,cm^2$)

Table 3.2　Characteristics of the communication circuit and its FPGA implementation

Parameter	Value
Model	Actel IGLOO AGLN125
Logic cell utilization	2,509 of 3,072 (81.7%)
Internal memory	2 kB
I/O supply voltage	3.3 V
Core supply voltage	1.5 V
Clock frequency	16 MHz
Data rate	4 Mbps

purpose of performance analysis, the communication nodes were first connected together with normal twisted wires; afterward the same experiment was repeated using conductive yarns for the interconnect. Signals flowing in the communication lines were analyzed and measured with an oscilloscope.

3.4.1 Communication at the MAC layer

Figure 3.10(a) shows the signals on the line, including both the TX (sender node) and RX (receiver node) signals, when two sensors are connected with normal twisted wire and communicating at 2 Mbps data rate. Figure 3.11 shows more details at 9 Mbps. The signal level at the output is between 0 and 1.5 V. RX and TX signals can be seen separately in Figure 3.10(b) and (c). This figure also shows the complete MAC layer

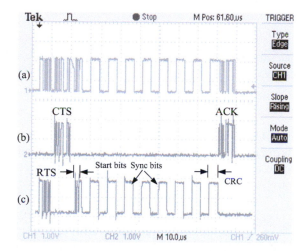

Figure 3.10   Encapsulated data packet in MAC layer: (a) line, (b) signals generated by receiver, (c) signals generated by transmitter

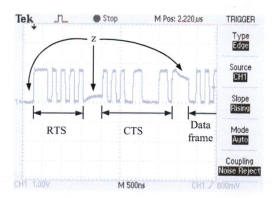

Figure 3.11   Signals over the line

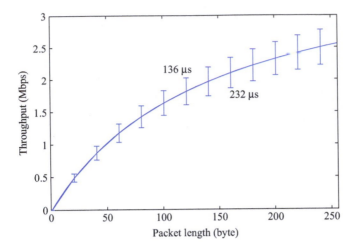

Figure 3.12   Node-to-node throughput as a function of packet length for a 4 Mbps data rate and T_{idle} in the range (136, 232 µs)

message including RTS, CTS, ACK and data packet frames with synchronization and start bits (waveform c). The transmitted packet includes 15 bytes all with 0x00 value (for this example) to highlight the synchronization bits in a packet including a long string of 0s. The trailer contains the CRC-8 checksum that is generated in the MAC layer.

It should be noted that the upper layers (network layer and above) do not implement any error-checking mechanism. In fact, extra error checking mechanisms are not necessary in the upper layers, because if any error is detected at the MAC level, the packet will be dropped before being handed over to the network layer. Furthermore, including other checking structures would increase the packet overhead. Assuming Poisson channels, the node-to-node or MAC level throughput is given by (3.3). Figure 3.12 depicts the throughput calculated with parameter values obtained from 20 times circuit operation and determined using (3.3) for $\lambda = 4$ Mbps. Obviously, throughput increases with packet length and reaches 2.55 Mbps at 256 bytes, which is 64% of the data rate. Throughput is mainly influenced by Tidle, which was measured to be 184 µs at 16 MHz on average ([136, 232 µs]).

3.4.2   Routing

To evaluate the routing performance of the circuits with the SRMCF protocol, a network containing three SNs and a BS, as shown in Figure 3.13, was used. The bold lines show the minimum cost path values between SN3 and BS defined during the network setup phase (as described in Section 3.5). The first experiment was designed to evaluate the effectiveness of the network in routing a packet sent from the BS to an SN. There are two intermediate nodes between the BS and the destination node SN3.

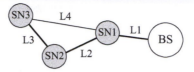

Figure 3.13   A network contains three SNs and BS

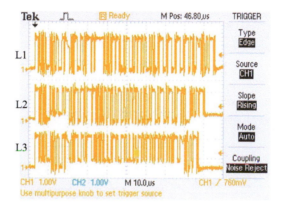

Figure 3.14   Routed packet at different points of the network from the BS to the destination SN

As will be mentioned in Section 3.5 and shown in Figure 3.14, the length of the packet sent from the BS to an SN is variable and decreases as it passes through intermediate nodes. In Figure 3.14, waveform L1 shows the packet after the BS, waveform L2 the packet after the first intermediate node and waveform L3 after passing the second node.

Another experiment was performed to evaluate the routing delay for transmissions from an SN to the BS node. Again, there are two intermediate nodes between the SN and BS. Figure 3.15 shows the packet Pa (generated by SN3 with a payload of 128 bytes) as it is routed through the network. As can be observed, the delay to the first intermediate node is almost 150 μs (packet serving time); the same holds for the second node. The end-to-end delay from SN3 to BS is 2.2 ms. The routing process time was measured to be in the range from 136 to 232 μs (with an average of 184 μs).

Figure 3.15 also shows that the signals are in a different polarity in the absence of communication between the SNs. Since NRZI is being used as the line coding scheme, the polarity of the signal is of no importance. The transmitted data indicated with an M in this figure shows two coexisting transmissions involving a common node. A time interval can be identified where SN3 is sending information to SN2, while a communication link is also active between SN2 and SN1. This shows the ability of the developed SNs to simultaneously send and receive data as described earlier.

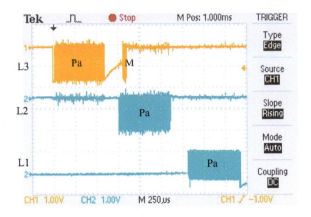

Figure 3.15   Data transmission from SN3 to BS observed at different points of the network (see Figure 3.13 for line identifiers)

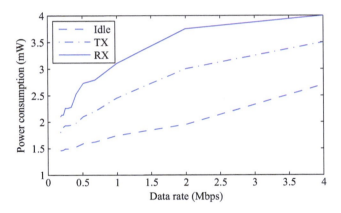

Figure 3.16   Power consumption in terms of data rate (VCC = 1.5 V)

3.4.3   Power consumption

The power consumption of the circuit as a function of data rate is depicted in Figure 3.16. The measurement of power consumption in idle mode indicates that the power expenditure for data rates below 512 kbps is 1.5 mW. It increases to 1.74 mW at 1 Mbps; 1.95 mW at 2 Mbps; and 2.7 mW at 4 Mbps. The increased power consumption in idle mode is due to the signal detector module, whose clock signal must be increased according to the data rate even in idle mode.

In order to operate at data rates above 4 Mbps, the internal PLL of the FPGA needs to be activated, which increases the power consumption. To attain the maximum data rate (9 Mbps), the PLL output frequency must be set to 70.993 MHz, increasing the power consumption to 7.92 mW.

The power consumption during communication increases to 3.5 mW at 4 Mbps and 9 mW at 9 Mbps in TX mode; in RX mode the power consumption is 4 mW at 4 Mbps and 10 mW at 9 Mbps. Usually wired systems consume less power than wireless counterparts. The values reported in [36] for Zigbee (30 mW, 250 kbps) and Bluetooth (100 mW, 3 Mbps) indicate that the proposed circuit consumes substantially less power than wireless alternatives.

3.5   The SRMCF routing protocol for sensor networks

This section presents SRMCF protocol, a reactive, energy-efficient routing protocol for WSN, which is used for both wired and wireless wearable sensor networks [37]. This protocol is based on source routing (SR) [38,39] concepts for ad hoc networks and minimum cost forwarding (MCF) [40] methods for heterogeneous WSN. Since the proposed concept combines SR with MCF, it is called the SRMCF protocol. SNs maintain no information about the network topology, but packets (from sensors to BS or vice-versa) always communicate over paths with minimum cost. In this approach, only the packets from the BS to SN include routing information [38,41]. The proposed protocol is intended for application scenarios where the availability of limited resources requires the energy consumption to be kept small. Experimental and simulation results show that the protocol reduces energy consumption and increases system lifetime, while preserving network performance. In the following subsections, the protocol is introduced and then compared with MCF. An implementation of the protocol is also presented. The contributions described in this section are the following:

1.   Design and implementation of the energy-efficient SRMCF protocol.
2.   Extensive comparative simulations of both routing protocols running with two different MAC protocols.

3.5.1   Supported message types

The routing algorithms for packets coming from the BS node and for packets generated by SNs are different. As a consequence, the header of data packets is also different. In both cases, data transmission is unicast and packets are sent over minimum cost paths. Intermediate nodes select the appropriate routing algorithm based on the packet type. In addition, SRMCF supports broadcast messages for network setup and failure recovery.

This protocol supports two kinds of broadcast messages: cost advertisement and cost request. Whenever the BS starts up or any SN gets a new cost value, they send a cost advertisement message to inform neighboring nodes of their current cost value. Nodes can broadcast cost request messages whenever a new cost value is needed. This occurs when a sensor turns on or a failure happens.

3.5.2   Network setup

The setup phase has two steps. In the first one, the cost field is set up: all nodes determine their cost values for communicating with the BS. In the second step, the BS node creates the routing table. In the setup phase of the network, each node is

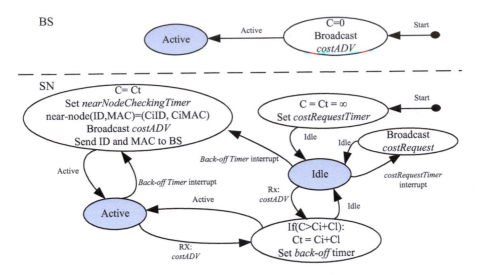

Figure 3.17   Network setup: determination of node cost values and initialization of BS

assigned a minimum cost value, together with the ID and MAC addresses of the adjacent node on the minimum cost path to the BS (hereafter called the near node). Figure 3.17 shows the finite state machines (FSMs) that describe the operation of SNs and BS in the first step of the setup. In this figure, C, Ci, Cl and Ct are, respectively, the current node cost, the node cost in the received message, the link cost and the temporary node cost. The setup procedure is discussed in the following subsections.

3.5.2.1   Determination of node cost value

This step is similar to the minimum cost forwarding back-off algorithm [40]. Each node sets its initial cost value to ∞ and the BS sets its own to zero. Then the BS starts broadcasting its cost value. A receiver node compares its own cost value with the received cost (including measured or calculated cost value of the sender plus the cost of the link between the sender and the receiver). If the received value is lower than the node's current cost value, then the node updates its current cost to the new value, waits, and then broadcasts the new value. The waiting time increases linearly with the link cost value [40]. This process goes on until all nodes set their cost values to the minimum.

The back-off algorithm decreases the number of cost advertisement messages significantly, as most of the nodes will broadcast their cost value only once. SRMCF utilizes a similar method, but here the nodes hold a unique ID.

3.5.2.2   Routing table creation

The routing table in the BS holds information about all optimum paths (i.e., paths with minimum cost value) between the BS node and all other nodes. During the first

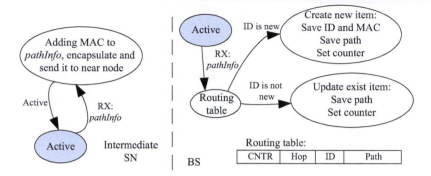

Figure 3.18   Routing table creation in the BS with collaboration of sensor nodes

phase, the cost value of each node changes (from the default value of ∞ to the cost determined in that phase). When a node changes its cost value, it sends a packet with its ID and MAC address to its near node. The receiver adds its own MAC address to the received payload and sends it to its own near node along the optimum path. Eventually, the BS receives a packet that includes the ID and MAC addresses of the source node, and the MAC of all the nodes in the minimum path between the source node and BS. It then saves the ID and MAC addresses in the row of the routing table corresponding to that particular source node. Figure 3.18 shows the FSMs that describe the initialization of the routing table in the BS.

In this way, SNs and BS collaborate in the creation of the routing table. In fact, the routing table is created without running any algorithm or calculation on the BS and without any information about the network topology. Note that the same procedure is performed during normal network operation whenever the cost value of a node changes. This may happen when a link or node failure occurs, or when a node gets a cost advertisement message with a lower value than its own previous cost.

3.5.3   Link and node failure recovery

In any kind of network, faults occur frequently and unexpectedly [42,43]. Recovering from a link or node failure involves updating the total cost (node + link costs) value field and the corresponding minimum-cost paths stored in the BS. In order to detect failures, each node has a timer, which is restarted whenever the node gets a message from its near node. Each node monitors the activity of its near node. Any message received from that neighbor will reset the timer. If there is no message for a specified time period, the node sends a query message to check the reachability of its near node. If there is no reply, the node initiates the recovery procedure by broadcasting cost request messages periodically.

Figure 3.19(a) shows a part of the network with established links and corresponding minimum cost paths (bold lines). In Figure 3.19(b), node 1 has failed, and nodes 17, 13, 18, 14 and 16 must find new minimum cost paths. Figure 3.19(c) shows the

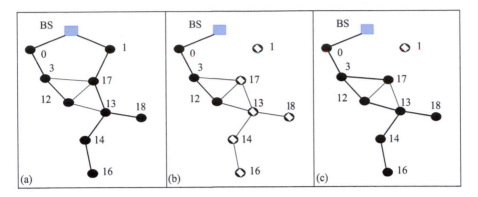

*Figure 3.19   Failure recovery: (a) initial situation, (b) failure of node 1,
(c) recovery from failure complete*

network after recovery. Node 13 selects node 12 as its near node, because the new cost value of node 17 is higher than that of node 12. In general, the cost values of nodes which are closer to the BS than the failing node are not affected.

The failure recovery mechanism of SRMCF is mainly local, with changes restricted to the place of occurrence without affecting other parts of the network, a behavior that enhances energy efficiency. In addition, data communication is not disrupted during failure recovery, as happens when a periodic setup phase is used with MCF [44].

3.5.4   Analytical comparison of SRMCF and MCF protocols

To increase the network performance and reduce energy consumption, SRMCF generates less packets than MCF for transmitting the same amount of data. This section compares SRMCF with MCF. In addition, average packet header length and routing table size are examined. Let the set $N = \{n_1, n_2, \ldots, n_m\}$ denote the WSN nodes. If n_i is within the transmission range of n_j with reliable communication, then $l_{i,j}$ denotes the link between these nodes. If the propagation channel between nodes is reciprocal and nodes are equipped with the same transmitter and receiver circuits, then $l_{i,j} = l_{j,i}$. In this case, let L denote the set of such reciprocal links. Then, we can model the network as an undirected graph $G = (N, L)$. In the following, we assume that G is connected. Between each pair of nodes, there are many possible paths. Let $P_{i,j}^k$ denote the kth path between n_i and n_j. For random networks with n nodes, both the maximum and the average number of hops in paths between nodes increase as $\Theta(\sqrt{N})$ [45,46].

Let $C_N(i)$ define the communication cost value of node n_i and $C_L(i, j)$ the cost value of link $l_{i,j} = l_{j,i}$. Let $C_{i,j}^k$ represent the cost for sending a message from node nj to ni over the kth path. Then

$$C_{i,j}^k = \sum_{r,s \in P_{i,j}^k} C_L(r, s) + \sum_{n \in P_{i,j}^k} C_N(n) \tag{3.4}$$

Since we are assuming reciprocal links, we have $C_{i,j}^{k} = C_{j,i}^{k}$, so the cost of communicating between two randomly selected nodes is independent of the direction. Whatever the metric used to measure the cost, there is at least one path with minimum cost between two arbitrary nodes:

$$C_{i,j} = \min\{C_{i,j}^{k}|k \in P_{i,j}\} \tag{3.5}$$

Obviously, the minimum cost value depends on the cost metric. If energy is used as the cost metric, (3.5) specifies a path that results in minimum energy consumption during communication; if the cost metric is the hop count, (3.5) specifies a shortest path. The SRMCF protocol uses the cost-field method described in [40] for calculating all $C_{i,j}$. Let T_M^n and T_M^s (M stands for MCF protocol) denote the total number if packets generated by source nodes and sink node, respectively, when using the MCF protocol. Since the average path length increases as $\Theta(\sqrt{N})$, the total throughput of the nodes is

$$T_M^n = h\sqrt{N}\lambda_n N \tag{3.6}$$

Here $h\sqrt{N}$ denotes the average hop count and h is a positive value ($\sqrt{N} > h$) that depends on the network topology. If the sink node communicates with $1/m$ of the SNs, then

$$T_M = T_M^N + T_M^S = h\sqrt{N}\lambda_n N + \frac{\lambda_n}{m}N, \qquad m > 1, \tag{3.7}$$

where T_M denotes to total number of packets generated when using the MCF protocol. For the SRMCF protocol, under the same conditions, the total number of generated packets T_S (S stands for SRMCF protocol) is

$$T_S = h\sqrt{N}\lambda_n N + \frac{\lambda_n}{m}\sqrt{N}. \tag{3.8}$$

From (3.7) and (3.8), we have

$$T_M = T_S + \frac{\lambda_n}{m}(N - h\sqrt{N}). \tag{3.9}$$

If the term $N - h\sqrt{N}$ is positive, (3.9) implies that the SRMCF protocol generates fewer packets than MCF. Because of $\sqrt{N} > h$, in any kind of node arrangement, $N - h\sqrt{N}$ is positive. Therefore, with the SRMCF protocol, each communication requires fewer packets than with MCF and the energy consumption is lower as well. Less packet generation leads to less collision and higher throughput as well with SRMCF.

3.5.5 Packet header length

SRMCF uses variable-length headers for packets directed from BS to SNs. In addition, the packet header length changes along the path. Let h_{fix} represent the packet header length (in bytes) used in packets from an SN to the BS. The header length for packets in the other direction is

$$h_{size} = h_{fix} + \alpha H, \tag{3.10}$$

where α is the length of the MAC address and H is the number of hops between the current node and the destination. The value of H is decremented as the packet progresses along the path from BS to SN. In fact, the packet header length at different nodes is an arithmetic progression with common difference α. Considering the average path length ($h\sqrt{N}$) and the change of the packet header length, the average packet header size for SRMCF is

$$h_{avg} \approx h_{fix} + \alpha \left(\frac{1 + \left[h\sqrt{N} \right]}{2} \right) \tag{3.11}$$

For $h\sqrt{N} \gg 1$, the expression becomes

$$h_{avg} \approx h_{fix} + \frac{\alpha h\sqrt{N}}{2} \tag{3.12}$$

Equation (3.12) shows that, for networks with a large number of nodes, eliminating unnecessary path information for packets from the BS to an SN almost decreases the average header length by a factor of two.

3.5.6   Routing table size

The BS keeps a routing table with the optimum path to reach each SN. Obviously, the size of aforementioned table depends on the number of SNs and on how they are distributed in the network. Because of the variable size path entry in the table, there is no fixed relation between the number of nodes and routing table size. To calculate the approximate size of the table, suppose that the average path size is $h\sqrt{N}$. Each entry of the routing table includes fixed-size elements and a variable-length path. Therefore,

$$T_{size} \approx N \left(T_{fix} + \alpha(h\sqrt{N}) \right), \tag{3.13}$$

where T_{size} is the total routing table size in bytes. For a large number of nodes, the fixed part of the entry will be much smaller than the variable part. Then,

$$T_{fix} \ll \alpha(h\sqrt{N}) \Rightarrow T_{size} \approx \Theta(N\sqrt{N}). \tag{3.14}$$

Equation (3.14) indicates that the routing table size grows like $N\sqrt{N}$, a fact that should be considered in the design of the BS. The variable entry size increases the complexity of managing the table, which must support queries by node, periodic updates of the expiration counter associated with each entry, and node insertion and deletions. If the BS has enough memory to keep a large table, using a table with a fixed entry size will be simpler and reduces processing time. In this case, the length of each entry has to be large enough to store the biggest path. Assuming that the maximum path size is much larger than the fixed-size elements, we have

$$T_{size} = N(T_{fix} + P_{max}). \tag{3.15}$$

Although the BS usually has more resources than the SN, it is still useful to compress the routing table. An approach to compression is based on the idea that, if an SN is recorded in the routing table, then its near node is also in the table. Therefore, if a pointer that refers to the location of the near node is used instead of the path information, the BS can generate the path by accumulating the path information. In this case, the routing table size is

$$T_{size} = N(T_{fix} + i), \quad\quad\quad (3.16)$$

where i is the size of the index into the routing table. This method is able to significantly compress the routing table.

3.5.7 Simulation and experimental results

This section presents results from simulations of the SRMCF and MCF protocols. The results of SRMCF for various scenarios are compared with the values obtained from simulations of the MCF protocol.

3.5.7.1 Routing table and packet header size

With SRMCF, packets generated by the BS have variable length, which depends on the number of nodes in the path between the BS and the destination nodes. Both SRMCF and MCF have a fixed 5-byte header for packets generated by the SNs. The implementation assumes that all sensors are from the same vendor. Therefore, the first three octets of MAC address, which identify the organization that issued the address, are the same for all nodes. It is not necessary to include them in the routing path, making the packet header more compact (three octets per each node address). Figure 3.20 shows the average packet header size obtained by simulating random networks with 100–1,000 nodes. The figure also shows the calculated average header size given by (3.12). The results confirm that the average header size only increases by a factor of 1.6 when the number of nodes increases ten times. Using a MAC address length $\alpha = 3$ and $h_{fix} = 5$, and using the simulation results for $N = 100$ to 1,000 (20 samples for

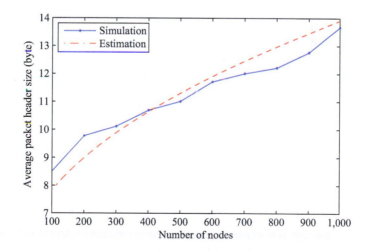

Figure 3.20   Average initial packet header size

each network size), allows us to solve (3.12) for h, obtaining $h = 0.281$. The average header size is then

$$h_{avg} \approx 5 + 0.281\sqrt{N}. \tag{3.17}$$

The dashed line in Figure 3.20 shows that h_{avg} agrees with the simulation values. Each node hands over the packets generated by BS to the next node with α bytes less in the packet header. With a fixed packet header (BS to node), the header size lies between 9.5 and 24.5 bytes (for networks with 100–1,000 nodes). Therefore, the SRMCF protocol achieves a significantly smaller average packet header size. It should be noted that both protocols have a relatively small header size in relation to the overall packet size (256 bytes) used in these simulations.

The average size of routing table for the uncompressed approach characterized by (3.13) was measured for randomly generated networks with 100–1,000 nodes (20 samples for each network size). Figure 3.21 depicts those results together with the values estimated from (3.13), (3.15) and (3.16). For this set of simulations, $T_{fix} = 5$ and the value of h obtained by interpolating the experimental results is 0.305. The simulation results agree closely with the calculated size given by (3.13). The simpler method of using only fixed-sized entries is acceptable for small or more compact networks, but the table size becomes very significant for larger ones: for 1,000 nodes, the size calculated from the average maximum path length is 34.9 kB, which is 1.49 times larger than the approach that uses variable-length entries. The use of compressed tables would allow significant savings in memory at the cost of more complex search and insertion algorithms: the table size for a 1,000 node network is predicted to be 8 kB by (3.16), a fixed value independent of network topology.

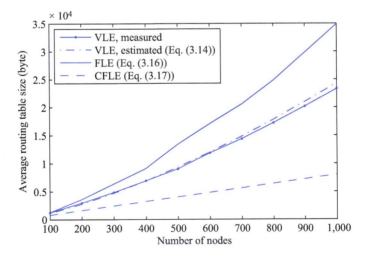

Figure 3.21   Average size of routing tables built using three different methods: VLE (variable-length entries), FLE (fixed-length entries), CFLE (compressed fixed-length entries)

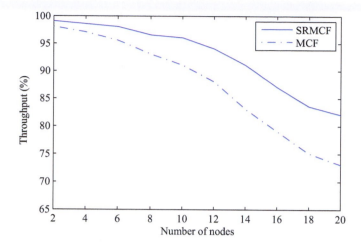

Figure 3.22   Total throughput in terms of the number of the nodes

3.5.7.2   Throughput in wired networks

The performance of both SRMCF and MCF protocols in a simulated wired network is evaluated and compared. The simulations assume that RTS/CTS handshaking at the MAC layer is used and that the node-to-node data rate is 10 Mbps. Each node periodically generates packets with 127 bytes resulting in an average traffic per node of 250 kbps. Figure 3.22 shows the total throughput of randomly generated networks (10 networks for a given number of the nodes) as a function of the number of the nodes. For small networks, both protocols have high throughput: 99% for SRMCF and 98% for MCF with two nodes. Throughput decreases as the number of the nodes increases, especially with MCF. For networks with 20 nodes, SRMCF has 83% throughput, but MCF achieves just 73% throughput, which is 10% less than SRMCF.

The power consumption of the communications module is mainly related to network activity, i.e., the number of generated and routed packets. In order to estimate the impact of the protocols on power consumption, the number of routed packet over generated packets was determined. Figure 3.23 depicts the results of the ratio of routed packets to generated packets obtained from the same networks as in the previous simulation. The results show that MCF needs to route more packets than SRMCF for the same number of data packets generated by the SNs. On the other hand, SRMCF has higher throughput. So, SRMCF consumes less power to hand over the same number of packets.

3.5.7.3   Network of sensors on textile

Figure 3.24 depicts a network of SNs and BS embedded in textile, which is used for data acquisition from the lower limb. For the network of Figure 3.24, CRC errors have been found to be extremely rare, because of the short range of communication over wires. So, for this experiment, the nodes have been set to not send ACK messages to

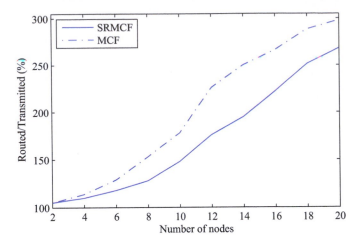

Figure 3.23   *Ratio of routed packets to generated packets as a function of the number of nodes*

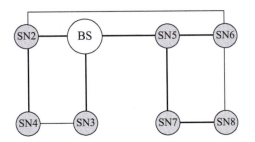

Figure 3.24   *A network of SNs embedded in textile used for data acquisition from the lower limb*

sender, but the other configuration parameters and sampling rate are the same as in previous measurements.

Figure 3.25 shows the average number of the packets saved in buffers as a function of packet length in BS, SN2 and SN5 while generating 20.8 kbps of data in each node (16 kbps by the sEMG, 2.4 kbps by the accelerometer and 2.4 kbps by the gyroscope). The service rate (μ) is

$$\mu = \frac{drate}{P_l} + T_{idle}, \tag{3.18}$$

where *drate* is data rate (4 Mbps), P_l is packet length and T_{idle} is the average time to process the received packet (184 μs for SNs and 0 s for BS). The results show that the average number of the packets in the buffers is always small. Therefore, the packet

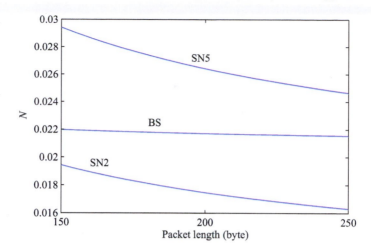

Figure 3.25   Average number of the packets in buffers with 20.8 kbps traffic per node

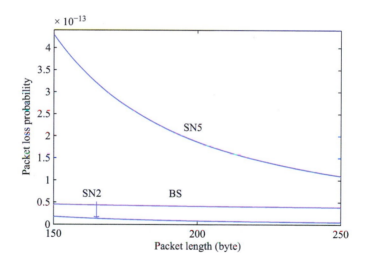

Figure 3.26   Packet loss probability with 20.8 kbps traffic per node

loss probability due to buffer overflow shown in Figure 3.26 is very small. So, in this specific case, the network is able to collect data from the SNs with high reliability.

The system performance under high traffic with the same configuration was also evaluated. For that, each SN generates 640 kbps, which is almost 31 times more than the previous value. Figures 3.27 and 3.28 depict the average number of packets in the buffer and packet loss probability, respectively. The performance of the network

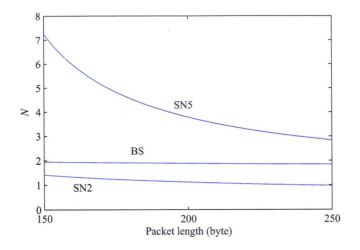

Figure 3.27   Average number of the packets in buffers with 640 kbps traffic per node

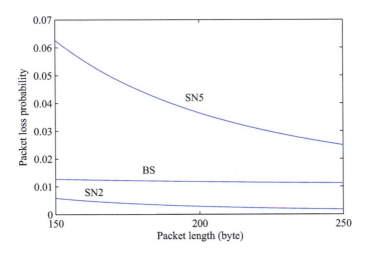

Figure 3.28   Packet loss probability with 640 kbps traffic per node

decreases especially for small packet lengths, as expected. So, to improve network performance with high data traffic, increasing the length of the packet can be a solution. This could be achieved by packing more sensor data in a single packet.

The average end-to-end delay for both low and high traffic (20.8, 80, 160, ..., 640 kbps) in terms of packet length is portrayed in Figure 3.29. In low traffic conditions, the delay increases linearly with packet length. In high traffic, the behavior is different because of the effect of Tidle. It always affects the network performance, but in high traffic with small packets, the effect is more significant and causes a notable increase in the end-to-end delay. Nevertheless, in all conditions, the average delay does not exceed 4.3 ms.

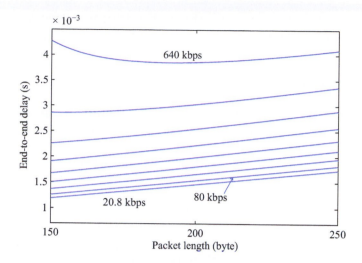

Figure 3.29   Average end-to-end delay

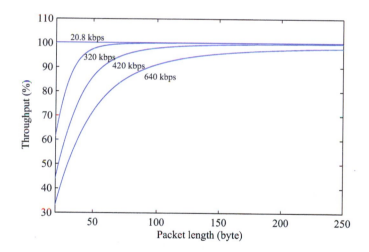

Figure 3.30   Throughput of the network in terms of the packet length

The SN-to-BS throughput (as a percentage of the total generated traffic) of the network in terms of packet length for different traffic conditions is shown in Figure 3.30. Obviously, throughput increases with packet size. Nevertheless, the throughput improvement does not have a linear relation with packet length, e.g., for 640 kbps, the throughput increases from 35% to 90% for packet lengths of 25 B and 100 B, respectively, which represents an increase of 2.57× in throughput. However, throughput improves only 1.07× (from 90% to 96%) when the packet length increases from 100 B to 250 B. In any case, it can be concluded that throughput improves as the packet length increases or traffic decreases and that it is better to use long packets.

3.6   Conclusion

In this chapter, a BAN system for low-power wearable applications has been described. The wearable unit is composed of a base and SNs, all connected to each other with conductive yarns forming a mesh topology with the base node at the center. The communication circuits described here support full-duplex communication. Design and implementation of the physical and MAC layers on an FPGA-based hardware prototype have been described. The measurement results obtained from the prototype show that the proposed circuits can work up to 18 Mbps which is high enough to satisfy most of BAN application requirements. The ability of the circuit to detect low quality signals in the communication lines, low power operation, high number of sensors, high data rate and reliable structure make it a suitable system for use in wearable applications with conductive yarns.

The proposed BAN system uses the SRMCF routing protocol, which has also been described. The adoption of source-based routing in SRMCF leads to an improvement of the performance over the MCF protocol. SRMCF is energy-efficient, reactive and does not require routing tables at every SN. Unlike MCF, it applies the minimum cost forwarding method in both directions. The absence of link and node failure control in MCF hampers its use in practical applications. The proposed failure recovery mechanism for SRMCF solves this limitation without affecting data communications. Both the theoretical analysis and simulation experiments described in this work show that SRMCF has both higher throughput and smaller energy consumption than MCF.

References

[1]   Xing J., Zhu Y. 'A Survey on Body Area Network'. *5th Intl. Conf. on Wireless Communications, Networking and Mobile Computing (WiCOM)*; 2009. pp. 1–4.

[2]   Yoo H.J., Hoof C.V. *Bio-Medical CMOS ICs*. 1st ed. New York: Springer; 2011.

[3]   Yuce M.R., Khan J.Y. *Wireless Body Area Networks*. 1st ed. Boca Raton, FL: CRC Press, Taylor & Francis Group; 2012.

[4]   Ullah S., Higgins H., Braem B., *et al.* 'A Comprehensive Survey of Wireless Body Area Networks: On PHY, MAC, and Network Layers Solutions'. *Journal of Medical System*. 2012;36:1065–1094.

[5]   Seyedi M., Kibret B., Lai D., Faulkner M. 'A Survey on Intrabody Communications for Body Area Network Applications'. *IEEE Transactions on Biomedical Engineering*. 2013;60(8):2067–2079.

[6]   Dangi K., Panda S. 'Challenges in Wireless Body Area Network—A Survey'. *Intl. Conf. on Optimization, Reliability, and Information Technology (ICROIT)*; 2014. pp. 204–207.

[7] Nakad Z., Jones M., Martin T. 'Fault Tolerant Networks for Electronic Textiles'. *Intl. Conf. on Communication in Computing*; 2004. pp. 100–106.

[8] Lee H.S., Park C.B., Noh K.J., Sunwoo J., Choi H., Cho I.Y. 'Wearable Personal Network Based on Fabric Serial Bus Using Electrically Conductive Yarn'. *Journal of Electronics and Telecommunications Research Institute (ETRI)*. 2010;32(5):713–721.

[9] Mizuno F., Hayasaka T., Tsubota K., Wada S., Yamaguchi T. 'Development of a Wearable Computer System with a Hands-Free Operation Interface for the Use of Home Health Caregiver'. *Journal of Technology and Health Care*. 2005;13(4):293–300.

[10] Cao H., Leung V., Chow C., Chan H. 'Enabling Technologies for Wireless Body Area Networks: A Survey and Outlook'. *IEEE Communications Magazine*. 2009;47(12):84–93.

[11] Patel M., Wang J. 'Applications, Challenges, and Prospective in Emerging Body Area Networking Technologies'. *IEEE Journal of Wireless Communications*. 2010;17(1):80–88.

[12] Chen M., Gonzalez S., Vasilakos A., Cao H., Leung V.C.M. 'Body Area Networks: A Survey'. *Journal of Mobile Networks and Applications*. 2010;16(2):171–193.

[13] Reddy G.P., Reddy P.B., Reddy R. 'Body Area Networks'. *Journal of Telematics and Informatics*. 2013 12;1(1):36–42.

[14] Cavallari R., Martelli F., Rosini R., Buratti C., Verdone R. 'A Survey on Wireless Body Area Networks: Technologies and Design Challenges'. *IEEE Communications Surveys & Tutorials*. 2014;16(3):1635–1657.

[15] Movassaghi S., Abolhasan M., Lipman J., Smith D., Jamalipour A. 'Wireless Body Area Networks: A Survey'. *IEEE Communications Surveys & Tutorials*. 2014;16(3):1658–1686.

[16] Pantelopoulos A., Bourbakis N. 'A Survey on Wearable Biosensor Systems for Health Monitoring'. *30th IEEE Intl. Conf. on Engineering in Medicine and Biology Society (EMBS)*; 2008. pp. 4887–4890.

[17] Ullah S., Khan P., Ullah N., Saleem S., Higgins H., Kwak K.S. 'A Review of Wireless Body Area Networks for Medical Applications'. *International Journal of Communications, Network and System Sciences*. 2009;2(8): 797–803.

[18] Wang H., Teng Q., Zhong X., Sweeney P. 'Using the Middle Tier to Understand Cross-Tier Delay in a Multi-Tier Application'. *IEEE Intl. Symp. on Parallel & Distributed Processing (IPDPS)*; 2010. pp. 1–9.

[19] Jovanov E., Milenkovic A. 'Body Area Networks for Ubiquitous Healthcare Applications: Opportunities and Challenges'. *Journal of Medical Systems*. 2011;35(5):1245–1254.

[20] Boulis A., Smith D., Miniutti D., Libman L., Tselishchev Y. 'Challenges in Body Area Networks for Healthcare: The MAC'. *IEEE Communications Magazine*. 2012;50(5):100–106.

[21] Post E.R., Orth M. 'Smart Fabric, or Wearable Clothing'. *1st Intl. Symp. on Wearable Computers*; IEEE; 1997. pp. 167–168.

[22]   Wade E., Asada H. 'Conductive Fabric Garment for a Cable-Free Body Area Network'. *IEEE Journal of Pervasive Computing*. 2007;6(1):52–58.

[23]   Yoo J., Lee S., Yoo H. 'A 1.12 pJ/b Inductive Transceiver with a Fault-Tolerant Network Switch for Multi-Layer Wearable Body Area Network Applications'. *IEEE Journal of Solid-State Circuits*. 2009;44(11):2999–3010.

[24]   Cho N., Yan L., Bae J., Yoo H.J. 'A 60 kb/s to 10 Mb/s Adaptive Frequency Hopping Transceiver for Interference-Resilient Body Channel Communication'. *IEEE Journal of Solid-State Circuits*. 2009;44(3):708–717.

[25]   Shikada K., Wang J. 'Development of Human Body Communication Transceiver Based on Impulse Radio Scheme'. *2nd IEEE CPMT Symp. Japan*; 2012. pp. 1–4.

[26]   Ogasawara T., Sasaki A.I., Fujii K., Morimura H. 'Human Body Communication Based on Magnetic Coupling'. *IEEE Transactions on Antennas and Propagation*. 2014;62(2):804–813.

[27]   Nemati E., Deen M., Mondal T. 'A Wireless Wearable ECG Sensor for Long-Term Applications'. *IEEE Communications Magazine*. 2012;50(1):36–43.

[28]   Lymberis A., Gatzoulis L. 'Wearable Health Systems: From Smart Technologies to Real Applications'. *2006 International Conference of the IEEE Engineering in Medicine and Biology Society*, New York, NY, 2006, pp. 6789–6792. doi: 10.1109/IEMBS.2006.260948

[29]   Gatzoulis L., Iakovidis I. 'Wearable and Portable eHealth Systems'. *IEEE Engineering in Medicine and Biology Magazine*. 2007;26(5):51–56.

[30]   Zambrano A., Derogarian F., Dias R., *et al.* 'A Wearable Sensor Network for Human Locomotion Data Capture'. *9th Intl. Conf. on Wearable Micro and Nano Technologies for Personalized Health (pHealth)*; 2012.

[31]   Gebali F. *Analysis of Computer and Communication Networks*. 1st ed. New York: Springer Science+Business Media, LLC; 2008.

[32]   Ilyas M., Mahgoub I. *Handbook of Sensor Networks: Compact Wireless and Wired Sensing Systems*. Boca Raton, FL: CRC Press LLC; 2005.

[33]   Konrad P. *The ABC of EMG: A Practical Introduction to Kinesiological Electromyography*. Noraxon INC; 2005 Apr. Available in https://www.noraxon.com/wp-content/uploads/2014/12/ABC-EMG-ISBN.pdf.

[34]   Libero IDE v9.1. User's Guide; 2010. Available in https://www.microsemi.com/document-portal/doc_download/130848-libero-ide-v9-1-user-s-guide.

[35]   IGLOO nano Low Power Flash FPGAs; 2015. Document number: DS0110, Revision 19. Available in http://www.microsemi.com/document-portal/doc_download/130695-igloo-nano-low-power-flash-fpgas-datasheet.

[36]   Pantelopoulos A., Bourbakis N.G. 'A Survey on Wearable Sensor-Based Systems for Health Monitoring and Prognosis'. *IEEE Transactions on Systems, Man, and Cybernetics, Part C: Applications and Reviews*. 2010;40(1):1–12.

[37]   Derogarian F., Ferreira J.C., Tavares V.M.G. 'A Routing Protocol for WSN Based on the Implementation of Source Routing for Minimum Cost Forwarding Method'. *5th Intl. Conf. Sensor Technologies and Applications (SENSORCOMM)*; ThinkMind; 2010. pp. 85–90.

[38] Zhong Y., Yuan D. 'Dynamic Source Routing Protocol for Wireless Ad Hoc Networks in Special Scenario Using Location Information'. *ICCT Intl. Conf. on Communication Technology Proceedings*; vol. 2; 2003. pp. 1287–1290.

[39] Garcia J.E., Kallel A., Kyamakya K., Jobmann K., Cano J.C., Manzoni P. 'A Novel DSR-Based Energy-Efficient Routing Algorithm for Mobile Ad-Hoc Networks'. *58th IEEE Intl. Conf. on Vehicular Technology*; vol. 5; 2003. pp. 2849–2854.

[40] Ye F., Chen A., Lu S., Zhang L. 'A Scalable Solution to Minimum Cost Forwarding in Large Sensor Networks'. *10th Intl. Conf. on Computer Communications and Networks*; 2001. pp. 304–309.

[41] Jetcheva J.G., Johnson D.B. 'Adaptive Demand-Driven Multicast Routing in Multi-Hop Wireless Ad Hoc Networks'. *2nd ACM Intl. Symp. on Mobile Ad hoc Networking & Computing*; MobiHoc '01. ACM; 2001. pp. 33–44.

[42] Yu M., Mokhtar H., Merabti M. 'Fault Management in Wireless Sensor Networks'. *IEEE Journal of Wireless Communications*. 2007;14(6):13–19.

[43] Paradis L., Han Q. 'A Survey of Fault Management in Wireless Sensor Networks'. *Journal of Network and Systems Management*. 2007;15:171–190.

[44] Henderson D.W., Torn S. 'Verification of the Minimum Cost Forwarding Protocol for Wireless Sensor Networks'. *IEEE Conf. on Emerging Technologies and Factory Automation (ETFA)*; 2006. pp. 194–201.

[45] Alazzawi L., Elkateeb A., Ramesh A. 'Scalability Analysis for Wireless Sensor Networks Routing Protocols'. *22nd Intl. Conf. on Advanced Information Networking and Applications, Workshops (AINAW)*; 2008. pp. 139–144.

[46] Santivanez C., McDonald B., Stavrakakis I., Ramanathan R. 'On the Scalability of Ad Hoc Routing Protocols'. *21th IEEE Intl. Conf. on Computer and Communications Societies (INFOCOM)*; vol. 3; 2002. pp. 1688–1697.

Chapter 4

Implementation study of wearable sensors for human activity recognition systems

Hamed Rezaie[1] and Mona Ghassemian[2]

This chapter addresses a number of activity recognition methods for a wearable sensor system. Three methods for data transmission, namely 'stream-based', 'feature-based' and 'threshold-based' scenarios are applied to study the accuracy against energy efficiency of transmission and processing power that affects the mote's battery lifetime. The impact of variation of sampling frequency and data transmission rate on energy consumption of motes is also analysed for each method. This study leads the authors to propose a cross-layer optimisation of an activity recognition system for provisioning acceptable levels of accuracy and energy efficiency.

4.1 Introduction

Human activity recognition (HAR) systems to automatically recognition of physical activities are used in rehabilitation centres to monitor the activity of daily living (ADL). A low-cost non-invasive solution for a continuous all-day and anyplace activity monitoring system can be realised using a wearable activity recognition system. In a wearable HAR system, multiple wearable sensor nodes are used to form a self-managing wireless body area network (WBAN). There are, however, arising challenges essential to be dealt with developing portable HAR systems, among which, network lifetime and detection accuracy can be highlighted as the most important ones.

In this chapter, we survey state-of-the-art in wearable HAR systems with an application-oriented approach, namely monitoring daily activity to support medical diagnosis, for rehabilitation, or to assist patients with chronic impairments to advance traditional medical methods such as multiple sclerosis, diabetes, emotion & mood, heart related treatments, osteoarthritis, pulmonary, obesity, autism and fall detection, as well as the sports and entertainment, social, military, child, game sectors that attract a great deal of interest.

[1] School of Computer Sciences and Engineering, Shahid Beheshti University, Tehran, Iran
[2] British Telecom PLC, Adastral Park, Ipswich, UK

Application requirements including sensory information, activity types and performance requirements (such as energy consumption, processing, accuracy, obtrusiveness and user flexibility) provide a baseline for the existing solutions to be discussed in this chapter. A general architecture is first presented along with a description of the main components of any HAR system.

The current state of HAR systems requires further modality for evaluating existing solutions based on the new emerging technologies. We provide a comprehensive background to the HAR procedures, designing, implementing and evaluating HAR systems. Sequence of techniques of the HAR chain [1], i.e. signal processing, pattern recognition and machine learning techniques will be described in details. We highlight open problems and ideas that require further work.

4.2   Application requirements

The HAR can be applied in variety of fields, especially for medical, military, sport and entertainment. For instance, patients with multiple sclerosis [2–4], diabetes [5–7], emotional problems [8,9], heart diseases [10–12], osteoarthritis [13–15], pulmonary [16–19], obesity [20–24], social [25–28], autism [29–33], oedema [34] and Parkinson disease [35,36] are often required to follow a well-defined exercise routine as part of their treatments [3]. Therefore, recognising activities such as sitting, lying, walking, jogging, eating or cycling becomes quite useful to provide feedback to caregivers about patients' behaviour. Patients with dementia and other mental pathologies or social disorders [4] can be monitored to detect abnormal activities to prevent undesirable consequences at early stages.

While the mentioned applications mainly refer to the elderly and patients with special needs, such applications are also demanded for infant and young children monitoring as well as assessment of infant movements [37–39].

In military scenarios, precise information on the soldiers' activities along with their locations, health conditions using body-fixed sensors [40,41], is highly beneficial for their performance and safety. Impact of training patterns on injury incidences [42] as well as energy expenditure estimation during daily military routines [43] is also helpful to support decision making in both combat and training scenarios.

Sport activity recognition systems support sport centres and athletics by monitoring their performance and health conditions by using smart garments during training assessment [44]. Other applications include classification of team sport activities using a wearable tracking device [45], sport coach online activity matching using wireless sensor network [46], swimmers' wearable integrated monitoring system [47] and innovative training machine learning approach for sport activity recognition from inertial data [48].

4.2.1   Sensory information

Sensors can be placed on the body or in the environment to detect the human activities or to benefit from a combination of them both. In this chapter, while we list all the applicable sensors, we focus on the body-worn sensors and their measurement circuit, characteristics and hardware requirements and limitations.

4.2.1.1 Privacy

Traditionally, researchers used vision sensors for activity recognition [49,50]. However, the use of camera devices is intrusive and disruptive [51] and can violate the privacy of the users [52] and in some cases against their cultural or religious beliefs. There exist solutions for privacy-preserving such as visualisation of the foreground masks only [53] and human body identification using frame differencing approach [54]; however, other challenges such as outdoor usability and user acceptance limit the camera-based systems.

With the advancements in micro-sensor technology, low-power and low-range wireless communication technologies and inertial sensor systems [55], developing a low-cost, effective and privacy-aware alternative for HAR systems becomes possible.

4.2.1.2 Sensor hardware and measurement circuits

A sensor is a transducer that receives and responds to a signal or stimulus. A sensor instrumental model that receives an observable variable X from the measurand is shown in Figure 4.1. X is related to the measurand in some *known* way (i.e. measuring mass)

Then, the sensor generates a signal variable that can be manipulated (i.e. processed, transmitted or displayed). In Figure 4.1, the generated signal is processed (i.e. amplified, noise filtering, quantised, etc.) and passed to a system to be displayed, recorded for further decisions by monitoring or control systems in place.

4.2.1.3 Sensor characteristics

Most of the sensor control systems are closed-feedback loop systems and the reliability of the sensor data plays an important role in the overall recognition results. In real-world HAR applications, the choice of sensing equipment and the variability in sensor characteristics are crucial decisions which should be made according to the specific scenario requirements, namely technology availability, practicality, usage duration, accuracy, precision, privacy and security.

Hardware challenges such as hardware errors, complete failures, sensor drift, sensor hysteresis, saturation, dead band, repeatability and ambient challenges including changes in the operating temperature, obstacles and moving objects are important parameters that affect overall system performance [57,58].

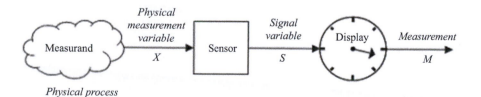

Figure 4.1   A simple instrument model [56]

In addition, in some applications, the wearable sensors should be fixed in specific places. However, portable devices containing sensors may be used in different ways or placed on different locations on the body [59]. Furthermore, different users with different body sizes and heights affect the orientation and positioning suggested by the HAR systems which demand methods such as accurate calibration.

4.2.1.4   Types of sensor for HAR systems

Applications requirements dictate the type of sensors for HAR systems. Inertial sensors including 3-dimentional gyroscope and tri-axial accelerometer, magnetometer, force sensitive resistor, compass, strain, barometric pressure sensors provide the information about the body movements as well as stroke numbers (hands & feet). HAR systems can also benefit from bio-signal sensors that capture skin temperature, blood pressure and heart rhythm, breathing rate and sweat index to assist when HAR is mainly used for healthcare applications. Further information can be recorded by microphone, camera, proximity sensor, GPS, radio signal measurements placed on the body to improve the recognition system. Table 4.1 presents a list of different sensor combinations applied by different applications in the literature. The majority of the applications using HAR systems apply accelerometer sensors to detect the activities.

The last three rows in Table 4.1 suggest the use of alternative sensor devices besides accelerometers. Capacitive sensors placed on the throat [66] can measure the distance of the electrode worn around the neck to indicate different activities, namely

Table 4.1   Sensor types applied in the literature for the HAR systems

Applications	Sensors					Ref.
	Accelero-meter	Gyroscope	Magneto-meter	Bio-signal sensors	Other sensor types	
Multiple sclerosis	✓					[2]
	✓	✓				[3,4]
Diabetes	✓	✓	✓			[5]
	✓					[6]
	✓	✓				[7]
Emotion	✓					[8]
	✓			Heart rate, temperature, pulse rate		[9]
Heart related	✓			Heart rate		[10,11]
	✓			ECG, skin temperature		[12]
Osteoarthritis	✓					[13–15]

(Continued)

Table 4.1   *(Continued)*

Applications	Sensors					Ref.
	Accelero-meter	Gyroscope	Magneto-meter	Bio-signal sensors	Other sensor types	
Pulmonary	✓					[16–19]
Obesity	✓					[20]
	✓	✓	✓			
	✓	✓	✓		Microphone	[21]
	✓					[22]
	✓					[23,24]
Social	✓				Microphone, proximity sensor	[25]
	✓				Compass, RSSI	[26]
	✓				Microphone, GPS	[27,28]
Sport	✓					[44]
	✓	✓				[45–48]
	✓				Heart rate, breathing rate	[47]
Military	✓			Heart rate		[40–43]
	✓					[41]
Child	✓					[37,38]
	✓				Barometric pressure	[39]
Autism	✓					[29–33]
Oedema	✓	✓	✓		Force sensitive resistor	[34]
Parkinson	✓					[35]
	✓	✓				[36]
Game	✓					[60]
Fall prevention	✓	✓				[61,62]
Fall detection	✓					[63,64]
	✓	✓				[65]
Chewing, swallowing, speaking, sighing					Capacitive sensor	[66]
General HAR					Force sensors	[67]
					Strain sensor	[68]

swallowing, chewing, speaking and singing. Force sensors can be placed on the tight to show the muscle's tension which indicates walking activity [67].

4.2.1.5   Number of sensors and sensor placement

The positioning and number of the sensors are well studied by Atallah *et al.* [69] and Yang *et al.* [70] which reported the impact of the wrong placement or the orientation of the sensors as well as changes in the position of sensors during activities. Therefore, the detection accuracy was decreased using smaller subsets of active sensors and the positioning depends on the type and level of activities under study. Studies conducted by Kale *et al.* [71] and Mannini *et al.* [72] highlighted the impact of the nodes misplacement with respect to the degree of rotation and confirm that -15 to $+15$ degree misplacements in each axis could be tolerated in HAR systems.

Zappi *et al.* [73] focused on selecting the minimum set of sensors according to their contributions to classification accuracy. The proposed dynamic sensor selection method showed the trade-off between the classification accuracy and battery lifetime by minimising the number of sensor nodes necessary to obtain a given classification accuracy for the HAR system. This method suggested selecting a set of active nodes to increase network lifetime may not be an efficient solution for HAR systems.

4.2.1.6   Adjusting sampling rate

Once the body sensor network design is done, other configurations such as sampling rate should be set on each sensor. French *et al.* [74] focused on the impact of sampling rate on the accuracy and discussed the energy consumption trade-off. Their reported analysis considered different sampling rate selection strategies including a baseline uniform sampling strategy, and a probability-based sampling when a transition occurring. This method, however, required a prior dataset in order to calculate transition probabilities.

The impact of three parameters, namely sampling rate, transmission rate and nodal processing were investigated by Rezaie and Ghassemian [75] to report on the wearable sensors lifetime and detection accuracy. The outcomes of this research suggested that among these parameters, the most significant one is sampling rate that impacts on the overall active time of sensor nodes as a result of the sleep periods. These experiments showed that optimum sampling rates should vary in different activity levels.

Further to the selection of an optimum sampling rate, event-driven applications such as HAR, smart homes mobility monitoring [76] and traffic control systems, can benefit from adjusting the sampling rate based on the collected data. The trade-off for adaptive mechanisms is addressed in Section 4.2.7 in details.

Rezaie and Ghassemian [77] proposed a feedback controller algorithm to adaptively change the sampling rate based on the level of activities. The proposed feedback controller algorithm is implemented and evaluated with respect to two parameters: the accuracy of detection and the energy consumption. Their method nearly doubled the activity recognition system lifetime and showed only 0.41% accuracy reduction.

4.2.1.7   Sensor limitations

Power consumption affects the size of the battery and sensor nodes, accordingly which affects the users' quality of experience (discussed in Section 4.2.6) when charging the device batteries are expected to be done by them [76]. Additionally, sensor nodes require sufficient computational power as well as memory space to store both for software components and data.

Hardware size for wearable sensors is also a limiting factor to make the body-worn devices less visible. The more complicated the algorithms get to ensure the accuracy and privacy, the higher the computational power is demanded. Sensor weight, especially if worn more than one, becomes another constraint which also affects the quality of the experience of the users.

4.2.2   Data collection phase

The discussed sensor characteristics, types, configurations and limitations, and the design of the body sensor network with respect to the number of nodes and their location on the body are the pre-requirements for the data collection phase.

Obtaining an ethical approval is the first point of actions once the data collection protocol is defined to be able to recruit the case studies. The ethical approval defines the data collection protocol by the related authority and data ownership and case study rights. Sensitive user information should be retrieved considering users' privacy and should be collected and encrypted before transmission [78].

Labelling the collected data is another crucial factor in the data collection protocol. In supervised HAR tasks, the collection of annotated or 'ground truth labelled' training data is a challenge. Ground truth annotation is an expensive and time-consuming task, as the annotator has to perform the annotation in real time [79].

In stationary and laboratory settings, annotation can often be obtained by *post hoc* labelling based on video recordings [80,81] manually or by a custom software considering the case study would give consent to be videoed. In daily life settings, ground truth annotation is a far more difficult problem. Different techniques such as daily self-recall methods [82], experience sampling [83] and reinforcement or active learning – all of which involve the user are being investigated. When few labelled training samples are available, semi-supervised [84], unsupervised [85] or knowledge transfer [80,81,86,87], learning techniques can be used [88].

In the HAR system, data collection phase is performed for gathering information and measuring activities from a variety of individuals following a special procedure to get a complete an accurate picture of subject's activities model. This phase can be performed either in a controlled or uncontrolled experiment [89] as described in the following:

4.2.2.1   Controlled/supervised

A very common way of collecting data in activity recognition is a controlled experiment where the researcher sets an experimental setup or pre-defined protocol for collecting data of each activity under study. The controlled or supervised experiments require two groups: a training set (an experimental group) and a testing set (a control group). The training set is a group of subjects that should perform any sequence and type of activity pre-defined by the researcher in the experimental protocol. Each activity is labelled (annotated) with the respective activity name/tag, which can be done either manually or through custom software.

The dataset collected from the training set is used to evaluate the HAR algorithm. The testing set is allowed to perform the activities without following the pre-defined protocol. The dataset collected from the testing set is used to evaluate the classification model. In principle, the data collected from the testing set should not have any activity labels. Due to the algorithm validation process, the data labelling can also be used to train the dataset.

4.2.2.2   Uncontrolled/unsupervised

In the uncontrolled or unsupervised data collection method, the users perform the unlimited number of activities in natural environment, such as clinical or home. The sensor is attached to one or multiple locations of the body. After collecting a reasonable amount of data for all activities, extensive data analysis is performed and the activity recognition algorithm is evaluated for certain activities. Due to the unlabelled dataset, machine learning techniques are used for activity classification and identification.

Limited number of datasets is collected in uncontrolled situations with elderly or patients. Most of the research work done in HAR systems is being targeted to elderly users in different contexts of healthcare and assisted living situations which are considered to be uncontrolled environments.

To validate the experiments, techniques such as process mining can be applied [90,91]. Process mining technologies can infer process models as workflows, specifically designed to be understood by experts, enabling them to improve the model with the data flow perspective.

The number of subjects and their physical characteristics are among the main factors in data collection methods. To achieve generalisation and a subject independent recognition classification model, each of data collection protocols should include a large number of subjects with diverse characteristics in terms of gender, age, height, weight and health conditions to ensure flexibility to recognise new subjects' activity without the need of collecting additional training data. However, large society for data collection is a time-consuming, costly and laborious task [92]. Machine learning techniques and transfer learning [93] (discussed in subsection 4.3.5) can help overcome this problem.

4.2.2.3   Controlled vs. uncontrolled

While the controlled data collection set up results in a clearer dataset for the researchers for the segmentation, training and evaluation, it limits the activity range and does not capture unpredicted moves as in a natural environment [9,10,18,19,22,44] setup.

Foerster *et al.* [94] demonstrated 95.6% of accuracy for activities in a controlled data collection experiment, but in an uncontrolled environment, the accuracy degrades to 66%.

Table 4.2   List of related literature with the number of training and test cases

Application	No. of subjects	Training-environment/ Test-environment	Ref.
Multiple sclerosis	14 subjects for test 11 subjects for control	Medical centre/Medical centre	[2]
	27 subjects for test 18 subjects for control	Medical centre/Medical centre	[3]
	21 subjects for test 12 subjects for control	Medical centre/Medical centre	[4]
Diabetes	17 subjects for test 21 subjects for control	Not specified/Not specified	[5]
	17 subjects for test 15 subjects for control	Medical centre/Medical centre	[7]
Social	2 subjects for train 3 subjects for test	Laboratory/Laboratory	[27]
Military	15 subjects for train 18–24 subjects for test	Laboratory/Field	[40]
	8 subjects for test 12 subjects for control	A controlled environment/ A natural environment	[43]

Each of data collection types or a combination of both can be employed for the training and testing the HAR systems as listed in Table 4.2. Furthermore, to check the quality of performing the activities, the tests can run by both a test group and a control group in some of the applications which are highlighted in the first column.

4.2.3   Activity set

States of locomotion (e.g. stand, walk, sit and lie) or gestures (e.g. open/close door, open/close drawers, drink cup and clean table) can be recognised by HAR systems. The level of activities can be categorised either based on the frequency of occurrence or their rate:

1. *Frequency:* Activities or gestures such as such as walking, running, rowing and biking are performed periodically. Sliding window segmentation and frequency-domain features are generally used for classification of such activities. Other activities or gestures occur sporadically, interspersed with other activities or gestures. Segmentation plays a key role to isolate the subset of data containing the gesture. For static the HAR system deals with the detection of static postures or static pointing gestures [88].

2. *Activity rate* [95]:
    i.   *Very low-level activities* (aka zero-displacement activities): Activities such as standing, sitting, lying. To detect these activities with a reasonable accuracy, wrist and ear worn sensors are in general applied.
    ii. *Low-level activities*: For this group, the waist sensor is selected by all classifiers as the one providing maximal precision and recall between this activity group and others. This group of activities is relatively varied, including eating, reading and socialising where body positions and motions can differ significantly.
    iii. *Medium level activities*: For this group, the chest and wrist sensor provide the best precision rates. The result is not surprising as the activities include walking and housework involving wiping tables and vacuuming. Recall is high for these sensors as well as the arm sensor and the ear worn sensor (especially from the Bayesian classifier).
    iv. *High-level activities* (aka strong displacement activities): These activities are picked up mostly by the ear worn sensor as it measures the change in body posture while walking and running. The ear, arm and knee sensors also perform well.
    v.   *Transitional activities*: As these activities involve sitting (from standing) and lying down (also from standing), the waist, chest and knee sensors reflect the parts of the body that are moving most. The ear sensor also gives good rates for both precision and recall over all the classifiers.

A summary of the activity types, sensor placements are listed in Table 4.3. The sampling frequency rate of 50 Hz is considered to be sufficient for all the activity levels. More discussion on the sampling rate is presented in subsection 4.2.1.6.

Table 4.3   A summary of the activity types and sensor placements

Activity levels	Frequency	Activity types	Sensor position
Very low-level activity	Static	Standing, sitting, lying down	Wrist and ear
Low-level activity	Periodic Sporadic	Brushing, eating Socialising, reading	Waist
Medium level activity	Periodic Sporadic	Walking level, walking upstairs, walking downstairs Housework involving wiping tables and vacuuming	Chest and wrist
High-level activity	Sporadic	Running slowly, rowing, biking	Ear, arm and knee
Transitional activity	Sporadic	Sitting-to-standing, sitting-to-lying, standing-to-sitting, level walking-to-stair walking, stair walking-to-level, walking, lying-to-sitting	Waist, chest, knee and ear

4.2.4   Energy consumption

In wearable HAR scenarios, extending the battery life ensures the system reliability and quality of experience, especially for continuous health monitoring that is compelled to deliver critical information.

The user quality of experience highly relies on the number of times the HAR system requires to be charged. However, some HAR analysis does not include energy expenditures, as a result of sensing, processing, communication and visualisation tasks. In the following subsections, we describe the sources of energy consumption and the strategies to reduce the energy consumption.

4.2.4.1   Sources of energy consumption

Sensor nodes are autonomous devices equipped with heavily integrated sensing, processing and communication capabilities. Therefore, there are three main sources of energy consumption in wireless sensors:

1. Sensing: A sensing subsystem for data acquisition from the physical surrounding environment.
2. Processing: A processing subsystem for local data processing and storage.
3. Transmission: A wireless communication subsystem for data transmission.

The battery source supplies the energy needed by the device to perform tasks. However, it consists of limited energy resource and frequent recharging requirements decreases HAR system user's quality of experience.

4.2.4.2   Energy-efficient design

There are mainly four strategies to achieve the energy-efficient design, namely:

1. Battery design: The first one is to improve the energy efficiency of the existing energy storage technologies, lithium-ion batteries and supercapacitors. Liu *et al.* [96] designed a novel structure of $ZnCo_2O_4$-urchins-on-carbon-fibers matrix to fabricate energy storage device that exhibits a reversible lithium storage capacity of 1,180 mAh/g even after 100 cycles.
2. Scavenging energy: The second approach relies on scavenge energy from the environment or human activities, such as the lower-limb motion and vibration, body heat and respiration, which have great potentials to become sustainable power sources for wearable sensors. The MIT Media Lab [97] realised an unobtrusive device that scavenged energy from heel-strike of the user with shoe-mounted piezoelectric transducers. Leonov [98] developed a hidden thermoelectric energy harvester of human body heat. It was integrated into clothing as a reliable powering source. Lombriser *et al.* [99] proposed an online activity recognition algorithm that was run on a miniaturised wireless sensor platform, namely the SensorButton consisting of a hybrid power supply circuitry.
3. Power management approaches: dynamic voltage scaling (DVS), battery energy optimal techniques and system energy management techniques are other solutions for improving the energy efficiency [100]. DVS technique reduces energy

consumption by varying the CPU frequency on the fly when low-power sleep is not an option due to application and/or environmental constraints [101].

4. Data-driven approaches: Such solutions are complementary ways to save energy in a smart node. In fact, data sensing can impact on energy consumption (i) because the sensing subsystem is power hungry [102], (ii) because sampled data have strong correlation (spatial or temporal [103], so that there is no need to communicate redundant information. Therefore, data-driven techniques have been designed to reduce the amount of sampled data by keeping the sensing accuracy within an acceptable level for the application. This approach mainly relies on reducing sampling rate [104–106] and data compression and redundancy elimination [107].

In addition, since all sensors may not be necessary simultaneously, turning off some of them or reducing their sampling/transmission rate is very convenient to save energy [73,108,109]. Rezaie and Ghassemian [75] have investigated on the impact of three parameters, namely sampling rate, transmission rate and nodal processing in the wearable sensors lifetime and detection accuracy and suggested that among these parameters, the most significant one to be sampling rate. Further experiments by Rezaie and Ghassemian [77] showed that optimum sampling rates should be varied in different activity levels to maintain the required detection accuracy depending on activity types. Data-driven approaches come usually with a computational cost that must be taken into consideration.

4.2.5 Processing

Another important point of discussion is where the recognition task should be performed, whether in the sink, the sensor, the network or offline.

1. Offline processing: For behavioural monitoring or pattern analysis over longer periods of time, offline data analysis and classification are sufficient [109]. However, for real-world applications, such as gesture-based input, real-time signal processing and classification are required.

2. In-sink processing: A sink node (aka server node) is often considered to be powered by main through a USB, or an adapter, hence expected to support huge processing, storage, and energy capabilities, allowing to incorporate more complex methods and models. However, HAR systems running on a mobile device should reduce communication energy consumption. Therefore, raw data should not have to be continuously transmitted to a sink node for processing.

3. In-node processing: The system can become more robust and responsive for real-time medical decision making when not transmitting over unreliable and error-prone wireless communication links. Furthermore, in-node processing can be considered more scalable since the server load is reduced by the locally performed feature extraction and classification computations. However, implementing HAR in portable devices (such as wristbands and pedometers) are

challenging due to their resource constraints with respect to processing, storage and energy. Therefore, feature extraction and learning methods should be carefully chosen to guarantee a reasonable response time and battery lifetime.

4. In-network processing: In-network processing solutions [73,110–112] move the processing complexity from the sensor node to the server side but also process information before it reached the sink node. This distributes the processing and energy consumption loads but demands a variety of functions related to security, performance and customisation to be implemented in online monitoring scenarios.

4.2.6   Obtrusiveness and user's quality of experience

To implement a practical and usable HAR system, a minimum number of sensors should be required to be worn by the user that do not often interact with the application. Minimising the number of sensors required to recognise activities not only improves comfort but also reduces complexity and energy consumption as less amount of data needs be processed. On the other hand, the higher data sources become available, the richer the information gets by extracting from the measured attributes. This highlights a new trade-off that a number of published works have investigated it:

Brigante *et al.* [113] developed a highly compact and lightweight wearable system for motion caption by careful selection of IMUs and optimised layout design. A highly integrated microsystem was designed for cardiac electrical and mechanical activity monitoring, which assembled the multi-sensor module, signal processing electronics and powering unit into a single platform with a flexible substrate [114].

There are systems which require the user to wear four or more accelerometers [92,115,116] or carry a heavy rucksack with batteries and recording devices [117]. Such configurations would not meet the quality of experience for continuous activity recognition.

However, a system with rather unobtrusive hardware such as a sensing platform that can be worn as a sport watch is presented by Maurer *et al.* [118]. Centinela [92] only requires a strap that is placed on the chest and a cellular phone. Finally, the systems introduced by Berchtold *et al.* [119] and Reddy *et al.* [120] recognise activities using a cellular phone only.

To set a minimum number of nodes while maintaining the accuracy level, Maurer *et al.* [118] also explored different subsets of features and sensors (accelerometers), as well as different sensor placements. They concluded that all sensors should be used together in order to achieve the maximum accuracy level. Bao *et al.* [116] carried out a similar experience, placing accelerometers on the individual's hip, wrist, arm, ankle, thigh and combinations of them and suggested the use of only two accelerometers (i.e. either wrist and thigh, or wrist and hip) to for HAR which would address the users' quality of experience. Rezaie and Ghassemian [77] also investigated the use of different number of nodes and placements and showed that using four nodes with lower sample rates results in higher accuracy compared to using less number of sensor nodes with a higher sampling rate.

4.2.7   Trade-offs

Designers of HAR systems also face challenges associated with the trade-off between accuracy, latency and processing and transmission power [75,104]. While for some HAR applications such as gesture recognition, low-latency classification and immediate feedback may be required, for others such as behavioural monitoring, this may be less critical [121]. Furthermore, among research that targets trade-off between detection accuracy and energy consumption, each study focused on only one particular challenge.

Yan *et al.* [104] proposed an activity-based strategy for continuous activity recognition with focus on both the sampling frequency and the classification features that were adapted in real time. The presented results showed that in an ideal condition activity-based strategy an energy savings of 50% was achieved.

Zappi *et al.* [73] focused on selecting the minimum set of sensors according to their contributions to classification accuracy. The proposed dynamic sensor selection method showed the trade-off between the classification accuracy and the battery lifetime by minimising the number of motes necessary to obtain a given classification accuracy for activity recognition. This method was tested by recognising manipulative activities of assembly-line workers in a car production environment.

French *et al.* [74] focused on the impact of sampling frequency. They evaluated different selective sampling strategies including a baseline uniform sampling strategy, one that samples over the distribution of duration times of activities, and one that samples based on the probability of a transition occurring.

Ghasemzadeh *et al.* [122] formulated coverage problem in the context of activity monitoring. Their method focused on the minimum number of sensor nodes that produced full activity coverage set. This solution eliminated redundant sensor nodes while maintaining an acceptable accuracy for the HAR system. Zappi *et al.* [73] focused on selecting the minimum set of sensors according to their contributions to the recognition accuracy. Their proposed dynamic sensor selection method showed the trade-off between the classification accuracy and battery lifetime by minimising the number of required sensor nodes to obtain a given classification accuracy for HAR. This method was tested by recognising manipulative activities of assembly-line workers in a car production environment.

Increasing the processing power or sending more frequent updates to ensure the level of the required accuracy typically decreases HAR system lifetime. One possible solution to this problem was presented Lu *et al.* [123] to introduce a central component in the experimental setup to aggregate, process and fuse the information extracted from different sensors applying an in-network processing solution.

Fallahzadeh *et al.* [124] presented an adaptive compressed sensing (CS) technique for wearable HAR systems, where a coarse-grained HAR module was applied to tune the CS parameters adaptively to reduce the sensing and transmission cost. To achieve the reduced cost, this technique required a prior dataset to be available to adjust the sampling rate.

Rezaie and Ghassemian [77] proposed a feedback controller algorithm to adaptively change the sampling rate based on the level of activities. The proposed feedback controller algorithm is implemented and evaluated with respect to two parameters: the accuracy of detection and the energy consumption. Their method nearly doubled the activity recognition system lifetime and showed only 0.41% accuracy reduction.

Except feedback controller algorithm, adaptive CS technique [124] and selective sampling technique [74], the sampling rate would either be considered fixed during all activities or variable based on a look-up table [104]. The feedback controller algorithm and sampling rate reduction techniques allowed sensor nodes to switch to the sleep mode between the sampling intervals. However, the effective sensor selection techniques [73,122] selected the sensor nodes to be switched to the sleep mode based on the prior dataset information when they did not contribute in activity recognition process. The requirement for prior dataset demanded a data collection and analysis to set the primary parameters. The feedback controller technique [77] did not require such information to set the primary parameters; instead the variation of the amplitude of the sensed signals was used. The requirements for prior dataset and HAR information polling from base station to retrieve the activity in place were considered to be strong assumptions in the discussed adaptive techniques. The feedback controller technique did not require either the use of prior dataset or the HAR information polling to adjust the sampling rate. Therefore, the proposed feedback controller technique should be counted as a more promising technique for real-life implementations. Table 4.4 presents a summary of the critical evaluation of different HAR algorithms [77] discussed in this subsection.

Table 4.4   Comparison of different HAR techniques [77]

Techniques	Reference	Sampling rate	Processing complexity	Transmission rate	Prior dataset requirements	AR info polling
Lossy compression techniques	Fallahzadeh *et al.* [124]	Adaptive (8–25 Hz)	Very low	Adaptive (8–25 Hz)	Yes	No
Sampling rate reduction techniques	Yan *et al.* [104]	Variable (5–100 Hz)	Moderate	Local processing	Yes	Yes
	French *et al.* [74]	Adaptive (0.06–6 Hz)	Very low	Adaptive (0.06–6 Hz)	Yes	No
Effective sensors selection techniques	Ghasemzadeh *et al.* [122]	Fixed (22 Hz)/ sleep	High	Fixed (22 Hz)/ Sleep	Yes	No
	Zappi *et al.* [73]	Fixed (N/A)/ sleep	High	Fixed (N/A)/ Sleep	Yes	No
Feedback controller algorithm	Rezaie and Ghassemian [77]	Adaptive (0.2–12 Hz)	Very low	Adaptive (0.2–12 Hz)	No	No

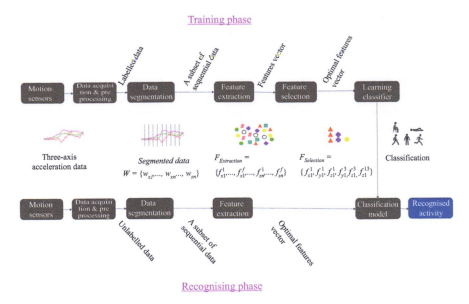

Figure 4.2   HAR system framework

4.3   Recognition architecture

In this section, we present two phases of the HAR system framework; training and recognition in details and describe functional stages as depicted in Figure 4.2. The training phase comprises components for data acquisition, synchronisation, segmentation, feature extraction and selection, training and classification, decision fusion and performance evaluation. Once the training phase to detect the activities is completed, same functionalities (i.e. data acquisition, preprocessing and data segmentation) are performed over similar data collected from person under study. The selected features in the training phase are extracted in the recognition phase to classify the activities based on the classification technique used in the training phase.

4.3.1   Sensor data acquisition and preprocessing

Raw data are collected from several sensors attached to different locations on the body as well as sensors placed in the environment for HAR systems. Since some sensors can provide multiple values (e.g. an acceleration sensor provides a 3D acceleration typically referred to as x-, y- and z-direction), or multiple sensors are jointly sampled, vector notation is used to describe the sensor's output:

$$S_i = (d^1, d^2, d^3, \ldots, d^t), \quad \text{for } i = 1, \ldots, k, \tag{4.1}$$

where k denotes the number of sensors and d^i the multiple values at a time t. The sampling rate of different sensors are not necessarily the same (e.g. GPS is 5 Hz and acceleration at 25 Hz or more), hence for a multivariate time series. Further-more, sensors can change/adapt their sampling frequency for power saving or due to requirements of the operating system. Therefore, t differs across S_i, and synchronisation across multimodal sensor data becomes a challenge. Moreover, raw sensor data can be corrupted by artefacts caused by a variety of sources (e.g. physical activity, sensor malfunction and AC line electromagnetic interference).

Collected data needs to be preprocessed, i.e. synchronised and the artefacts to be filtered, to prepare the acquired data for feature extraction. It is important to note that this preprocessing is supposed to be generic and not dependent on a particular person; that is, it should not depend on anything but the data itself. The synchronisation stage transforms the raw multivariate and nonsynchronous time series data into a preprocessed time series D':

$$D' = \begin{pmatrix} d'^1_1 & \cdots & d'^t_1 \\ \vdots & \cdots & \vdots \\ d'^1_n & \cdots & d'^t_n \end{pmatrix} = (d'_1, \ldots, d'_n)^T,$$

where d'_i corresponds to one dimension of the preprocessed time series, n to the number of total data dimensions and t to the number of samples.

In additional to synchronisation function, preprocessing stage differs in different sensors. For instance, acceleration and gyroscope signals may require calibration, unit conversion, normalisation, resampling, synchronisation or signal-level fusion [125]. Physiological signals, such as electro-oculography, typically require preprocessing algorithms for denoising or baseline drift removal [126].

4.3.2 Data segmentation

Segmenting a continuous sensor stream is a difficult task. Human activities are often a consecutive set of activities are not clearly separated by pauses. The data segmentation stage identifies those segments of the preprocessed data streams that are likely to contain information about activities (also often referred to as activity detection or 'spotting'). Information on activity segments not only is useful for classification but also can be used to reduce the sampling rate or turn off the HAR system to save power when limited or no activity is sensed.

Each data segment $W_i = (t_1, t_2)$ is defined by its start time t_1 and end time t_2 within the time series. n set of segments W is defined by the segmentation stage, each contains a potential activity.

$$W = \{W_1, \ldots, W_n\}.$$

Often, the exact boundaries of an activity are difficult to define an exact activity. A drinking activity, for instance, might start with reaching for a full glass from the table, holding the glass, and end after sipping or after putting the glass back on the table. In the literature [88], various solutions addressed the segmentation problem, namely segmentation using a sliding window, energy-based segmentation, rest-position segmentation, the use of additional sensor modality to segment data of a sensor of another modality, and the use of external context sources. We describe some of the segmentation methods, components and related works in more details in the following subsections:

4.3.2.1   Sliding window

In the sliding window methods, a window is moved over the time series data to 'extract' a data segment that is then used in subsequent HAR processing stages. The delay of the HAR system is proportional to the window size, i.e. larger window sizes result in longer delay for a new segment to be available for processing [127]. Sliding window algorithms are simple, intuitive and online algorithms and therefore are popular for medical applications [128].

Besides the delay, the window size reflects on a trade-off between segmentation precision and computational load. The larger the window size, the less frequently all subsequent stages of the HAR system are executed. While this reduces computational load, the less accurately the segmentation borders can be defined. A sliding window algorithm can start with a small fixed subsequence of time series and adds new data points until the fit-error for the potential segment is greater than the threshold which is defined by the user. Although they are simple and online, sliding window algorithms might cause poor results in some cases [129]. They work with a complexity of $O(nL)$, where L is the average length of a segment [130] and n the number of segments.

4.3.2.2   Energy-based segmentation

Energy-based segmentation method employs the fact that different activities are performed with different intensities which are directly translated to different energy levels of the recorded sensor signals. The energy E of a signal S is calculated as:

$$E = \int_{-\infty}^{\infty} |S(t)|^2 dt.$$

By thresholding on E, data segments for similar activities can be identified [131]. In this method, a predefined rest position between each activity [20,132] needs to be assumed, particularly for gesture-based HAR problems that involve discrete activities or gestures. Whenever the rest position is detected by the HAR system, a segment border is assumed [133]. For whole-body activity recognition, the rest position can be a certain posture; for the recognition of gestures, a defined hand position can be used.

An adaptive sliding window technique supports natural movements, based on naturally occurring pauses, such as the turning point of arms [134].

4.3.2.3   Additional sensors and contextual sources

Collected sensor data can be segmented using information recorded from additional sensors during the activities. For instance, long-term acceleration data recorded on a mobile phone accelerometer can be segmented using GPS traces [135] or feet step sounds recorded using the internal microphone of the mobile phone [136]. Similarly, segmentation can be performed using external context sources (i.e. sensors external to the recording device), such as PIR location tracking [76] or a self-recorded activity recording such as meetings or questionnaires. Such solutions clearly demand additional hardware instruments, transmitting and processing the data which can be counted as their downside.

4.3.2.4   Top-down

Top-down methods [78] (aka iterative end-point fits [137]) are used to break time series data into many segments by splitting data at the best location. They start by dividing the data into two halves until all segments have approximation errors below the user-defined threshold. These algorithms work recursively and their computational complexity is $O(n^2 K)$, where K is the number of segments for n-length time series [130].

4.3.2.5   Bottom-up

Bottom-up segmentation methods start the segmentation process with the finest possible approximation which is $n/2$ segments for n-length time series [128] and then merge a pair of adjacent subsequence of time series to create larger segments. This process continues until the calculated cost value reaches a stopping criterion. Similar to the sliding window algorithm, the computational complexity for this algorithm is $O(nL)$ [130].

Keogh *et al.* [128] proposed a two-level segmentation procedure to a time series, i.e. the Sliding Window and Bottom-Up (SWAB). While SWAB supports and online behaviour of sliding window algorithm, it maintains the superiority of bottom-up. SWAB uses sliding window algorithm to create a single segment that is then moved into the pre-allocated buffer to apply bottom-up approach. This process continues as long as new data arrives. The complexity for SWAB is slightly higher than the bottom-up approach while the performance is reported as good as the bottom-up algorithm [130].

4.3.3   Feature extraction

Extracting important and useful information is a challenging issue for continuous HAR [55]. Two approaches have been proposed to extract features from segmented time series data: Intuitional and statistical. Intuitional approaches take into account prior knowledge about the area of study, whereas statistical methods use quantitative characteristics of the data to extract features. Wavelet analysis has also been used to derive the so-called time-frequency features [138]. Brief descriptions of the three feature extraction approaches are described in the followings:

4.3.3.1   Intuitive approach

Intuitional approaches utilise knowledge retrieved from the area of study, namely the body model or the events:

1. Body model is calculated from a 3D skeleton using multiple sensors' placement on the subjects' body [134,139]. Encoding prior knowledge increases robustness across subjects and can lead to higher performance [139]. Polynomial features that describe signal trends such as mean, slope and curvature are used for trajectories of limbs [81].

2. In event-based [16,61–65] applications such as eye movements, features are extracted from saccades, fixations or blinks, as well as features describing the characteristics of repetitive eye movement sequences [126].

4.3.3.2   Statistical approach

In statistical methods, a range of different approaches has been used to obtain features from segmented data, with some researchers deriving features directly from the time-varying acceleration signal [26–28,38,44,140,141] and others from a frequency analysis [11,14,33,39,41,115,142,143].

1. Time-domain approach: Time-domain features include mean, median, variance, root mean square (RMS), standard deviation, amplitude, inter-axis correlation coefficient, skewness, kurtosis, range [144–147]. Bouten *et al.* [146], applied the integral method to offer estimation of energy expenditure using an inertial sensor and used the total integral of modulus of accelerations. This metric is referred to the time integrals of the module of accelerometer signals (Equation (4.4)):

$$IMA_{tot} = \int_{t=1}^{N} |a_x| dt + \int_{t=0}^{N} |a_y| dt + \int_{t=0}^{N} |a_z| dt, \tag{4.4}$$

where a_x, a_y, a_z denote the orthogonal components of accelerations, t denotes time and N represents the window length. Figo *et al.* [125] also used other time-domain features such as zero-crossings, correlation-coefficient and RMS.

2. Frequency-domain approach: Discrete Fourier transform [148] is used to compute frequency spectrum of the discrete data signal x [149].

    i.   Power spectral density (PSD) is one of the most important frequency-domain features used for HAR. This feature has been used by Nham *et al.* [149] to recognise activities such as walking, cycling, running and driving. PSD can be computed as the squared sum of its spectral coefficients normalised by the length of the sliding window (N):

$$P(f) = \frac{1}{N} \sum_{i=0}^{N-1} a_i^2 + b_i^2, \tag{4.5}$$

with $a_i = x_i \cos\left(\frac{2\pi f i}{N}\right)$ and $b_i = x_i \sin\left(\frac{2\pi f i}{N}\right)$,

where f denotes fth Fourier coefficient in the frequency domain.

    ii. The peak frequency is also a frequency domain feature that represents the frequency corresponding to the highest computed PSD over the sliding window and is used in several studies related to HAR [125,149,150].

    iii. Entropy feature is widely used in HAR [116] as a frequency domain feature and helps discriminate between activities that have the same PSD but different patterns of movement [150]. Entropy can be formulated as presented in Formula 6 in the following:

$$H(f) = \frac{1}{N} \sum_{t=1}^{N} c_i \log(c_i), c_i = \frac{\sqrt{a_i^2 + b_i^2}}{\sum_{k=0}^{N-1} \sqrt{a_i^2 + b_i^2}}. \tag{4.6}$$

iv. The *DC* component is another important feature used in HAR [150] which represents the PDS at frequency $f = 0$ Hz. It can be formulated as the squared sum of its real spectral coefficients normalised by the length of the sliding window:

$$DC = \frac{1}{N}\sum_{i=0}^{N-1} a_i^2. \tag{4.7}$$

4.3.3.3 Wavelet approach

With wavelet analysis, the segmented signal is decomposed into a series of coefficients, which carries both spectral and temporal information which enables localised temporal instances identification when there is a change in frequency characteristics of the original signal [151].

In particular, the original time-domain signal sampled with maximum frequency f, is initially decomposed into a coarse approximation and detail information by low-pass filtering ($[0, f/2]$) and high-pass filtering ($[f/2, f]$), respectively [152].

At high frequency (shorter time intervals or lower levels), the wavelets can capture discontinuities, ruptures and singularities in the original data. At low frequency (longer time intervals or higher levels), the wavelet characterises the coarse structure of the data to identify the long-term trends. Thus, the wavelet analysis allows us to extract the hidden and significant temporal features of the original data.

Previous activity classification studies have used wavelet analysis to derive only a small number of features. In contrast, Wang *et al.* [153] used wavelet packet analysis to derive 33 features from a triaxial accelerometer signal. With wavelet packet analysis, the detail coefficients are split into a further approximation and detail coefficients. This allows additional information to be extracted from the original signal. The features suggested by Wang *et al.* [153] involved summing the squares of the detail and wavelet packet approximation coefficients across different levels. In addition, they calculated standard deviations and RMS values of detail and wavelet packet approximation coefficients different levels.

Tamura *et al.* [154] used wavelet features for classification of accelerometer data. With this approach, the accelerometer signal is decomposed using the wavelet transform and the features defined as signal power measurements, calculated as the sum of the squared detail coefficients at levels 4 and 5. Sekine *et al.* [155] suggested another set of wavelet features for classification of accelerometer data. Again, there are two features, the first being the total of the summations of the detail signal at levels 6 and 7. This quantity is divided by the number of steps (N), which is obtained by counting the number of times the signal, reconstructed from levels 6 and 7, changes sign. For the second feature, the total of the summations of the detail signal from levels 4 to 7 is normalised against the sum of the squares from the original signal.

4.3.4 Feature selection

A basic property of a unique feature is to contain useful information about the different classes in the data. When a feature is conditionally independent of the class labels, it becomes irrelevant [156]. Feature relevance [157] provides a measurement of the feature's usefulness in discriminating the different classes. Here the issue of relevancy of a feature has to be raised, i.e. how do we measure the relevancy of a feature to the data or the output. Therefore, if a feature is to be relevant, it can be independent of the input data but cannot be independent of the class labels, i.e. the feature that has no influence on the class labels can be discarded.

Inter feature correlation plays an important role in determining unique features. For practical applications the underlying distribution is unknown and is measured by the classifier accuracy. Due to this, an optimal feature subset may not be unique because it may be possible to achieve the same classifier accuracy using different sets of features. Several publications [157–161] have presented various definitions and measurements for the relevance of a variable.

Generally, feature selection approaches can be divided into three folds: filters, wrappers and embedded methods which differ from three aspects [158], i.e. search strategies, evaluation criterion definition (e.g. relevance index or prediction of classifiers), and evaluation criterion estimation (e.g. statistical test or cross-validation/performance bounds).

4.3.4.1   Filter approach

Filter methods [162] use variable ranking techniques as the principal criterion for variable selection by ordering their relevance indices. There are various heuristics to design relevance indices for filters, including univariate prediction error rate (i.e. evaluate the relevance of a feature as how accurate the prediction is using only itself), correlation-based (e.g. Pearson coefficient and signal to noise ratio), distances between distributions (K-L divergence [163] and Jeffreys-Matusita distance [164]), information theory (mutual information (MI) [165] and minimum description length [166]), decision trees (C4.5 [167] and CART [168]) and Relief (a class of filters incorporating sample relations into feature selection [169]).

The downside of filter approach is that the relevance index is calculated based solely on a single feature without considering the values of other features, hence assuming orthogonality between features which usually is not true in practice. However, filters are efficient and proved to be more robust to overfitting theoretically [170].

On the other hand, they differ in how to use data to evaluate the usefulness of a single feature. Heuristics other than decision trees and Relief are global, i.e. they do not account for distances between samples. Relief makes use of the local information of an area of the feature space to calculate the average usefulness of a feature since features can be relevant to the target in some area. Decision trees divide the feature space hierarchically to investigate the relevance of features at different stages. Such local information will be helpful when the data domain is complicated (e.g. in image analysis).

There is no general guideline for choosing the most appropriate relevance index for a problem. However, the performance of relevance indices is related to the type of data (binary, integer or continuous) and prior information about the data distribution, according to their definitions and properties.

Ranking methods are used due to their simplicity and good success is reported for practical applications. A suitable ranking criterion can be used to score the variables and a threshold to remove variables below the threshold.

4.3.4.2   Wrapper approach

Instead of ranking every single feature, wrappers search to rank through all possible subsets of features and explore the MI between features, as first proposed by Mallat and Hwang [151]. After choosing a classifier, wrappers evaluate the prediction performance either by cross-validation or by theoretical performance bounds. The wrapper approach [160,161] uses a predictor as a black box and the predictor performance as the objective function to evaluate the variable subset. Since evaluating all subsets leads to an NP-hard problem and is prone to overfitting, suboptimal subsets are found by employing search methods that find a subset heuristically. Such methods broadly classified into Sequential Selection Algorithms and Heuristic Search Algorithms. First we will look at sequential selection algorithms followed by the heuristic search algorithms:

The sequential selection algorithms start with an empty set (full set) and add features (remove features) until the maximum objective function is obtained. To speed up the selection, a criterion is chosen which incrementally increases the objective function until the maximum is reached with the minimum number of features. The heuristic search algorithms evaluate different subsets to optimise the objective function. Different subsets are generated either by searching around in a search space or by generating solutions to the optimisation problem.

4.3.4.3   Embedded approach

Embedded approaches are based on the selected features during the learning process of a specific classifier. In contrast to wrappers incorporate the feature selection as part of the training process which is a limiting factor, hence the feature selected by one embedded method might not be suitable for others.

Embedded methods [171] aim to minimise the computation time taken up for reclassifying different subsets which are performed by wrapper methods. The methods to pursue such approximate solutions can be categorised as: (1) greedy search based on the gradient between the empirical risk and the weight indicators such as least angle regression [172], Recursive feature elimination [173] and decision trees and (2) relaxation of the integrality restriction on weight indicators by gradient descent regarding the bounds for generalisation errors [174] or incorporating proper priors for the weights (Joint Classifier and Feature Optimisation [175]) inclusion of a sparsity term in the minimisation problem (for linear models).

4.3.5   Classification methods

Once features have been derived to characterise a window of sensor data, they are used as input to a classification algorithm with different degree of complexities. We describe a range of classifications methods from simple threshold-based schemes to more advanced algorithms, such as hidden Markov models, neural networks and machine learning techniques (such as Supervised learning approaches, Semi-supervised learning, Unsupervised learning), transfer learning and deep learning (DL) in the following:

4.3.5.1   Threshold-based classification

A derived feature is being compared to a predetermined fixed threshold to determine whether a particular activity is being performed in threshold-based classification. Threshold-based classification has been used successfully to:

1. Identify postural transitions using data on the change in segmental angles derived from either accelerometer [176] or gyroscopes [176,177].
2. Differentiate between static postures, such as standing, bending, sitting and lying, using angles derived from two tri-axial accelerometers placed on combinations of the pelvis/trunk [178–180], lower limb segments [179,181–183] and chest [176,184].
3. Differentiate between static postures and dynamic activity by using a feature which quantifies variation in the acceleration signal [185–187].

Sekine *et al.* [188,189] and Nyan *et al.* [190] were able to use wavelet-based features and applied threshold-based classification to differentiate between three different gaits to detect falls. A fall can be considered an extreme instance of a postural transition where heuristic features are used to detect this abnormal behaviour occurs during a fall activity.

4.3.5.2   Machine learning

Machine learning tools are applied as a classification tool in HAR systems which learn to recognise and associate patterns in the input features with each activity. The associated patterns are to be discovered from a set of given examples, so-called training set. The examples in the training set may or may not be labelled. As discussed in subsection 4.2.2, labelling data is a time-consuming process because it may require an expert to examine the examples and assign a label based upon their experience which makes such data mining applications expensive and time-consuming [191–193].

There are mainly two machine learning approaches, namely *supervised* and *unsupervised* learning [194,195], which deal with labelled and unlabelled data, respectively. Supervised approaches are more commonly used in HAR systems since the HAR system should return a label such as walking, sitting, running, etc. Discriminating activities in a completely unsupervised context is rather challenging. A semi-supervised learning approach allows part of the data to be unlabelled. The three types of learning mechanisms are briefly described in the followings:

Supervised learning approaches
In machine learning algorithms every instance of dataset is represented by using the same set of features. If instances are labelled, then the learning scheme is known as supervised (aka classification for discrete-class problems). In this subsection, a summary of different supervised learning approaches applied for HAR systems is presented.

Sensor data can either be stored in a non-volatile memory [118], while a researcher supervises the collection process and manually labels the activities, or a mobile application allows the user to select the activity to be performed from a list [92] and matched each sample to an activity label, and then stored in the server. Some of the commonly used supervised learning applied in the HAR system field are summarised as follows:

1. *Decision trees* build a hierarchical model in which attributes are mapped to nodes and edges represent the possible attribute values. Accurate classification rules [195,196] exist to automate the process and create a compact set of rules from the root to a leaf node. C4.5 is perhaps the most widely used decision tree classifier and is based on the concept of information gain to select the attributes that should be placed in the top nodes [167]. Decision trees can be evaluated in $O(\log n)$ for n attributes, and usually generate models that are easy to understand by humans. A binary decision tree classifier [197] based on acceleration thresholds is simplified to recognise the physical activity based on waist accelerations in order to fit the embedded system.

2. The *Bayesian classifier* is based on the estimated conditional probabilities of the signal patterns available from each activity class. The probability of a new unknown pattern having been generated by a specific activity can be estimated directly. The *Bayesian Network* (BN) [198] classifier and *Naïve Bayes* (NB) [199] (which is a specific case of BN where the input features are assumed to be independent of each other) are the principal exponents of this family of classifiers. A key issue in BN is the topology construction, as it is necessary to make assumptions on the independence among features. For instance, the NB classifier assumes that all features are conditionally independent given a class value, yet such assumption does not hold in many cases. As a matter of fact, acceleration signals are highly correlated, as well as physiological signals such as heart rate, respiration rate and ECG amplitude. Although the assumption of feature independence is often violated, the Bayesian approach is popular due to its simplicity and ease of implementation [26,37,61,68].

3. *Support Vector Machines* (SVM) [200], is a popular machine learning that is broadly used in HAR [33,38,39,45,48,63] although it does not provide a set of rules understandable by humans. Instead, knowledge is hidden within the model, which may hinder the analysis and incorporation of additional reasoning.

4. *Artificial Neural Networks* (ANN) [201] replicate the behaviour of biological neurons in the human brain, propagating activation signals and encoding knowledge in the network links. Besides, ANNs have been shown to be universal function approximators. The high computational cost and the need for large amount of training data are two common drawbacks of neural networks which may not be feasible for mobile applications.

5. *k-nearest neighbour* (kNN) classification schemes [194,196] construct a multi-dimensional feature space, in which each dimension corresponds to a different feature. The feature space is first populated with all training data points for each activity. Unknown windows of sensor data are represented in the feature space

and the k-nearest neighbours of training data is identified. The majority of the k-nearest neighbours classifies to a given activity. The value of k typically varies from 1 to a small percentage of the training data and is selected using trial and error, or ideally using cross-validation procedures [9,39].

6. *Ensembles of classifiers* combine the output of several classifiers to improve classification accuracy and by applying majority voting (where the majority class is accepted), stacked generalisation (which trains the base classifiers and then uses their predictions as data to a new learning stage), or boosting (which assigns weights to the training patterns to combine the performance of weak classifiers) [194,195]. Ensembles of classifiers have higher computational complexity as several models need to be trained and evaluated [9].

Semi-supervised learning

As discussed, labelling is time-consuming and difficult in some scenarios, particularly in highly dynamic activities. A semi-supervised approach requires part of the data to be labels [84,191–193,202]. There are not yet any standard algorithms or methods for semi-supervised learning and instead each system implements its own approach. For instance, for a naturalistic data collection procedure, subjects are to perform activities without the researchers' supervision. When the subject does not follow the sequence of activity types correctly, some data would be unlabelled, hence the HAR model be trained would by means of semi-supervised learning.

The semi-supervised approach has implementation limitations with respect to the accuracy, cost and obtrusiveness and cannot yet be considered as a solution for HAR system: En-co-training [203] report no substantial improvement in the classification accuracy in HAR context. Ali *et al.* [192] designed a system to recognise finger gestures with a laparoscopic gripper tool from the worn sensor gloves with two biaxial accelerometers. The system was intended for a very specific purpose but not suitable for recognising daily activities thereby limiting its applicability to context-aware applications. Finally, the system proposed by Huynh *et al.* [202] required 11 sensors to recognise 8 ambulation and daily activities which introduce high obtrusiveness.

Online implementation of learning would open the possibility to use the unlabelled data collected in the production stage to improve the recognition performance.

Unsupervised learning

Unsupervised learning techniques [194,196] can be used for the analysis and interpretation of wearable sensor unlabelled data which allows implementation for the real-world user. Unsupervised approaches can identify unusual events from clusters of related patterns in the feature space which is valuable for a fall detection system, where there are typically no available training data. Unsupervised approach can also be applied to determine the cluster structure to help labelling process of the clusters, followed by a supervised learning layer which largely reduces the cost of labelling large datasets. By combining unsupervised with supervised approaches, it is possible to develop off-the-shelf systems which can be trained by the user with only occasional input.

Despite the usefulness of unsupervised learning approaches in the HAR systems, limited studies are conducted: The potential of unsupervised learning techniques in activity monitoring was first demonstrated by Van and Cakmakci [204] by defining a feature space from a number of simple time-domain features obtained from two thigh-mounted accelerometers to recognise one of seven defined activities. A Kohonen Self-Organising feature Map [205] which can be considered an array of discrete nodes or neurons, used to group of the original data to a much lower dimensional feature space. Once the original data have been grouped, they can be labelled with minimal user input and then a supervised classification layer added to recognise unlabelled sensor data. Nguyen *et al.* [206] performed a single subject study demonstrating the potential of unsupervised clustering for the recognition of both usual and unusual events. Data collected from a waist-mounted accelerometer was applied as input to a combined algorithm of hidden Markov models and Gaussian mixture models to perform data segmentation and clustering without prior knowledge. The degree of similar activities clustered together was demonstrated where optimal results were obtained using 'raw features' in comparison to other time-domain features.

4.3.5.3   Transfer learning

Applying the knowledge gained in one problem to a different but related problem is called transfer learning [207,208]. Different types of transfer learning are possible using machine learning methods for HAR systems [87,93,209]. Considering the problem of activity recognition using motion sensors, HAR algorithms require substantial amounts of labelled training data yet need to perform well under real scenarios. As a result, researchers work on methods to identify and utilise subtle connections between activity recognition datasets or to perform transfer-based activity recognition. To build a mapping between the two domains, web knowledge is used as a bridge to help link the different label spaces by Hu *et al.* [210]. Another application of transferring learned knowledge of activities [211] is based on modelling activities using structural, temporal and spatial features. This method not only avoids the tedious task of collecting and labelling huge amounts of data in the target space, but also the information learned in previous feature spaces.

4.3.5.4   Deep learning

Deep learning (DL) techniques have been successfully applied in recognition tasks [212,213] which allow an automatic extraction of features without any domain knowledge. The well-known DL models include convolutional neural network, deep belief network and auto-encoders [214]. Depending on the usage of label information, the DL models can be learned in either supervised or unsupervised manner.

A HAR framework is proposed by Plötz *et al.* [215] that employs PCA and deep belief networks for feature learning and is evaluated by means of recognition experiments on four publicly available HAR datasets. Automatically estimated features outperformed classic heuristic features for all the considered analysed HAR tasks. To automate feature extraction for the HAR task [214] builds a new deep architecture to investigate the multichannel time series data which employs the convolution and pooling operations to capture the salient patterns of the sensor signals at different time scales. All identified salient patterns are systematically unified among multiple channels and finally mapped into the different classes of human activities.

To extract human activity features without any domain knowledge, Zeng *et al.* [216] proposed to capture the local dependencies and scale-invariant features of activity signals. Thus, variations of the same activity can be effectively captured through the extracted features that outperform the state-of-the-art methods.

4.3.6   Evaluating HAR systems

In order to evaluate a method for activity recognition, we typically run experiments using a previously recorded dataset which consists of sensor data and a set of annotated ground truth labels, and a recognition method is used to infer which sequence of activities best explains the sensor data. To determine the quality of the inferred sequence, we compare it with the ground truth labels. Instead of individually comparing each inferred label with its corresponding ground truth label, an evaluation metric is generally used which provides with a number that represents the quality of recognition.

Some standard metrics used in HAR systems are: accuracy, recall, precision, *F*-measure, Kappa statistic and ROC curves. Evaluation metrics are preferably compact (i.e. a single number) which is useful for comparing the overall performance of a method, but do not provide any detail about the strengths and weaknesses of the model. Furthermore, different applications have different requirements and so no single metric is likely to be applicable to all applications.

In this section, we present several evaluation metrics and indicate their applicability. We start with a typical pattern recognition evaluation metric and discuss its limitations and present a number of measures for dealing with these limitations.

4.3.6.1   Performance metrics

We assume data is discretised using time-slices of constant length, this allows us to create a confusion matrix by aligning the inferred sequence of labels to the ground truth sequence of labels.

To get a better understanding of the errors made by the HAR system, the full confusion matrix is usually included for presenting the results. However, confusion matrices take up a lot of space in a paper and they only show that one activity was confused with another. An example of such a confusion matrix using three activities is given in Table 4.5 [217].

Using the values of the confusion matrix, we can calculate a commonly used measure in the pattern recognition community known as accuracy. The accuracy measure indicates the percentage of correctly classified time-slices out of all time-slices which can be presented as follows:

$$\text{Accuracy} = \frac{\sum_{i=1}^{Q} TP_i}{n},\tag{4.8}$$

with Q being the number of activities or classes and n being the total number of time-slices in the dataset (see Table 4.5). A problem with the accuracy measure is that it does not take differences in the frequency of activities into account. These differences in frequency can correspond to how often a particular activity is performed (e.g. sleeping is generally done once a day, while toileting is done several times a day), or to the

Table 4.5   Confusion Matrix showing the true positives TP, total number of ground truth time-slices (NG) and total number of inferred labels (NI) for each class

Ground Truth \ Inferred	Class 1	Class 2	Class 3	
1	TP_1	ε_{12}	ε_{13}	NG_1
2	ε_{21}	TP_2	ε_{23}	NG_2
3	ε_{31}	ε_{32}	TP_3	NG_3
	NI_1	NI_2	NI_3	n

number of time-slices an activity takes up (e.g. a sleeping activity generally takes up considerably more time-slices than a toileting activity). Not incorporating this class imbalance results in an evaluation metric which is biased towards a recognition method that favours the recognition of frequent classes [218,219].

Activity recognition datasets are generally imbalanced, meaning certain activities occur more frequently than others. To deal with the effects of class imbalance, we need to calculate an evaluation metric for each activity and normalise it according to the total number of time-slices used for that particular activity. However, the total number of time-slices for each activity in the ground truth often differs from the total number of time-slices in the sequence that was inferred by the recognition method.

Taking the total number of correctly recognised time-slices as a percentage of the total number of inferred and ground truth time-slices leads to two measures that are known in the information retrieval community as precision and recall. The conventional use of precision and recall assumes a two-class problem (i.e. a positive class and a negative class), and calculates the measures with respect to the number of true positives. If dealing with a multi-class problem, the precision and recall need to be calculated for each class separately by considering one class as the positive class and all other classes as negative classes. Instead of using the precision and recall values for all activities as a metric, the average precision and recall over all activities can be considered.

Precision and recall are often combined into a single measure known as the F-measure, which is a weighted average of the two measures. These three measures can be calculated using the values from the confusion matrix as follows:

$$\text{Precision} = \frac{1}{Q}\sum_{i=1}^{Q}\frac{TP_i}{NI_i},$$

$$\text{Recall} = \frac{1}{Q}\sum_{i=1}^{Q}\frac{TP_i}{NG_i},$$

$$F\text{-Measure} = \frac{2 - \text{Precision} - \text{Recall}}{\text{Precision} - \text{Recall}}$$

where *NG* and *NI* are the total number of ground truth time-slices and the total number of inferred labels for each class, respectively. Note that in calculating these metrics the precision and recall of each activity is weighted as equally important. While the *F*-measure presents a compact metric for comparing the performance of a model, the precision and recall show the performance with respect to the inferred sequence and the ground truth. Therefore, all these measures express the performance in terms of the percentage of correctly classified time-slices. However, these measures do not provide us any information about the type of errors that the HAR system makes.

4.3.6.2   Error analysis

Most activity recognition problems include a NULL class or garbage class which corresponds to any time-slices for which no annotation is available or in which activities that we are not interested in, are being performed. An activity recognition method has to determine the start and end time of each activity and therefore also has to determine at which points of time the NULL class is being performed. However, the NULL class is generally one of the most difficult activities to recognise. It is composed of all activities we are not interested in and therefore it is not a very clearly defined class.

Types of errors
In time series data, we can distinguish several error categories [220], we distinguish between four categories: substitution errors, occurrence errors, timing errors and segmentation errors:

1. Substitution errors S occur if one activity is incorrectly recognised as another activity. If the ground truth label and the inferred label do not match and neither of the two labels is the NULL activity. This means that an activity was being performed (according to the ground truth), but the recognition method substituted it with another activity. This type of error corresponds to the non-diagonal elements of the confusion matrix.

2. Occurrence errors happen if the recognition method confuses whether an activity has occurred or not. Occurrence errors can occur in two forms, an insertion error I is made when a ground truth segment that is a NULL activity is recognised as an actual activity (non-NULL activity) in the inferred sequence. A deletion error D is made when a ground truth segment of an activity is recognised as a NULL activity.

3. Timing errors refer to the cases in which the activity is correctly recognised, but the start and end time of the activity is not correctly recognised. There are two types of timing errors: an overfilled error O occurs when in the ground truth, the activity ended and a NULL activity started, while in the inferred sequence the activity continued. An under filled U error occurs when in the ground truth, the activity continued, but in the inferred sequence a NULL activity started.

4. Segmentation errors correspond to the cases in which either a single segment of an activity is recognised as multiple segments, or multiple segments are recognised as a single segment. Segmentation errors can either be a fragmentation error F or a merging error M. A fragmentation error means the activity is correctly

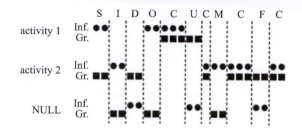

Figure 4.3   Examples of the different error types: circles denote inferred (inf.) and squares denote ground truth (Gr.) [217]

recognised but several time-slices of NULL class were inserted, rather than the activity being recognised as a single segment. Merging means several segments of activities were recognised as a single segment and were in that sense merged together.

Although the importance of these errors can differ among applications, for most applications substitution errors and occurrence errors are most important. Examples of the different error types are depicted in Figure 4.3, where the substitution error in the beginning leads to activity 2 to be mistakenly inferred as activity 1 followed by an insertion error makes the HAR system infer no activity (NULL) as activity 2.

Error metrics
For each of the errors, the number of time-slices with error should be counted and presented as a percentage of the ground truth or the inferred sequence to calculate a number of error metrics. For activity i, the error metrics are as follows:

$$S_G^* = \frac{1}{Q}\sum_{i=1}^{Q}\frac{\sum_{j=1}^{Q}S(i,j)}{NG_i} \qquad S_I^* = \frac{1}{Q}\sum_{i=1}^{Q}\frac{\sum_{j=1}^{Q}S(i,j)}{NI_i}$$

$$D^* = \frac{1}{Q}\sum_{i=1}^{Q}\frac{D_i}{NG_i} \qquad I^* = \frac{1}{Q}\sum_{i=1}^{Q}\frac{I_i}{NI_i}$$

$$U^* = \frac{1}{Q}\sum_{i=1}^{Q}\frac{U_i}{NG_i} \qquad O^* = \frac{1}{Q}\sum_{i=1}^{Q}\frac{O_i}{NI_i}$$

$$F^* = \frac{1}{Q}\sum_{i=1}^{Q}\frac{F_i}{NG_i} \qquad M^* = \frac{1}{Q}\sum_{i=1}^{Q}\frac{M_i}{NI_i}$$

with Q being the total number of classes. Each metric corresponds to the percentage of the total number of time-slices that the error occurs and is averaged over all classes. A summary of the error metrics is listed in Table 4.6.

Table 4.6   Compact result matrix showing all the error metrics

	Error metrics		
			Avg.
	Ground	Infer.	
Correct	Recall	Precision	*F*-measure
Substitution	S_G^*	S_I^*	*Sub**
Occurrence	D^*	I^*	*Occ**
Timing	U^*	O^*	*Tim**
Segmentation	F^*	M^*	*Seg**

The resulting values (denoted by the superscript in Table 4.6) account for the class imbalance and are averaged over all classes to present the error using a single number [217]. The weighted average between the two values can be used by the *F*-measure to calculate the precision and recall.

4.4   Communication platforms

Low power/low range wireless communication technologies and standards enable WBANs to collect and transmit the HAR system information to be monitored and managed by the authorised entities in the same network or remotely connected ones. Each WBAN consists of a number of sensing devices with processing and communication capabilities. In this section, we discuss the challenges and impairments of wireless media for the related wireless technologies which are also in place for HAR systems. We categorise the wireless communication technologies into three tiers; first tier connects the sensors to a gateway located on the user body or in the vicinity area, second tier relays the collected data to fixed access networks available to the user and the last tier forwards the information to the authorised users (see Figure 4.4). First-tier and second-tier communication protocols and standards are presented which connect and network the sensed data to the concerned authorities for necessary actions. Furthermore, Tactile Internet [221] concept with high reliability and low latency demand for relaying information to sensors and actuators can be considered as the networking platform for HAR systems.

4.4.1   Tier-1

Intra-WBAN communication includes the network interaction of nodes in Tier-1 and their respective transmission ranges ($\sim$2 metres) in and around the human body. Figure 4.4 illustrates WBAN communication within a WBAN and between the WBAN and its multiple tiers. In Tier-1, a variety of sensors are used to forward body signals to a personal server (PS) or a dual-stack gateway that bridges Tier-1 to Tier-2 or Tier-1 to Tier-3. Wireless standards which can be applied in the Tier-1 communications should be low power and low range. Suitable wireless communication standard for on-body

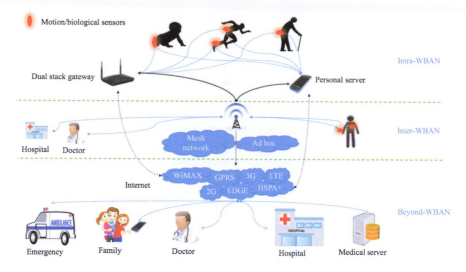

Figure 4.4   Communication architecture of an HAR system

healthcare applications including HAR systems are BLE [222], IEEE 802.15.6 [223] and IEEE 802.15.4j [224]. However, the IEEE 802.15.6 radio is unavailable to be employed within WBAN applications.

1. Bluetooth low energy (BLE) [222] was originally introduced under the name Wibree by Nokia in 2006. In 2010, the Bluetooth Special Interest Group merged Wibree into the Bluetooth standard as a part of the 4.0 core specification. BLE expands the functionality and applicability of Bluetooth and makes it a suitable choice for health monitoring systems. BLE involves several changes compared to traditional Bluetooth. It operates in the spectrum band 2,402–2,480 MHz, divided into 40×2 MHz channels instead of 79×1 MHz channels in Bluetooth. In BLE, three advertising channels are dedicated to broadcast messages, using frequencies 2,402, 2,426 and 2,480 MHz to mitigate interference from other technologies working in the same frequency band. BLE employs a frequency hopping mechanism that reduces the risk of eavesdropping on transmitted packets. In BLE, timing requirement in frequency hopping is more relaxed due to the longer period of each channel. The physical layer data rate is 1 Mbps. Security requirements are covered by advanced encryption standards, pairing to create shared secrets, and bonding to enable trusted device pair and device authentication.
2. IEEE 802.15.6 [223] defines a MAC layer that supports several PHY layers, such as narrowband (NB) with frequencies 400, 800 and 900 MHz, ultra-wideband with frequencies 2.3 and 2.4 GHz, and human body communication with 10–50 MHz, while the data rate varies from 75.9 kbps to 15.6 Mbps. Selecting a proper PHY layer with accompanying frequency band should be influenced by the application requirements and limitations. With 802.15.6, sensor nodes are organised in a one- or two-hop star topology, communicating to a single

coordinator or hub. In a two-hop topology, special nodes with relay capability are supposed to be placed in order to forward the data from sensor nodes towards the coordinator. The IEEE 802.15.6 standard divides the time into beacon periods or super frames with the equal length. The coordinator defines boundaries of the super frame that is separated into a number of slots, used for data transmission. Beacons are transmitted periodically for synchronisation purposes among all sensor nodes [225]. The IEEE 802.15.6 supports three security levels with different security properties, protection levels, and frame formats, which are known as (i) unsecured communication level (low-security level), (ii) authentication level (medium security level) and (iii) authentication and encryption (high-security level).

3. In 2013, the IEEE 802.15 Task Group 4j introduced an amendment to the IEEE 802.15.4 standard extending its PHY layer and MAC layer to support Medical Body Area Network services operating in 2,360–2,400 MHz frequency band [224]. The IEEE 802.15.4j standard is backward compatible with IEEE 802.15.4 so that the current IEEE 802.15.4 implementations can be reused. It must be noted that although the IEEE 802.15.4j PHY supports the medical band, there are no special improvements or provisions in the MAC to suit the requirements of a WBAN.

A summary of data rates supported by intra BAN wireless technologies are presented in Table 4.7 which confirms the appropriateness of these solutions for HAR systems with data rate ranging between 1 and 10 kbps depending on the sampling frequency.

4.4.2 Tier-2

Inter-WBAN communication covers communications between the PS and one or more access points (APs). The APs can be considered as part of the infrastructure (mesh-based), or even be placed strategically in a mobile system (*ad hoc*-based) to handle emergency situations. Tier-2 communication aims to interconnect WBANs with various networks, which can easily be accessed in daily life [226]. The more technologies supported by a WBAN, the integration process with other applications becomes more possible.

The paradigms of inter-WBAN communication as shown in Figure 4.4 are divided into two subcategories as follows:

1. Mesh-based architecture – The architecture is used in most WBAN applications as it facilitates dynamic deployment in a limited space such as a hospital as well as providing centralised management and security control. The APs can act as mesh clients and servers to relay the data to the database server related to its application.

2. *Ad hoc*-based architecture – In this architecture, multiple APs transmit information inside medical centres. The APs in this architecture form a mesh construction that enables flexible and fast deployment, allowing for the network to easily

Table 4.7 BAN wireless communication standards and technologies

Standard	Topology	No. of nodes	Frequency	Data rate	Tx power consumption	Tx range	Protocol stack size	Stronghold	Applications
Wi-Fi	p2p, p2mp	2,007	2.4/5 GHz	up to 600 Mbps	Transmit Power: 18 dBm ~180 mA	300 m	100KB+	High data rate	Internet browsing, PC networking, file transfers
Bluetooth	Piconet, Scatternet	7	2.4 GHz	3 Mbps	Transmit Power: +8.5 dBm ~40 mA	30 m	100KB+	Interoperability, cable replacement	Wireless USB, handset, headset
Bluetooth 4.0 (BLE)	Star	undefined	2.4 GHz	1 Mbps	Transmit Power: <11 dBm <10 mA	~ 50 m	250 KB	Low cost, Ability to run for years on standard coin-cell batteries	Healthcare, fitness, security and home entertainment industries
802.15.4/ 802.15.4j	P2P, Star, Cluster Tree, Mesh	+65,000	868/915 MHz 2.4 GHz	40 Kbps 250 Kbps	Transmit Power: 0 dBm 17.4 mA	100 m	34 KB/ 14 KB	Long battery life, low cost	Remote control, battery-operated products, sensors
802.15.6	Star (One hop, two-hop extendable)	255	2.4 GHz, 800 MHz, 900 MHz, 400 MHz	75.9 Kbps (narrowband) up to 15.6 Mbps	Transmit Power: −10 dBm 2.4 mA	<100 m	N/A	Low cost, high reliability, ultra-low power and short-range wireless communication in or around the human body	Medical (Wearable Health Monitoring) Non-Medical (Real-Time Streaming, Entertainment, Sport, military applications)
802.11ah	Single-hop	8,191	Sub-1 GHz	150 Kbps up to 78 Mbps	–	100–1,000 m	N/A	Ultra-low power, long-range coverage, backward compatibility with 802.11	Smart sensors and meters, Backhaul aggregation, Extended range hotspot and cellular offloading

expand, provides larger radio coverage due to multi-hop dissemination and supports patient mobility. The coverage range of this configuration is much larger compared to the mesh-based architecture and therefore facilitates movement around larger areas. In fact, this interconnection extends the coverage area of WBANs from 2 to 100 m, which is suitable for both short- and long-term setups [226].

In a health monitoring system, measurements are usually forwarded from a WBAN through a gateway towards the cloud. The gateway is used to bridge two different network technologies, from low-power to high-power wireless network, or from a wireless network to a wired network. The energy-hungry wireless networks such as IEEE 802.15.4 [227], Bluetooth [228], IEEE 802.11 and IEEE 802.11ah (aka low-power WiFi) [229] which can provide more reliable and secure way of data communication are not applicable for HAR systems.

Available wireless communication standard for inter-BAN communication are:

1. IEEE 802.15.4 [227] defines Physical (PHY) and medium access control (MAC) layers for LPWNs. It provides three frequency bands of 868 MHz, 915 MHz and 2.4 GHz, with a data rate of 250 kbps. The 2.4 GHz Industrial, Scientific and Medical (ISM) band is available worldwide and is therefore most commonly used. IEEE 802.15.4 defines two topologies: star topology (all sensor nodes communicate directly with the coordinator (single-hop)), and peer-to-peer topology (any sensor node can communicate with an other sensor node), where star topology is more common in health monitoring applications. ZigBee [230] is an open specification that complements the IEEE 802.15.4 standard with network and security layers, as well as application profiles. ZigBee supports mesh topology, where each node can communicate with any other node, through a single- or multi-hop, by relaying the transmission through multiple additional nodes. The network then can spread out over a larger area. To secure transmitted data, ZigBee networks use the advanced encryption standard (AES) encryption algorithm, which is one of the most secure, robust and reliable algorithms that encrypts 128-bit blocks of data, using multiple substitution and permutation operations.

2. IEEE 802.15.1 or Bluetooth [228], is designed and implemented for short-range wireless communication. It supports different frequency bands, such as 13.56 MHz, 2.4 GHz and 2.5 GHz, with the data rate 1–2.1 Mbps. Two types of topologies have been defined: Piconet and Scatternet. A Piconet is formed by the master node and one or more Bluetooth devices as slaves. A clock is set by master node in order to obtain synchronisation, and frequency hopping is applied to reduce the probability of interference. Slaves have point-to-point communication with their master node. However, a master node can either unicast or multicast to slaves within the Piconet. A Scatternet is a collection of some Piconet. A Bluetooth unit can be a member of different Piconet, i.e. it can be a slave in many Piconet.

3. IEEE 802.11 or WiFi that operates in the 2.4 and 5 GHz bands is the most popular wireless technology for indoor environments. The main features of WiFi are: high

data rate, easy deployment, low cost and high power consumption (compared with LPWNs). Due to the increasing demands for ubiquitous devices with low-power consumption, low power WiFi (IEEE 802.111ah) [229] was designed. This radio that works at 780, 868, 915 and 950 MHz is a potential solution for relaying data from body sensors towards the cloud. IEEE 802.11ah is scalable, supporting more than 8,000 devices, and it brings seamless connectivity with WiFi. In general, WiFi enables security via WiFi Protected Access that includes both access control and privacy for the communication.

Further information about these wireless standards in Tier-1 and Tier-2 is listed in Table 4.7.

4.4.3   Tier-3

Unlike Tier-2, the design of Tier-3 communication is for use in metropolitan areas. A gateway such as an AP can be used to link the connection between Tier-2 and Tier-3, as shown in Figure 4.4. The beyond-BAN tier communications can enhance the application and coverage range of an HAR system a step further by enabling authorised healthcare personnel (e.g. doctor, nurse or patient relatives) to remotely access a patient's medical information by means of cellular network or the Internet. A database is also an important component of the 'beyond-BAN' tier. This database maintains the user's profile and medical history. According to user's service priority and/or doctor's availability, the doctor may access the user's activity information as needed. At the same time, automated notifications can be issued to his/her relatives based on this data via various means of telecommunications. Details of a number of standard/under standardisation beyond-BAN communication technologies are listed in Table 4.8.

Table 4.8   Example beyond-BAN communication technologies

	SIGFOX	LoRa	Clean state	NB LTE-M Rel. 13	LTE-M Rel. 12/13	EC-GSM Rel. 13	5G
Range	<13 km	<11 km	<15 km	<15 km	<11 km	<15 km	<15 km
Max. coupling loss	160 dB	157 dB	164 dB	164 dB	156 dB	164 dB	164 dB
Spectrum	Unlicensed 900 MHz	Unlicensed 900 MHz	Licensed 7–900 MHz	Licensed 7–900 MHz	Licensed 7–900 MHz	Licensed 8–900 MHz	Licensed 7–900 MHz
Bandwidth	100 Hz	<500 KHz	200 KHz or dedicated	200 KHz or shared	1.4 MHz or shared	2.4 MHz or shared	shared
Data rate	>100 bps	>10 Kbps	>50 Kbps	>150 Kbps	>1 Mbps	10 Kbps	>10 Kbps
Battery life	>10 years	>10 years	>10 years	>10 years	>10 years	>10 years	>10 years

4.5   HAR open problems

In this chapter, we have highlighted a variety of challenges in each stage of HAR systems. Besides common research challenges in recognition systems (such as intra-class variability, inter-class similarity and the NULL class problem), there are s specific challenges with respect to the recognition algorithms, hardware and communication technologies to HAR systems that we summarise in this section:

4.5.1   Recognition challenges

1. **Definition and diversity of activities**: Clear definition of physical activities may seem trivial at first but human activity is highly complex, blur and diverse. Each activity can be performed in many different ways and depending on different scenarios, subjects and contexts. Two examples of developed definitions are: the ADLs index as a tool in elderly care [231] and the comprehensive compendium of physical activity [232].

2. **Class imbalance**: In the presence of considerable class imbalance, modelling different activity classes such as long-term behavioural monitoring, only a few activities occur often, such as sleeping or working, while most activities occur rather infrequently, such as taking a sip of a drink [233]. In general pattern recognition, class imbalance is often addressed by recording additional training data. This challenge can be either overcome by oversampling (i.e. duplicating) a smaller class size to equal the bigger class size [234] or collecting additional training data to ensure equal class distributions.

3. **Ground truth annotation**: Ground truth annotation is an expensive and time-consuming task, as the annotator has to perform the annotation in real time [79] or to skim through the raw sensor data and manually label all activity instances *post hoc* [80,81]. Researchers have investigated different techniques to address this problem, including daily self-recall methods [82], experience sampling [83] and reinforcement or active learning – all of which involve the user. If only a few labelled training samples are available, semi-supervised [84], unsupervised [85] or transfer learning techniques (as discussed in Section 4.3.5.3) [81,86,87] can be used.

4. **Data collection protocols**: Using standard datasets is crucial for reproducible research. The research community in HAR systems does not follow an agreed data collection protocol to collect rich and thus more general-purpose datasets of human physical activity. Data collection has a diverse requirement, such as high data quality, large numbers of modalities or sensors, long-term recordings, or large numbers of participants. Furthermore, researchers are faced with a trade-off between unobtrusiveness and ease of use of the sensors; the time required to prepare, conduct and maintain experiments; and the logistics and costs for participants, user's quality of experience, and the equipment(s). A number of publicly available datasets [235,236] to the research community have published in which researchers worldwide were invited to participate.

5. **Composite and overlapping activities**: When an activity (such as playing tennis) is composed of several instances of activities, additional uncertainty and complexity is introduced. Blanke *et al.* [81] provide an overview of this topic and propose a solution through several layers of inference. Furthermore, concurrent and overlapping activities [237] report great research opportunities in this field since generally a person can be walking while brushing their teeth, or watching TV while having lunch.

6. **Crowd-HAR**: The majority of the HAR systems work for a single or limited number of subjects. Although information from social networks has been shown to be effective to recognise human behaviours [238], recognising collective physical activity patterns is an open research challenge. If we could gather activity patterns from a significant sample of people in certain area (e.g. a city, a state or a country), that information could be used to estimate levels of sedentarism, exercise habits/addictions and even predict health conditions in a target population.

7. **Predicting future activities**: Published works have estimated activities and build behaviour routines [239]. Based on this information, the system could predict what the user is about to do and use this information for different applications where permitted by the user. For instance, if the user is going to have lunch, related advertisements on restaurants nearby may be sent to the user.

4.5.2   Hardware challenges

1. **Sensor characteristic variations**: This sensor changes may be internal or external. Hardware errors, complete sensor failures or sensor drift are internal causes. External causes may include changes in the operating temperature, humidity or loose placement of the sensors on the user's body [57,58]. Furthermore, portable devices containing sensors, such as mobile phones, may be used in different ways or carried at different locations on the body [59]. Sensor displacement and changes in sensor orientation may be detected when obvious differences in the recorded signals are observed [240]. Overcoming hardware (such as signal drift) and placements errors over time are more difficult to identify.

2. **Trade-off points in HAR system design**: Highly miniaturised embedded sensors for data recording typically have limited processing power which impacts on runtime. As discussed in subsection 4.2.6, there is the trade-off between accuracy, system latency and processing power. While for some HAR systems low-latency classification and immediate feedback may be required, for other applications this may be less critical.

3. **Sensor node heterogeneity**: Sensor nodes have different storage capacities, processing capabilities, power consumption limits, wireless technologies and QoS requirements. As per application requirements, a broad range of sensor devices can be exploited which makes heterogeneity factor quite challenging. The radio communication challenges will be addressed in subsection 4.5.3 separately.

4. **Synchronisation and calibration**: Since distributed sensor devices do not share a common power source, accurate calibration between wireless sensor nodes is

required. Tight synchronisation is needed when measuring a delay between two sensor nodes with two different clocks. The synchronisation between different clocks has a tremendous impact on the accuracy of the delay measurement.

4.5.3 Communication challenges

1. **Transmission power**: While higher transmission power would increase the range, the temperature rises of on-body sensors, radio interference between adjacent nodes, and energy consumption should be minimised [241] to improve the user's quality of experience.
2. **Power resource limitations**: Energy and device lifetime of WBANs is strictly limited due to the small size hardware and battery size. Due to the limitation of available resources in WBANs, WBAN nodes bound to fail because of the unavailablility of battery power, memory and bandwidth. Different solutions across all layers should consider the limited resources and exchange information (cross-layer design) to enhance resource usage. Alternative/supplementary power scavenging solutions are exploited to increase the system lifetime [96–98].
3. **Antenna design**: Certain constraints of HAR system due to the size and weight demanded for wearable devices in shape of antenna, material, size, malicious RF environment and installation location should be addressed as open challenges.
4. **Reliability**: Some of the WBAN standards such as the IEEE 802.15.6 standard allowed the deployment of dynamic channel hopping, which assists the network to minimise interference from other narrowband transmitters. Additionally, the IEEE 802.15.6 standard aims to eliminate interference by shifting beacon transmission. Also, in cases where the required levels of reliability cannot be achieved through a one-hop star topology, the use of relays is allowed [242]. The IEEE 802.15.6 standard also allows a one-hop or two-hop communication. While multi-hop transmission can increase system reliability, the energy consumption increases [243].
5. **Challenges of routing**: Postural body movements result in node mobility, frequent changes in topology and possible environmental obstacles which increase error in HAR as a result of dynamism in WBANs. Furthermore, the link quality variation between sensor nodes in WBANs follows a function of time due to various body movements [241]. Some body parts and clothing result in signal blockage that RF attenuation. However, considering short range between WBAN sensor nodes and limited energy resources, most communications can be performed on a single-hop basis, hence relaying is not a common scenario.

4.6 Summary

In this chapter, the current state of HAR systems and the procedures are presented and the application requirements and parameters for the evaluation existing solutions are described. Different aspects of the HAR design, implementation and evaluation HAR systems from sensor operation, data collection protocols, signal processing and pattern recognition, down to machine learning techniques as well as the communication technologies. The performance trade-offs such as energy consumption, accuracy, reliability and privacy should be reflected for selection of the HAR system based on the application requirements such as accuracy, energy efficiency, reliability and data privacy. We conclude the chapter by highlighting the open problems and ideas that require further work.

References

[1]   O. D. Lara, and M. A. Labrador, "A survey on human activity recognition using wearable sensors," IEEE Communications Surveys & Tutorials, vol. 15, no. 3, pp. 1192–1209, 2013.

[2]   Y. Baram, and A. Miller, "Auditory feedback control for improvement of gait in patients with Multiple Sclerosis," Journal of the Neurological Sciences, vol. 254, no. 1, pp. 90–94, 2007.

[3]   R. I. Spain, M. Mancini, F. B. Horak, and D. Bourdette, "Body-worn sensors capture variability, but not decline, of gait and balance measures in multiple sclerosis over 18 months," Gait & Posture, vol. 39, no. 3, pp. 958–964, 2014.

[4]   I. Carpinella, D. Cattaneo, and M. Ferrarin, "Quantitative assessment of upper limb motor function in Multiple Sclerosis using an instrumented Action Research Arm Test," Journal of Neuroengineering and Rehabilitation, vol. 11, no. 1, pp. 1, 2014.

[5]   B. Najafi, D. Horn, S. Marclay, R. T. Crews, S. Wu, and J. S. Wrobel, "Assessing postural control and postural control strategy in diabetes patients using innovative and wearable technology," Journal of Diabetes Science and Technology, vol. 4, no. 4, pp. 780–791, 2010.

[6]   B. Najafi, D. G. Armstrong, and J. Mohler, "Novel wearable technology for assessing spontaneous daily physical activity and risk of falling in older adults with diabetes," Journal of Diabetes Science and Technology, vol. 7, no. 5, pp. 1147–1160, 2013.

[7]   B. Najafi, J. S. Wrobel, G. Grewal, R. A. Menzies, T. K. Talal, M. Zirie, and D. G. Armstrong, "Plantar temperature response to walking in diabetes with and without acute Charcot: the Charcot Activity Response Test," Journal of Aging Research, vol. 2012, Article ID 140968, pp. 1–5, 2012.

[8]   Z. Zhu, H. Satizabal, U. Blanke, A. Perez-Uribe, and G. Troester, "Naturalistic recognition of activities and mood using wearable electronics," IEEE Transactions on Affective Computing, vol. 7, no. 3, pp. 272–285, 2015.

[9]   A. Zenonos, A. Khan, G. Kalogridis, S. Vatsikas, T. Lewis, and M. Sooriyabandara, "HealthyOffice: mood recognition at work using smartphones and wearable sensors," In Pervasive Computing and Communication Workshops (PerCom Workshops), 2016 IEEE International Conference on, pp. 1–6, IEEE, 2016.

[10] Y. Wen, R. Yang, and Y. Chen, "Heart rate monitoring in dynamic movements from a wearable system," In Medical Devices and Biosensors, 2008.

ISSS-MDBS 2008. 5th International summer school and symposium on, pp. 272–275, IEEE, 2008.

[11] L. Galway, S. Zhang, C. Nugent, S. McClean, D. Finlay, and B. Scotney, "Utilizing wearable sensors to investigate the impact of everyday activities on heart rate," In international conference on smart homes and health telematics, pp. 184–191, Springer Berlin Heidelberg, 2011.

[12] H. Solar, E. Fernández, G. Tartarisco, G. Pioggia, B. Cvetković, S. Kozina, M. Luštrek, and J. Lampe, "A non invasive, wearable sensor platform for multi-parametric remote monitoring in CHF patients," Health and Technology, vol. 3, no. 2, pp. 99–109, 2013.

[13] P. E. Taylor, G. J. Almeida, T. Kanade, and J. K. Hodgins, "Classifying human motion quality for knee osteoarthritis using accelerometers," In Engineering in medicine and biology society (EMBC), 2010 annual international conference of the IEEE, pp. 339–343, IEEE, 2010.

[14] K.-H. Chen, P.-C. Chen, K.-C. Liu, and C.-T. Chan, "Wearable sensor-based rehabilitation exercise assessment for knee osteoarthritis," Sensors, vol. 15, no. 2, pp. 4193–4211, 2015.

[15] J. T. Matthews, G. J. Almeida, E. A. Schlenk, R. Simmons, P. Taylor, and R. R. Da Silva, "Usability of a virtual coach system for therapeutic exercise for osteoarthritis of the knee," In IROS workshop on motivational aspects of robotics in physical therapy, Vilamoura, Portugal, 2012.

[16] B. G. Steele, B. Belza, K. C. Cain, J. Coppersmith, S. Lakshminarayan, J. Howard, and J. K. Haselkorn, "A randomized clinical trial of an activity and exercise adherence intervention in chronic pulmonary disease," Archives of Physical Medicine and Rehabilitation, vol. 89, no. 3, pp. 404–412, 2008.

[17] L. Sewell, S. J. Singh, J. E. Williams, R. Collier, and M. D. Morgan, "Can individualized rehabilitation improve functional independence in elderly patients with COPD?," CHEST Journal, vol. 128, no. 3, pp. 1194–1200, 2005.

[18] B. G. Steele, L. Holt, B. Belza, S. Ferris, S. Lakshminaryan, and D. M. Buchner, "Quantitating physical activity in COPD using a triaxial accelerometer," CHEST Journal, vol. 117, no. 5, pp. 1359–1367, 2000.

[19] B. M. de Blok, M. H. de Greef, N. H. ten Hacken, S. R. Sprenger, K. Postema, and J. B. Wempe, "The effects of a lifestyle physical activity counseling program with feedback of a pedometer during pulmonary rehabilitation in patients with COPD: a pilot study," Patient Education and Counseling, vol. 61, no. 1, pp. 48–55, 2006.

[20] O. Amft, H. Junker, and G. Troster, "Detection of eating and drinking arm gestures using inertial body-worn sensors," In Wearable Computers, 2005. Proceedings. Ninth IEEE International Symposium on, pp. 160–163, IEEE, 2005.

[21] O. Amft, M. Kusserow, and G. Tröster, "Probabilistic parsing of dietary activity events," In 4th International workshop on wearable and implantable body sensor networks (BSN 2007), pp. 242–247. Springer Berlin Heidelberg, 2007.

[22] A. G. Bonomi, G. Plasqui, A. H. Goris, and K. R. Westerterp, "Improving assessment of daily energy expenditure by identifying types of physical

activity with a single accelerometer," Journal of Applied Physiology, vol. 107, no. 3, pp. 655–661, 2009.

[23]   S. E. Crouter, K. G. Clowers, and D. R. Bassett, "A novel method for using accelerometer data to predict energy expenditure," Journal of Applied Physiology, vol. 100, no. 4, pp. 1324–1331, 2006.

[24]   C. Bouten, K. Westerterp, M. Verduin, and J. Janssen, "Assessment of energy expenditure for physical activity using a triaxial accelerometer," Medicine and Science in Sports and Exercise, vol. 23, no. 1, pp. 21–27, 1994.

[25]   H. Hung, G. Englebienne, and J. Kools, "Classifying social actions with a single accelerometer," In Proceedings of the 2013 ACM international joint conference on Pervasive and ubiquitous computing, pp. 207–210, ACM, 2013.

[26]   A. Matic, V. Osmani, A. Maxhuni, and O. Mayora, "Multi-modal mobile sensing of social interactions," In Pervasive computing technologies for healthcare (PervasiveHealth), 2012 6th International conference on, pp. 105–114, IEEE, 2012.

[27]   E. Miluzzo, N. D. Lane, S. B. Eisenman, and A. T. Campbell, "CenceMe-injecting sensing presence into social networking applications," In European conference on smart sensing and context, pp. 1–28, Springer Berlin Heidelberg, 2007.

[28]   E. Miluzzo, N. D. Lane, K. Fodor, *et al.*, "Sensing meets mobile social networks: the design, implementation and evaluation of the CenceMe application," In Proceedings of the 6th ACM conference on Embedded network sensor systems, pp. 337–350, ACM, 2008.

[29]   J. A. Kientz, G. R. Hayes, T. L. Westeyn, T. Starner, and G. D. Abowd, "Pervasive computing and autism: assisting caregivers of children with special needs," IEEE Pervasive Computing, vol. 6, no. 1, pp. 28–35, 2007.

[30]   G. Paragliola, and A. Coronato, "Intelligent Monitoring of Stereotyped Motion Disorders in Case of Children with Autism," In Intelligent environments (IE), 2013 9th International conference on, pp. 258–261, IEEE, 2013.

[31]   M. S. Goodwin, S. S. Intille, W. F. Velicer, and J. Groden, "Sensor-enabled detection of stereotypical motor movements in persons with autism spectrum disorder," In Proceedings of the 7th International conference on Interaction design and children, pp. 109–112, ACM, 2008.

[32]   T. Westeyn, K. Vadas, X. Bian, T. Starner, and G. D. Abowd, "Recognizing mimicked autistic self-stimulatory behaviors using HMMS," In Wearable computers, 2005. Proceedings of Ninth IEEE international symposium on, pp. 164–167, IEEE, 2005.

[33]   M. S. Goodwin, M. Haghighi, Q. Tang, M. Akcakaya, D. Erdogmus, and S. Intille, "Moving towards a real-time system for automatically recognizing stereotypical motor movements in individuals on the autism spectrum using wireless accelerometry," In Proceedings of the 2014 ACM international joint conference on pervasive and ubiquitous computing, pp. 861–872, ACM, 2014.

[34]   R. Fallahzadeh, M. Pedram, R. Saeedi, B. Sadeghi, M. Ong, and H. Ghasemzadeh, "Smart-Cuff: A wearable bio-sensing platform with activity-sensitive information quality assessment for monitoring ankle edema," In Pervasive computing and communication workshops (PerCom workshops), 2015 IEEE international conference on, pp. 57–62. IEEE, 2015.

[35]   A. Weiss, S. Sharifi, M. Plotnik, J. P. van Vugt, N. Giladi, and J. M. Hausdorff, "Toward automated, at-home assessment of mobility among patients with Parkinson disease, using a body-worn accelerometer," Neurorehabilitation and Neural Repair, vol. 25, no. 9, pp. 810–818, 2011.

[36]   M. Bahrepour, N. Meratnia, Z. Taghikhaki, and P. J. Havinga, Sensor fusion-based activity recognition for Parkinson patients, InTech 2011, pp. 171–190, 2011.

[37]   J. Goto, T. Kidokoro, T. Ogura, and S. Suzuki, "Activity recognition system for watching over infant children," In RO-MAN, 2013 IEEE, pp. 473–477, IEEE, 2013.

[38]   D. Gravem, M. Singh, C. Chen, J. Rich, J. Vaughan, K. Goldberg, F. Waffarn, P. Chou, D. Cooper, and D. Reinkensmeyer, "Assessment of infant movement with a compact wireless accelerometer system," Journal of Medical Devices, vol. 6, no. 2, pp. 021013, 2012.

[39]   Y. Nam, and J. W. Park, "Child activity recognition based on cooperative fusion model of a triaxial accelerometer and a barometric pressure sensor," IEEE Journal of Biomedical and Health Informatics, vol. 17, no. 2, pp. 420–426, 2013.

[40]   T. Wyss, and U. Mäder, "Recognition of military-specific physical activities with body-fixed sensors," Military Medicine, vol. 175, no. 11, pp. 858–864, 2010.

[41]   D. Minnen, T. Westeyn, D. Ashbrook, P. Presti, and T. Starner, "Recognizing soldier activities in the field," In 4th International workshop on wearable and implantable body sensor networks (BSN 2007), pp. 236–241, Springer Berlin Heidelberg, 2007.

[42]   T. Wyss, L. Roos, M.-C. Hofstetter, F. Frey, and U. Mauder, "Impact of training patterns on injury incidences in 12 Swiss Army basic military training schools," Military Medicine, vol. 179, no. 1, pp. 49–55, 2014.

[43]   T. Wyss, and U. Mäder, "Energy expenditure estimation during daily military routine with body-fixed sensors," Military Medicine, vol. 176, no. 5, pp. 494–499, 2011.

[44]   G. Andreoni, P. Perego, M. C. Fusca, R. Lavezzari, and G. C. Santambrogio, "Smart garments for performance and training assessment in sport," In Wireless Mobile Communication and Healthcare (Mobihealth), 2014 EAI 4th International conference on, pp. 267–270, IEEE, 2014.

[45]   D. W. Wundersitz, C. Josman, R. Gupta, K. J. Netto, P. B. Gastin, and S. Robertson, "Classification of team sport activities using a single wearable tracking device," Journal of Biomechanics, vol. 48, no. 15, pp. 3975–3981, 2015.

[46]   A. Horst, "Sport Coach: Online activity matching using wireless sensor network," Master's thesis, University of Twente, 2010.

[47]   I. WITKOWSKA, "S&D. Integration of sport and design for innovative systems. Application to a swimmer wearable integrated monitoring system for innovative training," PHD thesis, Politecnico di Milano, Italy, 2014.

[48]   L. Minetto, "Machine Learning approach to sport activity recognition from inertial data," Master's thesis, University of Padova, Italy, 2015.

[49]   A. Pentland, "Looking at people: sensing for ubiquitous and wearable computing," IEEE Transactions on Pattern Analysis and Machine Intelligence, vol. 22, no. 1, pp. 107–119, 2000.

[50]   D. M. Gavrila, "The visual analysis of human movement: a survey," Computer Vision and Image Understanding, vol. 73, no. 1, pp. 82–98, 1999.

[51]   X. Hong, C. Nugent, M. Mulvenna, S. McClean, B. Scotney, and S. Devlin, "Evidential fusion of sensor data for activity recognition in smart homes," Pervasive and Mobile Computing, vol. 5, no. 3, pp. 236–252, 2009.

[52]   M. Boyle, C. Edwards, and S. Greenberg, "The effects of filtered video on awareness and privacy," In Proceedings of the 2000 ACM conference on computer supported cooperative work, pp. 1–10, ACM, 2000.

[53]   Zhang C, Tian Y, Capezuti E: Privacy preserving automatic fall detection for elderly using RGBD cameras. In Proceedings of the 13th International Conference on Computers Helping People with Special Needs. Edited by: Miesenberger K, Karshmer A, Penaz P, Zagler W. Linz: Springer-Verlag Berlin; 2012:625–633.

[54]   Liu CL, Lee CH, Lin PM: A fall detection system using k-nearest neighbor classifier. Expert System Application 2010, 37: 7174–7181.

[55]   N. C. Krishnan, C. Juillard, D. Colbry, and S. Panchanathan, "Recognition of hand movements using wearable accelerometers," Journal of Ambient Intelligence and Smart Environments, vol. 1, no. 2, pp. 143–155, 2009.

[56]   J. G. Webster, and H. Eren, Measurement, instrumentation, and sensors handbook: electromagnetic, optical, radiation, chemical, and biomedical measurement, CRC press, 2014.

[57]   H. Bayati, J. d. R. Mill, and R. Chavarriaga, "Unsupervised adaptation to on-body sensor displacement in acceleration-based activity recognition," In Wearable computers (ISWC), 2011 15th Annual International symposium on, pp. 71–78, IEEE, 2011.

[58]   K. Kunze, and P. Lukowicz, "Dealing with sensor displacement in motion-based onbody activity recognition systems," In Proceedings of the 10th international conference on ubiquitous computing, pp. 20–29, ACM, 2008.

[59]   U. Blanke, and B. Schiele, "Sensing location in the pocket," Ubicomp Poster Session, pp. 2, 2008.

[60]   A. Whitehead, H. Johnston, K. Fox, N. Crampton, and J. Tuen, "Homogeneous accelerometer-based sensor networks for game interaction," Computers in entertainment (CIE), vol. 9, no. 1, pp. 1, 2011.

[61] M. Nyan, F. E. Tay, and E. Murugasu, "A wearable system for pre-impact fall detection," Journal of Biomechanics, vol. 41, no. 16, pp. 3475–3481, 2008.

[62] T. Tamura, T. Yoshimura, M. Sekine, M. Uchida, and O. Tanaka, "A wearable airbag to prevent fall injuries," IEEE Transactions on Information Technology in Biomedicine, vol. 13, no. 6, pp. 910–914, 2009.

[63] T. Zhang, J. Wang, L. Xu, and P. Liu, "Fall detection by wearable sensor and one-class SVM algorithm," Intelligent Computing in Signal Processing and Pattern Recognition, pp. 858–863: Springer, 2006.

[64] J. Chen, K. Kwong, D. Chang, J. Luk, and R. Bajcsy, "Wearable sensors for reliable fall detection," In Engineering in medicine and biology society, 2005. IEEE-EMBS 2005. 27th Annual international conference of the, pp. 3551–3554, IEEE, 2006.

[65] Q. Li, J. A. Stankovic, M. A. Hanson, A. T. Barth, J. Lach, and G. Zhou, "Accurate, fast fall detection using gyroscopes and accelerometer-derived posture information," In Wearable and implantable body sensor networks, 2009. BSN 2009. Sixth international workshop on, pp. 138–143, IEEE, 2009.

[66] J. Cheng, O. Amft, and P. Lukowicz, "Active capacitive sensing: exploring a new wearable sensing modality for activity recognition," In International conference on pervasive computing, pp. 319–336. Springer Berlin Heidelberg, 2010.

[67] P. Lukowicz, F. Hanser, C. Szubski, and W. Schobersberger, "Detecting and interpreting muscle activity with wearable force sensors," In International conference on pervasive computing, pp. 101–116, Springer Berlin Heidelberg, 2006.

[68] C. Mattmann, O. Amft, H. Harms, G. Troster, and F. Clemens, "Recognizing upper body postures using textile strain sensors," In Wearable computers, 2007 11th IEEE international symposium on, pp. 29–36, IEEE, 2007.

[69] L. Atallah, B. Lo, R. King, and G.-Z. Yang, "Sensor positioning for activity recognition using wearable accelerometers," IEEE Transactions on Biomedical Circuits and Systems, vol. 5, no. 4, pp. 320–329, 2011.

[70] A. Y. Yang, R. Jafari, S. S. Sastry, and R. Bajcsy, "Distributed recognition of human actions using wearable motion sensor networks," Journal of Ambient Intelligence and Smart Environments, vol. 1, no. 2, pp. 103–115, 2009.

[71] N. Kale, J. Lee, R. Lotfian, and R. Jafari, "Impact of sensor misplacement on dynamic time warping based human activity recognition using wearable computers," In Proceedings of the conference on wireless health, pp. 1–8, ACM, 2012.

[72] A. Mannini, A. M. Sabatini, and S. S. Intille, "Accelerometry-based recognition of the placement sites of a wearable sensor," Pervasive and Mobile Computing, vol. 21, pp. 62–74, 2015.

[73] P. Zappi, C. Lombriser, T. Stiefmeier, E. Farella, D. Roggen, L. Benini, and G. Tröster, "Activity recognition from on-body sensors: accuracy-power trade-off by dynamic sensor selection," Wireless sensor networks, pp. 17–33: Springer, 2008.

[74] B. French, D. P. Siewiorek, A. Smailagic, and M. Deisher, "Selective sampling strategies to conserve power in context aware devices," In Wearable computers, 2007 11th IEEE international symposium on, pp. 77–80, IEEE, 2007.

[75] H. Rezaie, and M. Ghassemian, "Implementation study of wearable sensors for activity recognition systems," Healthcare Technology Letters, vol. 2, no. 4, pp. 95–100, 2015.

[76] A. Jafari, M. Shirali, and M. Ghassemian, "A testbed evaluation of MAC layer protocols for smart home remote monitoring of the elderly mobility pattern," In Proceedings of the International Conference on Medical and Biological Engineering, CMBEBIH 2017, Springer, 2017, pp. 568–575.

[77] H. Rezaie and M. Ghassemian, "An Adaptive Algorithm to Improve Energy Efficiency in Wearable Activity Recognition Systems," IEEE Sensors Journal, 2017, doi: 10.1109/JSEN.2017.2720725.

[78] A. Avci, S. Bosch, M. Marin-Perianu, R. Marin-Perianu, and P. Havinga, "Activity recognition using inertial sensing for healthcare, wellbeing and sports applications: a survey," In Architecture of computing systems (ARCS), 2010 23rd International conference on, pp. 1–10, VDE, 2010.

[79] A. Bulling, J. A. Ward, and H. Gellersen, "Multimodal recognition of reading activity in transit using body-worn sensors," ACM transactions on applied perception (TAP), vol. 9, no. 1, pp. 2, 2012.

[80] D. Roggen, K. Forster, A. Calatroni, *et al.*, "OPPORTUNITY: towards opportunistic activity and context recognition systems," In World of wireless, mobile and multimedia networks & workshops, 2009. WoWMoM 2009. IEEE international symposium on a, pp. 1–6. IEEE, 2009.

[81] U. Blanke, and B. Schiele, "Remember and transfer what you have learned-recognizing composite activities based on activity spotting," In Wearable computers (ISWC), 2010 International symposium on, pp. 1–8, IEEE, 2010.

[82] K. Van Laerhoven, D. Kilian, and B. Schiele, "Using rhythm awareness in long-term activity recognition," In Wearable computers, 2008. ISWC 2008. 12th IEEE international symposium on, pp. 63–66, IEEE, 2008.

[83] A. Kapoor, and E. Horvitz, "Experience sampling for building predictive user models: a comparative study," In Proceedings of the SIGCHI conference on human factors in computing systems, pp. 657–666, ACM, 2008.

[84] M. Stikic, D. Larlus, S. Ebert, and B. Schiele, "Weakly supervised recognition of daily life activities with wearable sensors," IEEE Transactions on Pattern Analysis and Machine Intelligence, vol. 33, no. 12, pp. 2521–2537, 2011.

[85] T. Huynh, M. Fritz, and B. Schiele, "Discovery of activity patterns using topic models," In Proceedings of the 10th international conference on ubiquitous computing, pp. 10–19, ACM, 2008.

[86] V. W. Zheng, D. H. Hu, and Q. Yang, "Cross-domain activity recognition," In Proceedings of the 11th international conference on ubiquitous computing, pp. 61–70, ACM, 2009.

[87]   T. Van Kasteren, G. Englebienne, and B. J. Kröse, "Transferring knowledge of activity recognition across sensor networks," In International conference on pervasive computing, pp. 283–300. Springer Berlin Heidelberg, 2010.

[88]   A. Bulling, U. Blanke, and B. Schiele, "A tutorial on human activity recognition using body-worn inertial sensors," ACM Computing Surveys (CSUR), vol. 46, no. 3, pp. 33, 2014.

[89]   H. Ali, E. Messina, and R. Bisiani, "Subject-Dependent Physical Activity Recognition Model Framework with a Semi-Supervised Clustering Approach," In Modelling symposium (EMS), 2013 European, pp. 42–47. IEEE, 2013.

[90]   S. Sadiq, M. Orlowska, W. Sadiq, and C. Foulger, "Data flow and validation in workflow modelling," In Proceedings of the 15th Australasian database conference, pp. 207–214, Australian Computer Society, Inc., 2004.

[91]   Fernandez-Llatas, Carlos, Aroa Lizondo, Eduardo Monton, Jose-Miguel Benedi, and Vicente Traver. "Process mining methodology for health process tracking using real-time indoor location systems." Sensors 15, no. 12 (2015): 29821–29840.

[92]   O. D. Lara, A. J. Pérez, M. A. Labrador, and J. D. Posada, "Centinela: a human activity recognition system based on acceleration and vital sign data," Pervasive and mobile computing, vol. 8, no. 5, pp. 717–729, 2012.

[93]   D. Cook, K. D. Feuz, and N. C. Krishnan, "Transfer learning for activity recognition: a survey," Knowledge and Information Systems, vol. 36, no. 3, pp. 537–556, 2013.

[94]   F. Foerster, M. Smeja, and J. Fahrenberg, "Detection of posture and motion by accelerometry: a validation study in ambulatory monitoring," Computers in Human Behavior, vol. 15, no. 5, pp. 571–583, 1999.

[95]   L. Atallah, B. P. Lo, R. C. King, and G.-Z. Yang, "Sensor Placement for Activity Detection Using Wearable Accelerometers," In Body sensor networks (BSN), 2010 International conference on, pp. 24–29. IEEE, 2010.

[96]   B. Liu, X. Wang, B. Liu, Q. Wang, D. Tan, W. Song, X. Hou, D. Chen, and G. Shen, "Advanced rechargeable lithium-ion batteries based on bendable $ZnCo_2O_4$-urchins-on-carbon-fibers electrodes," Nano Research, vol. 6, no. 7, pp. 525–534, 2013.

[97]   N. S. Shenck, and J. A. Paradiso, "Energy scavenging with shoe-mounted piezoelectrics," IEEE Micro, vol. 21, no. 3, pp. 30–42, 2001.

[98]   V. Leonov, "Thermoelectric energy harvesting of human body heat for wearable sensors," IEEE Sensors Journal, vol. 13, no. 6, pp. 2284–2291, 2013.

[99]   C. Lombriser, N. B. Bharatula, D. Roggen, and G. Tröster, "On-body activity recognition in a dynamic sensor network," In Proceedings of the ICST 2nd international conference on Body area networks, pp. 17–33, 2007.

[100] A. A. Abidi, G. J. Pottie, and W. J. Kaiser, "Power-conscious design of wireless circuits and systems," Proceedings of the IEEE, vol. 88, no. 10, pp. 1528–1545, 2000.

[101] P. Moinzadeh, K. A. Mechitov, R. Shiftehfar, T. F. Abdelzaher, G. A. Agha, and B. F. Spencer Jr, Dynamic Voltage Scaling Techniques for Energy

Efficient Synchronized Sensor Network Design, Technical report, University of Illinois at Urbana Champaign, 2011.

[102] C. Alippi, G. Anastasi, M. Di Francesco, and M. Roveri, "Energy management in wireless sensor networks with energy-hungry sensors," IEEE Instrumentation & Measurement Magazine, vol. 12, no. 2, pp. 16–23, 2009.

[103] M. C. Vuran, Ö. B. Akan, and I. F. Akyildiz, "Spatio-temporal correlation: theory and applications for wireless sensor networks," Computer Networks, vol. 45, no. 3, pp. 245–259, 2004.

[104] Z. Yan, V. Subbaraju, D. Chakraborty, A. Misra, and K. Aberer, "Energy-efficient continuous activity recognition on mobile phones: an activity-adaptive approach," In Wearable computers (ISWC), 2012 16th International symposium on, pp. 17–24, IEEE, 2012.

[105] X. Qi, M. Keally, G. Zhou, Y. Li, and Z. Ren, "AdaSense: adapting sampling rates for activity recognition in Body Sensor Networks," In Real-time and embedded technology and applications symposium (RTAS), 2013 IEEE 19th, pp. 163–172. IEEE, 2013.

[106] S. Bosch, R. Marin-Perianu, P. Havinga, and M. Marin-Perianu, "Energy-Efficient Assessment of Physical Activity Level Using Duty-Cycled Accelerometer Data," Procedia Computer Science, vol. 5, pp. 328–335, 2011.

[107] H. Ghasemzadeh, N. Amini, R. Saeedi, and M. Sarrafzadeh, "Power-aware computing in wearable sensor networks: an optimal feature selection," IEEE Transactions on Mobile Computing, vol. 14, no. 4, pp. 800–812, 2015.

[108] D. Gordon, J. Czerny, T. Miyaki, and M. Beigl, "Energy-efficient activity recognition using prediction," In Real-time and embedded technology and applications symposium (RTAS), 2013 IEEE 19th, pp. 163–172, IEEE, 2013.

[109] Y. Wang, J. Lin, M. Annavaram, Q. A. Jacobson, J. Hong, B. Krishnamachari, and N. Sadeh, "A framework of energy efficient mobile sensing for automatic user state recognition," In Proceedings of the 7th international conference on Mobile systems, applications, and services, pp. 179–192, ACM, 2009.

[110] O. Baños, M. Damas, H. Pomares, and I. Rojas, "Activity recognition based on a multi-sensor meta-classifier," In International work-conference on artificial neural networks, pp. 208–215. Springer Berlin Heidelberg, 2013.

[111] O. Banos, M. Damas, H. Pomares, and I. Rojas, "On the use of sensor fusion to reduce the impact of rotational and additive noise in human activity recognition," Sensors, vol. 12, no. 6, pp. 8039–8054, 2012.

[112] H. Ghasemzadeh, V. Loseu, and R. Jafari, "Structural action recognition in body sensor networks: distributed classification based on string matching," IEEE Transactions on Information Technology in Biomedicine, vol. 14, no. 2, pp. 425–435, 2010.

[113] C. M. Brigante, N. Abbate, A. Basile, A. C. Faulisi, and S. Sessa, "Towards miniaturization of a MEMS-based wearable motion capture system," IEEE Transactions on Industrial Electronics, vol. 58, no. 8, pp. 3234–3241, 2011.

[114] Y. Chuo, M. Marzencki, B. Hung, C. Jaggernauth, K. Tavakolian, P. Lin, and B. Kaminska, "Mechanically flexible wireless multisensor platform

for human physical activity and vitals monitoring," IEEE transactions on biomedical circuits and systems, vol. 4, no. 5, pp. 281–294, 2010.

[115]　E. M. Tapia, S. S. Intille, W. Haskell, K. Larson, J. Wright, A. King, and R. Friedman, "Real-time recognition of physical activities and their intensities using wireless accelerometers and a heart rate monitor," In Wearable computers, 2007 11th IEEE international symposium on, pp. 37–40, IEEE, 2007.

[116]　L. Bao, and S. S. Intille, "Activity recognition from user-annotated acceleration data," In International Conference on Pervasive Computing, pp. 1–17, Springer Berlin Heidelberg, 2004.

[117]　J. Parkka, M. Ermes, P. Korpipaa, J. Mantyjarvi, J. Peltola, and I. Korhonen, "Activity classification using realistic data from wearable sensors," IEEE Transactions on Information Technology in Biomedicine, vol. 10, no. 1, pp. 119–128, 2006.

[118]　U. Maurer, A. Smailagic, D. P. Siewiorek, and M. Deisher, "Activity recognition and monitoring using multiple sensors on different body positions," In Wearable and implantable body sensor networks, 2006. BSN 2006. International Workshop on, pp. 99–102, IEEE, 2006.

[119]　M. Berchtold, M. Budde, D. Gordon, H. R. Schmidtke, and M. Beigl, "Actiserv: activity recognition service for mobile phones," In Wearable computers (ISWC), 2010 International symposium on, pp. 1–8, IEEE, 2010.

[120]　S. Reddy, M. Mun, J. Burke, D. Estrin, M. Hansen, and M. Srivastava, "Using mobile phones to determine transportation modes," ACM Transactions on Sensor Networks (TOSN), vol. 6, no. 2, pp. 13, 2010.

[121]　K. Van Laerhoven, and E. Berlin, "When else did this happen? Efficient subsequence representation and matching for wearable activity data," In Wearable computers, 2009. ISWC'09. International symposium on, pp. 101–104, IEEE, 2009.

[122]　H. Ghasemzadeh, E. Guenterberg, and R. Jafari, "Energy-efficient information-driven coverage for physical movement monitoring in body sensor networks," IEEE Journal on Selected Areas in Communications, vol. 27, no. 1, pp. 58–69, 2009.

[123]　H. Lu, J. Yang, Z. Liu, N. D. Lane, T. Choudhury, and A. T. Campbell, "The Jigsaw continuous sensing engine for mobile phone applications," In Proceedings of the 8th ACM conference on embedded networked sensor systems, pp. 71–84. ACM, 2010.

[124]　R. Fallahzadeh, J. P. Ortiz, and H. Ghasemzadeh, "Adaptive compressed sensing at the fingertip of Internet-of-Things sensors: an ultra-low power activity recognition," In 2017 Design, Automation & Test in Europe Conference & Exhibition (DATE), 2017, pp. 996–1001.

[125]　D. Figo, P. C. Diniz, D. R. Ferreira, and J. M. Cardoso, "Preprocessing techniques for context recognition from accelerometer data," Personal and Ubiquitous Computing, vol. 14, no. 7, pp. 645–662, 2010.

[126]　A. Bulling, J. A. Ward, H. Gellersen, and G. Troster, "Eye movement analysis for activity recognition using electrooculography," IEEE Transactions on Pattern Analysis and Machine Intelligence, vol. 33, no. 4, pp. 741–753, 2011.

[127] T. Huynh, and B. Schiele, "Analyzing features for activity recognition," In Proceedings of the 2005 joint conference on Smart objects and ambient intelligence: innovative context-aware services: usages and technologies, pp. 159–163, ACM, 2005.

[128] E. Keogh, S. Chu, D. Hart, and M. Pazzani, "An online algorithm for segmenting time series," In Data mining, 2001. ICDM 2001, Proceedings IEEE international conference on, pp. 289–296, IEEE, 2001.

[129] H. Shatkay, and S. B. Zdonik, "Approximate queries and representations for large data sequences," In Data engineering, 1996. Proceedings of the twelfth international conference on, pp. 536–545, IEEE, 1996.

[130] M. R. Maurya, R. Rengaswamy, and V. Venkatasubramanian, "Fault diagnosis using dynamic trend analysis: a review and recent developments," Engineering Applications of Artificial Intelligence, vol. 20, no. 2, pp. 133–146, 2007.

[131] E. Guenterberg, S. Ostadabbas, H. Ghasemzadeh, and R. Jafari, "An automatic segmentation technique in body sensor networks based on signal energy," In Proceedings of the fourth international conference on body area networks, pp. 21–27, 2009.

[132] C. Lee, and Y. Xu, "Online, interactive learning of gestures for human/robot interfaces," In Robotics and automation, 1996. Proceedings, 1996 IEEE international conference on, vol. 4, pp. 2982–2987, IEEE, 1996.

[133] A. D. Wilson, and A. F. Bobick, "Realtime online adaptive gesture recognition," In Pattern recognition, 2000. Proceedings. 15th International conference on, vol. 1, pp. 270–275, IEEE, 2000.

[134] A. Zinnen, C. Wojek, and B. Schiele, "Multi activity recognition based on body model-derived primitives," In International symposium on location-and context-awareness, pp. 1–18, Springer Berlin Heidelberg, 2009.

[135] D. Ashbrook, and T. Starner, "Using GPS to learn significant locations and predict movement across multiple users," Personal and Ubiquitous Computing, vol. 7, no. 5, pp. 275–286, 2003.

[136] H. Lu, W. Pan, N. D. Lane, T. Choudhury, and A. T. Campbell, "SoundSense: scalable sound sensing for people-centric applications on mobile phones," In Proceedings of the 7th International conference on mobile systems, applications, and services, pp. 165–178, ACM, 2009.

[137] R. O. Duda, and P. E. Hart, Pattern Classification and Scene Analysis. Wiley, New York, 1973.

[138] S. J. Preece, J. Y. Goulermas, L. P. Kenney, and D. Howard, "A comparison of feature extraction methods for the classification of dynamic activities from accelerometer data," IEEE Transactions on Biomedical Engineering, vol. 56, no. 3, pp. 871–879, 2009.

[139] A. Zinnen, U. Blanke, and B. Schiele, "An analysis of sensor-oriented vs. model-based activity recognition," In Wearable computers, 2009. ISWC'09. International symposium on, pp. 93–100. IEEE, 2009.

[140] J. Yin, Q. Yang, and J. J. Pan, "Sensor-based abnormal human-activity detection," IEEE Transactions on Knowledge and Data Engineering, vol. 20, no. 8, pp. 1082–1090, 2008.

[141]   A. S. A. Nisha, and S. Dhatchayini, "Feature extraction and wearable sensors based patient monitoring system," International Journal of Advanced Research in Electrical, Electronics and Instrumentation Engineering, vol. 3, no. 4, pp. 8498–8508, 2014.

[142]   U. Jensen, H. Leutheuser, S. Hofmann, *et al.*, "A wearable real-time activity tracker," Biomedical Engineering Letters, vol. 5, no. 2, pp. 147–157, 2015.

[143]   I. Cleland, B. Kikhia, C. Nugent, *et al.*, "Optimal placement of accelerometers for the detection of everyday activities," Sensors, vol. 13, no. 7, pp. 9183–9200, 2013.

[144]   K. Altun, B. Barshan, and O. Tunçel, "Comparative study on classifying human activities with miniature inertial and magnetic sensors," Pattern Recognition, vol. 43, no. 10, pp. 3605–3620, 2010.

[145]   J. Farringdon, A. J. Moore, N. Tilbury, J. Church, and P. D. Biemond, "Wearable sensor badge and sensor jacket for context awareness," In Wearable computers, 1999. Digest of papers. The third international symposium on, pp. 107–113, IEEE, 1999.

[146]   C. V. Bouten, K. T. Koekkoek, M. Verduin, R. Kodde, and J. D. Janssen, "A triaxial accelerometer and portable data processing unit for the assessment of daily physical activity," IEEE Transactions on Biomedical Engineering, vol. 44, no. 3, pp. 136–147, 1997.

[147]   F. Attal, S. Mohammed, M. Dedabrishvili, F. Chamroukhi, L. Oukhellou, and Y. Amirat, "Physical human activity recognition using wearable sensors," Sensors, vol. 15, no. 12, pp. 31314–31338, 2015.

[148]   A. V. Oppenheim, and R. W. Schafer, Discrete-time signal processing. Pearson Higher Education, 2010.

[149]   B. Nham, K. Siangliulue, and S. Yeung, "Predicting mode of transport from iphone accelerometer data," Machine Learning Final Projects, Stanford University, 2008.

[150]   J. Ho, "Interruptions: using activity transitions to trigger proactive messages," PhD Thesis, Massachusetts Institute of Technology, 2004.

[151]   S. Mallat, and W. L. Hwang, "Singularity detection and processing with wavelets," IEEE Transactions on Information Theory, vol. 38, no. 2, pp. 617–643, 1992.

[152]   S. G. Mallat, "A theory for multi-resolution signal decomposition: the wavelet representation," IEEE Transactions on Pattern Analysis and Machine Intelligence, vol. 11, no. 7, pp. 674–693, Jul. 1989.

[153]   N. Wang, E. Ambikairajah, N.H. Lovell, and B.G. Celler, "Accelerometry-based classification of walking patterns using time-frequency analysis," in Proc. 29th Annu. Conf. IEEE Eng. Med. Biol. Soc., Lyon, France, 2007, pp. 4899–4902.

[154]   T. Tamura, M. Sekine, M. Ogawa, T. Togawa, and Y. Fukui, "Classification of acceleration waveforms during walking by wavelet transform," Methods in Informative Medicine, vol. 36, pp. 356–359, 1997.

[155]   M. Sekine, T. Tamura, T. Togawa, and Y. Fukui, "Classification of waist acceleration signals in a continuous walking record," Medical Engineering Physics, vol. 22, pp. 285–291, 2000.

[156]   M. H. Law, M. A. Figueiredo, and A. K. Jain, "Simultaneous feature selection and clustering using mixture models," IEEE Transactions on Pattern Analysis and Machine Intelligence, vol. 26, no. 9, pp. 1154–1166, 2004.

[157]   R. Kohavi, and G. H. John, "Wrappers for feature subset selection," Artificial Intelligence, vol. 97, no. 1, pp. 273–324, 1997.

[158]   I. Guyon, and A. Elisseeff, "An introduction to variable and feature selection," Journal of Machine Learning Research, vol. 3, no. Mar, pp. 1157–1182, 2003.

[159]   P. Langley, "Selection of relevant features in machine learning," In Proceedings of the AAAI fall symposium on relevance, vol. 184, pp. 245–271, 1994.

[160]   A. L. Blum, and P. Langley, "Selection of relevant features and examples in machine learning," Artificial Intelligence, vol. 97, no. 1, pp. 245–271, 1997.

[161]   G. H. John, R. Kohavi, and K. Pfleger, "Irrelevant features and the subset selection problem," In Machine learning: proceedings of the eleventh international conference, pp. 121–129, 1994.

[162]   O. Banos, M. Damas, H. Pomares, and I. Rojas, "Novel method for feature-set ranking applied to physical activity recognition," International Conference on Industrial, Engineering and Other Applications of Applied Intelligent Systems, pp. 637–642, 2010.

[163]   S. Kullback, and R. A. Leibler, "On information and sufficiency," The Annals of Mathematical Statistics, vol. 22, no. 1, pp. 79–86, 1951.

[164]   S. M. Davis, D. A. Landgrebe, T. L. Phillips, P. H. Swain, R. M. Hoffer, J. C. Lindenlaub, and L. F. Silva, "Remote sensing: the quantitative approach," McGraw-Hill International Book Co., 1978. 405 p., vol. 1, 1978.

[165]   T. M. Cover, and J. A. Thomas, Elements of information theory. John Wiley & Sons, 2012.

[166]   J. Rissanen, "Modeling by shortest data description," Automatica, vol. 14, no. 5, pp. 465–471, 1978.

[167]   J. R. Quinlan, C4.5: programs for machine learning. Elsevier, 2014.

[168]   L. Breiman, J. Friedman, C. J. Stone, and R. A. Olshen, Classification and regression trees. CRC press, 1984.

[169]   K. Kira, and L. A. Rendell, "The feature selection problem: traditional methods and a new algorithm," In AAAI, vol. 2, pp. 129–134. 1992.

[170]   A. Y. Ng, "On feature selection: learning with exponentially many irrelevant features as training examples," In Proceedings of the Fifteenth International Conference on Machine Learning, pp. 404–412, 1998.

[171]   G. Chandrashekar and F. Sahin, "A survey on feature selection methods," Computers & Electrical Engineering, vol. 40 no. 1, pp. 16–28, 2014.

[172]   B. Efron, T. Hastie, I. Johnstone, and R. Tibshirani, "Least angle regression," The Annals of Statistics, vol. 32, no. 2, pp. 407–499, 2004.

[173]   I. Guyon, J. Weston, S. Barnhill, and V. Vapnik, "Gene selection for cancer classification using support vector machines," Machine Learning, vol. 46, no. 1–3, pp. 389–422, 2002.

[174] J. Weston, S. Mukherjee, O. Chapelle, M. Pontil, T. Poggio, and V. Vapnik, "Feature selection for SVMs," In Advances in Neural Information Processing Systems, pp. 668–674, 2001.

[175] B. Krishnapuram, L. Carin, and A. Hartemink, "1 Gene expression analysis: joint feature selection and classifier design," Kernel Methods in Computational Biology, pp. 299–317, 2004.

[176] B. Najafi, K. Aminian, A. Paraschiv-Ionescu, F. Loew, C. J. Bula, and P. Robert, "Ambulatory system for human motion analysis using a kinematic sensor: monitoring of daily physical activity in the elderly," IEEE Transactions on Biomedical Engineering, vol. 50, no. 6, pp. 711–723, 2003.

[177] B. Najafi, K. Aminian, F. Loew, Y. Blanc, and P. A. Robert, "Measurement of stand-sit and sit-stand transitions using a miniature gyroscope and its application in fall risk evaluation in the elderly," IEEE Transactions on Biomedical Engineering, vol. 49, no. 8, pp. 843–851, 2002.

[178] J. Boyle, M. Karunanithi, T. Wark, W. Chan, and C. Colavitti, "Quantifying functional mobility progress for chronic disease management," In Engineering in medicine and biology society, 2006. EMBS'06. 28th Annual international conference of the IEEE, pp. 5916–5919. IEEE, 2006.

[179] K. Culhane, G. Lyons, D. Hilton, P. Grace, and D. Lyons, "Long-term mobility monitoring of older adults using accelerometers in a clinical environment," Clinical Rehabilitation, vol. 18, no. 3, pp. 335–343, 2004.

[180] M. Uiterwaal, E. Glerum, H. Busser, and R. Van Lummel, "Ambulatory monitoring of physical activity in working situations, a validation study," Journal of Medical Engineering & Technology, vol. 22, no. 4, pp. 168–172, 1998.

[181] H. Busser, J. Ott, R. Van Lummel, M. Uiterwaal, and R. Blank, "Ambulatory monitoring of children's activity," Medical Engineering & Physics, vol. 19, no. 5, pp. 440–445, 1997.

[182] J. Bussmann, Y. Van de Laar, M. Neeleman, and H. Stam, "Ambulatory accelerometry to quantify motor behaviour in patients after failed back surgery: a validation study," Pain, vol. 74, no. 2, pp. 153–161, 1998.

[183] M. Makikawa, and H. Iizumi, "Development of an ambulatory physical activity memory device and its application for the categorization of actions in daily life," Medinfo, vol. 8, pp. 747–750, 1994.

[184] K. Aminian, P. Robert, E. Buchser, B. Rutschmann, D. Hayoz, and M. Depairon, "Physical activity monitoring based on accelerometry: validation and comparison with video observation," Medical and Biological Engineering and Computing, vol. 37, no. 3, pp. 304–308, 1999.

[185] M. Mathie, A. Coster, N. Lovell, and B. Celler, "Detection of daily physical activities using a triaxial accelerometer," Medical and Biological Engineering and Computing, vol. 41, no. 3, pp. 296–301, 2003.

[186] D. Maxwell, "Addressing the challenge of quantifying free-living activity – the activPALTM professional," In Proceedings recent advances in assistive technology and engineering (RAATE) 2002 conference, pp. 18–19, 2002.

[187] P. H. Veltink, H. J. Bussmann, W. De Vries, W. J. Martens, and R. C. Van Lummel, "Detection of static and dynamic activities using uniaxial

accelerometers," IEEE Transactions on Rehabilitation Engineering, vol. 4, no. 4, pp. 375–385, 1996.

[188]   M. Sekine, T. Tamura, T. Togawa, and Y. Fukui, "Classification of waist-acceleration signals in a continuous walking record," Medical Engineering & Physics, vol. 22, no. 4, pp. 285–291, 2000.

[189]   M. Sekine, T. Tamura, T. Fujimoto, and Y. Fukui, "Classification of walking pattern using acceleration waveform in elderly people," In Engineering in medicine and biology society, 2000. Proceedings of the 22nd Annual international conference of the IEEE, vol. 2, pp. 1356–1359. IEEE, 2000.

[190]   M. Nyan, F. Tay, K. Seah, and Y. Sitoh, "Classification of gait patterns in the time–frequency domain," Journal of Biomechanics, vol. 39, no. 14, pp. 2647–2656, 2006.

[191]   M. Stikic, D. Larlus, and B. Schiele, "Multi-graph based semi-supervised learning for activity recognition," In Wearable computers, 2009. ISWC'09. International symposium on, pp. 85–92, IEEE, 2009.

[192]   A. Ali, R. C. King, and G.-Z. Yang, "Semi-supervised segmentation for activity recognition with multiple Eigen spaces," In Medical devices and biosensors, 2008. ISSS-MDBS 2008. 5th International summer school and symposium on, pp. 314–317, IEEE, 2008.

[193]   D. Guan, W. Yuan, Y.-K. Lee, A. Gavrilov, and S. Lee, "Activity recognition based on semi-supervised learning," In Embedded and real-time computing systems and applications, 2007. RTCSA 2007. 13th IEEE international conference on, pp. 469–475, IEEE, 2007.

[194]   S. Theodoridis, and K. Koutroumbas, Pattern Recognition, 4th edition. Boston, MA: Academic Press, 2009.

[195]   A. R. Webb, Statistical pattern recognition. West Sussex: John Wiley & Sons, 2003.

[196]   R. O. Duda, P. E. Hart, and D. G. Stork, Pattern classification. New York: John Wiley & Sons, 2012.

[197]   M. Ermes, J. Pärkkä, J. Mäntyjärvi, and I. Korhonen, "Detection of daily activities and sports with wearable sensors in controlled and uncontrolled conditions," IEEE Transactions on Information Technology in Biomedicine, vol. 12, no. 1, pp. 20–26, 2008.

[198]   P. Antal, "Construction of a classifier with prior domain knowledge formalised as Bayesian network," In Industrial electronics society, 1998. IECON'98. Proceedings of the 24th Annual conference of the IEEE, vol. 4, pp. 2527–2531, IEEE, 1998.

[199]   H. Zhang, "The optimality of naive Bayes," AA, vol. 1, no. 2, pp. 3, 2004.

[200]   C. Cortes, and V. Vapnik, "Support-vector networks," Machine Learning, vol. 20, no. 3, pp. 273–297, 1995.

[201]   S. I. Gallant, "Perceptron-based learning algorithms," IEEE Transactions on Neural Networks, vol. 1, no. 2, pp. 179–191, 1990.

[202]   T. Huynh, and B. Schiele, "Towards less supervision in activity recognition from wearable sensors," In Wearable computers, 2006 10th IEEE international symposium on, pp. 3–10, IEEE, 2006.

[203]   A. Blum, and T. Mitchell, "Combining labeled and unlabeled data with co-training," In Proceedings of the eleventh annual conference on computational learning theory, pp. 92–100, ACM, 1998.

[204]   K. Van Laerhoven, and O. Cakmakci, "What shall we teach our pants?" In Wearable computers, the fourth international symposium on, pp. 77–83, IEEE, 2000.

[205]   S. M. Guthikonda, "Kohonen Self-Organizing Maps," Weittenberg University [Online], 2005. Available: http://www.shy.am/wpcontent/uploads/2009/01/kohonen-self-organizing-maps-shyamguthikonda.pdf

[206]   A. Nguyen, D. Moore, and I. McCowan, "Unsupervised clustering of free-living human activities using ambulatory accelerometry," In Engineering in medicine and biology society, 2007. EMBS 2007. 29th Annual international conference of the IEEE, pp. 4895–4898, IEEE, 2007.

[207]   R. Raina, A. Y. Ng, and D. Koller, "Constructing informative priors using transfer learning," In Proceedings of the 23rd international conference on Machine learning, pp. 713–720, ACM, 2006.

[208]   R. Caruana, "Multitask learning," Learning to learn, vol. 28, no. 1, pp. 95–133, 1998.

[209]   Z. Zhao, Y. Chen, J. Liu, Z. Shen, and M. Liu, "Cross-people mobile-phone based activity recognition," In Twenty-second international joint conference on artificial intelligence, pp. 2545–2550, 2011.

[210]   D. H. Hu, and Q. Yang, "Transfer learning for activity recognition via sensor mapping," In IJCAI proceeding of international joint conference on artificial intelligence, vol. 22, no. 3, pp. 1962–1967, 2011.

[211]   P. Rashidi, and D. J. Cook, "Activity recognition based on home to home transfer learning," In AAAI'10: Proceedings of the 24th conference on artificial intelligence, pp. 45–52, AAAI press, 2010.

[212]   U. Bagct, and L. Bai, "A comparison of Daubechies and Gabor wavelets for classification of MR images," In Signal processing and communications, 2007. ICSPC 2007. IEEE international conference on, pp. 676–679, IEEE, 2007.

[213]   Y. Tang, R. Salakhutdinov, and G. Hinton, "Robust Boltzmann machines for recognition and denoising," In Computer vision and pattern recognition (CVPR), 2012 IEEE conference on, pp. 2264–2271, IEEE, 2012.

[214]   J. Yang, M. N. Nguyen, P. P. San, X. Li, and S. Krishnaswamy, "Deep Convolutional Neural Networks on Multichannel Time Series for Human Activity Recognition," In twenty-fourth international joint conference on artificial intelligence, pp. 3995–4001, 2015.

[215]   T. Plötz, N. Y. Hammerla, and P. Olivier, "Feature learning for activity recognition in ubiquitous computing," In twenty-fourth international joint conference on artificial intelligence, pp. 1729–1734, 2011.

[216]   M. Zeng, L. T. Nguyen, B. Yu, O. J. Mengshoel, J. Zhu, P. Wu, and J. Zhang, "Convolutional neural networks for human activity recognition using mobile sensors," In mobile computing, applications and services (MobiCASE), 2014 6th International conference on, pp. 197–205, IEEE, 2014.

[217] T. L. van Kasteren, H. Alemdar, and C. Ersoy, "Effective performance metrics for evaluating activity recognition methods," In Proceedings of the 23rd international conference on architecture of computing systems (ARCS), 2010, Italy, 2011.

[218] T. Van Kasteren, A. Noulas, G. Englebienne, and B. Kröse, "Accurate activity recognition in a home setting," In Proceedings of the 10th international conference on ubiquitous computing, pp. 1–9, ACM, 2008.

[219] T. L. van Kasteren, G. Englebienne, and B. J. Kröse, "Human activity recognition from wireless sensor network data: benchmark and software," Activity recognition in pervasive intelligent environments, pp. 165–186: Springer, 2011.

[220] D. Minnen, T. Westeyn, T. Starner, J. Ward, and P. Lukowicz, "Performance metrics and evaluation issues for continuous activity recognition," Performance Metrics for Intelligent Systems, vol. 4, no. 1, pp. 4–11, 2006.

[221] G. P. Fettweis, "The tactile internet: applications and challenges," IEEE Vehicular Technology Magazine, vol. 9, no. 1, pp. 64–70, 2014.

[222] C. Gomez, J. Oller, and J. Paradells, "Overview and evaluation of bluetooth low energy: an emerging low-power wireless technology," Sensors, vol. 12, no. 9, pp. 11734–11753, 2012.

[223] "IEEE Std 802.15.6-2012, IEEE standard for local and metropolitan area networks – Part 15.6: Wireless Body Area Networks," February, 2012.

[224] IEEE Computer Society, LAN/MAN Standards Committee, Institute of Electrical and Electronics Engineers, and IEEE-SA Standards Board. IEEE Standard for Local and Metropolitan Area Networks. Part 15.4, Amendment 4, Part 15.4, Amendment 4, 2013. http://ieeexplore.ieee.org/servlet/opac?punumber=6471720.

[225] S. Ullah, M. Mohaisen, and M. A. Alnuem, "A review of IEEE 802.15.6 MAC, PHY, and security specifications," International Journal of Distributed Sensor Networks, vol. 9, no. 4, pp. 950704, 2013.

[226] M. Chen, S. Gonzalez, A. Vasilakos, H. Cao, and V. C. Leung, "Body area networks: a survey," Mobile Networks and Applications, vol. 16, no. 2, pp. 171–193, 2011.

[227] I. W. Group, "IEEE Standard for Local and Metropolitan Area Networks – Part 15.4: Low-Rate Wireless Personal Area Networks (LR-WPANs)," IEEE Std, vol. 802, pp. 4–2011, 2011.

[228] Specification, Bluetooth. "Version 1.1," See http://www.opensearch.org/Specifications/OpenSearch/1.1 (2001).

[229] T. Adame, A. Bel, B. Bellalta, J. Barcelo, and M. Oliver, "IEEE 802.11 AH: the WiFi approach for M2M communications," IEEE Wireless Communications, vol. 21, no. 6, pp. 144–152, 2014.

[230] ZigBee, Alliance. "Zigbee specification," ZigBee document 053474r13 (2006).

[231] S. Katz, T. D. Downs, H. R. Cash, and R. C. Grotz, "Progress in development of the index of ADL," The Gerontologist, vol. 10, no. 1, pp. 20–30, 1970.

[232]   B. E. Ainsworth, W. L. Haskell, S. D. Herrmann, *et al.*, "2011 Compendium of Physical Activities: a second update of codes and MET values," Medicine and Science in Sports and Exercise, vol. 43, no. 8, pp. 1575–1581, 2011.

[233]   U. Blanke, B. Schiele, M. Kreil, P. Lukowicz, B. Sick, and T. Gruber, "All for one or one for all? Combining heterogeneous features for activity spotting," In Pervasive computing and communications workshops (PERCOM Workshops), 2010 8th IEEE international conference on, pp. 18–24, IEEE, 2010.

[234]   A. Bulling, C. Weichel, and H. Gellersen, "Eyecontext: recognition of high-level contextual cues from human visual behaviour," In Proceedings of the SIGCHI conference on human factors in computing systems, pp. 305–308, ACM, 2013.

[235]   "2011 Activity Recognition Challenge," http://www.oportunity-project. eu/challenge.

[236]   "Datasets for Human Activity Recognition from Tim Van Kasteren's"; https://sites.google.com/site/tim0306/datasets.

[237]   R. Helaoui, M. Niepert, and H. Stuckenschmidt, "Recognizing interleaved and concurrent activities: A statistical-relational approach," In Pervasive computing and communications (PerCom), 2011 IEEE international conference on, pp. 1–9, IEEE, 2011.

[238]   N. D. Lane, Y. Xu, H. Lu, A. T. Campbell, T. Choudhury, and S. B. Eisenman, "Exploiting social networks for large-scale human behavior modeling," IEEE Pervasive Computing, vol. 10, no. 4, pp. 45–53, 2011.

[239]   S. Lee, H. X. Le, H. Q. Ngo, H. I. Kim, M. Han, and Y.-K. Lee, "Semi-Markov conditional random fields for accelerometer-based activity recognition," Applied Intelligence, vol. 35, no. 2, pp. 226–241, 2011.

[240]   K. Kunze, P. Lukowicz, H. Junker, and G. Tröster, "Where am i: recognizing on-body positions of wearable sensors," In International symposium on location-and context-awareness, pp. 264–275, Springer Berlin Heidelberg, 2005.

[241]   A. Maskooki, C. B. Soh, E. Gunawan, and K. S. Low, "Opportunistic routing for body area network," In Consumer communications and networking conference (CCNC), 2011 IEEE, pp. 237–241, IEEE, 2011.

[242]   S. Ullah, H. Higgins, B. Braem, *et al.*, "A comprehensive survey of wireless body area networks," Journal of Medical Systems, vol. 36, no. 3, pp. 1065–1094, 2012.

[243]   B. Braem, B. Latre, I. Moerman, C. Blondia, and P. Demeester, "The wireless autonomous spanning tree protocol for multihop wireless body area networks," In Mobile and ubiquitous systems: networking & services, 2006 Third annual international conference on, pp. 1–8. IEEE, 2006.

Chapter 5

Electromagnetic characterisation of textile materials for the design of wearable antennas and systems

*Caroline Loss[1], Marco Rossi[3], Sam Agneessens[3],
Ricardo Gonçalves[2], Hendrik Rogier[3], Pedro Pinho[2,4],
and Rita Salvado[5]*

The accurate characterisation of textile materials to be used as a dielectric substrate in wearable systems is fundamental. However, little information can be found on the electromagnetic properties of regular textiles. Woven, knits and nonwovens are inhomogeneous, highly porous, compressible and easily influenced by the environmental hygrometric conditions, making their electromagnetic characterisation difficult. For these reasons, there is no standard method to measure the dielectric properties of textiles. This chapter presents a survey on the evolution of flexible antennas and the textile materials used to manufacture them. Besides, it gives an overview of several methods used to characterise the permittivity of textile materials. Furthermore, it presents and applies a resonator-based experimental technique to characterise textile materials. This experimental technique is based on the theory of resonant methods and consists in calculating the electromagnetic parameters of the material under test (MUT), at a single frequency, by measuring the shift in frequency and the value of Q-factor of one resonator board with a microstrip patch antenna. To validate the experimental characterisation method, four textile antennas have been designed to resonate in the 2.45 GHz ISM band. This bandwidth (BW) also supports the Wireless Local Area Network (WLAN), Bluetooth and Short Range Communication Systems (SRCS) applications. All antennas operated in the targeted frequency band and showed excellent agreement between simulated and measured parameters, supporting the validation of this method. The resonator-based experimental technique proved to be an efficient, simple, easy and fast technique for the characterisation of electromagnetic properties

[1]FibEnTech Research Unit, Universidade da Beira Interior, Portugal
[2]Instituto de Telecomunicações – Aveiro, Campus de Universitário de Santiago, Portugal
[3]Department of Information Technology, Ghent University/iMinds, Belgium
[4]Instituto Superior de Engenharia de Lisboa, Rua Conselheiro Emídio Navarro, Portugal
[5]LabCom, Universidade da Beira Interior, Portugal

of textile materials for the development of wearable antennas and body-centric communication.

5.1   Introduction

In current days, the Internet of Things (IoT) scenario boosts the concept of wearable health-monitoring systems [1–3]. The development of smart objects for IoT applications include wearable products to be identifiable, to communicate and to interact with their environment. In this framework, wearable technology has been addressed to enable the person, mainly through clothes and wearables, to communicate with and be part of this technological network.

The success of such wireless wearable systems is intrinsically related to the performance and stability of the antenna included in the system [4]. Moreover, in systems for on-body applications, the integration of antennas into clothing has been challenging, because antennas are conventionally built on rigid substrates, hindering their efficient and comfortable integration into the garment [5]. Since 2001, when Salonen *et al.* [6] proposed the first prototype of a flexible antenna made of fabrics, the use of textile materials to develop wearable antennas has increased exponentially.

When designing antennas, the knowledge of the electromagnetic properties of the component materials is essential to correctly dimension the antennas and ensure their high stable performance [7]. Specific electrically conductive textiles are available on the market and have been successfully used in wearable antennas [8]. Also, conventional textile fabrics have been applied as dielectric substrates. However, little information can be found about the electromagnetic properties of these regular textiles. Indeed, textile materials are highly porous, compressible, anisotropic [9–11] and easily influenced by the environmental hygrometric conditions [8,11,12], making their electromagnetic characterisation difficult. For these reasons, there is no standard method to measure the dielectric properties of conventional textiles. Therefore, the main objectives of this chapter are to review the electromagnetic methods applied to characterise the dielectric properties of textile fabrics and leathers, and to present a new experimental technique for the characterisation of the dielectrics for wearable antennas.

Section 5.2 provides a description and an overview of the textile materials, considering both conductive and dielectric ones, which have been used to design wearable antennas. It also presents a survey about the evolution of the flexible antennas and the textile materials used to manufacture them. Section 5.3 explains the methods applied to characterise the textile material and was divided into two sub-sections: (1) Resonant methods and (2) Non-resonant methods. Furthermore, Section 5.4 presents the resonator-based experimental technique to characterise the dielectric materials. Finally, Section 5.5 draws the main conclusions.

5.2   Textile materials for the design of wearable antennas

In the broad context of wireless body sensor networks (WBSN) for healthcare, the design of wearable antennas targets several applications, such as ubiquitous

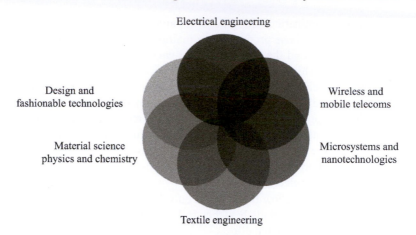

Figure 5.1   Convergent fields for the development of smart clothing

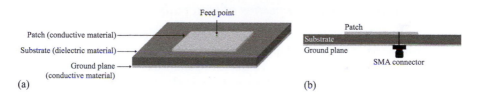

Figure 5.2   Layout of microstrip patch antenna (a) front view and (b) cross section

monitoring, communication and energy harvesting and storage. Nowadays, after several attempts, diverse research areas converge to the development of smart clothes for healthcare applications, as illustrated in Figure 5.1, integrating new sensors and actuators, low-power circuits and self-sustainable wireless sensors networks [14–16].

The integration of electronics into textiles has started a new era for the apparel industry. In a near future, wearable antennas will be part of everyday clothing, transforming the garment in an interface for wireless communication actions. To achieve a low profile and unobtrusive integration of the antenna into the garment, antennas have to be thin, lightweight, robust and easy to maintain. Moreover, they must be low cost for manufacturing and commercialising [8,17]. In this way, planar antennas have been proposed for wearable applications, because this type of antenna topology combines all these characteristics, and is also adaptable to any surface [10,11]. Such antennas are usually formed by assembling conductive and dielectric layers [7], as shown in Figure 5.2. Furthermore, planar antennas, such as the microstrip patch antenna, radiate perpendicularly to a ground plane, which shields the antenna radiation, ensuring that the human body is exposed only to a very small fraction of the radiation.

To develop planar textile antennas, the knowledge of the properties of textile materials is crucial as well as insight into the manufacturing techniques for assembling the layers, such as laminating with glue, seam and adhesive sheets. Fabrics are planar fibrous materials whose properties are mainly determined by the properties of the component fibres and the structure of the yarns and/or of the fabric. They are porous materials, in which the density of the fibres, air volume and size of the pores determine general behaviour, for instance, air permeability and thermal insulation. Also, fabrics are flexible and consist of compressible materials, whose thickness and density might change with changes in pressures. Moreover, the main orientation of the fibres and/or yarns introduces an intrinsic planar anisotropy of general properties. In addition, fibres are constantly exchanging water molecules with the surroundings, which affects their morphology and properties. All these features must be considered when characterising the electromagnetic properties of the textile materials. Moreover, as these features are somehow difficult to control in real applications of textiles, it would be very important to study how they influence the behaviour of the antenna in order to minimise any unwanted effects.

5.2.1 Electromagnetic properties of materials

The response of the material to electromagnetic fields is determined by the constitutive parameters of the material: permittivity (ε), permeability (μ) and conductivity (σ). The relationship between these constitutive parameters and the electromagnetic fields is described through Maxwell's equations [18]. These parameters also determine the spatial extent to which the electromagnetic field can penetrate the material at a given frequency. For the design of antennas, the permittivity is the key parameter to characterise the substrate material and the conductivity characterises the conductive components. The next subsections will discuss the conductive and dielectric textile materials that have been used to develop wearable antennas.

5.2.1.1 Conductive materials

Generally, the fabrics are insulating materials. However, there are materials with high electrical conductivity, which incorporate fibres, filaments or coatings of metals or conductive polymers [19–24]. There is a large range of commercially available electrotextiles (e.g., Less EMF Inc. or Shieldex Trading Inc.) with high conductivity, enabling the development of antennas with acceptable performance.

Fabrics are planar materials and, therefore, their electrical behaviour may be quantified by the surface resistance (R_s) and characterised by the surface resistivity (ρ_s). The surface resistance, whose unit is (Ω), is the ratio of a DC voltage (U) to the current flowing between electrodes that are in contact with the same face of the material under test (MUT) in a specific configuration, as shown in Figure 5.3. The DC surface resistance is the ratio of the DC voltage drop per unit length (L) to the surface current (I_s) per unit width (D). DC surface resistance is thus a property of the material, not depending on the configuration of the electrodes used for the measurement [25]. The results are given in (Ω/sq) [26].

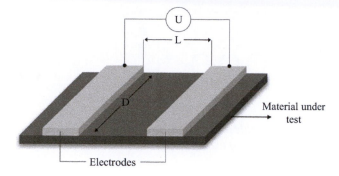

Figure 5.3   Basic setup for surface resistance measurement

The conductivity of the fabric, whose unit is Siemens per metre (S/m), is related to the surface resistance by (5.1), where t is the thickness of the fabric:

$$\sigma = \frac{1}{(\rho_s \cdot t)}. \tag{5.1}$$

The surface resistance is usually given by the manufacturer and may be measured by standard methods, such as ASTM Standard D 257-99 [26], AATCC Test Method 76-2011: Electrical Surface Resistivity of Fabrics [27], ASTM D4496: 2004 Standard Test Method for DC Resistance or Conductance of Moderately Conductive Materials [26], among others [28,29]. Despite the existence of several standard methods, an accurate characterisation of highly conductive fabrics demands specific techniques, such as for instance the ones based on transmission lines and waveguide cavities [9,26,27].

For a good textile antenna performance, it is recommended to use conductive textiles with surface resistance $\leq 1\ \Omega$/sq, in order to minimise the losses of the superficial waves. Moreover, the surface resistance must be homogeneous. In 2006, a study of metallic coated textiles [32] concluded that the yarns that are on the back face, under the other yarns (due to the interlacing of warp and weft, see Figure 5.4), are not covered by the metallic alloy of the coating, causing a discontinuity in the flow of electric current, increasing the electrical resistance of the material.

Flexibility and elasticity are other requirements of the textile materials used to develop wearable antennas, in order to withstand the deformations caused by the bending angles of the human body. However, it is important to note that a very large deformability, as typically occurs in knits, decreases the geometric precision of the antenna [8]. Furthermore, the electric resistance of the knits varies with the direction of the deformation [13,32]. This phenomenon is due to the different number of contact points established between the conductive yarns and the length of the electrical route. So when knits are deformed in the wales direction, the electrical resistance exhibits little changes, since in this direction the laces have little elasticity. However, when knits are deformed in the courses direction, the electrical resistance

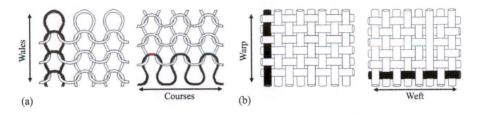

Figure 5.4   *Schema of textile directions in (a) example of Jersey knit – wales and courses and in (b) example of plain weave – warp and weft*

increases significantly, because in this direction the laces are very extensible, as shown in Figure 5.4.

Other factors, as for instance the float, which is the length of the conductive yarns laying on the surface of the woven, also influence the electrical behaviour of the material. The side with the longest floats of conductive electric wires exhibits lower resistance than the other side [31]. Likewise, a higher density of conductive wires also causes lower resistance [32]. Furthermore, the humidity content in the material is also an important factor to consider when determining the electrical resistance of textile materials, because the presence of moisture in the fibres significantly decreases the electrical resistance [9,13,31].

5.2.1.2   Dielectric materials

In the dielectric materials, the conductivity is very small or null. For this reason, the electromagnetic behaviour of low-conductive materials is mainly determined by the permittivity and the permeability. Permittivity describes the interaction of the material with the electric field applied on it, whereas permeability describes the interaction of the material with the magnetic field applied on it [18].

The permittivity is a complex value that generally depends on frequency, temperature and moisture. Furthermore, permittivity is usually expressed as a relative value, ε_r, given by (5.2), where ε_0 is the permittivity of vacuum, equal to 8.854×10^{-12} F/m [33].

$$\varepsilon = \varepsilon_0 \varepsilon_r = \varepsilon_0 (\varepsilon_r' - j\varepsilon_r'). \tag{5.2}$$

The relative permittivity is often called dielectric constant. The real part of the relative permittivity (ε_r') is a measure of how much energy from an external field is stored in the material. The imaginary part of the relative permittivity (ε_r'') is a measure of how dissipative a material is to an external electric field. The ratio between the real and the imaginary part of the relative permittivity is the loss tangent (tan δ) – often called dissipation factor (D_f), is expressed by

$$\tan \delta = \frac{\varepsilon_r''}{\varepsilon_r'}. \tag{5.3}$$

Table 5.1   *Dielectric constant of different*
materials

Reference	Material	ε_r
[33]	Alumina	10.1
[34]	FR-4	4.50
[35]	Paper	7
	Glass	5–10

When designing antennas, the key parameter of the performance of the dielectric substrate is the relative permittivity as well as the loss tangent. As a reference, Table 5.1 presents values of the dielectric constant of diverse materials.

As textile materials are anisotropic materials, their characterisation also depends on the electric field's orientation. This anisotropy is fully described by a permittivity tensor, although in most practical applications like the ones surveyed in this paper, one specific component of this tensor is enough to characterise the behaviour of the textile material for a specific application.

The dielectric behaviour of textile materials depends on frequency, temperature and humidity [11,12,16,29–31], the properties of the component fibres and the structure of the yarns and/or of the fabrics, and on the fibre packing density in the fibrous material [13,38]. Also, textile materials incorporate some difficulties due to their inhomogeneity and instability, being a challenge in terms of accurate characterisation [39], as discussed in Section 5.4.

As pointed in the review presented in [8], for the development of wearable antennas, several conventional textile fabrics have been applied as dielectric substrate, exhibiting very low ε_r and $\tan \delta$, which reduce the surface wave losses and improve the impedance bandwidth (BW) of the antenna [6,7,15,34]. Surface waves are connected to the guided wave propagation within the substrate. Hence, by reducing the dielectric constant, the contribution of the spatial waves increases, and consequently the impedance BW of the antenna increases, allowing the development of antennas with high gain and acceptable efficiency [35,36]. A sufficiently wideband and efficient planar textile antenna is realised by selecting a substrate with a low-dielectric constant (preferably ≤ 4) and a low-loss tangent ($< 10^{-2}$) [43].

The thickness of the substrate is critical in the design of the antenna. As almost all fabrics have the same permittivity, the thickness of the substrate is generally the parameter that determines the BW of the antenna [7]. Besides, the variation of the thickness also influences the resonance frequency [43]. In addition, the substrate thickness also affects the geometric dimensioning of the antenna, which means that a thicker substrate with lower permittivity value (between 1 and 2) results in a larger patch, whereas a thinner substrate with high permittivity results in a smaller patch [36,37]. For a fixed relative permittivity, the substrate thickness may be chosen to maximise the BW of the planar antenna. Therefore, the choice of the thickness of the dielectric material is a compromise between efficiency and BW of the antenna [43]. The influence of the thickness on the BW of the antenna may be explained by (5.4), where Q is the antenna quality factor:

$$BW \sim 1/Q. \tag{5.4}$$

The Q-factor is influenced by the space wave (Q_{rad}) losses, the conduction Ohmic (Q_c) losses, the surface waves (Q_{sw}) and dielectric (Q_d) losses as shown in 5.5 [7]:

$$\frac{1}{Q} = \frac{1}{Q_{rad}} + \frac{1}{Q_c} + \frac{1}{Q_d} + \frac{1}{Q_{sw}}. \tag{5.5}$$

For thin substrates ($h \ll \lambda$, where h is the thickness of the material and λ is the wavelength of the frequency), the quality factor associated with radiation (Q_{rad}) is usually the dominant factor and is inversely proportional to the height of the substrate [7]. Therefore, increasing the height of the substrate lowers the Q-factor (Q_t). As the Q factor decreases with an increased spacing between the patch and the ground planes of the antenna, a thicker substrate yields a larger antenna BW [43].

The textile materials are always establishing a dynamic equilibrium with the temperature and relative humidity of the environmental air. However, the amount of water absorbed by a material to reach this equilibrium depends on the type of molecular structure and on the type of chemical components of the fibres, resulting in very different moisture contents in the textile materials [13]. The moisture content changes the electromagnetic properties of the textile material. Being porous, textiles have many air cavities, and this presence of air yields a textile permittivity that approaches the value of 1. In fact, the relative permittivity of textile materials is usually between 1 and 2, while the permittivity of water is approximately 78 at 2.45 GHz and 25°C. In this way, the higher dielectric constant of water dominates the moistened material causing an increase in their dielectric constant and loss tangent [37].

Hence, when water is absorbed by the textile material of the antenna, it dramatically changes the parameters of the antenna. The absorption of water by the dielectric substrate reduces the resonance frequency [37,41] and the BW of the antenna [44]. Besides, when the fibres absorb water their volume changes. This effect, commonly known as swelling, directly influences the dimensional stability of some fabrics, since the increased diameter of the fibres results in the shrinkage of yarns and of the fabrics [13]. This shrinkage of the fibres directly influences mechanical and geometrical stabilities of the antenna, which are essential requirements to preserve its performance [32].

5.2.2 *Brief survey on textile materials used in wearable antennas*

Wearable antennas are a recent research subject [17]. One of the first proposals on the subject was a Planar Inverted F Antenna (PIFA) for dual-band operation, built on a flexible unspecified substrate, presented in 2001 by Salonen *et al.* [6]. It was intended to be placed in the sleeve of clothing and operate at Global System for Mobile

(GSM) (900 MHz) and Bluetooth (2.4 GHz) frequency, although the lower band was not achieved, the antenna still showed good performance, even in human body presence, around the 2.4 GHz band. Later in 2003, they presented an antenna built on a textile substrate intended for Wireless Local Area Network (WLAN) applications [41], where results are claimed to be acceptable. Furthermore, in 2004 Salonen *et al.* [45] proposed a Global Positioning System (GPS) antenna with circular polarisation, in which they have experimented with five different synthetic fabric materials as dielectric substrates. The conductive parts were made of copper tape. The dielectric synthetic materials used were: (1) Vellux®, which is a 5 mm thick fabric covered on both surfaces with thin layers of plastic foam; (2) synthetic felt, which is a 4 mm thick nonwoven in which fibres are looser on the surface than in the centre; (3) Delinova 200®, which is a strong fabric made of polyamide Cordura® fibres laminated with Gore-Tex membrane, weighing about 370 g/m^2 and having a thickness of 0.5 mm; (4) fleece, which is a very soft polyester fabric with 4 mm thickness, commonly used in sportswear; (5) upholstery fabric, which is composed of three fabric layers bound together resulting in a thin (1.1 mm) fabric of polyester and acrylic that has firmness. The ε_r, of the five fabrics was measured by a cavity perturbation method, at 1.575 GHz, and the values ranged between 1.1 and 1.7. Among the studied fabrics, the one made of high tenacity polyamide fibres (Cordura®) was pointed out as the more interesting fabric for the development of a flexible antenna, because of its constant thickness and its high resistance. These properties yield more stable geometric dimensions of the antenna.

More recent demonstrations on wearable antennas for personal area networks (PANs) to operate in the 2.45 GHz industrial, scientific and medical (ISM) band and for GPS applications are presented in [42,46]. In these examples, antennas for wearable protective clothing intended for professional use under rough conditions are presented and their behaviour in various practical scenarios is discussed. High performance aramid fabric that can withstand high temperatures is applied as a substrate, while conductive textiles, like Shieldit® and Flectron®, are used for the antenna patch and ground planes. These antennas have shown acceptable performance, even in a real environment with human-body presence and when subjected to bending and deformations.

In 2006, Locher *et al.* [32] have built four purely textile wearable patch antennas for Bluetooth applications. They have used three electrical conductive fabrics: (1) a nickel-plated woven fabric (with plating thickness about 250 nm applied on the fabric surface); (2) a silver-plated knitted fabric; (3) a silver–copper–nickel-plated woven fabric. Fabric (3) is the one preferred for building textile antennas with geometric precision, as it is woven and not knitted and its electric surface resistance was more homogeneous than the one of Fabric (1). For the dielectric substrate, they used two types of fabrics: (1) woollen felt of 1.050 g/m^2 with a thickness of 3.5 mm and (2) polyamide spacer fabric, of 530 g/m^2, with a thickness of 6 mm. The felt was dimensionally more stable and harder to bend, whereas the spacer fabric was lighter and more elastic owing to its knitting-based structure. The dielectric properties were measured by a transmission line method, at a frequency of 2.4 GHz, obtaining as results for the felt: $\varepsilon_r = 1.45$ and $\tan \delta = 0.02$, and for the spacer fabric: $\varepsilon_r = 1.14$ and the loss tangent was negligible. The four different antennas produced have shown good performance and could satisfy the Bluetooth specifications, even when subjected to bending effects. However, the antennas lose their circular polarisation when subject to bending.

In the same year, Tronquo *et al.* [44] presented rectangular-ring textile antennas for body area networks (BAN) that are circularly polarised, covering a BW of more than 190 MHz. For the conductive antenna patch and ground plane they applied Flectron®, which is a thin copper-plated fabric with low-surface resistance, lesser than 0.1 Ω/sq. For the dielectric substrate they relied on a fleece fabric of 2.56 mm thickness. Its dielectric properties were measured by testing antennas and they obtained a dielectric constant of $\varepsilon_r = 1.25$.

In 2007, Zhu and Langley [47] developed a dual-band coplanar patch antenna integrating electromagnetic bandgap (EBG) material, to operate at the 2.45 and 5.8 GHz wireless bands. The conductive parts were made of Zelt® fabric whereas the dielectric substrate was a thin felt, with 1.1 mm thickness, and with $\varepsilon_r = 1.30$ and $\tan \delta = 0.02$.

The performance of the textile antennas presented earlier can be improved with integrated solutions, if diverse techniques such as for instance multiple-input–multiple-output (MIMO) are considered [48], or, as shown in [49], by introducing a low-noise amplifier (LNA) in a garment to achieve an active integrated antenna, increasing the sensitivity and the gain of the overall system.

Declercq *et al.* [50] showed another integrated solution consisting of an aperture-coupled antenna on a textile and foam substrate, with a flexible solar cell for tracking and monitoring solutions. Instead of integrating a LNA to increase the wearable antenna performance, Zhu and Langley [47,51] developed a dual-band coplanar patch antenna, to operate in the 2.45 and 5.8 GHz wireless bands, in which they integrated EBG to reduce body presence effects and increase antenna gain. As shown in Table 5.1, the conductive parts were made of Zelt® fabric, while the dielectric substrate was thin felt with $\varepsilon_r = 1.30$ and $\tan \delta = 0.02$. They proved that the introduction of the 3×3 arrays EBG with the coplanar patch could reduce the radiation towards the body by 10 dB, while increasing the antenna gain in 3 dB.

The wearable health-monitoring systems are not only endowed with sensing, processing, actuation and communication abilities, but also with energy harvesting and storage applications, emerging as a solution to the challenges of ubiquitous monitoring of people in several contexts. In the energy harvesting field, Gonçalves *et al.* [52] presented a dual-band textile antenna for electromagnetic energy harvesting, operating at GSM 900 and digital selective calling (DSC) 1,800 frequency bands, for WBAN application. For the dielectric substrate a Cordura® fabric was considered, with 0.5 mm thickness, $\varepsilon_r = 1.9$ and $\tan \delta = 0.0098$; and for the conductive parts of the antenna an e-textile Zelt® was applied. In the numerical simulation, the obtained gain is about 1.8 dBi and 2.06 dBi allied with 82% and 77.6% radiation efficiency for the lowest and highest frequency bands, respectively.

In order to improve flexibility in the antenna design and to achieve compact dimensions for easier antenna integration into clothing, in 2015 [53] Agneessens *et al.*

designed and manufactured a circular quarter-mode textile antenna, using substrate integrated waveguides (SIW) technology, operating in the ISM band. The patch and ground planes were made of conductive e-textile copper-plated polyester plain woven fabric, with low-surface resistance 0.18 Ω/sq at 2.45 GHz, whereas the dielectric substrate was a flexible closed-cell expanded rubber protective foam, with 3.7 mm thickness, $\varepsilon_r = 1.495$ and tan $\delta = 0.016$. Furthermore, this paper studies the influence of the human body on the antenna behaviour, obtaining gains equals 3.8 and 4.2 dBi in an on-body and free space scenario, respectively. Previously, in 2014 [54], these authors had already proposed another antenna design, a dual-band antenna for a body-worn application operating in the ISM frequency bands 2.4 and 5.8 GHz, using the same materials described above. In this SIW technology, several wearable antennas were implemented with conductive fabrics and diverse flexible substrates, including three-dimensional (3D) knit and cork [55–57].

In the radio frequency identification (RFID) context, the miniaturisation of the tag technology brought significant advances in this field, improving their functionality and applicability. Nowadays, the size of the RFID tag depends on the constraints of the antenna. For this reason, the design of suitable antennas continues to be challenging. Based on this framework, some authors have been investigating textile antennas for commercial advertisement proposes, such as brand names and logotypes. Elmahgoub *et al.* [58] in 2010 proposed two RFID tags based on the merge of logos from The University of Mississippi (TUM) and Tampere University of Technology (TUT). The authors applied four different substrates: (1) polyethylene terephthalate film, (2) thin film, (3) paper and (4) fabric, and applied three manufacturing techniques: (A) manual cutting, (B) etching and (C) screen print, in order to verify the efficiency of the designed logo. In the textile fabric case, despite that the woven fabric used as dielectric substrate remains unspecified, the most promising results in terms of conductivity and performance were achieved by screen printing with silver ink. Using the materials described above (3 and C), two tags have been proposed: one operating at 866 MHz and another at 915 MHz. Both tags performed well when compared with simulation, and long read distances were achieved of 11.2 and 7 m, respectively.

Not only limited to RFID tags, Tak and Choi in 2015 [59], developed a Luis Vuitton logo antenna for integration into a handbag. This dual-band antenna, for ISM 2.45 and 4.5 GHz, was designed and fabricated using five layers of the sheepskin-leather substrate with 0.7 mm thickness for each layer (3.5 mm in total), $\varepsilon_r = 2.5$ and tan $\delta = 0.035$. For the conductive parts, Zell® fabric (from Shieldex Trading USA), with $R_s = 0.02$ Ω/sq and 0.1 mm thickness was used. The antenna is essentially a microstrip patch antenna, in which the patch assumes the form of the Louis Vuitton logo. It achieves dual-band operation due to its characteristic structure. The size of the logo was obtained through optimisation in order to match its input impedance in the ISM bands. The radiation pattern is expected to be directive. However, due to the large losses of the dielectric, the efficiencies fall short at 15.2% at 2.49 GHz, and at 41.4% at 4.52 GHz, yielding gains of −0.29 and 3.05 dBi, respectively. Nevertheless, it is reasonable to consider this antenna for IoT applications, which usually comprise short distances of communication.

Besides the use of conductive fabrics, several other ways to manufacture the conductive parts of the antenna have been explored, such as embroidering and screen printing. Matthews and Pettit presented in [60] three types of antennas, which are integrated into clothing: a broadband wire dipole, a bowtie and a spiral antenna, operating in frequencies from 100 MHz to 1 GHz. They have tested different materials (textiles and others), different frames and manufacturing techniques. Among the tested conductive materials there are conductive ribbon, conductive paint and ink, conductive nylon fabric (that is also adhesive on the back face), phosphor bronze mesh fabric (also adhesive on the back face), conductive thread, liquid crystal polymer (LCP) and copper coated fabric. The phosphor bronze mesh, LCP and copper coated fabric have the advantage that the antennas can be directly soldered onto. In some antennas, a conducting epoxy was used to bond the materials, but this results in some lack of robustness. In terms of radio frequency (RF) performance of the designed antennas, the spiral antenna, where the spiral is embroidered with conductive thread, performed worse than any other antenna and was clearly lossy. Overall, based on RF performance, they concluded the most attractive materials to design wearable antennas were the textile fabrics: the conductive nylon and the copper coated fabrics.

In [61] the stability and efficiency of wearable and washable antennas are discussed for textile antennas in which the conductive parts were screen printed with conductive ink. These antennas have shown acceptable performance. The combination of screen printing with a breathable thermoplastic polyurethane (TPU) coating ensured that performance was maintained even after several wash cycles.

An embroidered technique was applied in [47,51] to sew conductive fibres into polymer and fabric substrates. It was proved that by increasing the density of the embroidering stitching, the conductivity of the conductive section increases as well as the accuracy of the fabricated prototypes, yielding better agreement with the simulations. Dipole, spiral and microstrip patch antennas were fabricated with this technique. They yielded very good RF performance when compared to the corresponding rigid copper structures.

Furthermore, in [5] the performance is analysed of an embroidered antenna, made by varying the number of stitches and the directions of the stitches. Despite all embroidered antennas exhibited good results, the result closest to the simulation was obtained by an antenna with the embroidery stitch direction parallel to the feed line, which homogenises the current flow. Also, in this study the authors concluded that a higher number of stitches make the current flow less continuous, due to the higher number of breaks and air gaps in the embroidery, reducing the conductivity of the patch.

A promising way to produce conductive elements is integrating conductive wires into the textiles through the 3D weaving technique. This technique may contribute to improving the mechanical robustness and to eliminating the influence of the glue and/or seam [62].

Table 5.2 summarises the several textile materials that have been applied to develop wearable antennas.

Table 5.2   *Summary of the textile materials that have been used to design of wearable antennas*

Reference	Application	Dielectric material				Conductive material
		Material	Thickness (mm)	ε_r	tan δ	
[6]	GSM (900 MHz) and Bluetooth (2.4 GHz)	Unspecified textile fabric	0.236	3.29	0.0004	–
[41]	WLAN (2.4 GHz)	Fleece fabric	3	1.04	–	Knitted copper fabric
[45]	GPS (1.5 GHz)	Cordura®	0.5	Between 1.1 and 1.7	–	Copper tape
[32]	Bluetooth (2.4 GHz)	Polyamide spacer fabric	6	1.14	Negligible	Silver–copper–nickel-plated woven fabric
[32]	Bluetooth (2.4 GHz)	Woollen felt	3.5	1.45	0.02	Silver–copper–nickel-plated woven fabric
[44]	190 MHz	Fleece fabric	2.56	1.25	–	Flectron®
[44], [51]	WLAN (2.45 and 5.8 GHz)	Felt	1.1	1.30	0.02	Zelt®
[50]	ISM (900 MHz)	Polyurethane protective foam	11	1.16	0.01	Flectron®
[52]	GSM 900 (900 MHz) and DCS1800 (1800 MHz)	Cordura®	0.5	≈1.9	0.009	Zelt®
[53]	2.45 GHz	Closed-cell expanded rubber protective foam	3.7	1.49	0.016	Flectron®
[59]	ISM (2.45 GHz and 4.5 GHz)	Sheepskin leather	0.7	2.5	0.035	Zell
[61]	ISM (2.4 GHz)	Cotton/polyester	2.808	1.6	0.02	Flectron®/conductive ink
[5]	GSM 900 (900 MHz) and DCS1800 (1800 MHz)	Cordura®	0.5	≈1.9	0.009	Silverspam conductive yarn

5.3   Methods for the electromagnetic characterisation of dielectric textiles

This section presents an overview of several methods to characterise the dielectric properties of textile materials. Also, the results obtained through these methods are compared and their suitability for the development of textile antennas is evaluated. The goal of the electromagnetic characterisation is to measure the dielectric constant of the specimen for a specific frequency and field orientation [33]. The methods used to determine the electromagnetic properties of the materials are generally subdivided into two main categories: resonant and non-resonant methods [18]. Each of these categories includes several procedures. The following subsections describe the ones that have already been proposed and applied to characterise textile materials.

5.3.1   Non-resonant methods

The non-resonant methods mainly include procedures based on reflection and transmission/reflection measurements. As the reflection-based techniques, the dielectric properties of the MUT are extracted based on the reflection of the electromagnetic waves in free space by the sample [63]. In transmission/reflection methods, the dielectric properties are calculated based on the reflection from and transmission through the sample [18]. For the characterisation of textile materials, several non-resonant methods have been tested, such as transmission lines [64–68], metallic and dielectric waveguides [30] and free space [11,61,62].

5.3.1.1   Parallel plate method

The parallel plate method is the oldest way to measure the dielectric properties in fibre materials [71]. In this method, the MUT is placed between two parallel plates, as shown in Figure 5.5. In this case, the structure creates a capacitor whose capacity is measured by a LCR metre[a]. Owing to the simplicity of the procedure and the power measured by the LCR, this method is limited to the maximum 1 MHz frequency. Details about

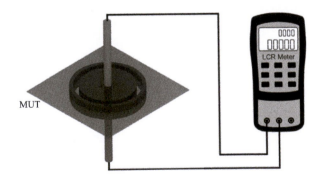

Figure 5.5   Parallel plate method

[a]LCR meter is an electronic equipment used to measure the inductance (L), capacitance (C) and resistance (R) of an electronic component.

the theoretical equations can be found in [12,13]. In [72], nine textile fabrics were characterised at 200 kHz, under controlled ambient conditions to avoid the influence of moisture. The obtained results are summarised in the following Table 5.3.

5.3.1.2 Planar transmission lines methods

Planar transmission lines methods are the most common methods applied to characterise textile materials. Based on the scattering (S) parameters, the advantage of this method is that it can also be applied to determine conductive properties, as presented in [14,15]. The planar transmission lines are subdivided into three types: microstip, coplanar and stripline ones, as illustrated in Figure 5.6.

As shown in Figure 5.6(d), the stripline consists of upper and lower grounding planes, and a central conductive line. The dielectric MUT is placed between the grounding planes and the central line. The advantage of this structure of transmission line is that the radiation losses are negligible. In [64], a stripline prototype was built to characterise a denim fabric, 100% cotton, at 2.45 GHz. For the conductive parts Flectron® fabric was applied and to design the transmission line they used an estimated $\varepsilon_r = 1.2$. As this stripline required $h = 2.45$ mm, five layers of denim fabric were superposed. The measurements with the vector network analyser (VNA) yielded $\varepsilon_r = 2.117$ and $\tan \delta = 0.01$. To validate this method, a microstrip patch antenna for 2.45 GHz was designed and manufactured using the same materials. The antenna has shown good performance and 6.16 dBi of gain was obtained.

In [66], Mantash *et al.* proposed a comparison between a stripline and an open stub resonator, described in [74], to characterise felt and denim fabric. For the stub resonator, a copper tape was used for the conductive parts. Measuring at a nonspecified single frequency, for the denim they extract $\varepsilon_r = 1.6$ and $\tan \delta = 0.05$, and for felt $\varepsilon_r = 1.22$ and $\tan \delta = 0.016$. As to the broadband measurements, the authors have used a stripline fixture resonator for the dielectric material. The results between 1 and 6 GHz have shown a range of values from $\varepsilon_r = 1.215$ to $\varepsilon_r = 1.225$ for the felt material, and from $\varepsilon_r = 1.6$ to $\varepsilon_r = 1.65$ for the denim fabric. Comparing the obtained results from the two methods, both are acceptable. In order to validate the results, two patch antennas for 2.45 GHz were proposed, using the permittivity value extracted at a single frequency, and Shieldit® Super fabric ($R_s \leq 0.5$ Ω/sq) for the conductive parts. These antennas have shown a good agreement between the results and the simulated parameters.

The propagation characteristics of two microstip transmission lines with different lengths were measured in [67]. The lines were made using Nora® fabric, with $R_s = 0.03$ Ω/sq, for the conductive parts, and an acrylic fabric with 0.5 mm of thickness for the dielectric substrate. Knowing the length difference and measuring the S-parameters between 3 and 10 GHz, the permittivity value for the acrylic fabric was calculated: $\varepsilon_r = 2.6$. In order to validate the method, two UWB (3.1–10.6 GHz) wearable antennas were designed. The measured antenna parameters have shown to be in agreement with simulation estimations. Besides, the antennas have been tested for transmission of UWB pulses into the human body in order to evaluate reflectivity, which can be used in diagnosis applications. Moreover, they have proven to be reliable when in close contact with the human body, proving the usefulness of the development of such antennas in textiles.

Table 5.3 Summary of non-resonant methods to characterise textile materials and leather

Reference	Frequency application	Material	Thickness (mm)	ε_r	tan δ	Method
[72]	200 kHz	100% Cotton, twill weave	0.62	2.231	0.0366	Parallel plate
		100% Cotton, plain weave	0.48	2.077	0.0314	
		100% Wool plain weave	0.42	1.865	0.0079	
		Wool, twill weave fabric	0.64	2.053	0.0076	
		Wool, plain weave fabric	1.26	1.670	0.0073	
		Wool + Polyamide, twill weave	1.47	1.529	0.0053	
		100% Polyester (PES) plain weave	0.36	1.748	0.0044	
		Viscose + PES twill weave	0.52	1.707	0.0079	
		100% Polyester plain weave	0.08	2.122	0.0035	
[64]	2.45 GHz	100% Cotton, Denim by Santista	0.49	2.117	0.01	Transmission line
[66]	Non-specified	Felt fabric	4.0	1.22	0.016	Stub resonator
	single frequency	Denim woven	–	1.6	0.05	
	1–6 GHz	Felt fabric	4.0	1.215–1.225	0.016	Stripline
		Denim woven	–	1.6–1.65	0.05	
[67]	UWB (3.1–10.6 GHz)	100% Polyacrylonitrile fabric	0.5	2.6	–	Two-lines method
[65]	2.45 GHz	98% PAR 2% carbon woven 1	0.6	1.57	0.007	Matrix-pencil two-line method
		98% PAR 2% carbon woven 2	0.4	1.91	0.015	
		100% PP nonwoven	3.60	1.18	0.025	
		Fleece fabric	2.56	1.25	0.007	
[69]	8–11 GHz	E-Glass G7628	0.210	4.8–5.0	0.003–0.11	Free space
		E-Glass G880	0.152	3.84–4.0	0.003–0.11	
		Kevlar K141	0.254	3.97–4.05	0.003–0.11	
		Kevlar K151	0.254	3.88–3.04	0.003–0.11	
[11]	330 GHz	Denim woven	0.8	2.73	0.073	Free space:
		Textile 1	0.45	2.74	0.031	angular-invariant approach
		Textile 2	0.25	2.72	0.068	
		Textile 3	0.7	3.66	0.042	
		Textile 4	0.25	2.54	0.066	
		Wool fabric from scarf	1.6	3.18	0.15	
		Stockinet fabric	0.65	3.22	0.08	
		Satin fabric	0.2	2.74	0.075	
		Natural leather	1.15	3.4	0.127	
		Artificial leather	0.95	3.104	0.079	
		Artificial leather	1	3.008	0.08	
		Artificial leather	0.6	2.17	0.066	
		Artificial leather	0.8	3.02	0.084	
		Artificial leather	0.6	2.49	0.085	

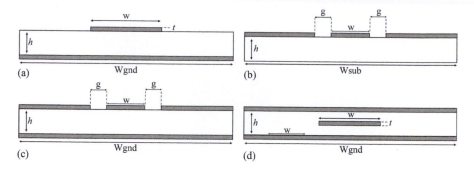

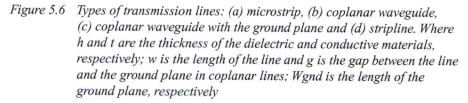

Figure 5.6   *Types of transmission lines: (a) microstrip, (b) coplanar waveguide, (c) coplanar waveguide with the ground plane and (d) stripline. Where h and t are the thickness of the dielectric and conductive materials, respectively; w is the length of the line and g is the gap between the line and the ground plane in coplanar lines; Wgnd is the length of the ground plane, respectively*

In [65], the authors have combined the two-lines method for microstrip lines with the matrix-pencil technique, in order to reduce the perturbations in the parameters of the transmission lines. They have characterised four different materials: (1) woven fabric, 98% aramid and 2% carbon, with $h = 0.60$ mm and $\varepsilon_r = 1.57$; (2) woven fabric, 98% aramid and 2% carbon, with $h = 0.40$ mm and $\varepsilon_r = 1.91$; (3) nonwoven polypropylene fabric with $h = 3.60$ mm and $\varepsilon_r = 1.18$; (4) fleece with $h = 2.56$ mm and $\varepsilon_r = 1.25$. For the conductive parts, copper sheet and Flectron® fabric were used to manufacture the transmission lines. The authors concluded that the lines using copper sheet enabled to estimate the loss tangent of the substrates under test, given that the losses in the copper sheet are much smaller than in the Flectron®. To validate this method, the authors designed three textile antennas for 2.45 GHz, using the tested materials (1, 2 and 3) as a dielectric substrate. For the conductive parts, copper sheets and Flectron® fabric were used, and both copper and Flectron®-based antennas exhibited good results.

Despite the acceptable results, the transmission lines require a complex sample preparation, due to the mechanical instability of the fabrics, such as fraying and deformation in/after the cut process. Furthermore, in the transmission lines, the measured values are influenced by the conditions of some variables, as for example, the type of e-textile or conductive metal that is used, the glue/adhesive sheet, the connector and the manufacturing technique to make the probe, which can lead to non-repeatability of the measurements and introduces errors in the final values.

5.3.1.3   Free space methods

This method typically consists of placing the MUT between two horn antennas, as shown in Figure 5.7. By measuring the S-parameters – transmission (S_{11}) and reflection (S_{21}) coefficients – of the antennas with a VNA, the dielectric constant is estimated. The main advantage of the free space method is that it is contactless and

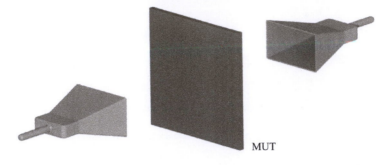

MUT

Figure 5.7 Free space measurement system

non-destructive. The main drawback of this method is the required precise calibration of the horn antennas.

In [69], this method was used to measure the dielectric properties of textile composites. The results of these characterisations can be found in Table 5.3. Furthermore, in [11] the authors present a variation of this method, based on an angular-invariant approach. Due to the typical non-regular surface of the textiles and leathers, in this angular-invariant approach based on Rayleigh scattering, the authors have measured the scattering parameters of the antenna by varying the angle of the material under the test. All materials tested in this work have shown variations in the complex permittivity as a function of the different angles of incidence, due to irregular thickness, surface roughness and texture of the materials. Also, the free space method can be used to characterise the conductive textiles [70].

5.3.2 Resonant methods

Resonant methods usually provide higher accuracy and sensibility than non-resonant methods for low-loss materials [10], even though they only allow material characterisation for a single frequency. They include the resonator method and the resonance-perturbation method. The resonator method is based on the fact that the resonant frequency and quality factor of a dielectric resonator with given dimensions are determined by its permittivity and permeability. These methods are usually applied to measure low-loss dielectric materials. The resonance-perturbation method is based on resonant perturbation theory. For a resonator with given electromagnetic boundaries, when part of the electromagnetic boundary condition is changed by introducing a sample, its resonant frequency and quality factor (Q-factor) will also change. Measuring these shifts in the frequency and in the Q-factor it is possible to extract the permittivity value, as presented in [18].

The Q-factor [18] is a parameter often used to describe an electromagnetic material, according to (5.6)–(5.8):

$$Q_e = \frac{\varepsilon_r'}{\varepsilon_r''} = \frac{1}{\tan \delta_e},$$

$$(5.6)$$

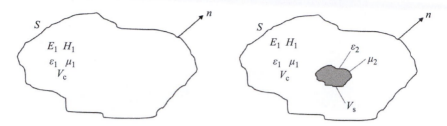

Figure 5.8 *Cavity perturbation (a) original cavity (b) perturbed cavity after the material insertion. E_1 and H_1 are the electric and magnetic fields, respectively; ε_1 and μ_1 are the permittivity and permeability of the cavity; V_c is the volume of the cavity; ε_2 and μ_2 are the permittivity and permeability of the material and V_s is the volume of the sample*

$$Q_m = \frac{\mu_r'}{\mu_r''} = \frac{1}{\tan \delta_m},\tag{5.7}$$

where Q_e is the electric quality factor and Q_m is the magnetic quality factor. Based on this, it is possible to calculate the total quality factor (Q) of the material:

$$\frac{1}{Q} = \frac{1}{Q_e} + \frac{1}{Q_m}.\tag{5.8}$$

5.3.2.1   Cavity perturbation methods

The most common resonance techniques are those based on resonant cavities formed by a rectangular or circular waveguide. The MUT is inserted into the resonant cavity and their electromagnetic properties are calculated from the changes in the resonant frequency and in the quality factor of the cavity caused by the introduction of the material [18,75–77]. This phenomenon is illustrated in Figure 5.8 and the theoretical equations can be found in [18].

These methods are difficult to apply as in every measurement the sample has to be placed in the same position. Figure 5.9 shows some examples of resonant cavities. Also, the cavity needs to be dismantled and reassembled every time a new sample is tested [63], consuming time and maybe introducing errors on the measurements. Moreover, the cavity resonator methods measure the permittivity in the plane of the sample and, therefore, cannot determine any anisotropy in the measured plane [10].

In [77], Kumar and Smith present the electromagnetic characterisation of yarns and fabrics using a cylindrical split cavity resonator. In this work, the averaged result obtained for five yarns with different diameters, $\varepsilon_r = 3.1141$, is compared to the averaged result for three fabrics made with these yarns $\varepsilon_r = 2.7851$. As the textile materials are a mixture of nylon fibres and air, the authors concluded that fabrics exhibit higher porosity and, for this reason, a lower permittivity value than the yarns measured stand-alone.

Furthermore, [73] presents the dielectric characterisation of nine textiles, using a closed cavity resonator. The authors have measured the textiles in all planes (x, y and z).

*Figure 5.9   Examples of resonant cavities: (a) circular resonator cavity
(b) rectangular resonator cavity*

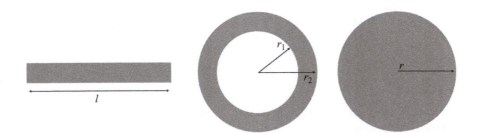

*Figure 5.10   Types of microstrip resonators: (a) straight ribbon resonator
(b) ring resonator and (c) circular resonator, where l is the length of
straight ribbon resonator; r_1 and r_2 are the inner and outer radius of
the resonator ring, respectively; and r is the radius of the circular
resonator*

The obtained results showed the influence of different alignments of the material with
the electromagnetic field. The anisotropy caused by the different density of yarns in
the warp and weft directions is clearly seen.

5.3.2.2   Microstrip ring resonator method

The microstrip resonator methods are based on the resonance-perturbation method.
The sample of the MUT is placed over the microstrip resonator affecting its resonant
frequency and quality factor. Measuring the shift in the frequency and in the Q-factor
enables to extract the permittivity values [78]. The microstrip resonator methods are
subdivided into three types: straight ribbon resonator, ring resonator and circular
resonator, as shown in Figure 5.10.

The microstrip ring resonator does not have open ends, decreasing the radiation
loss and consequently increasing the quality factor. For this reason, the ring resonator
is more accurate and sensible than the straight and circular resonators. Details about
the theoretical equations of microstrip resonator can be found in [18]. For the characterisation of the textile materials, the procedure of the microstrip ring resonator
method only involves attaching the conductive parts, made of copper foil, to the textile dielectric probe under test. In [79], the characterisation of a "Bakhram" textile

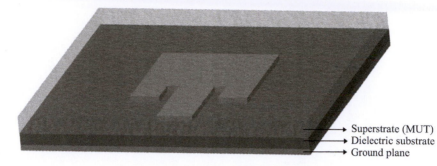

Superstrate (MUT)
Dielectric substrate
Ground plane

Figure 5.11   Setup up of dielectric material covering a patch antenna

fabric at two different frequencies (831.940 and 1.6890 MHz) is presented. After the measurements, the permittivity and loss tangent extracted values were $\varepsilon_r = 2.031$ and $\tan \delta = 0.0038$, and $\varepsilon_r = 1.965$ and $\tan \delta = 0.0024$, respectively. To validate the method, the authors designed a textile patch antenna to resonate at 831.940 MHz, also made of copper foil in the conductive parts. Despite the deviation in the measured frequency, the authors considered the characterisation method well-suited for the characterisation of textile materials, considering this deviation as part of the design and manufacturing process of the antenna.

5.3.2.3   Microstrip patch sensor

The microstrip patch sensor consists of a patch antenna covered with the dielectric MUT, called superstrate, as shown in Figure 5.11. As for the microstrip ring resonator method, this technique relies on the resonance-perturbation method, where the calculation of the permittivity value is based on the shift in frequency caused by the introduction of the superstrate. More information about theoretical equations can be found in [80–82].

In [81], this method is presented to measure solids and liquids. In [82], the method is used to test six nonwoven materials at 1.9 GHz. All tested nonwoven materials were manufactured by the stitching method and the results of their electromagnetic characterisation are presented in Table 5.4. This non-destructive method can be a solution for a quick, easy and low-cost characterisation procedure.

5.3.2.4   Microstrip patch radiator method

In [63], Sankaralingam and Bhaskar proposed a novel microstrip patch radiator method, that consists of designing a patch antenna using an estimated permittivity value found by literature review. After manufacturing the antenna and measuring its S_{11}, the real value of the dielectric constant is calculated based on the shift of the resonant frequency. Six different textile materials were characterised and the extracted values can be found in Table 5.4. To validate this method, the authors designed some textile antennas for 2.45 GHz, using three tested materials as a dielectric substrate,

Table 5.4 Summary of resonant methods to characterise textile materials and leather

Reference	Frequency application	Material	Thickness (mm)	ε_r	tan δ	Method
[77]	9.8 GHz	100% Nylon 6.6 fabric	–	2.8230	0.02681	Resonant microwave cavity
			–	2.7522	0.02420	
			–	2.7801	0.02831	
[45]	GPS (1.5 GHz)	100% PA, Cordura® fabirc	0.5	Between 1.1 and 1.7	–	Cavity perturbation technique
[41]	WLAN (2.4 GHz)	Fleece fabric	3	1.04	–	
[79]	831. 940 MHz	"Bakhram" fabric	0.37	2.031	0.00038	Microstrip ring resonator method
	1.6890 GHz			1.965	0.0024	
[63]	2.45 GHz	100% washed cotton fabric	3.0	1.51	–	Microstrip patch radiator
		100% cotton, denim	2.84	1.67	–	
		65% PES 35% CO fabric	3.0	1.56	–	
		100% CO, fabric for Curtain	3.0	1.47	–	
		100% Polyester (PES)	2.85	1.44	–	
		100% CO, Bed sheet/ floor spread fabric	3.0	1.46	–	
[59]	ISM (2.45 GHz) and 4.5 GHz	Original cowhide leather	0.7	1.76	0.0009	Agilent 85070E dielectric measurement probe kit
		Original sheepskin leather	0.7	2.5	0.0035	
		Original oiled sheepskin leather	0.7	2.66	0.085	
		Original scratched cowhide leather	0.7	3.13	0.15	
		Original oiled cowhide leather	0.7	2.3	0.04	
[82]	1.9 GHz	PES + (LMF-PES), Stitched nonwoven	10	1.013	–	Microstrip patch sensor
		P84® + (LMF-PES), stitched Nonwoven	8	1.012	–	
		Kermel® + (LMF-PES), stitched Nonwoven	8	1.014	–	
		(PES + LMF-PES)/T, stitched Nonwoven, Thermal processed	1.4	1.175	–	
		(P84® + LMF-PES)/T, stitched Nonwoven, Thermal processed	4	1.036	–	
		(Kermel® + LMF-PES)/T, stitched Nonwoven, Thermal processed	4	1.050	–	

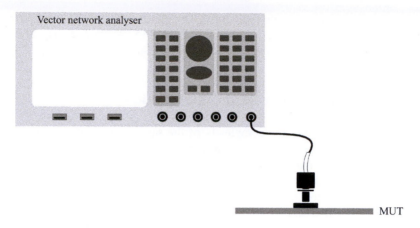

Figure 5.12   Schematic of the coaxial probe by Agilent measurement kit

and copper sheets for the conductive parts. Despite that these antennas exhibit good results, confirming the suitability of this method, the results can be influenced by the manufacturing technique. Indeed, the use of an interface to assemble the layers, such as glue or an adhesive sheet, and the inaccuracies when soldering the SubMiniature version A (SMA) connector to the microstrip patch radiator, can introduce errors in the final values.

5.3.2.5   Agilent 85070E dielectric measurement probe kit

The Agilent 85070E dielectric measurement probe kit, is an equipment by Agilent Technologies, available on the market to measure the dielectric properties of several types of materials [83]. In [59], the characterisation of textile and leather was performed using the open-ended probe method. In this method the permittivity value is calculated only from the S_{11} parameter from the coaxial probe, measured with the VNA or impedance analyser, as illustrated in Figure 5.12. On the one side, the advantages of this method are: non-destructive technique and quick and easy to perform. On the other side, this method is very expensive (price of probe kit + software) and a complex calibration of the probe is required, using standard materials such as distilled water or methanol.

In [59], five types of leather were characterised at 2.45 GHz: (1) original cowhide, $\varepsilon_r = 1.76$ and $\tan \delta = 0.0009$, (2) original sheepskin $\varepsilon_r = 2.5$ and $\tan \delta = 0.035$, (3) oiled sheepskin $\varepsilon_r = 2.66$ and $\tan \delta = 0.085$, (4) scratched cowhide $\varepsilon_r = 3.13$ and $\tan \delta = 0.15$, and (5) oiled cowhide $\varepsilon_r = 2.3$ and $\tan \delta = 0.04$. Among the characterised leathers, considering the operating frequency, leather (2) was chosen to develop a dual-band wearable antenna for 2.45 and 4.5 GHz applications. Due to the structure of the five-layered substrate, the simulated and measured results deviate. For this reason, the authors included a new simulation using four air gaps of 0.0875 mm thickness to correct the results. The size of the air gaps was obtained through the difference between the simulated substrate thickness (3.5 mm) and the measured thickness of the final substrate (3.85 mm). The gain and the radiation efficiency at the lower frequency (2.49 GHz) are −0.29 dBi and 15.2%, respectively, and at the higher frequency (4.52 GHz) 3.05 dBi and 41.4%, respectively.

5.4   Resonator-based experimental technique

To alleviate the problems and complexity of the sample preparation, the influence of the conductive material and the expensive equipment required in the methods described in the previous section, an experimental resonator-based technique was developed. It is based on the theory of resonance-perturbation, based on similar principles as the microstrip patch sensor technique presented in Section 5.3.6 [81,82]. However, the method outlined in this section relies on full-wave simulators rather approximating formulas for the extraction of the electromagnetic properties of the MUT, thus yielding an overall higher accuracy. More specifically, it consists of computing the electromagnetic parameters of the MUT by comparing the simulated and measured shifts in the resonance frequency and in the value of Q-factor of a microstrip patch antenna. These shifts are caused by the introduction of a superstrate on the patch.

Indeed, the presence of the superstrate will change the characteristics of the antenna, such as resonance frequency, as shown in Figure 5.13. This shift in the frequency ($f_{r1} - f_{r2}$) is caused by the difference of the electromagnetic wavelength of the uncovered antenna, a simulation during the design process and the measured antenna covered by a dielectric material of unknown ε_r. The amplitude of the frequency shift depends on the ε_r and on the thickness of the covering material.

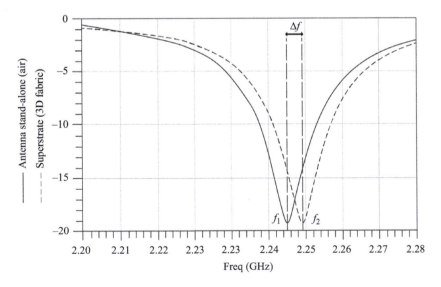

Figure 5.13   Measurement results from the patch antenna stand-alone and after the introduction of superstrate

The test equipment consists of a board, on which a microstrip patch antenna is implemented, and in a VNA to perform the measurements. The antenna is designed to operate at 2.25 GHz, thus in the proximity of the 2.45 GHz ISM frequency band, which is paramount for wearable systems. The antenna is realised on a rigid I-Tera MT40 high-frequency laminate, with $\varepsilon_r = 3.56$, $\tan \delta = 0.0035$ and thickness $h = 0.508$ mm, whereas the patch is etched in a copper layer with a thickness of 35 μm. It is fed by a coaxial line through the SMA connector. The designed antenna is shown in Figure 5.14 and its dimensions are given in Table 5.5.

The measurement setup of this resonator-based experimental technique is illustrated in Figure 5.15.

5.4.1 Experimental procedure

In order to extract the ε_r and $\tan \delta$ of the textile fabric under test, the following procedure is carried out:

(1) Prepare the textile samples, cutting a square 10×10 cm^2 (or sized enough to cover all the patch of the antenna) from the selvedge of the fabric and numerating its sides clockwise, as shown in Figure 5.16;

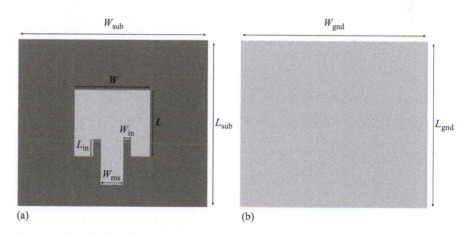

(a)   (b)

Figure 5.14   *Design of microstrip patch antenna resonating at 2.25 GHz (a) front and (b) rear*

Table 5.5   *Dimensions of microstrip patch antenna for 2.45 GHz*

Parameter	Dimension (mm)
W_{sub}, L_{sub}	100, 100
W, L	36.5, 36
W_{in}, L_{in}, W_{ms}	3, 10, 1.2
W_{gnd}, L_{gnd}	100, 100

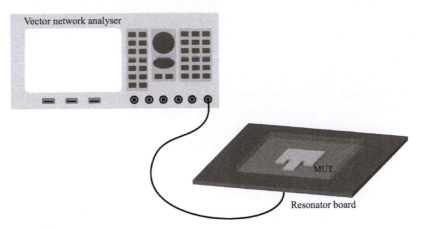

Figure 5.15   Setup of resonator-based experimental technique

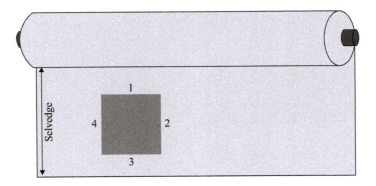

Figure 5.16   Scheme to cut and prepare the textile samples

(2)   Calibrate the VNA in the frequency range of interest by using the electronic calibration module;
(3)   Connect the microstrip patch antenna to the VNA;
(4)   Measure the S_{11} parameter of the antenna;
(5)   Save the data;
(6)   Put the textile sample on the resonator board and mark the board in order to place future samples in the exact same place;
(7)   Measure again the S_{11} parameter of the antenna, this time with the textile sample on the patch;
(8)   Save the data;
(9)   Repeat the previous steps (6, 7 and 8) four times by positioning the substrate sample such that each time one different side is aligned with the microstrip

feed of the antenna. Then, repeat the complete procedure by turning the sample upside-down. As a result, a total number of 8 measurements is acquired (2 faces × 4 positions);

(10)   Finally, by leveraging full-wave simulations of the antenna, extract the ε_r and $\tan \delta$ of the MUT by fitting the measured shift $(f_{r1} - f_{r2})$ in the antenna's resonance frequency and the measured variation in the Q-factor to the simulated ones, respectively.

The most advantageous feature of this experimental method is that it does not require sophisticated equipment, so it is fast and simple to apply. Moreover, in contrast to some of the aforementioned techniques, this method is non-destructive. As such, the MUT is not modified during the characterisation process, thus yielding values of the permittivity and loss tangent that are not influenced by the glue, the conductive material or the manufacturing of the measured structures. Finally, it allows considering the influence of the anisotropy of the textile structures, since the MUT can easily be placed along several orientations to measure its permittivity in different directions.

5.4.2   Measurement results

Eleven common textile materials applied in the fashion industry were characterised using the discussed experimental method. More specifically, two samples of each material were characterised according to the procedure outlined in Section 5.4.1. The extracted values of the permittivity ε_r and the loss tangent $\tan \delta$, reported in Table 5.6, were calculated following (5.9) and (5.10):

$$\varepsilon_r = \frac{\overline{\varepsilon_{rf1}} + \overline{\varepsilon_{rf2}}}{2}, \tag{5.9}$$

$$\tan \delta = \frac{\overline{\tan \delta_{f1}} + \overline{\tan \delta_{f2}}}{2}, \tag{5.10}$$

where $\overline{\varepsilon_{rf1}}$, $\overline{\varepsilon_{rf2}}$, $\overline{\tan \delta_{f1}}$ and $\overline{\tan \delta_{f2}}$ are the average values of ε_r and $\tan \delta$ extracted with the MUT positioned with the face-side (f_1) and the reverse-side (f_2) facing the board, respectively.

The thickness of the samples was measured using the Kawabata's evaluation system for fabrics, the KES-F – 3 Compressional Tester, under controlled environment conditions of 25°C and 65% of relative humidity.

5.4.3   Discussion of the results

In resonator methods, such as the experimental technique presented here, the roughness of the material may introduce some inaccuracies in the results due to the air gap between the sample and the resonator board. Even though the presence of this air gap can be accounted for during the characterisation process, it is important to analyse the properties of the material in order to understand whether the results are correct and the material is compatible with this kind of characterisation method.

Figure 5.17 presents the extracted dielectric constants obtained by positioning the sample with each of the two faces contacting the board. We notice that the permittivity

Table 5.6 Summary of characterised materials using the resonator-based experimental technique at 2.25 GHz

Sample	Manufacturer	Manufacturer reference	Composition	Thickness (mm)	ε_r	$\tan \delta$
3D Fabric I	LMA – Leandro Manuel Araújo Ltda. (Matosinhos, Portugal)	3003	100% Polyester	2.650	1.10	0.005
3D Fabric II		3013		3.068	1.10	0.006
3D Fabric III		3037		2.821	1.12	0.017
3D Fabric IV		3006		2.410	1.13	0.018
3D Fabric V		3015		4.140	1.11	0.004
Cordura I	B. W. Wernerfelt Group (Søborg, Danmark)	LTE1N184	Plain weave 100% PA 6.6, PU coated	0.503	1.58	0.008
Cordura II		LTE1N185	Plain weave 100% PA 6.6, TF coated	0.501	1.56	0.008
Neoprene I	Sedo chemical neoprene GmbH (Fürstenwalde/Spree, Germany)	N00S1	Neoprene laminated with jersey 100% Polyester	5.000	1.37	0.001
Neoprene II			Neoprene laminated with jersey 100% Polyester (face side) and 100% Nylon (reverse side)	3.095	1.30	0.001
Fake Leather I	–	–	Carded knit 100% polyester, PU coated	0.831	1.45	0.017
Fake Leather II		–		0.923	1.43	0.012

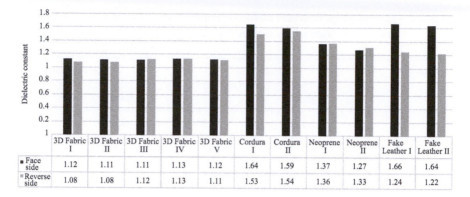

	3D Fabric I	3D Fabric II	3D Fabric III	3D Fabric IV	3D Fabric V	Cordura I	Cordura II	Neoprene I	Neoprene II	Fake Leather I	Fake Leather II
■ Face side	1.12	1.11	1.11	1.13	1.12	1.64	1.59	1.37	1.27	1.66	1.64
■ Reverse side	1.08	1.08	1.12	1.13	1.11	1.53	1.54	1.36	1.33	1.24	1.22

Figure 5.17   Results of the dielectric constant of textile materials when positioning face- and reverse-side in contact with the resonator board

values vary according to the position of the sample on the patch. In particular, the largest variation is observed for the coated textiles.

The surfaces of the textile materials were characterised by measuring the superficial geometric roughness (SMD) with Kawabata's evaluation system, KES-F – 4 Surface Tester. The SMD is measured by pulling a U-shaped steel wire with 0.5 mm diameter through the fabric length under a normal force of 10 gf. In this way, a height profile of the fabric along its length is obtained. The standard mean deviation of this height profile is considered as surface roughness and is called geometric roughness (SMD) [84]. This test was carried out in both the warp and the weft directions (five measurements per each direction), for both face- and reverse-side of the sample (20 measurements in total). Table 5.7 shows the average value of these measurements and the averaged results of the permittivity when positioning face- and reverse-side of the sample in contact with the resonating board.

We observe that for all samples the face of the material with higher SMD has lower permittivity value. Indeed, during the dielectric characterisation when positioning the rougher face in contact with the board more air is trapped on the surface of the probe, lowering the extracted value of ε_r.

5.4.4   Validation of the method

In order to validate the discussed method, four textile antennas were designed using the CST Microwave Studio 2011 full-wave simulator to resonate in the 2.45 GHz ISM band. This BW also supports the WLAN, Bluetooth and SRCS (802.15.4) applications. Then, the antennas were manufactured by applying a commercial e-textile (Pure Copper Polyester Taffeta Fabric, Ref. 1212, from Less EMF Inc., USA), with resistance $R_s = 0.18\ \Omega$/sq (at 2.45 GHz) and a thickness equal to 0.080 mm, for the patch and the ground plane. As for the dielectric substrate, the 3D Fabrics II and V, neoprene II and Fake Leather I were considered.

*Table 5.7   Surface roughness and dielectric constant positioning the sample face-
and reverse-side of the sample on the test board*

Sample	Face side		Reverse side	
	SMD	$\overline{\varepsilon_r}$	SMD	$\overline{\varepsilon_r}$
3D fabric I	5.320	1.12	9.130	1.08
3D fabric II	4.988	1.11	10.700	1.08
3D fabric III	5.743	1.11	2.578	1.12
3D fabric IV	3.523	1.13	3.140	1.13
3D fabric V	7.723	1.12	7.865	1.11
Cordura I	2.755	1.64	3.130	1.53
Cordura II	2.353	1.59	3.770	1.54
Neoprene I	3.440	1.37	3.795	1.36
Neoprene II	2.403	1.27	2.258	1.33
Fake Leather I	1.430	1.66	2.985	1.24
Fake Leather II	1.985	1.64	3.383	1.22

In order to limit the impact of production uncertainties on the performance of the antenna, a square patch was chosen and the patches were cut with a laser cutting machine (LC6090C CCD, from Jinan G. Weike Science & Technology Co. Ltd., Jinan, China). All four antennas were produced by assembling the components with a thermal adhesive sheet (Fixorete Losango, Ref. 252/001, from JAU Têxteis, Serzedo, Portugal). Finally, the S_{11} parameters of the patch antennas were measured. They are reported in Figure 5.18. Both the simulated and the measured resonance frequencies and gains of the antennas are reported in Table 5.8.

We notice from Figure 5.18 that all the antennas resonate in the ISM frequency band and that there is excellent agreement between the simulated and measured parameters. In particular, Antenna 4 exhibits the best performance. As for Antenna 3, the observed shift in the resonance frequency may be explained by the fact that, when applying the conductive layer on the knitted structure, the superficial pores of the material are "reduced," thereby decreasing the quantity of air in the substrate. As a result, the permittivity of the material increases and the antenna resonates at a lower frequency.

The difference between simulated and measured values of the resonant frequency of the Antennas 2 and 1 is small. It is probably due to the uncertainties introduced during the production. Indeed, after assembly with an adhesive sheet, the presence of the glue can increase the permittivity of the substrate, with a corresponding increase in the resonance frequency value. Also, the presence of the adhesive sheet may increase the superficial electrical resistance of the e-textile, since the glue introduces discontinuities in the electrical current flow. As a result, the magnitude of the resonance peak may decrease, as is observed for Antenna 2.

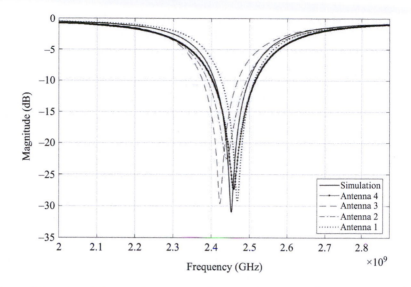

Figure 5.18   S$_{11}$ of the patch antennas designed for the validation of the method

Table 5.8   *Simulated and measured parameters of the antennas*

Prototype		Simulation parameters						Simulated results		Measured results	
	Dielectric substrate	Substrate size (mm)	Substrate height (mm)	Patch size – width × length (mm)	Feed position (mm)*		Resonance frequency (GHz)	Gain (dB)		Resonance frequency (GHz)	Gain (dB)
Antenna 1	3D Fabric II	80 × 80	3.068	52.5 × 52.5	15		2.45	7.76		2.46	7.73
Antenna 2	3D Fabric V	80 × 80	4.140	51.5 × 51.5	14		2.45	7.16		2.44	7.58
Antenna 3	Neoprene II	80 × 80	3.095	46.5 × 46.5	11		2.45	7.28		2.42	6.46
Antenna 4	Fake Leather I	80 × 80	0.092	48.1 × 48.1	4		2.45	0.61		2.45	2.24

*Vertical distance from the centre to the top edge of the patch.

5.5   Conclusions

This chapter has presented an overview of the textile materials commonly applied for wearable antennas and systems. Because off-the-shelf fabrics represent a common choice for antenna substrates, the characterisation of their electromagnetic properties is essential to carry out an optimal antenna design and achieve excellent antenna performance. Therefore, we have discussed a multitude of methods described in the literature to characterised textile materials.

Furthermore, a new resonator-based experimental technique was introduced to characterise several textile fabrics of interest. This method is based on the theory of resonance-perturbation, which consists in computing the electromagnetic parameters of the MUT, at a single frequency (2.25 GHz), by measuring the shift in the resonance frequency and in the value of the Q-factor of a microstrip patch antenna.

In general, the resonant methods yield higher accuracy than non-resonant ones and do not require complex sample preparation. Another advantage of the method is that it is only applied to the antenna substrate. Therefore, the procedure is not affected by arbitrary effects, such as, for example, the type of glue/adhesive sheet, the connector and the manufacturing technique of the probe, which can lead to the non-repeatability of the measurements and introduce variations in the final values.

Some physical properties of the textile materials influence their electromagnetic characteristics, as discussed in Section 5.2. During the dielectric characterisation of textile and other related materials using the resonator-based experimental technique, it was observed that when positioning the rougher face of the MUT in contact with the resonator board, the extracted ε_r values were lower than the ones extracted with the material upside-down. This effect is attributed to the amount of air that is confined between the textile and the resonator board because of the roughness of the fabric. Therefore, the experimental results presented in this chapter have mainly focussed on the roughness of the MUT and the consequent existence of pores on its surface.

Finally, to validate the analysis, four wearable antennas operating in the 2.45 GHz ISM band were designed and manufactured. All antennas operated in the targeted frequency band and showed excellent agreement between simulated and measured figures of merit. Therefore, the resonator-based experimental method discussed in this chapter proved to be an efficient, simple and fast technique for the characterisation of the electromagnetic properties of textile materials for wearable antennas and body-centric communication.

Acknowledgements

This work is funded by FCT/MEC through national funds and when applicable co-funded by FEDER-PT2020 in partnership agreement under the projects UID/EEA/50008/2013 and UID/Multi/00195/2013. The authors wish to thank the European COST Action IC1301 WiPE for the support given to this research; CAPES Foundation for the PhD grant, process n.9371-13/3 and LMA Leandro Manuel Araújo Ltda for supplying the 3D textile samples.

References

[1]   S. Warren and B. Natarajan, "Wireless Communication Technologies for Wearable Systems," in *Wearable Monitoring Systems*, New York: Springer, pp. 51–80, 2011.

[2]   A. Pantelopoulos and N. G. Bourbakis, "A Survey on Wearable Sensor-Based Systems for Health Monitoring and Prognosis," *IEEE Trans. Syst. man Cybern. – Part C Appl. Rev.*, vol. 40, no. 1, pp. 1–12, 2010.

[3]   D. Miorandi, S. Sicari, F. De Pellegrini, and I. Chlamtac, "Ad Hoc Networks Internet of Things: Vision, Applications and Research Challenges," *Ad Hoc Networks*, vol. 10, no. 7, pp. 1497–1516, 2012.

[4]   H. Khaleel, Ed., *Innovation in Wearable and Flexible Antennas*. Southampton, UK: WIT Press, 2015.

[5]   C. Loss, R. Gonçalves, C. Lopes, P. Pinho, and R. Salvado, "Smart Coat with a Fully-Embedded Textile Antenna for IoT Applications," *Sensors*, vol. 16, no. 6. 2016.

[6]   P. Salonen, M. Keskilammi, J. Rantanen, and L. Sydanheimo, "A Novel Bluetooth Antenna on Flexible Substrate for Smart Clothing," *IEEE Conf. Publ.*, vol. 2, pp. 789–794, 2001.

[7]   C. A. Balanis, *Antenna Theory: Analysis and Design*, 3rd ed. Hoboken, NJ: Wiley Interscience, 2005.

[8]   R. Salvado, C. Loss, P. Pinho, and R. Goncalves, "Textile Materials for the Design of Wearable Antennas: A Survey," *Sensors*, pp. 15841–15857, 2012.

[9]   J. Lilja and P. Salonen, "Textile Material Characterization for Softwear Antennas," in *Antennas and Propagation Society International Symposium (APSURSI)*, pp. 1–7, 2009.

[10] J. Baker-Jarvis, R. Geyer, J. Grosvenor, *et al.*, "Dielectric Characterization of Low-loss Materials: A Comparison of Techniques," *IEEE Trans. Dielectr. Electr. Insul.*, vol. 5, no. 4, pp. 571–577, 1998.

[11] B. Kapilevich, B. Litvak, M. Anisimov, D. Hardon, and Y. Pinhasi, "Complex Permittivity Measurements of Textiles and Leather in a Free Space: An Angular-Invariant Approach," *Int. J. Microw. Sci. Technol.*, pp. 1–7, 2012.

[12] M. Pourova, R. Zajicek, L. Oppl, and J. Vrba, "Measurement of Dielectric Properties of Moisture Textile," in *14th Conference on Microwave Techniques, 2008*, pp. 1–4, 2008.

[13] W. E. Morton and W. S. Hearle, *Physical properties of textile fibres*, 4th ed. Cambridge: Woodhead Publishing in Textiles, 2008.

[14] P. Taylor, A. Schwarz, L. Van Langenhove, and P. Guermonprez, "A Roadmap on Smart Textiles," *Text. Prog.*, vol. 42, no. 2, pp. 99–180, 2010.

[15] M. Catrysse, F. Pirotte, and R. Puers, "The Use of Electronics in Medical Textiles," in *Smart Textiles for Medicine and Healthcare*, Cambridge: Woodhead Publishing in Textiles, 2007, pp. 88–104.

[16] K. Cherenack and L. Pieterson, "Smart Textiles: Challenges and Opportunities," *J. Appl. Phys.*, vol. 112, no. 091301, 2012.

[17] B. Grupta, S. Sankaralingam, and S. Dhar, "Development of Wearable and Implantable Antennas in the Last Decade: A Review," *IEEE Conf. Publ.*, pp. 251–267, 2010.

[18] L. F. Chen, C. K. Ong, C. P. Neo, V. V. Varadan, and V. K. Varadan, Microwave Electronics: Measurement and Materials Characterization. Chichester: John Wiley & Sons Ltd., 2004.

[19] R. R. Bonaldi, E. Siores, and T. Shah, "Electromagnetic Shielding Characterization of Several Conductive Fabrics for Medical Applications," *J. Fiber Bioeng. Informatics*, vol. 2, no. 4, pp. 245–253, 2010.

[20] S. Brzeziński, T. Rybicki, I. Karbownik, *et al.*, "Textile Materials for Electromagnetic Field Shielding Made with the Use of Nano- and Micro-Technology," *Cent. Eur. J. Phys.*, vol. 10, no. 5, pp. 1190–1196, 2012.

[21] S. Brzeziński, T. Rybicki, I. Karbownik, *et al.*, "Textile Multi-Layer Systems for Protection against Electromagnetic Radiation," *Fibers Text. Eastern Eur.*, vol. 17, no. 2, pp. 66–71, 2009.

[22] S. Gimpel, U. Mohring, H. Muller, A. Neudeck, and W. Scheibner, "Textile-Based Electronic Substrate Technology," *J. Ind. Text.*, vol. 33, no. 3, pp. 179–189, 2004.

[23] V. Kaushik, J. Lee, J. Hong, *et al.*, "Textile-Based Electronic Components for Energy Applications: Principles, Problems, and Perspective," Namomaterials, vol. 5, no. 3, pp. 1493–1531, 2015.

[24] W. Zeng, L. Shu, Q. Li, S. Chen, F. Wang, and X. M. Tao, "Fiber-Based Wearable Electronics: A Review of Materials, Fabrication, Devices, and Applications," *Adv. Mater.*, vol. 26, pp. 5310–5336, 2014.

[25] W. A. Maryniak, T. Uehara, and M. A. Noras, "Surface Resistivity and Surface Resistance Measurements – Using a Concentric Ring Probe Technique," *Trek Appl. Note*, vol. 1005, pp. 1–4, 2003.

[26] "ASTM Standards D 257-99," in *Standard Test Methods for DC Resistance or Conductance of Insulating Materials*, 1999.

[27] "AATCC Test Method 76-2011," in *Electrical Surface Resistivity of Fabrics*, American Association of Textile Chemists and Colorists, 2011, pp. 1–3.

[28] "ISO 10965:2011," in *Textile Floor Coverings – Determination of Electrical Resistance*, 2011, pp. 1–5.

[29] "ISO 21178:2013," in *Light Conveyor Belts – Determination of Electrical Resistances*, pp. 1–20, 2013.

[30] R. Shawl, B. Longj, D. Werner, and A. Gavrin, "The Characterization of Conductive Textile Materials Intended for Radio Frequency Applications," *IEEE Antennas Propag. Mag.*, vol. 49, no. 3, pp. 28–40, 2007.

[31] D. Cottet, J. Gryzb, T. Kirstein, and G. Troster, "Electrical Characterization of Textile Transmission Lines," *IEEE Trans. Adv. Packag.*, vol. 26, no. 2, pp. 182–190, 2003.

[32] I. Locher, M. Klemm, T. Kirstein, and G. Tröster, "Design and Characterization of Purely Textile Patch Antennas," *EEE Trans. Adv. Packag.*, vol. 29, no. 4, pp. 777–788, 2006.

[33] J. Baker-Jarvis, M. D. Janezic, and D. C. DeGroot, "High-Frequency Dielectric Measurements," *IEEE Instrum. Meas. Mag.*, vol. 13, no. 3, pp. 24–31, 2010.

[34] J. S. Thorp, M. Akhtaruzzaman, and D. Evans, "The Dielectric Properties of Alumina Substrate for Microelectronic Packaging," *J. Mater. Sci.*, vol. 25, no. 9, pp. 4143–4149, 1990.

[35] J. Coonrod, "Understanding When to Use FR-4 or High Frequency Laminates," *Onboard Technology*, pp. 26–30, 2011.

[36] M. N. O. Sadiku, *Elements of Electromagnetics*, 4th ed. New Delhi: Oxford University Press, 2007.

[37] C. Hertleer, A. V. Laere, H. Rogier, and L. Van Langenhove, "Influence of Relative Humidity on Textile Antenna Performance," *Text. Res. J.*, vol. 80, no. 2, pp. 1–9, 2009.

[38] K. Bal and V. K. Kothari, "Measurement of Dielectric Properties of Textile Materials and Their Applications," *Indian J. Fibre Text. Res.*, vol. 34, no. June, pp. 191–199, 2009.

[39] U. C. Hasar, "A New Microwave Method for Electrical Characterization of Low-Loss Materials," *IEEE Microw. Wirel. Components Lett.*, vol. 19, no. 12, pp. 801–803, Dec. 2009.

[40] C. Hertleer, A. Tronquo, H. Rogier, and L. Van Langenhove, "The Use of Textile Materials to Design Wearable Microstrip Patch Antennas," *Text. Res. J.*, vol. 78, no. 8, pp. 651–658, Aug. 2008.

[41] P. Salonen and H. Hurme, "A Novel Fabric WLAN Antenna for Wearable Applications," *IEEE Conf. Publ.*, vol. 2, pp. 100–103, 2003.

[42] C. Hertleer, H. Rogier, L. Vallozzi, and L. Van Langenhove, "A Textile Antenna for Off-Body Communication Integrated Into Protective Clothing for Firefighters," *IEEE Trans. Antennas Propag.*, vol. 57, no. 4, pp. 919–925, 2009.

[43] S. Brebels, J. Ryckaert, C. Boris, S. Donnay, W. De Raedt, E. Beyne, and R. P. Mertens, "SOP Integration and Codesign of Antennas," *IEEE Trans. Adv. Packag.*, vol. 27, no. 2, pp. 341–351, 2004.

[44] A. Tronquo, H. Rogier, C. Hertleer, and L. Van Langenhove, "Applying textile materials for the design of antennas for wireless body area networks," *First Eur. Conf. Antennas Propagation, EuCAP*, no. October, pp. 1–5, 2006.

[45] P. Salonen, Y. Rahmat-samii, M. Schafhth, and M. Kivikoski, "Effect of Textile Materials on Wearable Antenna Performance: A Case Study of GPS Antenna," *IEEE Conf. Publ.*, vol. 1, no. June, pp. 459–462, 2004.

[46] C. Hertleer, H. Rogier, L. Vallozzi, and F. Declercq, "A Textile Antennas Based on High-performance Fabrics," in *2nd European Conference on Antennas and Propagation*, pp. 1–5, 2007.

[47] S. Zhu and R. Langley, "Dual-Band Wearable Antennas over EBG Substrate," *Electron. Lett.*, vol. 43, no. 3, 2007.

[48] L. Vallozzi, V. P. Torre, C. Hertleer, H. Rogier, M. Moeneclaey, and J. Verhaevert, "Wireless Comunications for Firefighters Using Dual-Polarized Textile Antennas Integrated in their Garment," *IEEE Trans. Antennas Propag.*, vol. 58, pp. 1357–1368, 2010.

[49] A. Dierck, F. Declercq, and H. Rogier, "Review of Active Textile Antenna Co-Design and Optimization Strategies," in *IEEE International Conference on RFID-Technologies and Applications*, pp. 194–201, 2011.

[50] F. Declercq, A. Georgiadis, and H. Rogier, "Wearable Aperture-Coupled Shorted Solar Patch Antenna for Remote Tracking and Monitoring Applications," in *5th European Conference on Antennas and Propagation*, pp. 2992–2996, 2011.

[51] S. Zhu and R. Langley, "Dual-band Wearable Textile Antennas over EGB Substrate," *IEEE Trans. Antennas Propag.*, vol. 57, pp. 926–935, 2009.

[52]   R. Gonçalves, N. Carvalho, P. Pinho, C. Loss, and R. Salvado, "Textile Antenna for Electromagnetic Energy Harvesting for GSM900 and DCS1800 Bands," in *Antennas and Propagation Society International Symposium (APSURSI) 2013*, pp. 1206–1207, 2013.

[53]   S. Agneessens, S. Member, S. Lemey, T. Vervust, and H. Rogier, "Wearable, Small, and Robust: the Circular Quarter-Mode Textile Antenna," *IEEE Antennas Propag. Lett.*, vol. 1225, no. c, pp. 1536–1225, 2015.

[54]   S. Agneessens and H. Rogier, "Compact Half Diamond Dual-Band Textile HMSIW On-Body Antenna," *IEEE Antennas Propag. Mag.*, vol. 62, no. 5, pp. 2374–2381, 2014.

[55]   M. Bozzi, A. Georgiadis, and K. Wu, "Review of Substrate-Integrated Waveguide Circuits and Antennas," *IET Microw. Antennas Propag.*, vol. 5, no. 8, p. 909, 2011.

[56]   O. Caytan, S. Lemey, S. Agneessens, *et al.*, "Half-Mode Substrate-Integrated-Waveguide Cavity-Backed Slot Antenna on Cork Substrate," *IEEE Antennas Propag. Lett.*, vol. 15, pp. 162–165, 2016.

[57]   S. Lemey, O. Caytan, D. Vande Ginste, P. Demeester, and H. Rogier, "SIW Cavity-Backed Slot (Multi-) Antenna Systems for the Next Generation IoT Applications," in *IEEE Topical Conference on Wireless Sensors and Sensors Networks (WISNet), 2016*, pp. 75–77, 2016.

[58]   K. Elmahgoub, T. Elsherbeni, F. Yang, A. Z. Elsherbeni, L. Sydänheimo, and L. Ukkonen, "Logo-Antenna Based RFID Tags for Advertising Application," *Appl. Comput. Electromagn. Soc. J.*, vol. 25, no. 3, pp. 174–181, 2010.

[59]   J. Tak, S. Member, J. Choi, and S. Member, "An All-Textile Louis Vuitton Logo Antenna," *IEEE Antennas Propag. Lett.*, vol. 1225, pp. 3–6, 2015.

[60]   J. C. G. Matthews and G. Pettitt, "Development of Flexible, Wearable Antennas," *IEEE Conf. Publ.*, pp. 273–277, 2009.

[61]   M. L. Scarpello, I. Kazani, C. Hertleer, H. Rogier, and D. Vande Ginste, "Stability and Efficiency of Screen-Printed Wearable and Washable Antennas," *IEEE Antennas Wirel. Propag. Lett.*, vol. 11, pp. 838–841, 2012.

[62]   L. Yao and Y. Qiu, "Design and Fabrication of Microstrip Antennas Integrated in Three Dimensional Orthogonal Woven Composites," *Compos. Sci. Technol.*, vol. 69, pp. 1004–1008, 2009.

[63]   S. Sankaralingam and G. Bhaskar, "Determination of Dielectric Constant of Fabric Materials and Their Use as Substrates for Design and Development of Antennas for Wearable Applications," *IEEE Trans. Instrum. Meas.*, vol. 59, no. 12, pp. 3122–3130, 2010.

[64]   A. Moretti, G. N. Malheiros-silveira, E. Hugo, and M. S. Gonçalves, "Characterization and Validation of a Textile Substrate for RF Applications," in *Microwave & Optoelectronics Conference (IMOC), 2011 SBMO/IEEE MTT-S International*, pp. 546–550, 2011.

[65]   F. Declercq, H. Rogier, and C. Hertleer, "Permittivity and Loss Tangent Characterization for Garment Antennas Based on a New Matirx-Pencil Two-Line Method," *IEEE Trans. Antennas Propag.*, vol. 56, no. 8, pp. 2548–2554, 2008.

[66] M. Mantash, A. C. Tarot, S. Collardey, and K. Mahdjoubi, "Investigation of Flexible Textile Antenna and AMC Reflectors," *Int. J. Antennas Propag.*, pp. 1–10, 2012.

[67] M. Klemm and G. Troster, "Textile UWB Antennas for Wireless Body Area Networks," *IEEE Antennas Propag.*, vol. 54, no. 11, pp. 3192–3197, 2006.

[68] R. Gonçalves, R. Magueta, P. Pinho, and N. B. Carvalho, "Dissipation Factor and Permittivity Estimation of Dielectric Substrate Using a Single Microstrip Line Measurement," *Appl. Comput. Electromagn. Soc. J.*, vol. 31, no. 2, pp. 118–125, 2016.

[69] A. S. A. Bakar, M. I. Misnon, D. K. Ghodgaonkd, *et al.*, "Comparison of Electrical Physical and Mechanical Properties of Textile Composites Using Microwave Nondestructive Evaluation," in *RF and Microwave Conference*, pp. 164–168, 2014.

[70] E. Hakansson, A. Amiet, and A. Kaynak, "Dielectric Characterization of Conducting Textiles Using Free Space Transmission Measurements: Accuracy and Methods for Improvement," *Synth. Met.*, vol. 157, pp. 1054–1063, 2007.

[71] W. L. Balls, "Dielectric Properties of Raw Cotton," *Nature*, vol. 158, pp. 9–11, 1946.

[72] J. Lesnikowski, "Dielectric Permittivity Measurement Methods of Textile Substrate of Textile Transmission Lines," *Electr. Rev.*, vol. 88, no. 3, pp. 148–151, 2012.

[73] J. Lilja, P. Salonen, P. D. Maagt, and N. K. Zell, "Characterization of Conductive Textile Materials for Softwear Antenna," in *Antennas and Propagation Society International Symposium (APSURSI) 2009*, pp. 1–4, 2009.

[74] O. Himdi and M. Lafond, "Printed Millimeter Antennas – Multilayer Technologies," in D. Lui, U. Pfeiffer, J. Gryzb, and B. Gaucher, (Eds). *Advanced Millimeter-Wave Technologies: Antennas, Packaging and Circuits*, Chichester: Wiley Interscience, 2009.

[75] D. Gershon, J. P. Calame, Y. Carmel, *et al.*, "Adjustable Resonant Cavity for Measuring the Complex Permittivity of Dielectric Materials," vol. 3207, no. 2000, pp. 12–15, 2007.

[76] U. Faz, U. Siart, T. F. Eibert, S. Member, and T. Hermann, "Electric Field Homogeneity Optimization by Dielectric Inserts for Improved Material Sensing in a Cavity Resonator," vol. 64, no. 8, pp. 2239–2246, 2015.

[77] A. Kumar and D. G. Smith, "Microwave Properties of Yarns and Textiles Using a Resonant Microwave Cavity," *IEEE Trans. Instrum. Meas.*, vol. 26, no. 2, pp. 95–98, 1977.

[78] G. Zou, P. Starski, and J. Liu, "High Frequency Characteristics of Liquid Crystal Polymer for System in a Package Application," in *8th Int. Symp. Adv. Packag. Mater.*, pp. 337–341, 2002.

[79] B. Roy and S. K. Choudhury, "Characterization of Textile Substrate to Design a Textile Antenna," in *International Conference on Microwave and Photonics (ICMAP), 2013*, pp. 1–5, 2013.

[80] I. J. Bahl and S. S. Stuchly, "Analysis of a Microstrip Covered with a Lossy Dielectric," *IEEE Trans. Microw. Theory Tech.*, vol. 28, no. 2, pp. 104–109, 1980.

[81]　M. Bogosanovich, "Microstrip Patch Sensor for Measurement of the Permittivity of Homogeneous Dielectric Materials," *IEEE Trans. Instrum. Meas.*, vol. 49, no. 5, pp. 1144–1148, 2000.

[82]　S. Hausman, Ł. Januszkiewicz, M. Michalak, T. Kacprzak, and I. Krucińska, "High Frequency Dielectric Permittivity of Nonwovens," *Fibers Text. Eastern Eur.*, vol. 14, no. 5, pp. 60–63, 2006.

[83]　Agilent Technologies, "Measuring Dielectric Properties Using Agilent's Materials Measurement Solutions." USA, pp. 1–4, 2014. Available at: http://www.academia.edu/6472263/Measuring_Dielectric_Properties_using_ Agilents_Materials_Measurement_Solutions

[84]　S. Kawabata, "Measurement of the Mechanical Properties of Fabrics," in *The Standarization and Analysis of Hand Evaluation*, 2nd ed., Osaka: The Textile Machinery Society of Japan, pp. 28–57, 1980.

Chapter 6

Human-movement identification using the radio signal strength in WBAN

Sukhumarn Archasantisuk[1], Takahiro Aoyagi[2],
Tero Uusitupa[3], Minseok Kim[4], and Jun-ichi Takada[5]

In this chapter, an intensive study of a novel human movement identification scheme using the radio signal strength in wireless body area network (WBAN) is presented. Since the WBAN channel characteristics are highly influenced by the human movement, the radio signal strength and its temporal variation can be used to determine the human movement without additional tools such as an accelerometer/gyroscope. This study included the process of developing the human movement identification system, performance assessment on different conditions including a vector size of the received signal levels, an antenna orientation, a classifier training algorithm, and a receiver location. It was found that the vector size of the received signal levels and the receiver location had strong impact on the identification accuracy. More than 80% of the identification accuracy can be achieved when using 30–40 received signal levels or the receiver location at thigh or upper arm. In addition, a feature selection method based on a correlation coefficient was used to remove redundant and less informative features. The classification results show that the comparable performance to the all feature vectors can be achieved by the subset feature vector with a lower computational cost.

6.1   Introduction

In a WBAN, wireless sensors are placed on/inside the human body to gather vital signs such as heart rate, blood pressure, body temperature, respiration rate, and oxygen saturation rate and then deliver the collected data to a central node, which is called

[1]Department of Human System Science, Graduate School of Decision Science and Technology, Tokyo Institute of Technology, Japan
[2]Department of Electrical and Electronic Engineering, School of Engineering, Tokyo Institute of Technology, Japan
[3]Department of Radio Science and Engineering, Aalto University School of Electrical Engineering, Finland
[4]Graduate School of Science and Technology, Niigata University, Japan
[5]Department of Transdisciplinary Science and Engineering, School of Environment and Society, Tokyo Institute of Technology, Japan

a coordinator. The coordinator can connect to an external network through a cellular network or a wide area network. The WBAN serves not only medical applications but also nonmedical applications. One important challenge in WBAN designs is that WBAN channels are affected by energy absorption, reflection, scattering, diffraction, and shadowing due to variations of different dielectric properties of human body tissue, such as the electrical conductivity and permittivity [1], and surrounding environment. Furthermore, since the sensor nodes are mounted on the human body, the WBAN channels often show dynamic characteristics as a result of a human movement. The WBAN channels were studied in a scenario-based approach and it was found that human motion is a dominant factor contributing to a great variation of the dynamic channels [2,3]. Time-varying channel fading was studied for seven different human actions using simulation [4]. The study showed that unintentional movements, such as breathing, caused a short-term fading, while a mid-term and a long-term channel fadings were caused by intentional movements. Many studies on the WBAN channel characteristics and channel models in dynamic scenarios can be found in [5–8].

A channel-fading phenomenon is generally not desired in any network since communication quality severely deteriorates under deep fading condition. However, since the body movements highly influence WBAN channel characteristics, these signal fluctuation patterns can be used to detect or identify the human movement. The information of human movements or human activities can be used in many different domains of applications, for example, to detect fall-related events [9], to assess a level of physical activities [10], to provide contextual knowledge for context-aware applications [11], to encourage fitness and healthy living against a sedentary lifestyle [12], to assist dependent-living elderly [13,14], to analyze motions in a sport training [15], and to support interactive games [16,17].

This chapter provides an intensive study of the human-movement identification using the radio signal strength in WBAN. Contents of this chapter include a summary of relevant studies in Section 6.2. A development process of the human-movement classifier, which is a tool used for identifying the human movement, is thoroughly explained in Sections 6.3–6.6. After that, the performance of the classifier is validated in Section 6.7. In addition, in Section 6.8, feature selection is performed to reduce computational overheads of the movement-identification process but still retains acceptable accuracy. Lastly, a summary of the development of the human-movement-identification system and the performance evaluation results is provided in Section 6.9.

6.2 Related works

Human-movement recognition and identification/classification has gained high attention from researchers in many disciplines, and hence there are many different approaches or methods that have been intensively studied, for instance, computer vision, micro Doppler signature, physiological sensors, or environmental sensing devices. These approaches try to achieve desired performance criteria of the human-movement identification, such as high accuracy rate, high robustness, and low latency.

Aiming at WBAN applications, we limit our attention to using body-mounted sensors or radio signal strength to identify the human movements. An accelerometer-equipped device receives the most recognition as a tool for the human-movement identification since it can directly measure acceleration of a body part to which it is attached. It was suggested in [18] that accelerometer data provided dominant features among a variety of physical sensing data, for example, an electrocardiogram signal, heart rate, respiratory rate, skin temperature, for the human-activity classification. In the paper, both time-domain and frequency-domain features were computed from the accelerometer data to classify seven activities including lying, sitting/standing, walking, Nordic walking, running, rowing, and cycling. These features achieved 82–86 percent of correct classification results. Recently, it was shown in [19] that using only time-domain features computed from 3 accelerometers placed at the wrist, chest, and ankle can accurately classify 12 human activities, which included sitting, standing, walking, running, cycling, Nordic walking, ascending stair, descending stair, vacuum cleaning, ironing clothes, jumping rope, and lying down, with 98 percent of accuracy rate. The accelerometer is a very comprehensive device for the human-movement-identification problem. However, an additional unit to measure the acceleration increases the complexity of a form factor for sensor nodes, and extra energy is required for this measurement unit.

In wireless communication, environmental change or a movement of sensor nodes causes a fluctuation footprint in received signal strength. The signal fluctuation pattern can thus be used to determine a certain context of the network. This approach enables a context identification system, for instance, the human-movement identification utilizing only the existing radio frequency (RF) signals. Therefore, costs of integrating a special sensing unit can be avoided. Based on the idea of utilizing the RF signals to identify the surrounding context of the wireless network, the RF signal fluctuation was used to detect movement in [20]. In addition, a simple velocity estimation method was also demonstrated in the paper. Later, Muthukrishnan *et al.* presented motion detection algorithms based on spectral and spatial analysis of wireless local area network signal [21]. Both algorithms were used to identify whether the user is moving or stationary. The algorithm, based on spectral analysis, performed better than the other algorithm that was based on spatial analysis. RF signal-based human activity identification was developed further to classify between four different human activities including crawling, lying, standing and walking, and to identify the empty environment [22]. In this study, the RF devices were fixed in positions on the hallway, where subjects performed the mentioned activities. Both frequency-domain and time-domain features were used by a k-nearest neighbor (k-NN) and a decision tree to identify the human activities or the empty hallway with 67.4–90.7 percent of accuracy rate. However, the classification accuracy deteriorated as the distance between RF devices and subjects increased.

RF-based human-movement identification using GSM cellular signal strength was introduced in [23]. In this paper, the sum of GSM signal strength and the number of detectable cells observed over some period of time were used by a neural network to identify between walking, boarding on a car, or stationary. This GSM-based approach achieved 89 percent of accuracy rate, but the prediction model had to be retrained if it was applied in different environment, such as rural and urban areas. More robust

GSM-based motion classification, which well performed across different environment, was then demonstrated in [24] using a hidden Markov model and automatic unsupervised calibration. Another GSM-based mobility detection can be found in [25], which introduced a number of features that can be extracted from GSM traces and obtained 85 percent of accuracy rate when classifying between walking, driving, or stationary. Mun *et al.* combined GSM signal strength and WiFi signal strength data to distinguish between dwelling state, walking, and driving, and obtained 83 percent of correct classification [26]. We can see from the abovementioned literature that the RF signal strength fluctuation caused by presence or motions of the human can be used to develop the human-movement identification system.

However, there has been little research considering the use of RF signal strength for the human-movement identification in WBAN. Ekure *et al.* performed a theoretical analysis of the feasibility of path-loss-based activity recognition in WBAN [27]. In the paper, the channel characteristics were numerically simulated from the average distance between a transmitter and a receiver for a variation of standing and sitting postures. Simple features such as difference from maximum, mean, and median path loss were used by a support vector machine (SVM) to identify the postures. The accuracy rate obtained from this method was 60–69 percent. Even though the classification result shown in this paper was not highly accurate, we can see that it was feasible to apply RF-based human-movement identification in WBAN. Performance comparison between RF-based and accelerometer-based human activity classification can be found in [28]. It was shown that the RF-based method obtained 75–96.67 percent of accuracy rate, while the accelerometer-based method obtained 68.3–91.7 percent of accuracy rate when both were used to classify between standing, walking, and jogging motion. Recently, Geng *et al.* applied the RF-based motion identification for emergency operation of fire fighters [29]. Seven candidate motions, namely, standing, walking, running, lying, crawling, climbing, and running upstairs, were included. Both frequency-domain and time-domain features were extracted from the received radio signal strength to classify the candidate motions. The paper showed that 82.72–90.42 percent of accuracy rate was achieved.

In the present study, we investigate the performance of RF-based human-movement identification considering only time-domain features. We also determine the dominant time-domain features for the human-movement identification problem. Frequency-domain features are avoided in this study due to its computational complexity. Furthermore, this study provides not only classification results between different dynamic movements but also investigates how well the RF-based approach can differentiate between static postures with unconscious movement, such as sitting, standing, or lying.

6.3　Human motion classification system

In this study, supervised learning was applied to construct a machine to identify human movements. The constructed machine is called a classifier since it will decide the movement class based on examples given during a training process. Figure 6.1

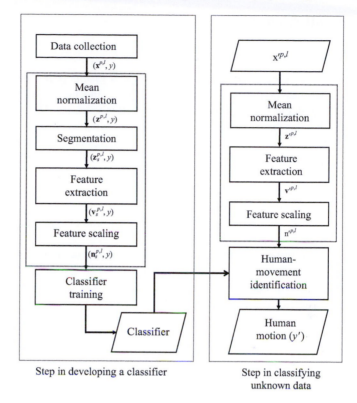

Figure 6.1   A schematic of the human-movement identification using the radio signal strength

illustrates the steps of developing the classifier for the human-movement identification using the radio signal strength and how to use the trained classifier to identify the human movement. In this section, we focus on the construction of the classifier, which is shown in the left side of the figure. The construction of the classifier starts with data collection. WBAN measurements are conducted to collect WBAN channels of several human movements. After that, the WBAN channel gains are pre-processed. Data preprocessing includes mean normalization, data segmentation, feature extraction, feature scaling as shown in the dashed-line rectangle. In general, the preprocessing step is to prepare input vectors for the classifier. The classifier is trained with different machine-learning techniques. After the classifier training, a validation process is then performed to estimate how well the classifier performs on unseen data—the data that are not used to train the classifier. The use of the classifier to identify the human movement is demonstrated in the right side of the figure. The segmentation is excluded when the trained classifier is used since we assume that a stream of received signal levels is processed in real-time.

6.4   Data collection

6.4.1   Measurement campaign

Measurements were conducted to collect the WBAN channels for several human movements. The measurements were performed in an anechoic chamber at Tokyo Institute of Technology. Two 3-port dielectric resonator antennas (DRAs) were used [30,31]. The DRA has three polarization modes, which can provide three independent channels. Three modes of the DRA are a monopole mode, denoted by M, and TEy_{111} and TEx_{111} modes, denoted by P(X) and P(Y), respectively, of which radiation patterns are similar to x-polarized and y-polarized microstrip antennas. The DRA was placed in a direction so that the directions of patch modes were parallel to human body, while the monopole was perpendicular to the human body [32]. A 3×3 MIMO channel sounder was used in the measurement so that nine channels were recorded concurrently [33].

Figure 6.2 illustrates a schematic of the measurement system. A signal generator generated a continuous wave (CW) at 2.45 GHz. The CW was then split into a reference line and a transmission line. To achieve the 3×3 MIMO channel measurement, a microcontroller controlled a switch (SW) to feed the signal to each DRA port, hence each transmitter can transmit in a time-division-multiplexing manner. The microcontroller also triggered a digital sampling oscilloscope (DSO). The DSO was equipped with four ports, where three ports were used for capturing the signal from receivers and amplifiers, and the other port was used to capture the reference signal. The frame rate of the DSO was 30 frames/s, which was sufficient to capture the body

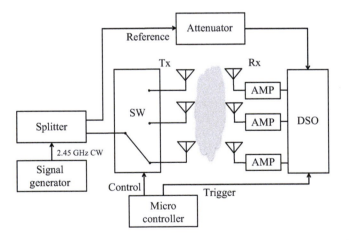

Figure 6.2   A schematic of the system used in the WBAN measurement. ©2016 IEICE. Reproduced, with permission, after Reference Archasantisuk S., Aoyagi T., Uusitupa T., Minseok K., Takada J. 'Human motion classification using radio signal strength in WBAN'. IEICE Transactions on Communications. 2016;99(3):592–601

movement. In other words, 30 snapshots of the 3×3 channel matrix were obtained in one second. The channel response is then obtained by comparing the generated signal and the received signal.

The DRAs were mounted on polystyrene foam, whose dimension was 10 cm $\times$ 10 cm $\times$ 3 cm (width $\times$ length $\times$ height). The polystyrene foam was stuck on a denim belt. The denim belt was then attached to the human body by velcro tape. Two subjects participating in this measurement campaign are Japanese males whose heights and weights are 165 cm, 72 kg and 178 cm, 65 kg, respectively. Both subjects performed five movements: standing still on their feet with arms at their sides; sitting on a chair with hands resting on their own lap; lying flat on the back with face up; walking on the spot at step frequency nearly 2 Hz; and running on the spot at step frequency nearly 4 Hz. For the walking and running movements, a second ticking sound was played to keep a constant speed. Subject 1 performed each movement for 30 s, while subject 2 performed each movement for 15 s. The WBAN channels were recorded in one-to-one link, where a transmitter position was fixed at the navel. The subjects performed each movement four times for measuring the WBAN channels of different four receiver positions. The receiver positions included ankle, thigh, upper arm, and wrist. Let $\mathbf{x}^{p,l} = (x_1^{p,l}, x_2^{p,l}, \ldots, x_N^{p,l})$ denote a time series of the path gains (dB), which is interchangeably referred to as the received signal levels for the rest of this chapter, where p is a pair of transmitting and receiving antennas where the first letter denotes the transmitting antenna and the latter letter denotes the receiving antenna, l is a location of the receiver on the human body, and N is the number of total samples of each measurement. The WBAN channels collected from this measurement campaign are denoted by $(\mathbf{x}^{p,l}, y)$, where y denotes a human-movement class. A summary of the WBAN measurement conditions is shown as follows:

$p \in \{$P(X)–P(X), P(X)–P(Y), P(X)–M, P(Y)–P(X), P(Y)–P(Y), P(Y)–M, M–P(X), M–P(Y), M–M$\}$
$l \in \{$ankle, thigh, upper arm, wrist$\}$
$y \in \{$lying, sitting, standing, running, walking$\}$
$N = 900$ for subject 1 and 450 for subject 2

6.4.2 Measurement results

Figure 6.3 shows path gains of the WBAN M–M channel from subject 1. Each subfigure shows path gains at different receiver locations including ankle, thigh, upper arm, and wrist. The sitting, standing, and lying were static postures with some unconscious movements; hence, the radio signal fluctuation was relatively small. Although involuntary movements, such as breathing, can cause the small fluctuation of the received signal levels as seen in the lying or standing case, the signal fluctuation was not so significantly compared to the fluctuation caused by walking or running movement. The median of the signal fluctuation ranges of standing movement across all receiver locations and all channels was 2.0 and 1.2 dB for subject 1 and 2, respectively. Similarly, the median of the signal fluctuation ranges of lying movement across all sensor locations and all channels was 1.0 and 1.2 dB for subject 1 and 2, respectively. However, for sitting, the large signal fluctuation was observed at the ankle during 17–25 s.

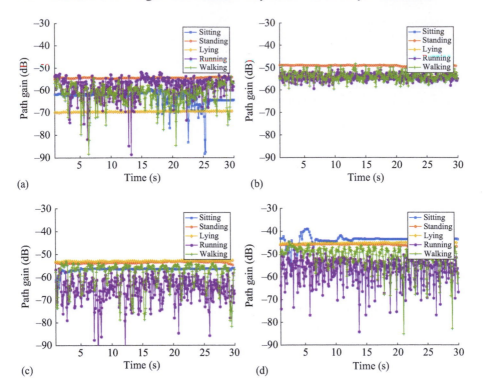

Figure 6.3   *Measurement results of subject 1: (a) WBAN path gains of M–M channel at ankle, (b) WBAN path gains of M–M channel at thigh, (c) WBAN path gains of M–M channel at upper arm, and (d) WBAN path gains of M–M channel at wrist*

In addition, the moderate signal fluctuation was observed at the upper arm during 1–7 s and at the wrist during 1–13 s as seen from the figure. The other channels also had the similar fluctuation in sitting movement. Furthermore, the WBAN channels at the ankle from subject 2 also moderately fluctuated. The WBAN channels of sitting movement with the moderate signal fluctuation were used since people tend to occasionally move their body during sitting, especially the limbs. On the other hand, walking and running, which were dynamic movements, had the rapid, strong, and constant signal fluctuation.

6.5   Data preprocessing

Data preprocessing is an important step to prepare data for training the classifier or for the classifier to determine an output class. The data preprocessing includes mean normalization, segmentation, feature extraction, and feature scaling.

6.5.1   Mean normalization

A reference path gain was obtained by averaging all path gains in a reference scenario. Standing movement was used as the reference scenario in this study. Relative path gains were obtained by normalizing the instantaneous path gains with the reference path gain. The relative path gains can partially remove a variation of the received signal levels from different body size and shape between the subjects. A time series of the relative path gains, denoted by $\mathbf{z}^{p,l} = (z_1^{p,l}, z_2^{p,l}, \ldots, z_N^{p,l})$, was obtained using (6.1), where $x_{\text{ref}}^{p,l}$ is the reference path gain.

$$z_i^{p,l} = x_i^{p,l} - x_{\text{ref}}^{p,l} \tag{6.1}$$

6.5.2   Segmentation

Data segmentation was performed using a windowing process, which is a technique to partition a large set of data into smaller subsets. A rectangular window was used to extract a segment of data within the window. Then, the window slid along the stream of data to extract other segments. There are two methods for the windowing process: adjacent and overlapping windowings. The adjacent windowing extracts disjoint segments of data from the dataset. The drawback of this method is that a small portion of data is obtained. The overlapping windowing, on the other hand, extracts segments of data by sliding the window less than the window size, thus resulting in overlapping data in different segments. Let $\mathbf{z}_s^{p,l} = (z_{s-W+1}^{p,l}, z_{s-W+2}^{p,l}, \ldots, z_s^{p,l})$ denote a segment of the relative path gains with the ending point at $z_s^{p,l}$, where $W \leq s \leq N$, and W is the window size. The overlapping windowing process was adopted in this work. To investigate the impact of the window size to the performance of the human-movement identification, different window sizes, which were 10, 20, 30, and 40 samples, were used and the window moved five steps from a current position to extract other segments of the relative path gains.

6.5.3   Feature extraction

Feature extraction is a step to provide a derived value, which is more informative than the raw value, to the classifier. Since WBAN devices have limited resources for computation, only time-domain features were considered. The candidate features adopted in this study were taken from those used in surface electromyography (sEMG) signal recognition for hand movements [34]. However, modifications of the feature computation were made to suit our input data, which were the RF signal strength.

6.5.3.1   Feature computation

Integrated received signal strength
The received signal strength implicitly provides information of the distance between the transmitter and the receiver. Integrated received signal strength (IRSS) is the summation of the signal strength over a windowing period.

$$\text{IRSS} = \sum_{j=1}^{W} z_j \tag{6.2}$$

Mean value

Since the received signal level could be used to estimate the distance between the transmitter and the receiver, thus a mean value (MV) is used to approximate the average distance between the transmitter and the receiver over a windowing period.

$$MV = \frac{1}{W}\sum_{j=1}^{W} z_j \tag{6.3}$$

Modified mean value1

A modified MV1 (MMV1) is a variation of the MV using a weighted window function.

$$MMV1 = \frac{\sum_{j=1}^{W} w_j z_j}{\sum_{j=1}^{W} w_j}$$

$$w_j = \begin{cases} 1 & \text{if } 0.25W \le j \le 0.75W \\ 0.5 & \text{otherwise} \end{cases} \tag{6.4}$$

Modified mean value2

This modified MV2 (MMV2) is another variation of the MV. The difference between MMV1 and MMV2 is the weighted window function. While MMV1 uses the simple weighted window function, MMV2 uses a continuous weighted function, which provides additional smoothness.

$$MMV2 = \frac{\sum_{j=1}^{W} w_j z_j}{\sum_{j=1}^{W} w_j}$$

$$w_j = \begin{cases} 1 & \text{if } 0.25W \le j \le 0.75W \\ 4j/W & \text{if } j < 0.25W \\ 4(W-j)/W & \text{if } j > 0.75W \end{cases} \tag{6.5}$$

Mean value slope

MV slope (MVSLP) is the difference of the MVs of two adjacent subsegments. This feature represents the variation of the average received signal strength.

$$MVSLP = \frac{1}{W-(W/2)}\sum_{j=(W/2)+1}^{W} z_j - \frac{1}{W/2}\sum_{j=1}^{W/2} z_j \tag{6.6}$$

Simple square integral

SSI is a commonly used feature in sEMG signal recognition to capture energy of the sEMG signal. In this study, SSI feature is used to capture size of the received RF signal.

$$SSI = \sum_{j=1}^{W} z_j^2 \tag{6.7}$$

Variance

Variance (VAR) of the received signal levels shows how much the instantaneous received signal levels deviate from the local mean.

$$\text{VAR} = \frac{1}{W-1} \sum_{j=1}^{W} (z_j - \mu)^2 \tag{6.8}$$

Root mean square

RMS is also known as a quadratic mean. It is used to measure the magnitude of the received signal levels regardless of its sign.

$$\text{RMS} = \sqrt{\frac{1}{W} \sum_{j=1}^{W} z_j^2} \tag{6.9}$$

Cumulative signal change

Cumulative signal change (CSC) is the cumulative difference between two adjacent received signal levels. This feature is used to distinguish between dynamic movements and static movements.

$$\text{CSC} = \sum_{j=1}^{W-1} |z_{j+1} - z_j| \tag{6.10}$$

Level crossing rate

LCR at threshold ρ is the number of times that the received signal level passes the value ρ in a positive direction. This feature provides an estimation of frequency domain properties with simpler computation than a frequency-domain feature that requires some transformation such as the fast Fourier transform function.

$$\text{LCR} = \sum_{j=1}^{W-1} [(z_j \leq \rho) \wedge (z_{j+1} > \rho)] \tag{6.11}$$

Slope sign change

Slope sign change (SSC) is another feature that estimates the frequency information of the received signal levels. SSC is defined as the number of times that the received signal levels change between a positive and a negative slope. A threshold value is set to disregard the background noise.

$$\text{SSC} = \sum_{j=2}^{W-1} [((z_j - z_{j-1}) \times (z_{j+1} - z_j) < 0) \wedge f(|z_j - z_{j-1}| + |z_{j+1} - z_j|)] \tag{6.12}$$

$$f(z) = \begin{cases} 1 & \text{if } z \geq threshold \\ 0 & \text{otherwise} \end{cases}$$

Willison amplitude

WAMP is defined as a number of times that the difference between two adjacent received signal levels is greater than a threshold value.

$$\text{WAMP} = \sum_{j=1}^{W-1} f(|z_{j+1} - z_j|)$$

$$f(z) = \begin{cases} 1 & \text{if } z \geq threshold \\ 0 & \text{otherwise} \end{cases}$$

(6.13)

Range

Range is defined as the difference between the local maximum and minimum received signal level in the same windowing period. This feature is sometimes called peak-to-peak amplitude.

$$\text{Range} = \max(\mathbf{z}) - \min(\mathbf{z})$$

(6.14)

Each feature is computed from the segment of the WBAN relative path gains. A feature vector is denoted by $\mathbf{v}_s^{p,l} = (v_{s,1}^{p,l}, v_{s,2}^{p,l}, \ldots, v_{s,D}^{p,l})$, where $v_{s,d}^{p,l}$ is the dth feature computed from $\mathbf{z}_s^{p,l}$ and D is the number of total features.

6.5.3.2 Optimal threshold selection

Each of the three features, namely, LCR, SSC, and WAMP, is computed according to a specific threshold value. LCR is used to measure the frequency of signal fading corresponding to a threshold. SSC is used to estimate the frequency of change from the improving channel to the deteriorating channel and vice versa. Similarly, WAMP measures the frequency of the change in the received signal amplitudes. Since these three features represent the frequency properties of the received signal, they are mainly used to discriminate static movements from dynamic movements and to distinguish between different dynamic movements. Therefore, the optimal threshold is selected from the ability to identify walking and running movements. In other words, the optimal threshold is the value that can maximize the identification rate of walking and running movements. A criterion used for measuring the rate of correct identification of a particular class is a true positive (TP) rate (TPR). The computation of the TPR is explained in detail in Section 6.7. To provide the common threshold value for different subjects, the optimal threshold is selected from the average TPR of all subjects' data.

The channel characteristics of WBAN vary depending on a sensor location and an antenna orientation. Therefore, the optimal threshold should be selected for each pair of the sensor location and antenna orientation. LCR counts the number of times that a signal level crosses the threshold value. For LCR, the following threshold values were considered: -15, -10, -5, and 0 dB. For SSC and WAMP, the threshold value was different from LCR since it was set to ignore very small signal fluctuation. The threshold values, which were 0.5, 1.0, 1.5, and 2.0 dB, were considered for SSC and WAMP. To evaluate the impact of these threshold values on the performance of the classifier, the classifier training and validation was performed using only one feature at a time. A decision tree was used to find the optimal threshold since it was

Table 6.1   *Optimal threshold values of LCR, SSC, and WAMP for individual sensor location and antenna pair*

Tx–Rx antenna	Threshold values (LCR, SSC, WAMP)			
	Ankle	**Thigh**	**Upper arm**	**Wrist**
P(X)–P(X)	−5, 1.5, 1.5	0, 1.5, 1.5	−5, 1.0, 0.5	−5, 1.0, 1.5
P(X)–P(Y)	0, 1.5, 1.5	0, 1.0, 1.5	−10, 1.0, 1.0	−10, 1.0, 1.0
P(X)–M	−5, 1.5, 1.5	0, 1.0, 0.5	−10, 1.0, 0.5	−15, 1.0, 1.0
P(Y)–P(X)	−10, 2.0, 1.0	−5, 2.0, 1.5	−10, 1.0, 1.0	−10, 1.5, 2.0
P(Y)–P(Y)	0, 2.0, 2.0	0, 1.0, 1.5	−5, 0.5, 1.5	0, 2.0, 1.0
P(Y)–M	0, 2.0, 2.0	−5, 2.0, 1.5	−15, 1.5, 1.0	−15, 1.5, 1.5
M–P(X)	−10, 2.0, 1.0	0, 1.5, 1.0	−10, 0.5, 0.5	−10, 1.0, 1.0
M–P(Y)	−5, 1.5, 0.5	0, 2.0, 1.5	−10, 1.0, 1.0	−5, 1.0, 1.5
M–M	−5, 0.5, 0.5	−5, 1.0, 0.5	−10, 0.5, 1.0	−10, 0.5, 1.5

the fastest among three algorithms considered in this work. The optimal threshold was the threshold value that can maximize the average TPR of running and walking movement. Table 6.1 summarizes the optimal threshold values of each feature for individual receiver location and antenna pair. These threshold values were used in the feature extraction to provide the feature vector for the classifier training and validation.

6.5.4   Feature scaling

Feature scaling is a method used to control a range of independent features. Some algorithms calculate the distance between feature vectors in the feature space. Features with a broad range of values can dominate other features with a small range, thus misleading the classification. Therefore, all features were scaled to the range of $[n_{min}, n_{max}]$ using (6.15), where $v_{min,d}^{p,l}$ is the minimum value of feature d, and $v_{max,d}^{p,l}$ is the maximum value of feature d with respect to conditions of p and l. A scaled feature vector is denoted by $\mathbf{n}_s^{p,l} = (n_{s,1}^{p,l}, n_{s,2}^{p,l}, \ldots, n_{s,D}^{p,l})$.

$$n_{s,d}^{p,l} = \frac{v_{s,d}^{p,l} - v_{min,d}^{p,l}}{v_{max,d}^{p,l} - v_{min,d}^{p,l}} \times (n_{max} - n_{min}) + n_{min} \tag{6.15}$$

6.6   Classifier training

Three algorithms, which appeared in the top ten machine-learning algorithms [35], were used to solve the human-movement identification problem. These three algorithms were k-NN, SVM, and decision tree. This section gives a brief explanation of how these three algorithms work. The parameter tuning used in this study for each algorithm is also given.

6.6.1 k-Nearest neighbor

A k-NN is one of the instance-based methods with lazy and competitive learning [36]. The k-NN does not need the explicit training process, which implies that the training phase of the k-NN is very simple and fast. However, it needs considerable memory space to store all training data for the classification. The k-NN can easily solve a classification problem by finding k closest instances in the feature space. The common class among k instances represents the class of an unknown object.

In this study, the Euclidean distance was used to compute the distance between instances. Three different k-NN sizes—the number of neighbors—including 1NN, 10NN, and 100NN, were examined. It was found that the 10NN performed the best in our problem; hence, the 10NN was used.

6.6.2 Support vector machine

A SVM is one of the most powerful techniques to solve a classification problem [37]. For nonlinearly seperable data, a kernel function is needed to map a feature vector into a higher dimensional vector, which tends to be more linearly seperable.

A radial basis function (RBF) kernel was used in this study. A penalty term is a parameter that trades off between the number of misclassifications and the margin between a hyperplane—the boundary that linearly separates data points into different classes—and the closest data points. The penalty term was set to 1 and the kernel scale in the RBF kernel was set to 3.6. Conventionally, the SVM is used to solve a binary classification problem. To apply the SVM to our problem, the one-against-one SVM was used.

6.6.3 Decision tree

A decision tree is a predictive model where properties of unknown instances are tested in nonterminal nodes. The unknown object travels down the tree according to a decision at each node until it reaches a terminal node, where the output class is labeled.

The decision tree was constructed using the CART algorithm [38], which a node split was based on the Gini's diversity index. The maximum number of the node splits was limited to 20 in this study. Note that the construction and classification of the decision tree can be performed without the feature scaling.

6.7 Classifier validation

A k-fold cross validation is suitable for a small data set since it efficiently utilizes all data for training the model. Therefore, the 5-fold cross validation was used to assess the performance of the classifier model.

6.7.1 Validation metrics

Table 6.2 demonstrates a confusion matrix in which the classification results are presented. Columns of the matrix contain the instances of an actual class. Rows of

Table 6.2   Confusion matrix

		Actual class	
		Target movement	**Other movements**
Predicted class	Target movement	True positive	False positive
	Other movements	False negative	True negative

the matrix contain the instances of a predicted class. Definitions of the terminologies used in the confusion matrix are as follows: TP refers to the instances of a target movement that are correctly classified; false positive (FP) refers to the instances of other movements that are classified as the target movement; false negative (FN) refers to the instances of the target movement that are classified as other movements; and true negative (TN) refers to the instances of other movements that are correctly classified.

An accuracy rate is the percentage of correctly classified outputs over all predictions. The accuracy rate indicates how well the classifier can identify the human movement in general. The accuracy rate is calculated as in (6.16):

$$\text{Accuracy rate} = \frac{TP + TN}{TP + FN + FP + TN} \times 100\% \qquad (6.16)$$

A TPR is sometimes referred to as recall or sensitivity. For any target output class, the TPR is used for measuring an ability to detect the positive condition. The TPR is applicable for only binary classification. Therefore, to apply this metric, the one-against-all evaluation was used to measure the TPR of each movement class. The TPR is calculated as shown in (6.17). In contrast to the accuracy rate, the TPR provides the indication of the classifier's ability to identify the target movement.

$$\text{TPR} = \frac{TP}{(TP + FN)} \times 100\% \qquad (6.17)$$

6.7.2   Validation results

Four conditions, which are the window size used in the segmentation, antenna orientation, classifier algorithm, and receiver location, were investigated. The parameters of these conditions are shown in Table 6.3. There were total 432 combinations of the classification parameters. The validation was independently performed for each combination of these parameters.

6.7.2.1   Window size of the segmentation

The effect of the different window sizes, which are 10, 20, 30, and 40 samples, on the classification results was examined. A cumulative distribution function (CDF) of the accuracy rate for each window size was computed. For each window size, there were 108 ($9 \times 3 \times 4$) classification results that were used to compute the CDF. Figure 6.4(a) shows the CDFs of the accuracy rate for four different window sizes

Table 6.3 Conditions and parameters for the validation of the human-movement identification

Condition	Parameters
Window size used in the segmentation	10, 20, 30, 40 samples
Tx–Rx antenna	P(X)–P(X), P(X)–P(Y), P(X)–M, P(Y)–P(X), P(Y)–P(Y), P(Y)–M, M–P(X), M–P(Y), M–M
Classifier algorithm	Support vector machine, k-nearest neighbor, decision tree
Receiver location	Ankle, thigh, upper arm, wrist

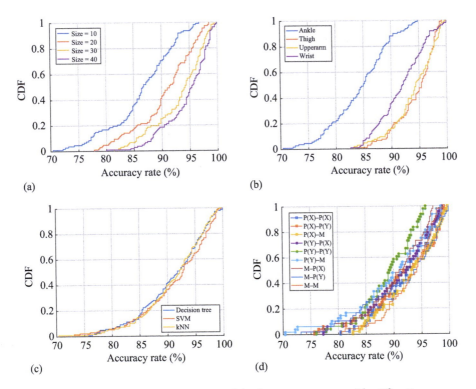

Figure 6.4 CDF of the accuracy rate of the human-movement identification: (a) CDF of the accuracy rate of different window sizes used in the data segmentation, (b) CDF of the accuracy rate of different receiver locations, (c) CDF of the accuracy rate of different classifier training algorithms, and (d) CDF of the accuracy rate of different pairs of Tx–Rx antennas

used in the data segmentation. Using the larger segment resulted in higher accuracy rate than using the smaller segment. As seen from the figure that only 30 percent of the classification results using the window size of 10 samples had the accuracy rate higher than 90 percent, but using the window size of 20 samples, it resulted in 57 percent of the classification results with the accuracy rate higher than 90 percent. Furthermore, 74 and 80 percent of the classification results achieved the accuracy rate greater than 90 percent when using the window size of 30 and 40 samples, respectively.

Table 6.4 shows the TPR of the classification results of M–M channel when the receiver locations are ankle and thigh. The classification results shown in this table are from the SVM classifier. It is noticed that the TPRs of running and walking movement were relatively low when the window size was 10 samples for both the ankle and the thigh location. The low TPRs were due to the misclassification of running movement as walking movement and vice versa. The TPRs can be improved by increasing the window size. As seen in the table that the TPRs of running movement were improved from 81.0 to 96.5 percent and from 79.9 to 98.4 percent for the ankle and the thigh location, respectively, when the window size was increased from 10 samples to 40 samples. Similarly, the TPRs of walking movement were improved from 69.8 to 95.3 percent and from 75.4 to 96.5 percent for the ankle and the thigh location, respectively. In addition, the TPR of sitting movement at the ankle location was also low since the received signal at the ankle location in the sitting scenario occasionally fluctuated. The TPR of sitting movement at the ankle location was also improved when the window size was increased. In summary, the window size used in the data segmentation had impact on the TPR. Using the larger window size resulted in the higher TPR for running and walking movement, and in some cases sitting movement as well, thus resulting in the higher accuracy rate.

6.7.2.2 Receiver location on human body

Impact of the receiver location to the performance of the classifier was also investigated. The location of the transmitter was fixed at the navel. The locations of the receiver were ankle, thigh, upper arm, and wrist. Therefore, there were 108 ($9 \times 3 \times 4$) classification results that were used to compute the CDF for each receiver

Table 6.4 TPR of the classification results of M–M channel at ankle and thigh location

Location	Window size	True positive rate (%)				
		Lie	Run	Sit	Stand	Walk
Ankle	10	99.3	81.0	69.4	98.5	69.8
	40	100	96.5	83.6	100	95.3
Thigh	10	100	79.9	100	98.5	75.4
	40	100	98.4	100	100	96.5

location. Figure 6.4(b) shows that the accuracy rate of the human movement identification at the ankle location was relatively low compared to the other locations. Both the thigh and upper arm location had the high accuracy rate, which was higher than the wrist location. Apart from the ankle location, the other locations produced more than 90 percent of the classification results with the accuracy rate higher than 85 percent. The low accuracy rate of the ankle location was attributed to the low TPR of sitting movement as seen in Table 6.4. These results showed that the human-movement identification using the received signal levels from the locations close to the body trunk had the higher accuracy rate than the received signal levels from the extremities such as the ankle.

6.7.2.3   Classifier algorithm

The classifier was trained by three algorithms, which were k-NN, SVM, and decision tree. The CDF of the accuracy rate of each algorithm was computed from 144 (4 $\times$ 9 $\times$ 4) classification results. Figure 6.4(c) shows the CDFs of the accuracy rate from different classifier algorithms. The classifier algorithms did not have much impact on the classification results. Although the SVM classifier had slightly higher accuracy rate than the k-NN and decision tree classifier, the difference of the accuracy rate was not much significant.

6.7.2.4   Pair of Tx–Rx antennas

Antenna orientation also affects the channel characteristics. Therefore, the results of the human-movement identification from different antenna orientations were investigated. The CDF of the accuracy rate of each Tx–Rx pair was computed from 48 (4 $\times$ 3 $\times$ 4) classification results. Figure 6.4(d) shows the CDFs of different pairs of Tx–Rx antennas. The classification results of the P(Y)–P(Y) and P(Y)–M had slightly lower accuracy rates than the other Tx–Rx pairs. The P(X)–M and M–M had the highest mean accuracy rates, which were 92.9 (SD=5.0) and 92.8 (SD=4.5) percent, respectively. In general, apart from the P(Y)–P(Y), more than 50 percent of the classification results of any Tx–Rx pairs had the accuracy rates higher than 90 percent.

6.8   Subset feature selection

The feature extraction introduced in Section 6.5.3 can provide more than 70 percent of the classification results with the accuracy rate higher than 90 percent when using the window size of 30 and 40 samples in the data segmentation. However, the feature vector contained very similar features, which some could be removed without affecting the classification results. This section explains about the feature-selection method for reducing the dimension of the feature vector to achieve lower computational cost.

6.8.1   Feature selection method

Feature selection is a procedure to remove redundant features from a feature vector. The feature selection makes more efficient use of the computational time and memory usage. The feature selection involved two steps. The first step was to partition the features into groups. The features in the same group are similar while the features from different groups are dissimilar. The measurement of the feature similarity used in this study was a correlation coefficient, which is a well-known measurement of the relation between two variables [39]. Next, the second step was to select a representative feature of each group. The representative features of all groups were combined to form a subset feature vector. Indices which were used to select the representative feature from each group were the classification accuracy [39] and the TPR.

The correlation coefficient of the features was computed from the features that were obtained by using the window size of 30 in the data segmentation. The correlation coefficient was computed across all pairs of Tx–Rx antennas. Figure 6.5 shows the correlation coefficient of the features at the ankle location. The correlation coefficient of the features from the other locations was similar to the ankle location; thus, it is not shown here. The features were divided into four groups according to the correlation coefficient. Group 1 contained IRSS, MV, MMV1, and MMV2. Group 2 contained RMS and SSI. Group 3 contained VAR, CSC, and Range. Lastly, Group 4 contained LCR, SSC, and WAMP.

MVSLP was not correlated to any other features. Hence, it was used in the subset feature vector. For the rest, only the representative feature of each group was selected to be in the subset feature vector. To select the representative feature, the features were computed from the received signal levels at the M–M channel and the decision tree was used for the classification. The M–M channel was used in the feature selection since it generally provided accurate classification results. The selection of the representative feature was done by training the classifier using only one feature

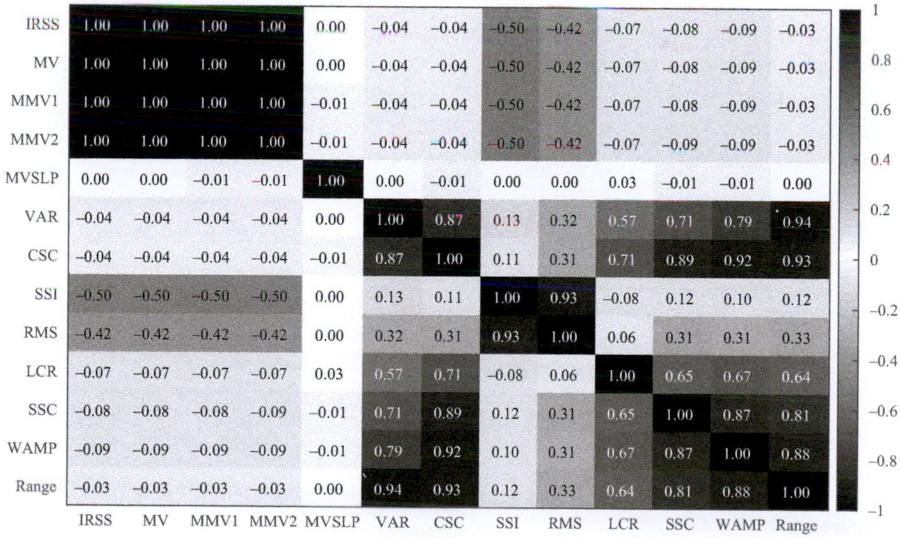

	IRSS	MV	MMV1	MMV2	MVSLP	VAR	CSC	SSI	RMS	LCR	SSC	WAMP	Range
IRSS	1.00	1.00	1.00	1.00	0.00	−0.04	−0.04	−0.50	−0.42	−0.07	−0.08	−0.09	−0.03
MV	1.00	1.00	1.00	1.00	0.00	−0.04	−0.04	−0.50	−0.42	−0.07	−0.08	−0.09	−0.03
MMV1	1.00	1.00	1.00	1.00	−0.01	−0.04	−0.04	−0.50	−0.42	−0.07	−0.08	−0.09	−0.03
MMV2	1.00	1.00	1.00	1.00	−0.01	−0.04	−0.04	−0.50	−0.42	−0.07	−0.09	−0.09	−0.03
MVSLP	0.00	0.00	−0.01	−0.01	1.00	0.00	−0.01	0.00	0.00	0.03	−0.01	−0.01	0.00
VAR	−0.04	−0.04	−0.04	−0.04	0.00	1.00	0.87	0.13	0.32	0.57	0.71	0.79	0.94
CSC	−0.04	−0.04	−0.04	−0.04	−0.01	0.87	1.00	0.11	0.31	0.71	0.89	0.92	0.93
SSI	−0.50	−0.50	−0.50	−0.50	0.00	0.13	0.11	1.00	0.93	−0.08	0.12	0.10	0.12
RMS	−0.42	−0.42	−0.42	−0.42	0.00	0.32	0.31	0.93	1.00	0.06	0.31	0.31	0.33
LCR	−0.07	−0.07	−0.07	−0.07	0.03	0.57	0.71	−0.08	0.06	1.00	0.65	0.67	0.64
SSC	−0.08	−0.08	−0.08	−0.09	−0.01	0.71	0.89	0.12	0.31	0.65	1.00	0.87	0.81
WAMP	−0.09	−0.09	−0.09	−0.09	−0.01	0.79	0.92	0.10	0.31	0.67	0.87	1.00	0.88
Range	−0.03	−0.03	−0.03	−0.03	0.00	0.94	0.93	0.12	0.33	0.64	0.81	0.88	1.00

Figure 6.5   Correlation coefficient of the features at ankle

at a time. For the group 1, 2, and 3, the accuracy rate of the classification results was used as the index to select the representative feature. For the group 4, the average TPR of walking and running movement was used. The accuracy rates of all features in the group 1 were almost the same. However, despite the similar accuracy rates, MV was selected since the computation was simpler than MMV1 and MMV2, and it was robust to some missing received signal levels in the case that a receiver cannot properly receive the transmitted signal. For the group 2, the accuracy rates of RMS and SSI were not significantly different. Therefore, RMS was selected for the same reason as MV that it was more robust to some missing received signal levels. For the group 3, the accuracy rate of CSC was higher than VAR and Range for all receiver locations. Therefore, CSC was selected from the group 3. Lastly, the average TPR of running and walking movement across all receiver locations when using SSC was 72.8 percent, which was higher than LCR and WAMP with the average TPR of 57.9 and 70.6 percent, respectively. Therefore, the representative feature of the group 4 was SSC. In summary, the subset feature vector contained five features, namely, MV, MVSLP, RMS, CSC, and SSC.

6.8.2   Evaluation of the subset feature vector compared to the all feature vector

The classification results of the M–M channel and the P(Y)–P(Y) channel were used to compare between the all feature vector and the subset feature vector. In this evaluation, two window sizes, which were 10 and 40 samples, were used in the data segmentation and the SVM was used for the classification.

Tables 6.5 and 6.6 provide the comparison of the accuracy rates between the all feature vector and the subset feature vector from the M–M channel and the P(Y)–P(Y) channel, respectively. The biggest difference between the all feature vector and the subset feature vector was at the ankle location, where the subset feature vector resulted in lower accuracy rate than the all feature vector by 3.4 and 3.3 percent for the M–M channel and the P(Y)–P(Y) channel, respectively. In addition, 1.7 percent of the decrease in the accuracy rate was observed from the M–M channel at the ankle using the window size of 10 samples and at the wrist using the window size of 40 samples.

Table 6.5   The accuracy rate of the classification results from the M–M channel

Rx location	Accuracy rate (%)			
	window size = 10 samples		window size = 40 samples	
	All feature	Subset feature	All feature	Subset feature
Ankle	83.6	81.9	95.1	91.7
Thigh	90.7	89.6	99.0	97.8
Upper arm	92.5	92.6	98.0	97.3
Wrist	90.7	90.0	97.3	95.6

Table 6.6   The accuracy rate of the classification results from the P(Y)–P(Y) channel

Rx location	Accuracy rate (%)			
	Window size = 10 samples		Window size = 40 samples	
	All feature	Subset feature	All feature	Subset feature
Ankle	75.4	74.8	87.7	84.4
Thigh	88.8	88.3	96.1	96.2
Upper arm	85.5	84.7	94.8	93.6
Wrist	86.4	86.8	94.8	94.0

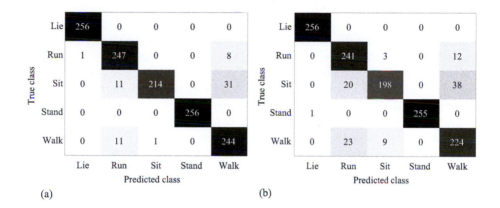

(a)   (b)

Figure 6.6   Confusion matrix of the classification results from the M–M channel at the ankle, where the window size is 40 samples: (a) confusion matrix of the result using the all feature vector and (b) confusion matrix of the result using the subset feature vector

However, apart from those mentioned above, the accuracy rates of the subset feature vector were comparable to the all feature vectors.

Figure 6.6 shows the confusion matrix of the classification results of the M–M channel at the ankle location using the all feature vector and the subset feature vector, for which the window size is 40 samples. Sitting and walking movement suffered from the low TPRs when using the subset feature vector. 14.8 percent of the sitting instances were classified as walking movement and 7.8 percent of the sitting instances were classified as running movement. The result was not entirely unexpected that the sitting instances were misclassified as the dynamic movement, such as walking, since the signal fluctuation due to unconscious movement was observed at the ankle location. As a result of removing some features, which helped distinguish between highly dynamic movement and semi-dynamic movement, more instances were misclassified for the subset feature vector. The misclassification of the walking instances as the running movement also increased from 4.3 to 9.0 percent due to the feature reduction.

In summary, the subset feature vector comprised five features, namely, MV, MVSLP, RMS, CSC, and SSC, showing about 60 percent of feature reduction in size compared to the all feature vector. The accuracy rate of the human-movement identification using the subset feature vector slightly decreased for the ankle location mainly due to the low TPR of sitting and walking movement. However, the accuracy rates of the subset feature vector were comparable to the accuracy rates of the all feature vector in most cases. Trade-off between the computational cost and the performance of the classifier must be well considered for a target application and hardware limitation.

6.9   Summary

In this chapter, the development of the human-movement identification using the radio signal strength in WBAN was discussed. Developing the human-movement identification system mainly consisted of three parts, which were data collection, data preprocessing, and classifier training. The WBAN measurements were conducted to collect the WBAN channels for static and dynamic movements. After that, the WBAN channel data underwent preprocessing procedures, which involved mean normalization, data segmentation, feature extraction, and feature scaling. However, the decision tree does not require the feature scaling. Therefore, the feature scaling was performed only for the k-NN and SVM. After the classifier was constructed, the validation was needed to assess the performance of the developed classifier. We examined various conditions for the human-movement identification system. It was found that a window size used in the data segmentation and a receiver location on human body had strong impact on the classification results. The thigh and upper arm location resulted in a higher accuracy rate than the ankle and wrist location. The low accuracy rate of the ankle and wrist location was the result of low TPR of the sitting movement. Since the received signal level of the sitting movement occasionally fluctuated at the ankle and wrist location, the sitting movement was sometimes misclassified as running or walking movement. For the window size used in the data segmentation, using a larger window size in the data segmentation showed a higher accuracy rate than using a smaller window size since a larger window size allowed a broader range of values for some features such as LCR, SSC, and WAMP, thus resulting in a higher TPR for walking and running movement than a smaller window size. On the other hand, the classification results of different Tx–Rx pairs showed slightly different accuracy rate. The P(Y)–P(Y) channel provided slightly lower accuracy rate than the other pairs due to the large number of walking instances classified as running instances and vice versa, and the large number of sitting instances classified as walking or running instances. In the end of this chapter, the correlation-based feature selection was performed to remove redundant features from the feature vector. The performance of the all feature vector and the subset feature vector was compared for the M–M channel and the P(Y)–P(Y) channel. The subset feature vector, which contained MV,

MVSLP, RMS, CSC, and SSC, was slightly inferior to the all feature vector. However, 60 percent reduction in the feature vector size was achieved.

The investigation results presented in this chapter support that the accurate human-movement identification using only the radio signal strength is feasible. This approach enables many context-aware applications in WBAN without additional installation of special hardware. However, it is worth noting that this human-movement identification approach is based on the WBAN channel characteristics, which are highly influenced by a sensor location and an antenna orientation; hence, the classifier is limited to use in the specific location and the orientation for which it is trained. Future works of this research include further measurements to construct highly reliable and robust classifier model and integration of the human-movement identification in communication design for WBAN.

References

[1] Wang J., Wang Q. *Body area communications: channel modeling, communication systems, and EMC.* Singapore: John Wiley & Sons; 2012.

[2] D'Errico R., Ouvry L. 'Time-variant BAN channel characterization'. *Proceedings of IEEE 20th International Symposium on Personal, Indoor and Mobile Radio Communications*; Tokyo, Japan. IEEE; 2009. pp. 3000–3004.

[3] D'Errico R., Ouvry L. 'A statistical model for on-body dynamic channels'. *International Journal of Wireless Information Networks.* 2010;17(3–4): 92–104.

[4] Zhen B., Aoyagi T., Kohno R. 'Reasons for time-varying fading of dynamic on-body channels'. *Proceedings of IEEE Wireless Communications and Networking Conference*; New South Wales, Australia. IEEE; 2010. pp. 1–5.

[5] Yazdandoost K., Sayrafian K. *Channel model for body area network (BAN).* IEEE P802.15-08-0780-09-0006; 2009.

[6] Smith D., Hanlen L., Zhang A., Miniutti D., Rodda D., Gilbert B. 'First and second-order statistical characterizations of the dynamic body-area propagation channel of various bandwidths'. *Annals of Telecommunications.* 2011;66(3):187–203.

[7] Aoyagi T., Minseok K., Takada J., Hamaguchi K., Kohno R. 'Numerical simulations for wearable ban propagation channel during various human movements'. *IEICE Transactions on Communications.* 2011;94(9):2496–2500.

[8] Uusitupa T., Aoyagi T. 'Analysis of dynamic on-body communication channels for various movements and polarization schemes at 2.45 GHz'. *IEEE Transactions on Antennas and Propagation.* 2013;61(12):6168–6179.

[9] Khan Z.A., Sohn W. 'Abnormal human activity recognition system based on R-transform and kernel discriminant technique for elderly home care'. *IEEE Transactions on Consumer Electronics.* 2011;57(4):1843–1850.

[10] Ainsworth B., Cahalin L., Buman M., Ross R. 'The current state of physical activity assessment tools'. *Progress in Cardiovascular Diseases.* 2015;57(4):387–395.

[11]   Tobón D.P., Falk T.H., Maier M. 'Context awareness in WBANs: a survey on medical and non-medical applications'. *IEEE Wireless Communications*. 2013;20(4):30–37.

[12]   Shuger S.L., Barry V.W., Sui X., *et al.* 'Electronic feedback in a diet-and physical activity-based lifestyle intervention for weight loss: a randomized controlled trial'. *International Journal of Behavioral Nutrition and Physical Activity*. 2011;8(41):1–8.

[13]   Virone G., Alwan M., Dalal S., *et al.* 'Behavioral patterns of older adults in assisted living'. *IEEE Transactions on Information Technology in Biomedicine*. 2008;12(3):387–398.

[14]   Wood A.D., Stankovic J.A., Virone G., *et al.* 'Context-aware wireless sensor networks for assisted living and residential monitoring'. *IEEE Network*. 2008;22(4):26–33.

[15]   Salazar A.J., Silva A.S., Borges C., Correia M.V. 'An initial experience in wearable monitoring sport systems'. *Proceedings of the 10th IEEE International Conference on Information Technology and Applications in Biomedicine*; Corfu, Greece. IEEE; 2010. pp. 1–4.

[16]   Darby J., Li B., Costen N. 'Activity classification for interactive game interfaces'. *International Journal of Computer Games Technology*. 2008: 1–8;Article ID 751268.

[17]   Zhang Z. 'Microsoft Kinect sensor and its effect'. *IEEE MultiMedia*. 2012;19(2):4–10.

[18]   Parkka J., Ermes M., Korpipaa P., Mantyjarvi J., Peltola J., Korhonen I. 'Activity classification using realistic data from wearable sensors'. *IEEE Transactions on Information Technology in Biomedicine*. 2006;10(1):119–128.

[19]   Arif M., Kattan A. 'Physical activities monitoring using wearable acceleration sensors attached to the body'. *PLoS One*. 2015;10(7):e0130851.

[20]   Woyach K., Puccinelli D., Haenggi M. 'Sensorless sensing in wireless networks: implementation and measurements'. *Proceedings of 4th International Symposium on Modeling and Optimization in Mobile, Ad Hoc and Wireless Networks*; Massachusetts, Unites States. IEEE; 2006. pp. 1–8.

[21]   Muthukrishnan K., Lijding M., Meratnia N., Havinga P. 'Sensing motion using spectral and spatial analysis of WLAN RSSI'. *Proceedings of the 2nd European Conference on Smart Sensing and Context*; Kendal, England. Springer-Verlag; 2007. pp. 62–76.

[22]   Sigg S., Scholz M., Shi S., Ji Y., Beigl M. 'RF-sensing of activities from non-cooperative subjects in device-free recognition systems using ambient and local signals'. *IEEE Transactions on Mobile Computing*. 2014;13(4):907–920.

[23]   Anderson I., Muller H. 'Context awareness via GSM signal strength fluctuation'. *Proceedings of the 4th International Conference on Pervasive Computing, Late Breaking Results*; Dublin, Ireland. Springer-Verlag; 2006. pp. 27–31.

[24]   Anderson I., Muller H. 'Practical context awareness for GSM cell phones'. *Proceedings of the 10th IEEE International Symposium on Wearable Computers*; Montreux, Switzerland. IEEE; 2006. pp. 127–128.

[25] Sohn T., Varshavsky A., LaMarca A., *et al.* 'Mobility detection using everyday GSM traces'. *The Proceedings of the 8th International Conference on Ubiquitous Computing*; California, United States. Springer; 2006. pp. 212–224.

[26] Mun M., Estrin D., Burke J., Hansen M. Parsimonious mobility classification using GSM and WiFi traces. *Presented at the Fifth Workshop on Embedded Networked Sensors*; Virginia, United States; 2008.

[27] Ekure I.N., Wang S., Zhou G. 'A theoretical analysis of path loss based activity recognition'. *Proceedings of IEEE 11th International Conference on Mobile Ad Hoc and Sensor Systems*; Pennsylvania, United States. IEEE; 2014. pp. 277–281.

[28] Fu R., Bao G., Pahlavan K. 'Activity classification with empirical RF propagation modeling in body area networks'. *Proceedings of the 8th International Conference on Body Area Networks*; Massachusetts, United States. ICST; 2013. pp. 296–301.

[29] Geng Y., Chen J., Fu R., Bao G., Pahlavan K. 'Enlighten wearable physiological monitoring systems: on-body RF characteristics based human motion classification using a support vector machine'. *IEEE Transactions on Mobile Computing*. 2016;15(3):656–671.

[30] Ishimiya K., Langbacka J., Ying Z., Takada J. 'A compact MIMO DRA antenna'. *Proceedings of 2008 International Workshop on Antenna Technology*; Chiba, Tokyo, IEEE; 2008. pp. 286–289.

[31] Ishimiya K., Chiu C.Y., Zhinong Y., Takada J. '3-Port MIMO DRAs for 2.4 GHz WLAN communications'. *IEICE Transactions on Communications*. 2016;99(9):2047–2054.

[32] Naganawa J., Takada J., Aoyagi T., Kim M. 'Antenna deembedding in WBAN channel modeling using spherical wave functions'. *IEEE Transactions on Antennas and Propagation*. 2017;65(3):1289–1300.

[33] Chen G., Naganawa J., Takada J., Kim M. 'Development of a tri-polarized dynamic channel sounder for wireless body area network'. *Proceedings of 9th International Symposium on Medical Information and Communication Technology*; Tokyo, Japan. IEEE; 2015. pp. 30–34.

[34] Phinyomark A., Limsakul C., Phukpattaranont P. 'A novel feature extraction for robust EMG pattern recognition'. *Journal of Computing*. 2009;1(1):71–80.

[35] Wu X., Kumar V., Quinlan J.R., *et al.* 'Top 10 algorithms in data mining'. *Knowledge and Information Systems*. 2008;14(1):1–37.

[36] Duda R.O., Hart P.E., Stork D.G. *Pattern classification*. 2nd ed. New York: Wiley-Interscience; 2001.

[37] Burges C.J. 'A tutorial on support vector machines for pattern recognition'. *Data mining and Knowledge Discovery*. 1998;2(2):121–167.

[38] Breiman L., Friedman J., Stone C.J., Olshen R.A. *Classification and regression trees*. Boca Raton, Florida: CRC Press; 1984.

[39] Mitra P., Murthy C., Pal S.K. 'Unsupervised feature selection using feature similarity'. *IEEE Transactions on Pattern Analysis and Machine Intelligence*. 2002;24(3):301–312.

Chapter 7

Cognitive radio and RF energy harvesting for medical WBANS

Fernando J. Velez[1], Raúl Chávez-Santiago[2],
Norberto Barroca[1], Luís M. Borges[1],
Jorge Tavares[1], and Ilangko Balasingham[2]

Wearable wireless medical sensors beneficially impact the healthcare sector, and this market is experiencing rapid growth. Globally, the digital health market is forecast to increase from $80 billion in 2015 to $200 billion in 2020. Medical body area networks (MBANs) improve the mobility of patients and medical personnel during surgery, accelerate the patients' recovery, while facilitating the remote monitoring of patients suffering from chronic diseases. Currently, MBANs are being introduced in unlicensed frequency bands, where the risk of mutual interference with other electronic devices can be high. Techniques developed during the evolution of cognitive radio (CR) can potentially alleviate these problems in medical communication environments. In addition, these techniques can help increase the efficiency of spectrum usage to accommodate the rapidly growing demand for wireless MBAN solutions and enhance coexistence with other collocated wireless systems. A viable architecture of an MBAN with practical CR features is proposed in this work. Additionally, ultra wideband (UWB) radio for the implementation of CR offers many advantages to MBANs, and some features of this technology can be exploited for effective implementation of CR. Conceptual aspects associated with energy harvesting and practical identification of spectrum opportunities for radio frequency (RF) energy scavenging motivates the options taken in the development of the protocols. The physical (PHY) and medium access control (MAC) layer aspects of the proposal are proposed in addition to their implementation challenges in the context of CR.

7.1 Introduction

Wireless communications have the potential to impact beneficially the medical practice through the development of ubiquitous health monitoring solutions [1,2]. This can

[1]Instituto de Telecomunicações and Departamento de Engenharia Electromecânica, Universidade da Beira Interior, Portugal
[2]Wireless Sensor Network Research Group, Intervention Centre, Oslo University Hospital, Norwegian University of Science and Technology, Norway

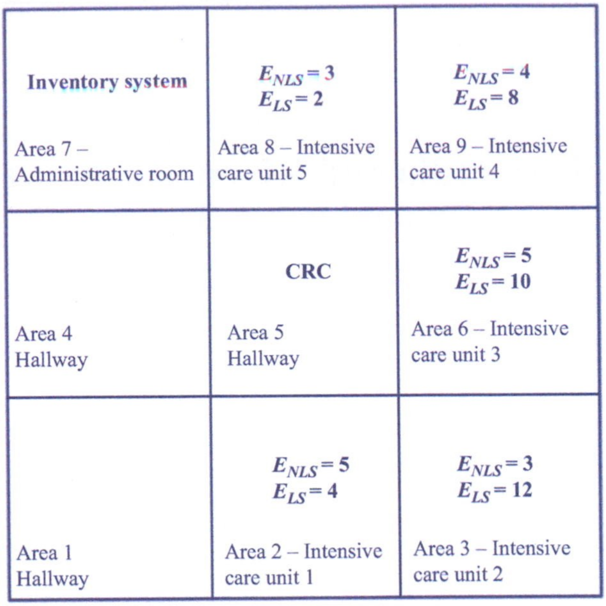

Inventory system	$E_{NLS} = 3$ $E_{LS} = 2$	$E_{NLS} = 4$ $E_{LS} = 8$
Area 7 – Administrative room	Area 8 – Intensive care unit 5	Area 9 – Intensive care unit 4
	CRC	$E_{NLS} = 5$ $E_{LS} = 10$
Area 4 Hallway	Area 5 Hallway	Area 6 – Intensive care unit 3
	$E_{NLS} = 5$ $E_{LS} = 4$	$E_{NLS} = 3$ $E_{LS} = 12$
Area 1 Hallway	Area 2 – Intensive care unit 1	Area 3 – Intensive care unit 2

*Figure 7.1   Layout of the hospital scenario considered for simulations in [17]
(adapted from [17])*

be achieved through the use of biomedical miniaturised sensors and actuators placed in, on, and around the human body, combined with small wireless intercommunicating radio transceivers for measuring, transmitting, and storing different physiological signals in real time. The interconnection of these wearable and implantable devices constitutes a Wireless Body Sensor Network (WBSN) that enables unobtrusive, continuous, and automated monitoring of users to support medical applications, which is the core of modern telemedicine systems, as shown in Figure 7.1.

According to [3], such a WBSN targeted to support medical applications is called medical BAN (medical body area network, MBAN) as shown in Figure 7.1. The MBAN technology has the potential to change healthcare delivery in ambulances, operation theatres, clinics, homes, and other pervasive scenarios. The benefits of unobtrusive and continuous monitoring and treatment include long-term trend analysis, prompt alerting of a carer to intervene in an emergency situation, regulation of treatment regimes, reduction of hospital stays, and improved patient well-being. As such, according to [3], the MBAN offers a paradigm shift in healthcare from managing illness to proactively managing wellness by focusing on prevention and early detection/treatment of diseases while augmenting the mobility of not only of the medical personnel during surgery but also patient's mobility, a fundamental factor for fast recovery after surgical procedures and interventions.

A dedicated frequency band in 2,360–2,400 MHz for body area networks (BANs) use on an opportunistic secondary basis has recently been designated in the

United States. Nevertheless, it is anticipated that a large number of wireless biomedical sensors will operate in unlicensed frequency bands too. In fact, the BAN IEEE 802.15.6 standard [4] has recommended the unlicensed 2.4 GHz industrial, scientific, and medical (ISM) frequency band in 2,400–2,500 MHz, and ultra wideband (UWB) in 3.1–10.6 GHz as substitute spectrum for wireless biomedical sensors. According to [3], IEEE 802.15.6 is optimised for low-power devices and operations on, in, or around the human body to serve a variety of medical applications. Wireless patient monitoring through MBANs, using low-cost wireless devices, will benefit a large number of patients in several hospitals at the same time, thereby reducing healthcare costs. Small transceivers compliant with the IEEE 802.15.4 (ZigBee) standard are commercially available for operation in unlicensed ISM bands and have been found suitable for health and fitness monitoring in confined indoor areas [5]. In addition, a large number of wireless local area networks (WLAN) based on the family of IEEE 802.11 standards also share the 2.4 GHz ISM band, making the coexistence of these different wireless devices challenging [6]. Similar interference scenarios may be expected in the UWB band [7].

According to [3], there are several dynamic use cases for CR in wireless medical networks (wireless body area networks, WBANs), in vehicular, outdoor, or indoor scenarios. Techniques to avoid mutual interference must be applied in such cases. In hospital scenarios, the coexistence problem is more critical in small areas like intensive care units (ICUs) and operating rooms (ORs) because electromagnetic interference (EMI) from wireless devices can disrupt the performance of non-communication medical equipment that is routinely present in such premises. CR is a promising technology that can ease the coexistence of wireless devices, while protecting the electronic medical equipment. Despite the recognised potential benefits of CR for BANs [8], this application has not been extensively investigated and just a few solutions have been proposed.

The authors from [3] mention that a new CR assisted medical telemetry paradigm is introduced in [9] that envisages transforming healthcare and medical telemetry through CR networks. Transmissions in the wireless medical telemetry services (WMTS) band and similar reserved channels, for patient's health-related data and life-critical communications, are hampered by interferences from adjacent digital television channels, and due to non-uniform access priority, as other transmissions such as utility telemetry transmissions and government-run radar sites also use the WMTS band. The authors from [9] show that with the CR technology, the WMTS frequencies can be dynamically utilised according to the high-priority users' activity patterns, and the Quality-of-Service (QoS) constraints of the patients' data, while providing a safe operation of sensitive medical equipment. CR enabled devices could be used to realise telemetry applications in ambulances and home environments as well as to utilise multimedia data.

The authors from [10] have proposed the architecture for an MBAN with CR features based on UWB radio technology. Using the UWB technology enhances the CR-MBAN implementation by implementing the cognitive capabilities in the body network controller (BNC) using different UWB modalities. This enables to convert a BNC into a cognitive radio controller (CRC) in a MBAN and the CRC manages the transmission parameters of CR clients (wearable sensors) for wireless access. Wearable sensors must have characteristics such as low cost, small in size, and low-power depletion and this can be attained through the use of an impulse radio-UWB (IR-UWB) radio interface for communications in the MBAN. Thus, according to IEEE 802.15.6 specifications, the communication links between the sensors and the BNC should be implemented with IR-UWB. The MBAN architecture proposed in [10] is equipped with frequency agility and frequency-domain spectrum shaping capabilities to avoid interference in scenarios where multiple devices are operating in common spectrum segments and in close proximity to each other, for instance, in a crowded medical centre.

Moreover, according to [9], the proposal to allocate the underused 2,360–2,400 MHz band for MBANs on a secondary basis will greatly enhance the QoS for life-critical monitoring applications. The IEEE 802.15 Task Group 4j has been set up to specifically develop standards for MBANs in the 2,360–2,400 MHz band by leveraging the existing IEEE 802.15.4 standard. Figure 7.1 shows both in-hospital and out-of-hospital solutions for using the 2,360–2,390 MHz band [8]. In general, the CR techniques can potentially mitigate the problems of interference and congestion in medical communication environments. Further, employing the CR techniques will help in increasing the efficiency of spectrum usage to accommodate the rapidly growing demand for wireless MBAN solutions and enhance coexistence with other collocated wireless systems. In addition, CR provides significant energy saving benefits in MBANs through the use of cognition (e.g., to know the battery level) and through better dynamic selection of spectrum, avoiding interference and thus, reducing necessary received power and hence, transmitted power [10].

Future improvements in radio frequency (RF) energy harvesting technology will facilitate the creation of a network with no need of dedicated transmitters as a reliable source of wireless energy power [11]. This can be accomplished by enabling the capture of electromagnetic energy from multiple available ambient RF energy sources, such as mobile base stations, TV and radio transmitters, microwave radios, and mobile phones, as discussed in Chapter 2. Moreover, since WBSN nodes are battery-operated, energy recharging is a possibility, avoiding the need for battery replacement. However, the service lifetime of the electronic components could be a major concern if there is no possibility to collect enough energy to generate the voltage needed to drive the sensor node. Medium access control (MAC) and routing protocols also play an important role in network performance [12]. As a consequence, choosing the best opportunities poses a high effect on the overall network performance, as well as on the energy consumption.

The WBAN MAC protocols are responsible for providing the mechanisms for resource management and allocation of the shared wireless channel. Compared to conventional WBANs, the MAC sub-layer of the WBAN architecture with CR capabilities must handle additional challenges, such as silent spectrum sensing periods and the need for high-priority access mechanisms, for the distribution of spectrum sensing and decision results [13]. Therefore, the new innovative MAC protocols must be designed regarding energy efficiency.

7.2 Hospital scenarios

CR has been identified as the enabling technology to tackle spectrum scarcity and interference in healthcare and medical telemetry by enabling dynamic utilisation of the WMTS frequency band, which comprises different parts of the spectrum, namely 608–614 MHz, 1,395–1,400 MHz, and 1,427–1,432 MHz [9]. A CR request-to-send/clear-to-send (RTS/CTS) protocol for e-health applications was proposed in [14], which adapts the transmit power of wireless devices operating in 2.4 GHz according to standardised EMI immunity constraints. The protocol effectively handles two different types of medical application traffic with different priorities. Through computer simulations it was demonstrated that this EMI-aware RTS/CTS protocol can reduce significantly the interference to protected non-communication medical devices in comparison to other MAC protocols like the one specified by IEEE 802.15.4. Finally, a CR solution for BAN based on UWB technology was proposed in [10]. UWB signals offer many advantages to BANs, and some features of this technology can be exploited for effective implementation of CR. Below, the two latter approaches, namely CR for BAN in 2.4 GHz and UWB bands, are described with more detail.

Although MBANs offer many benefits for healthcare, the surge in adoption rates across the healthcare sector will certainly create new interference scenarios with other collocated electronic systems. CR is a paradigm for opportunistic access of licensed (primary) parts of the electromagnetic spectrum by unlicensed (secondary) users that can provide solutions for these interference scenarios, and also enhance scalability. As discussed in [10], a cognitive radio user (CRU) can combine spectrum sensing and geolocation database access to determine occupancy, and dynamic reconfiguration of its transceiver parameters in order to avoid interference with primary users (PUs).

7.2.1 Cognitive radio solution in 2.4 GHz

In the hospital case study addressed in [14], two different types of traffic from two wireless e-health applications were considered to be handled by the CR system:

(1) Real-time non-critical telemedicine, which is used to transmit data that are not delay/loss-sensitive, for example, remote consultation, patient record transfers, and remote diagnosis.
(2) Hospital information system, which collects patient, technical, and facility data that are intended for better clinical decisions and to prevent patient complications. This system collects information with the aid of BANs and other wireless sensor networks (WSNs) located in the hospital.

In the CR context, the telemedicine system is treated as a PU and the hospital information system as a secondary user (SU).

The CR system consists of three components, namely an inventory system, a CRC, and CR clients. The inventory system is a database containing information about all the medical devices in the hospital premises. Information like location, activity status, and EMI immunity levels are stored in the database. The CRC is a computer that controls the transmission parameters of the CR clients, i.e., PUs and SUs. For this sake, the CRC uses the information in the inventory system to compute the appropriate transmit power for each CR client in order to avoid interference that exceeds the EMI immunity levels of non-communication medical devices located in the vicinity. The CR system operates using a dedicated control channel (DCC) and a data channel (DATC). Both channels are in the unlicensed spectrum, for example, the 2.4 GHz ISM band. Every CR client transmits its data through the CRC. The CRC can transmit/receive data from both channels simultaneously, whereas the CR clients can transmit just in one of the two channels at a time. The DCC is used to broadcast information to all the CR clients about their corresponding maximum power, P_{ctrl}, for transmitting RTS messages. Each CR client has a different P_{ctrl} value depending on its location, which is calculated as established in Section 6.3.2.1 from [15], and depends on $P_{NLS}(n)$ and $P_{LS}(n)$, the upper bounds on transmit power for non-life-supporting (NLS) medical device n and life-supporting (LS) medical device m, respectively.

Details for the computation of these transmitter power values in a frequency range of 800–2,500 MHz are given in [15], as well as details on the message exchange involved in the EMI-aware RTS/CTS protocol for channel access.

The performance of the EMI-aware RTS/CTS protocol in a single channel scenario consisting of hospital premises over 27 m^2 arranged in nine areas of equal size comprising a hallway, an administration room, and five ICUs, as shown in Figure 7.1, is also discussed in [15].

In this case study, the CRC was located at the centre of Area 5. Ten NLS and LS non-communication medical devices were located in the ICUs, and their corresponding EMI immunity levels (E_{NLS} and E_{LS}) are also given in Figure 7.1. The locations of the NLS and LS medical devices and the CRC were fixed, whereas the CR clients were mobile and uniformly distributed over the area. Details on the mobility, propagation model and interference probability were discussed in [15] as well. In the EMI-aware RTS/CTS protocol, interference occurs when the ON status of the medical devices is reported wrongly to the CRC. This probability of misdetection was set to 0.01 in [15].

7.2.2   Enhancement through the use of an additional channel

In [17], the EMI-aware RTS/CTS protocol proposed in [14] can be enhanced by including dual-band operation, as proposed by the authors from [16]. The use of an additional "emergency" channel (AEC) in a different frequency band that can serve as a control/data channel for potential interferers can reduce the outage probability. The recently allocated 2,360–2,400 MHz BAN frequency band and the 900 MHz ISM band (902–928 MHz) are suitable for allocation of the AEC. Through computer simulations, the performance of this MAC scheme was evaluated in [15] in terms of the outage probability. The comparison with the EMI-aware RTS/CTS protocol from [14] is shown in Figure 7.2.

Clearly, the use of an AEC reduced the outage probability, but the improvement is determined by the centre frequency of the AEC. Marginal improvement is obtained with an AEC in 2,360–2,400 MHz (AEC1), whereas significant improvement can be

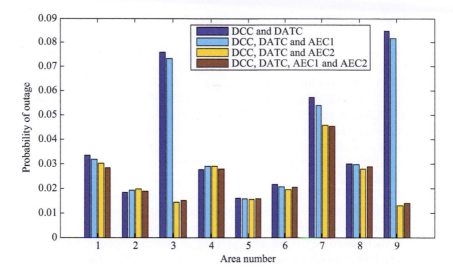

Figure 7.2   Outage probability over the nine areas of the hospital scenario for different multiband EMI-aware RTS/CTS protocols (extracted from [15])

obtained by using an AEC in the 900 MHz ISM band (AEC2). The simultaneous use of AEC1 and AEC2 does not provide further improvement.

7.3   CR solution in the UWB band

According to [10], besides large bandwidth that enables high data-rate transmission, UWB technology has other attractive characteristics for BAN implementation. UWB signals have an inherent noise-like behaviour due to their extremely low-maximum effective isotropically radiated power (EIRP) spectral density of −41.3 dBm/MHz. This makes UWB difficult to detect and increases its robustness against jamming, potentially rescinding the need for complex encryption algorithms in small, low-cost transceivers. Additionally, compared to single-band orthogonal frequency-division multiplexing (OFDM) whose symbols are continually sent on one frequency band, for multiband OFDM (MB-OFDM), symbols are interleaved over multiple sub-bands across both time and frequency. By interleaving the OFDM symbols across sub-bands in this manner, multiband UWB can maintain the power level associated with a single-band OFDM transmission yet the data throughput can be significantly increased. MB-OFDM UWB technology can achieve rates ranging from 53.3 to 480 Mbps over distances up to 10 m.

Furthermore, according to [10], UWB signals do not represent a threat to patients' safety and are not significant sources of interference to other medical devices. IR-UWB transceivers have a simple structure and very low-power consumption characteristics. These features definitely facilitate their miniaturisation for wearable biomedical sensors.

In fact, the authors from [10] have proposed the architecture for an MBAN with CR features based on UWB radio technology. The UWB technology enhances the CR-MBAN implementation by deploying the cognitive capabilities in the BNC using two UWB modalities, namely IR and MB-OFDM. This enables to convert a BNC into a CRC in an MBAN. The CRC manages the transmission parameters of CR clients (wearable sensors) for wireless access. According to [10], these wearable sensors must have the characteristics such as low cost, small in size, and low-power depletion and this can be accomplished through the use of an IR-UWB radio interface for communications in the MBAN. Thus, according to IEEE 802.15.6 specifications, the communication links between the sensors and the BNC should be implemented with IR-UWB. The MBAN architecture proposed in [10] is equipped with frequency agility and frequency-domain spectrum shaping capabilities to avoid interference in scenarios where multiple devices are operating in common spectrum segments and in close proximity to each other, for example, in a busy medical centre.

7.4   Wireless body area networks

7.4.1   Introduction

Finite battery lifetime has become a major concern over the past years. Hence, there is a need to seek innovative solutions enabling to power supply wireless nodes (e.g., WSN or WBAN nodes). Energy harvesting holds a promising future in the next generation of WBSNs. Since there are a variety of energy sources available for energy harvesting in the environment, the opportunities are vast. Table 7.1 presents the power density measurements for different energy harvesting sources, as discussed in [17].

Energy harvesting from RF electromagnetic holds a promising future for power supply of wireless electronic devices. Nowadays, RF energy is currently broadcasted from billions of radio transmitters (e.g., mobile communications base stations and television/radio stations), which can be collected from the ambient or from dedicated sources, enabling wireless charging of the low-power devices. Additionally, RF transmitters and receivers can be used when other potential intermittent energy scavenging sources (e.g., solar, vibration, and heat) are not available. Hence, by adding new energy harvesting capabilities to the sensor nodes, one provides a predictable and reliable power system that uses controlled broadcasted RF energy for wirelessly charging the battery (or supercapacitor) energy storage systems. Powercast offers several Powerharvester receivers [18] modules that have been designed for charging batteries, energy storage devices, and for direct power applications. The P2110 Powerharvester long range receiver module has the following characteristics:

- Low-RF input for longer range operation;
- RF scavenging range down to −11 dBm input power; and
- Frequency range from 850 to 950 MHz;

Table 7.1   *Power density and performance for different*
energy harvesting methods (extracted from [17])

Energy source	Power density and performance
Acoustic noise	$0.003 \ \mu W/cm^3$ @ 75 dB
	$0.96 \ \mu W/cm^3$ @ 100 dB
Temperature variation	$10 \ \mu W/cm^3$
Ambient RF	$1 \ \mu W/cm^2$
Ambient light	$100 \ mW/cm^2$ (direct sun)
	$100 \ \mu W/cm^2$ (illuminated office)
Thermoelectric	$60 \ \mu W/cm^2$
Vibration (micro generator)	$4 \ \mu W/cm^3$ (human motion – Hz)
	$800 \ \mu W/cm^3$ (machines – kHz)
Vibration (piezoelectric)	$200 \ \mu W/cm^3$
Airflow	$1 \ \mu W/cm^2$
Push buttons	$50 \ \mu J/N$
Shoe inserts	$330 \ \mu W/cm^2$
Hand generators	30 W/kg
Heel strike	$7 \ W/cm^2$

Table 7.2   *Amount of power harvested by the P2110 harvester using*
a patch antenna (extracted from [19])

Distance (m)	$P \ (\mu W)$	$I \ (\mu A)$	Recharge time (h)
1.52	1,925	1,604	42.24
3.05	386	322	210.50
4.57	189	158	429.40
5.49	131	109	618.50
6.10	102	85	797.50
7.62	50	41	1,639.00
9.14	19	16	4,353.00
10.67	5	4	15,517.00
10.97	1	1	70,019.00

- Configurable regulated output voltage up to 5.5 V;
- Received signal strength indicator (RSSI) and data output; and
- Interrupt available for sophisticated systems.

Table 7.2 shows the time to charge a battery with a capacity of 1,150 mAh using the P2110 module at a given distance, according to the experiments conducted in [19]. One advantage of collecting RF energy is that it is essentially "free." Besides, RF energy is universally present over an increasing range of frequencies and power levels, especially in highly populated urban areas. These radio waves represent a unique and widely available source of energy if it can be effectively and efficiently harvested. Moreover, the growing number of wireless transmitters is naturally increasing RF power density and availability. Dedicated power transmitters will enable engineered and predictable wireless power solutions.

Since the power consumption of wireless devices is decreasing and the sensitivity of passive RF harvesting receivers is increasing, the applications for wireless charging by means of RF-based wireless power and energy harvesting will continue to grow.

7.4.2   Architecture for WBSNs with CR capabilities

In the near future, energy harvesters will enable to supply all the nodes of a WBSN without the need of replacement of the primary source of energy (i.e., batteries). The next generation of CR networks will be supplied by renewable energy from natural resources, such as solar, wind, and RF energy [20]. This energy could be used overnight to increase the battery charge or to prevent power leakage. In a hazardous situation, if a battery or a solar-collector/battery package completely fails, harvested energy from radio waves can enable the system to transmit a wireless distress signal, whilst potentially maintaining critical functionalities [21].

MAC and routing protocols also play an important role in the network performance of the WBSN. The MAC protocols are responsible for managing the radio transmission and reception through the shared wireless link, whereas the routing protocols are responsible for the selection of the best path in order to send packets from one source to one single-destination or multi-destination. Hence, choosing the best ones has a high effect on the overall network performance as well as on the energy consumption. Driven by the intense usage of some frequency bands, while others are being liberated (e.g., white spaces left by analogue television switch-off) investigation on multi-hop CR networks has experienced an evolution in the latest years. The authors from [22] use graph theory for the routing algorithms, where the CR aspects are considered by assigning different colours to each considered frequency band. The developed algorithm is not computationally heavy and based on hop-count for the routing. However, as pointed out by the authors, they were not able to mitigate the interference between neighbours. In [23], the authors aim to minimise the interference between nodes. To achieve this objective, they propose the use of relays, with visible enhancements of the channel utilisation, whilst decreasing the energy consumption and delay.

Agile spectrum access can be accomplished by enabling an unlicensed user (i.e., SU) to adaptively adjust its operating parameters and exploit the spectrum which is unused by licensed users (i.e., PUs), in an opportunistic manner. Therefore, CR allows for SUs to seek and utilise "spectrum holes" in a time and location-varying radio environment as long as they do not cause interference to the PUs [24]. This opportunistic use of the spectrum leads to new challenges, making the network protocols to adapt to the varying available spectrum. The extreme flexibility of CR has significant implications in the design of network algorithms and MAC protocols at both local/access network and global inter-networking levels.

Therefore, new cross-layer algorithms can be envisaged which can adapt to the changes in the transmission link, based on the quality of the received signal, radio interference, radio node density, network topology, or traffic demand. As such, it may require an advanced control and management framework with support for cross-layer

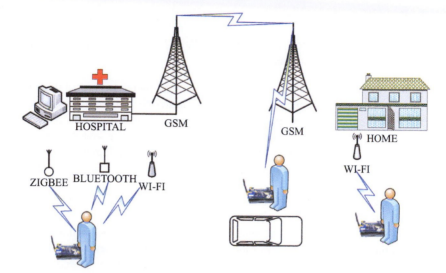

Figure 7.3 *BAN scenario with electromagnetic energy harvesting and CR*
capabilities

information and inter-node collaboration [25]. Figure 7.3 shows a typical WBSN scenario with CR. Depending on the spectrum availability, sensor nodes transmit their results from the sensing to the next hops and ultimately to the sink [24] in an opportunistic manner.

A dynamic spectrum access (DSA) entity will be used to identify the CR opportunities, in which nodes can be equipped with Powerharvester receivers [18], facilitating to collect energy by converting the energy from the RF waves to DC power. By using RF-based wireless scavenging devices, we intend to eliminate the cost of replacing batteries of wireless sensors as well as the service downtime caused by depleted batteries.

Future improvements in RF energy harvesting technology will allow for the creation of a network without the need of a dedicated transmitter to a single-sided system (the main characteristic of wireless power transfer). This is accomplished by enabling the capture of RF waves emitted by existing and commonly used ambient RF energy sources, such as mobile base stations, TV and radio transmitters, microwave radios, and mobile phones, as presented in Figure 7.3. Besides, future improvements in the performance of the electronic components will lead to a decrease of the power consumption, which results in faster charge times and more frequent broadcasts of data, enabling the creation of an always-connected ubiquitous wireless power sensor network (UWPSN).

Moreover, it is envisaged a system that cognitively seeks the best signal available from multiple frequency bands for collecting energy simultaneously finds the best transmission opportunities. An adaptive frequency hopping (AFH) algorithm containing a blacklist of the bands which contain interference caused by the same/other protocols can be implemented.

7.4.3   Topology aspects

Depending on the application, the scavenging WBSNs with CR capabilities may be applied to different scenarios, as follows:

- **Static networks:** In ad hoc networks, nodes send their readings to the gateway node in a multi-hop manner. In addition, the bandwidth availability and computing resources (e.g., hardware and battery power) are restricted. To overcome these limitations, joint optimisation between the MAC and PHY layers, to maximise energy efficiency, must be addressed. In the scenario from Figure 7.3, a CR node senses several channels simultaneously and chooses the best ones to transmit data to the receiver. Besides, by considering the PHY layer measurements, an RF-based scavenged device can be used, to power the sensor nodes based on local measurements (e.g., by using WiFi inside the hospital).

- **Mobile networks:** In static networks, node position can be determined once during initialisation. However, in a mobility scenario, since nodes frequently change position, some adjustments in the transmission power may be required (e.g., when nodes are close to each other, their transmission power can be lowered). The challenge in node mobility is how to cope with motion on different speeds, which determines the updates of the frequency of transmission power. Besides, our CR system must be capable to adapt to the frequent changes in the control and data channels that may vary in different clusters. This requires additional time and energy, as well as the availability of a rapid localisation service. Since frequent updates have a great impact on the network performance, the electromagnetic energy from RF signals can be collected by using available the spectrum opportunities.

- **Hybrid:** In a hybrid ad hoc WBSNs, nodes are mobile and stationary. Therefore, they are able to form connections at both MAC and network layers. When mobile nodes are inserted into the static network, in order to maintain the connectivity of the network, route information must be set up, which increases the drain of the power source. Hence, mobility-aware dynamic spectrum management CR solutions must be considered to overcome the challenges covered by this additional source of complexity. Spectrum sensing parameters include signal-to-noise ratio (SNR), frequencies available, and RF energy harvesting opportunities based on the energy detection performed at the PHY layer.

7.4.4   Hardware for the cognitive sensor node

The cognitive sensor node hardware of the WBSN is composed by four sub-systems: (i) communication, (ii) computational and storage, (iii) sensing and actuation, and (iv) power system. Figure 7.4 presents the general hardware architecture.

The description of each sub-system is as follows:

- **Communication:** The communication sub-system consists of a radio transceiver and an antenna that enables the wireless communications between neighbouring nodes.

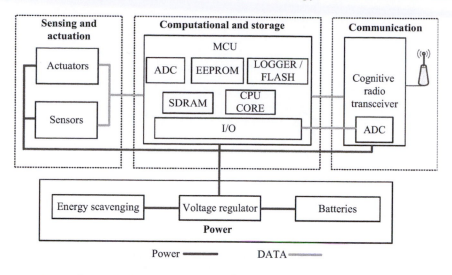

Figure 7.4   Hardware structure and sub-systems for CR sensor nodes

- **Computational and storage:** This sub-system allows for data processing and the management of the nodes functionalities.
- **Sensing and actuation:** The interfaces between the environment and the WBSN are the sensors and the actuators. Basic environmental sensors include, but are not limited to, light, temperature, humidity, pressure, acceleration/seismic, acoustic, magnetic, and sound. Basic environmental actuators include, but are not limited to, light-emitting diodes, speakers, and buzzers.
- **Power system:** The appropriate energy infrastructure to supply the nodes, includes the batteries and the energy scavenging systems, which allows for supporting the operation of the nodes from a few hours to months or years.

The inclusion of a CR transceiver in the communication sub-system is the main difference between the hardware structure of CR sensor and classic sensors.

7.5   Communication aspects of WBSNs with CR capabilities

This section investigates the specific cross-layer design aspects, between the PHY, MAC, and network layers of a CR sensor node.

7.5.1   PHY layer aspects

The PHY layer acts as a mediator between the data link layer and the PHY wireless environment. Since CRs need to sense the spectrum in order to find spectrum opportunities, the PHY layer is also responsible for spectrum sensing, reporting it to the microprocessor of the CR node. As presented in Figure 7.5, the PHY layer also aims

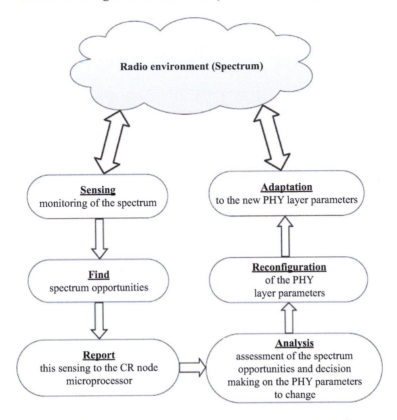

Figure 7.5   Tasks related to spectrum sensing at the PHY layer

at the reconfiguration of the transmission parameters according to the decisions from the microprocessor.

WBSNs which do not consider CR capabilities impose environmental and propagation constraints that must be taken into account in system design:

- Current drawn is 400 and 1 μA for the active and idle modes, respectively [26].
- Typically, the sensor nodes are placed near the human skin, which imposes restrictions to the transmitter power from the radio transceiver, as the radiation caused by the large wireless transmission power may possibly harm human health. The IEEE 802.15.6 group, which is focused in WBANs, already foresees and advises these thresholds. Safe transmitter power thresholds depend on the location of the sensor node relatively to the human body.
- The surrounding environment of WBSNs produces situations such as the phenomena of body shadow effect already identified by the authors of [13]. This effect is due to the propagation of a signal over the human body.
- The multi-path effect leads to interference due to the reflection of the radiated signal. This is caused by the ground and surrounding objects. In WBSNs the

values of reflection signal depend on the position of the sensor node in the human body [27].

- The antenna characteristics cause some negative impact on the overall performance of WBSNs. The coupling effect appears and depends much on the relative positions of the sensor nodes in the human body.

WBSNs without CR capabilities must be able to cope with the majority of the aforementioned effects in order to properly receive the low-power signal. Therefore, the use of CR capabilities in WBSNs must also consider these design requirements of the system in order to be efficient.

In addition to the environmental and propagation constraints, the CR transceiver hardware must be suitable to enable CR operations WBSNs. The ability of the CR to reconfigure the PHY layer parameters (i.e., modulation, channel coding, and transmitter power) is the main difference between the WSN and the cognitive radio sensor network (CRSN) PHY layer. This in situ reconfiguration does not require hardware replacement. To accomplish this reconfiguration, software defined radio (SDR) based RF front-end transmitters and receivers are needed. Special attention is needed in the use of SDRs due to the nodes' scarce power supply in WBSNs. Energy harvesting is a solution to complement the power supply requirements in the context of SDR in CR sensor nodes. A preferable solution is the development of SDRs specifically to energy-efficient CR sensor nodes. The spectrum sensing task of the CR sensor nodes is a challenging issue due to its limited processing capabilities. Spectrum sensing is a highly demanding signal processing task, since the radio signals in WBSNs are weak and with possible large background noise. Digital signal processing (DSP) hardware and algorithms can be added to the CR sensor node to achieve more efficient wideband spectrum opportunities sensing and detection. Furthermore, if conventional SDRs applied to CR sensor nodes are not considered, it is impossible to support different modulations schemes, waveforms or supporting wide-band spectrum sensing, due to the limited processing capabilities of the CR sensor node. Another design consideration aspect in WBSNs with CR capabilities is the development of transmission power and interference adaptive algorithms that cope with interference in the deployment of CR sensor nodes over the human body.

7.5.2 *Medium access control sub-layer aspects*

The MAC protocols are responsible to determine and change the operation mode of the radio transceiver, allowing for nodes to access the medium in a more fair and efficient manner. Compared to conventional WSNs, the MAC layer of a CRSN node must handle additional challenges such as silent spectrum sensing periods and the need for high-priority access mechanism for the distribution of spectrum sensing and decision results [5]. In WBSNs with CR capabilities, sensor nodes may use a control-channel-request-to-send and control-channel-clear-to-send (CRTS/CCTS) handshake mechanism to negotiate the access to the channel before transmitting packets [28]. The use of a CRTS/CCTS mechanism on a separated control channel packets allows for decreasing the number of collisions, whereas the DATA and ACK packets could be transmitted in a group of frequency bands, as shown in Figure 7.6.

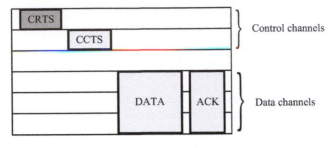

Figure 7.6 CRTS/CCTS/DATA/ACK handshake

Therefore, compared to conventional WSNs, the MAC layer of WBSNs with CR capabilities must address additional challenges regarding the coordination of DSA, as follows:

- **Spectrum sensing and decision results:** Cooperative sensing [29] and decision results are used to increase sensing accuracy and sharing efficiency. Therefore, nodes must share extra control information. The CR MAC protocol must include mechanisms to share information with higher priority.
- **Minimum overhead:** The wide range of MAC protocols for WBSNs use control packets. This type of packets can be received by all nodes within radio range of the sender, resulting in an increase of the power drain in a potentially large number of nodes. Since nodes are required to remain awake in order to receive control packets, the battery life can be significantly reduced. Therefore, it is envisaged a CRSN with a minimum exchange of control packets and without the need of having additional hardware requirements (i.e., an extra transceiver or GPS for synchronisation). This can be accomplished by using a block acknowledgement (BACK) policy feature [30]. This way every time a node accesses to the medium, it optimises the transmission time by reducing the amount of overhead, whilst increasing the channel capacity. The piggyback mechanism (PM) could also be used, in which a receiver station is allowed to piggyback a data frame to a sender station once if the receiver station has a frame to send to the sender [31].
- **Adaptive duty cycle:** Since the spectrum opportunities are time and location variant, self-adaptive mechanisms must be introduced. Therefore, based on the spectrum sensing measurements we can increase the duty cycle, which means more opportunities to cover multiple neighbours with one forwarding. Besides, since collecting RF energy to power the sensor nodes is foreseen based on the available energy and CR opportunities, a threshold can be implemented to send packets based on the existing energy scavenging opportunities versus interference metrics.

7.5.3 Network layer aspects

Cognitive networks and its design aspects are very challenging for the network layer design. In traditional wireless networks, the nodes use the same frequency to communicate with each other. Thus, routing is performed by considering only one link between neighbours. In CR networks, the network layer is responsible for choosing the best next hop as well as the frequency to use. This brings an extra load to the routing algorithm, since it must also take into account the readings from the spectrum to decide the routes for the packets to follow. Therefore, the first challenge is how to have available information on the spectrum usage. This information must be provided by an external (to the routing) entity that constantly monitors the communication bands and evaluates its interference or not. This can be performed by the node locally, using its own sensing capacities, can be provided by an external entity dedicated to that task, or by a combination of these two, in which neighbouring nodes exchange information about their spectrum sensing. The second challenge is how to combine the information gathered and choose the optimal path for the packets to be sent. Traditional approaches use hop-count, RSSI strength or QoS metrics to decide the best routes for the packets; however, in CR networks the frequency to be used must also be included in the algorithm. With the dynamic nature of the spectrum and constant frequency usage by other entities, the algorithm will have to adapt the paths to the current conditions of the channel. However, open research issues are still open, and must be addressed in order to have a complete network solution as follows:

- **Spectrum aware routing:** When nodes send packets through the network, spectrum sensing techniques must be addressed to opportunistically route data packets across paths avoiding spectrum congested areas. To achieve this goal, innovative awareness mechanisms (that facilitate to know about the presence, characteristics, and requirements of other wireless devices in the same area) that consider spectrum mobility and resource constraints must be employed to find the optimal traffic according to the available spectrum resources.
- **Adaptive and QoS routing:** Cross-layer mechanisms will be responsible for providing up-to-date local QoS information for the adaptive routing protocol. Hence, new techniques based on varying channel conditions must be taken into consideration for real-time communication in CRSNs.
- **Multi-hop routing maintenance and reparation:** In a multi-hop CR scenario the sudden appearance of a PU in a given location may impose an unusable channel for the SUs, leading to unpredictable route failures. Therefore, effective signalling mechanism must be addressed to restore the paths with minimal effect on the network performance.

7.6 Spectrum opportunities for RF energy harvesting

RF energy harvesting in WBANs can be accomplished by using a new generation of wearable antennas that allow for power supplying the sensor nodes [32]. Ubiquitously available RF sources, operating at different bands, are therefore exploited for RF electromagnetic energy harvesting purposes.

Based on power density measurements, we have been able to identify the best spectrum opportunities that may be considered in order to conceive multiband antennas for electromagnetic energy harvesting. The RF energy harvesting system

developed in the context of the PROENERGY-WSN project is presented as an example
of energy harvesting implementation [33], which consists of an impedance matching
circuit, rectifier, and the energy storage sub-system.

In order to seek the best spectrum opportunities for RF energy harvesting, field
trial measurements have been conducted in Covilhã, Portugal, by using the NARDA-
SMR spectrum analyser with a tri-axial measuring (isotropic) antenna, in both indoor
and outdoor environments.

7.6.1   *Average received power*

By analysing the power density measurements in 36 different locations, we have found
the recommendations for frequencies for RF energy harvesting. Besides, the identified
spectrum opportunities have been considered to conceive multi-band antennas. The
location for the measurements is shown in Figure 7.7. To determine the received
power, P_r, of the spectrum analyser, we multiply the power density, P_d, by the effective
receiving area of the antenna, A_e, and gain, $G = 1$, as follows:

$$P_d \ [\text{W/m}^2] = |E^2|/(120 \cdot \pi), \tag{7.1}$$

$$\overline{P_r \ [\text{dBm}]} = 10 \cdot \log\left(P_d \frac{\lambda^2 \cdot G}{4\pi}\right) + 30, \tag{7.2}$$

where E is the electric field and λ is the wavelength.

Figure 7.7   Locations of the measurements to identify spectral RF energy
harvesting opportunities in Covilhã, Portugal (outdoor measurements
are within the green balloons while the indoor measurements are
identified by the red balloons)

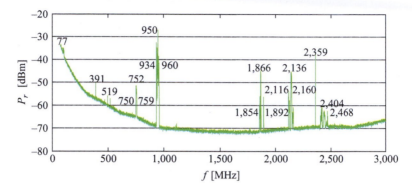

*Figure 7.8 Average received power as a function of the frequency for the university
scenario (indoor)*

To choose the best frequency bands for electromagnetic energy harvesting, we
have determined the average of each of the n values of the received power, P_{ri} [W]
in linear units, in five different locations, where n is the number of measurements
taken, for each frequency. The average received power, in dBm, is then calculated as
follows:

$$\overline{P_r \text{ [dBm]}} = 10 \cdot \log\left(\frac{\sum_{i=1}^{n} P_{ri}[W]}{n}\right) + 30. \tag{7.3}$$

7.6.2 Indoor opportunities

Figure 7.8 facilitates to identify the indoor spectrum opportunities as observed at
Universidade da Beira Interior in Covilhã.

The set of frequencies with high-energy available for harvesting comprises the
range from 934 to 960 MHz (GSM 900), 1,854 to 1,892 MHz (GSM 1800), 2,116
to 2,160 MHz (UMTS), 2,359 MHz (amateur, SAP/SAB applications, video), and
2,404 to 2,468 MHz (WiFi).

7.6.3 Outdoor opportunities

The location of public places in the outdoor scenario for the field trial results are
identified in Figure 7.7 as locations numbered 8, 9, 12, 13, 14, 21, and 22. The
corresponding values of the average received power are shown in Figure 7.9.

The set of frequencies with more energy available for harvesting are in the range
from 79 to 96 MHz (mobile/radio broadcast stations), 391 MHz (emergency broadcast
stations), 750 to 759 MHz (digital television broadcast stations), 935 to 960 MHz
(GSM 900 broadcast stations), 1,854 to 1,870 MHz (GSM 1800 broadcast stations),
and 2,115 to 2,160 MHz (UMTS broadcast stations).

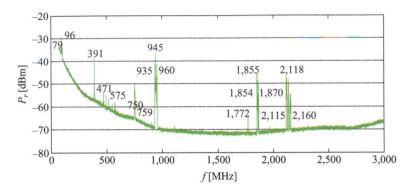

Figure 7.9   Average received power for the outdoor scenario

The conclusion from the above-reported measurements indicated that the GSM 900/1800 frequency bands appear to be the most promising bands for RF energy harvesting.

7.7   Innovative MAC protocols

In [34] the authors have shown that one fundamental reason for IEEE 802.15.4/4a MAC inefficiency is overhead, where the use of acknowledgement (ACK) control frames can decrease the bandwidth efficiency about 10%. In this work, we propose and analyse two innovative mechanisms to reduce overhead in IEEE 802.15.4: (i) concatenation and (ii) piggyback [35]. The main idea is to improve channel efficiency by aggregating several ACK responses into one single transmission (i.e., one single frame) like in the IEEE 802.11e standard. This aggregation of ACKs aims at reducing the overhead by transmitting less ACK control frames and by decreasing the time periods the transceivers should switch between different states.

The proposal of the Sensor Block Acknowledgement (SBACK-MAC protocol aims at increasing the throughput as well as decreasing the end-to-end delay, whilst providing a feedback mechanism for the receiver to inform the sender about how many transmitted (TX) frames were successfully received (RX). Our proposal also considers the use of the RTS/CTS mechanism in order to avoid the hidden terminal problem.

In IEEE 802.15.4 the protocol overhead impacts on end-to-end delay and throughput. In order to reduce end-to-end delay and increase throughput, we propose a new innovative MAC protocol that solves the above problems, along with the elimination of the backoff period repetitions, the SBACK-MAC protocol [36]. The main difference compared to IEEE 802.15.4 is related to the way that SBACK-MAC treats the ACK control frames. The SBACK-MAC allows the aggregation of several ACK responses into one special frame. The *BACK Response* will be responsible to confirm a set of data frames successfully delivered to the destination. This frame has the same length as an ACK frame in IEEE 802.15.4. Hence, an ACK control frame will not be received in response to every data frame sent/received. By decreasing the number of control frames exchanged in a wireless medium, it is possible to decrease not only the number of collisions but also the number of back-off periods (the time a node must wait before attempting to transmit/retransmit the frame) on each node. Moreover, in WSNs the length of control frames can be of the order of magnitude of the data frames. Since nodes are battery-operated, the transmission of such frames leads to energy decrease, whilst reducing the number of data frames that will be transmitted containing useful information (i.e., goodput).

The SBACK-MAC also considers the *ccaTime*. This way, clear channel assessment (CCA) nodes are able to determine the channel state (i.e., busy or idle), which allows for providing statistical information for the MAC sub-layer and upper layers. Moreover, since the CCA result is based on the obtained RSSI, transmission power control techniques could be used to estimate the minimum transmission power for sending each frame to a neighbouring node.

7.7.1 BACK mechanism with BACK Request

The version of the proposed SBACK-MAC protocol with *BACK Request* considers the exchange of two special frames, that is, *RTS ADDBA* and *CTS ADDBA*, where *ADDBA* stands for "Add Block Acknowledgement." After this successful exchange, data frames are transmitted from the transmitter to the receiver (e.g., 100 frames are sent during the active periods). Afterwards, by using the BACK Request primitive, the transmitter inquires the receiver about the total number of data frames that successfully reach the destination. In response, the receiver will send a special data frame, called BACK Response identifying the frames that require retransmission, and the BACK mechanism finishes.

Figure 7.10 presents the message sequence chart for the BACK mechanism.

The exchange of two types of special control frames used at the beginning and end of the BACK mechanism allows for mitigating the hidden-terminal and exposed-terminal problems like in IEEE 802.11e. The BACK mechanism also aims at reducing the power consumption by transmitting less ACK control frames and by decreasing

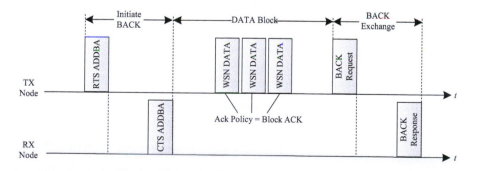
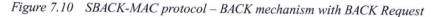

Figure 7.10   SBACK-MAC protocol – BACK mechanism with BACK Request

*Figure 7.11   Frame sequence for the IEEE 802.15.4 SBACK-MAC protocol with
BACK Request (concatenation)*

the time periods the transceivers should switch between different states. By using the BACK there is no need to receive an ACK for every DATA frame sent, as presented in Figure 7.11. Besides, during the data transmission there is no way to know how many frames have successfully reached the destination, except at the end of communication by using the *BACK Request/BACK Response* (we set *BE* equal to 0), as if there is no congestion.

When the SBACK-MAC with *BACK Request* is considered, the minimum average delay, D_{min}, in seconds, is given by

$$D_{min\_BACK} = \left(\overline{CW} + ccaTime + T_{RTS_{ADDBA}} + H_1\right) / n, \tag{7.4}$$

where $H_1 = T_{TA} + T_{CTS\_ADDBA} + n \times (ccaTime + T_{TA} + T_{DATA} + T_{TA} + T_{IFS}) + ccaTime + T_{TA} + T_{BRequest} + T_{TA} + T_{BResponse} + T_{IFS}$.

The maximum average throughput, S_{max}, in bits per second, is given by

$$S_{max\_BACK} = 8L_{DATA}/D_{min\_BACK}. \tag{7.5}$$

The meaning of the parameters is defined in Table 7.3.

By analysing equations (7.4) and (7.5), we conclude that by using the *BACK Request* primitive we allow several MAC Protocol Data Units (MPDUs) to be acknowledged by a single *BACK Request* frame. Therefore, compared with IEEE 802.15.4 in the basic access mode, there is no need to consider the use of individual ACK control frames. We also assume that the back-off period is equal to 0, as if there is no congestion (no activity in the shared medium). This is explained by the fact that a CTS can be only sent if there is no congestion at the receiver. The *BACK Response* contains the information about the reception of corresponding MPDUs by using a specific bitmap that is transmitted as an answer to an explicit transmitter request. This request is performed by the new *BACK Request* control frame. Both the *BACK Request* and *BACK Response* are transmitted at the same data rate (i.e., 250 kb/s).

7.7.2   Proposed scheme with no BACK Request

The version of the SBACK-MAC protocol with no *BACK Request* ("piggyback mechanism") also considers the exchange of the *RTS ADDBA* and *CTS ADDBA* frames at the beginning of the communication. However, at the end of the communication the BACK Request primitive is not transmitted. Therefore, the last aggregated data frame must include the information about the frames previously transmitted, as shown in Figure 7.12.

Table 7.3   IEEE 802.15.4 and SBACK-MAC typical parameters and values

Description	Symbol	Value
Backoff period duration	T_{BO}	320 μs
PHY SHR duration	T_{SHR}	160 μs
CCA detection time	T_{CCA}	128 μs
TX/RX or RX/TX switching time	T_{TA}	192 μs
Short interframe spacing (SIFS) time	T_{SIFS}	192 μs
Long interframe spacing (LIFS) time	T_{LIFS}	640 μs
ACK transmission time	T_{ACK}	352 μs
RTS transmission time	T_{RTS}	352 μs
CTS transmission time	T_{CTS}	352 μs
RTS ADDBA transmission time	T_{RTS_ADDBA}	352 μs
CTS ADDBA transmission time	T_{CTS_ADDBA}	352 μs
BACK Request transmission time	$T_{BRequest}$	352 μs
BACK Response transmission time	$T_{BResponse}$	352 μs
ACK wait duration time	T_{AW}	560 μs
DATA transmission time	T_{DATA}	576 μs
Time to set up radio to RX or TX states	$rxSetupTime$	1,792 μs
PHY length overhead	L_{H_PHY}	6 bytes
MAC overhead	L_{H_MAC}	9 bytes
DATA payload	L_{H_DATA}	3 bytes
DATA frame length	L_{FL}	18 bytes
ACK/RTS/CTS frame length	L_{ACK}	11 bytes
Number of TX frames	n	1 to 100
Data rate	R	250 kb/s

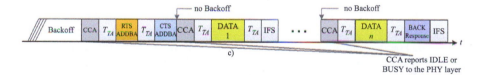

Figure 7.12   Frame sequence for the IEEE 802.15.4 SBACK-MAC protocol without BACK Request (piggyback)

The SBACK-MAC version with "piggyback" does not consider the use of the BACK Request primitive, as shown in Figure 7.12. As a consequence, the control overhead and the end-to-end delay are reduced, whilst increasing the throughput, by piggybacking the BACK information into the last data fragment. However, with this scheme, the system becomes less robust. If the last aggregated frame (DATA frame *n*) is lost, the destination does not know that an ACK needs to be sent back.

When the SBACK-MAC with no *BACK Request* is considered, the minimum average delay, $D_{\min\_Piggy}$, in seconds, is given by:

$$D_{\min\_Piggy} = \left(\overline{CW} + ccaTime + T_{RTS_{ADDBA}} + H_2\right)/n, \tag{7.6}$$

where $H_2 = T_{TA} + T_{CTS\_ADDBA} + (n - 1) \times (ccaTime + T_{TA} + T_{DATA} + T_{TA} + T_{IFS}) + ccaTime + T_{TA} + T_{DATA} + T_{TA} + T_{BResponse} + T_{IFS}$.

The maximum average throughput, S_{max_Piggy}, in bits per second, is given by

$$S_{max\_Piggy} = 8L_{DATA}/D_{min\_Piggy}. \tag{7.7}$$

The meaning of the parameters is defined in Table 7.3.

By analysing equations (7.6) and (7.7), we conclude that in the SBACK-MAC with no BACK Request, $(n-1)$ frames are transmitted (which corresponds to less than one IFS) like in the SBACK-MAC with *BACK Request*. However, since the last data frame includes the information about the total number of frames previously transmitted, the *BACK Response* is transmitted immediately after the reception of the last data frame. Therefore, there is no need to transmit the *BACK Request* frame.

7.7.3 *Modelling and simulation results*

The SBACK-MAC was evaluated by using the MiXiM simulation framework [37] from the OMNeT++ simulator. SBACKMAC throughput and end-to-end delay with and with no *BACK Request* have been compared against IEEE 802.15.4, by considering a 95% confidence interval; however, as it is too small, we decided not to plot it in the figures. Table 7.3 presents the MAC parameters considered for the network in our simulations. The performance analysis of the proposed schemes is conducted for the best-case scenario. Therefore, we are assuming that the channel is an ideal channel, with no transmission errors. During the active period, there is only one node that always has a frame to be sent. The other stations can only accept frames and provide acknowledgements.

Figure 7.13 presents the maximum average throughput and the minimum average delay versus the payload size for the SBACK-MAC protocol with and with no *BACK Request*. The discontinuity around 18 bytes is due to the use of SIFS and LIFS (i.e., MPDU less of equal than 18 bytes must be followed by a SIFS, whilst MPDU longer than 18 bytes must be followed by a LIFS). The number of transmitted frames, n, is 10 (i.e., for the SBACK-MAC, the frames are aggregated and transmitted in a burst).

It is observed that by increasing the payload size S_{max} also increases. This conclusion is valid for all the three presented mechanisms. For small frame sizes (i.e., data payload less than or equal to 18 bytes) by comparing the IEEE 802.15.4 with the SBACK-MAC protocol, with and with no *BACK Request*, S_{max} increases 17% and 25%, respectively. Moreover, by using the IEEE 802.15.4 basic access mode with DATA/ACK, the maximum achievable throughput is approximately 108.7 kb/s whereas, by using the SBACK-MAC with and with no *BACK Request*, the maximum achievable throughput is 118.1 and 123.2 kb/s, respectively. Results for D_{min} as a function of the payload size show that by using SBACK-MAC with and with no *BACK Request* for small frames sizes (i.e., data payload less or equal to 18 bytes) D_{min} decreases 17% and 25%, respectively. For larger frame sizes, by considering SBACK-MAC with and with no *BACK Request*, D_{min} decreases by 8% and 13%, respectively.

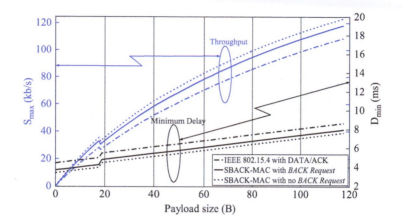

Figure 7.13   Maximum throughput and minimum delay versus payload size

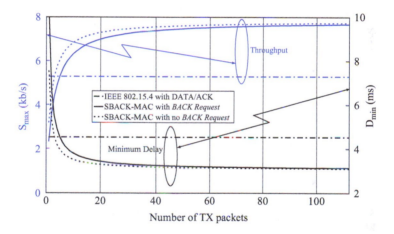

Figure 7.14   Maximum throughput and minimum delay versus number of TX frames

Figure 7.14 presents S_{max} and D_{min} as a function of the number of TX frames. A fixed payload size of 3 bytes (i.e., LDAT A = 3 bytes) is considered, since it is one of the values in the range from 1 to 18 bytes presented in Figure 7.13, by considering the worst throughput performance, when taking into account the BACK mechanism. Even for the shortest payload sizes, it is possible to improve the network performance by using the proposed BACK mechanisms.

It may be observed that when the number of TX frames is less than 4, the IEEE 802.15.4 standard through the basic access mode achieves higher throughput in comparison to SBACK-MAC (either with or with no *BACK Request*). Moreover, by considering the IEEE 802.15.4 standard in the basic access mode, S_{max} does not depend on the number of TX frames and achieves the maximum value of 5.2 kb/s.

In SBACK-MAC with and with no *BACK Request* (i.e., concatenation and piggyback), S_{max} increases by increasing the number of TX frames (i.e., the number of aggregated frames). For a number of TX frames equal to 18, by considering the SBACK-MAC with *BACK Request* (i.e., concatenation version) S_{max} is about 6.3 kb/s. This value corresponds to an increase of 21% in the throughput in comparison to the MAC protocol from the IEEE 802.15.4 standard in the basic access mode, whereas by considering the SBACK-MAC with no *BACK Request* (i.e., piggyback version), the achievable throughput is 6.8 kb/s, an increase of 30%. However, the difference in the throughput between the SBACK-MAC with and with no *BACK Request* tends to decrease by increasing the total number of TX frames (i.e., by aggregating more frames). We also conclude that, for more than 4 TX frames, SBACKMAC (with and with no *BACK Request*) delay is significantly shorter than for IEEE 802.15.4 in the basic access mode. The difference is mitigated by increasing the total number of TX frames (i.e., by aggregating more frames).

The previous results have shown that the use of the proposed BACK mechanisms enables to improve the network performance, whilst increasing the channel use optimisation. As a consequence, in the context of cognitive radio wireless sensor networks (CRWSN) the proposed BACK mechanisms can be applied in a scenario where the SU's flow alternates between sensing (i.e., CCA mechanism) and sending rapidly at short intervals [38], see Figure 7.15.

By considering that the primary user data flow (from TX_1 to RX_1) is bursty, there will be inactive periods between transmissions. These inactive periods are used by the SUs to send their own traffic. Since, by using the proposed BACK mechanism we decrease the end-to-end delay, whilst increasing throughput we intend to exploit the temporal opportunities (i.e., temporal "white spaces") [38] more efficiently, by decreasing the transmission times of SUs. Moreover, CRWSN nodes must perform sensing and make a decision about the channel state (i.e., busy or idle). This spectrum decision time can also be seen as overhead, leading to an increase in the energy consumption. Therefore, in the case of re-transmissions, the SBACK-MAC protocol with and with no *BACK Request*, is able to decrease the time periods before making

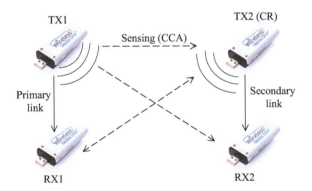

Figure 7.15   Signal and interference paths in a CR sensor network

CCA again, since there is no back-off phase between two consecutive data frames, which allows for decreasing the total overhead.

7.8   Conclusions

The growing use of radio communication technology for healthcare is triggering the deployment of BANs in medical premises mostly on a license-exempt secondary basis. This fact motivates further studies on interference mitigation techniques for medical communication environments. CR offers viable cost-effective and future-proof solutions addressing both scalability and coexistence issues.

The next generation of CR networks will be supplied by renewable energy from natural resources, where RF energy harvesting plays an important role. In a near future, RF energy will enable to power supply all the nodes without the need of replacement of the primary source of energy (i.e., batteries).

We have identified the spectrum opportunities for RF energy harvesting to power supply the wireless sensor nodes in real indoor/outdoor scenarios. The set of indoor/outdoor most promising frequency bands are 79–96 MHz (mobile/radio broadcast stations), 391 MHz (emergency broadcast stations), 750–758 MHz (digital television broadcast stations), 935–960 MHz (GSM 900 broadcast stations), 1,855–1,868 MHz (GSM 1800 broadcast stations), and 2,115–2,160 MHz UMTS broadcast stations). A scenario of using RF harvesting with dedicated devices had been analysed in the PROENERGY-WSN [33]. It enables future WBAN to operate without the need of replacing batteries.

In order to improve the efficiency of WBANs and reduce the overhead that affects WBANs communications, a new innovative SBACK-MAC protocol which aggregates several ACK responses (concatenation) into one special frame has been proposed. By using a concatenation mechanism the total overhead is decreased, since there is no back-off between two consecutive data frames.

Two innovative solutions have been proposed to improve the IEEE 802.15.4 performance. The first one considers the SBACK-MAC protocol in the presence of *BACK Request* (concatenation mechanism), while the second considers the SBACK-MAC in the absence of BACK Request (PM). The results presented here show that the proposed concatenation mechanisms considerably improve network performance in terms of throughput and end-to-end delay. In the context of CRWSNs, enhancements will be achieved because of the reduction of queue lengths of SUs as well as the channel utilisation.

DSA can be used to mitigate spectrum scarcity. This is accomplished by enabling an unlicensed user (i.e., SU) to adaptively adjust its operating parameters and exploit the spectrum which is unused by licensed users (i.e., PUs), in an opportunistic manner.

References

[1]   H. Alemdar and C. Ersoy, "Wireless sensor networks for healthcare: A survey," *Computer Networks*, vol. 54, no. (15), pp. 2688–2710, 2010.

[2]   E. Jovanov and A. Milenkovic, "Body area networks for ubiquitous health-care applications: Opportunities and challenges," *Journal of Medical Systems*, vol. 35, no. (5), pp. 1245–1254, 2011.

[3]   P. Rawat, K. D. Singh and J. M. Bonnin, "Cognitive radio for M2M and internet of things: A survey," *Computer Communications*, vol. 94, pp. 1–29, 2016.

[4]   *IEEE Standard for Local and Metropolitan Area Networks – Part 15.6: Wireless Body Area Networks*, IEEE Standard 802.15.6-2012, 2012.

[5]   H. Cao, V. Leung, C. Chow and H. Chan, "Enabling technologies for wireless body area networks: A survey and outlook," *IEEE Communications Magazine*, vol. 47, no. (12), pp. 84–93, 2009.

[6]   J.-H. Hauer, V. Handziski and A. Wolisz, "Experimental study of the impact of WLAN interference on IEEE 802.15.4 body area networks," *Wireless Sensor Networks*, vol. 5432, pp. 17–32, 2009.

[7]   R. Chávez-Santiago, A. Khaleghi, I. Balasingham and T. A. Ramstad, "Architecture of an ultra wideband wireless body area network for medical applications," in *Proc. 2nd Intl. Symp. on Applied Sciences in Biomedical and Communication Technologies (ISABEL)*, Bratislava, Slovak Republic, November 24–27, pp. 1–6, 2009.

[8]   J. Wang, M. Ghosh and K. Challapali, "Emerging cognitive radio applications: A survey," *IEEE Communications Magazine*, vol. 49, no. (3), pp. 74–81, 2011.

[9]   R. Doost-Mohammady and K. R. Chowdhury, "Transforming healthcare and medical telemetry through cognitive radio networks," *IEEE Wireless Communications*, vol. 19, no. (4), pp. 67–73, 2012.

[10] R. Chávez-Santiago, K. E. Nolan, O. Holland *et al.*, "Cognitive radio for medical body area networks using ultra wideband," *IEEE Wireless Communications – Special Issue on Cognitive Radio – A Practical Perspective*, vol. 19, no. (4), pp. 74–81, 2012.

[11] H. Jabbar, Y. S. Song and T. T. Jeong, "RF energy harvesting system and circuits for charging of mobile devices," *IEEE Transactions on Consumer Electronics*, vol. 56, no. (1), pp. 247–253, 2010.

[12] N. Barroca, J. M. Ferro, L. M. Borges, J. Tavares and F. J. Velez, "Electromagnetic energy harvesting for wireless body area networks with cognitive radio capabilities," in *Proc. of URSI Seminar of the Portuguese Committee*, Lisbon, Portugal, Nov. 2012.

[13] S. L. Cotton and W. G. Scanlon, "Characterization and modeling of the indoor radio channel at 868 MHz for a mobile bodyworn wireless personal area network," *IEEE Antennas and Wireless Propagation Letters*, vol. 6, no. (1), pp. 51–55, 2007.

[14] P. Phunchongharn, E. Hossain, D. Niyato and S. Carmolinga, "A cognitive radio system for e-health applications in a hospital environment," *IEEE Wireless Communications*, vol. 17, no. (1), pp. 20–28, 2010.

[15] Dariusz Wiecek and Fernando José Velez, "Case Studies for Advancing CR Deployment," in A. A. Medeisis and O. Holland (eds.), *Cognitive Radio Policy and Regulation: Techno-Economic Studies to Facilitate Dynamic Spectrum Access*, Chapter 6, A. Springer, London, UK, 2014, pp. 309–348.

[16] R. Chávez-Santiago, D. Jankūnas, V. V. Fomin and I. Balasingham, "A dual-band MAC protocol for indoor cognitive radio networks: An e-health case study," in *Proc. 8th Intl. Conf. on Body Area Networks (BodyNets)*, Boston, MA, September 30–October 2, 2013.

[17] F. Yildiz, D. Fazarro and K. Coogler, "The green approach: Self-power house design concept for undergraduate research," *Journal of Industrial Technology*, vol. 26, no. (2), pp. 1–10, 2010.

[18] http://www.powercastco.com, July 2018.

[19] A. M. Zungeru, L. Ang, S. R. S. Prabaharan and K. P. Seng, *Radio Frequency Energy Harvesting and Management for Wireless Sensor Networks*, in H. Venkataraman and G. Munteam (eds.), Green Mobile Devices and Networks – Energy Optimization and Scavenging Techniques, CRC Press, Boca Raton, Florida, 2012.

[20] T. Le, K. Mayaram, and T. Fiez, "Efficient far-field radio frequency energy harvesting for passively powered sensor networks," *IEEE Journal of Solid-State Circuits*, vol. 43, no. (5), pp. 1287–1302, 2008.

[21] G. Tech, "Air Power: New Device Captures Ambient Electromagnetic Energy to Drive Small Electronic Devices," http://www.rh.gatech.edu/news/68714/ambient-electromagnetic-energy-harnessed-small-electronic-devices, July 2018.

[22] X. Zhou, L. Lin, Jianping Wang and X. Zhang, "Cross-layer routing design in cognitive radio networks," *Wireless Personal Communications*, vol. 49, no. (1), pp. 123–131, 2009.

[23] M. Xie, W. Zhang, and K.-K. Wong, "A geometric approach to improve spectrum efficiency for cognitive relay networks," *IEEE Transactions on Wireless Communications*, vol. 9, no. (1), pp. 268–281, 2010.

[24] O. B. Akan, O. B. Karli, and O. Ergul, "Cognitive radio sensor networks," *IEEE Network*, vol. 23, no. (4), pp. 30–40, 2009.

[25] D. Raychaudhuria, X. Jinga, I. Seskara, K. Lea, and J. B. Evansb, "Cognitive radio technology: From distributed spectrum coordination," *Pervasive and Mobile Computing*, vol. 4, no. (3), pp. 278–302, 2008.

[26] E. Jovanov, D. Raskovic, J. Price, J. Chapman, A. Moore, and A. Krishnamurthy, "Patient monitoring using personal area networks of wire-less intelligent sensors," *Biomedical Sciences Instrumentation*, vol. 37, no. (1), pp. 373–378, 2001.

[27] A. Fort, J. Ryckaert, C. Desset and P. D. Doncker, "Ultra-wideband channel model for communication around the human body," *IEEE Journal on Selected Areas in Communications*, vol. 24, no. (4), pp. 927–933, 2006.

[28] H. Wang, H. Qin and L. Zhu, "A survey on MAC protocols for opportunistic spectrum access in cognitive radio networks," in *Proc. of International Conference on Computer Science and Software Engineering*, Wuhan, China, 2008, pp. 214–218.

[29] G. Ganesan and Y. G. Li, "Cooperative spectrum sensing in cognitive radio networks," in *Proc. of First IEEE International Symposium on New*

Frontiers in Dynamic Spectrum Access Networks, Baltimore, MD, USA, 2005, pp. 137–143.

[30]   O. Cabral, A. Segarra and F. J. Velez, "Implementation of IEEE 802.11e Block acknowledgement policies," *IAENG International Journal of Computer Science (IJCS)*, vol. 36, no. (1), pp. 85–93, 2009.

[31]   Y. Xiao, "IEEE 802.11 performance enhancement via concatenation and piggyback mechanisms," *IEEE Transactions on Wireless Communications*, vol. 4, no. (5), pp. 2182–2192, Sep. 2005.

[32]   J. S. Bellon, M. Cabedo-Fabres, E. Antonino-Daviu, M. Ferrando-Bataller and F. Penaran-da-Foix, "Textile MIMO antenna for Wireless Body Area Networks," in *Proc. of the 5th European Conference on Antennas and Propagation (EUCAP)*, Rome, Italy, Apr. 2011, pp. 428–432.

[33]   PROENERGY-WSN, "Prototypes for Efficient Energy Self-sustainable Wireless Sensor Networks," http://www.e-projects.ubi.pt/proenergy-wsn, July 2018.

[34]   M.S. Chowdhury, N. Ullah, Md. H. Kabir, P. Khan and K. S. Kwak, "Through-put, Delay and Bandwidth Efficiency of IEEE 802.15.4a Using CSS PHY," in *Proc. of The International Conference on Information and Communication Technology Convergence (ICTC)*, Jeju Island, Korea 2010.

[35]   *IEEE Standard for Local and Metropolitan Area Networks–Part 15.4: Wireless Body Area Networks, IEEE Standard 802.15.4*, 2011.

[36]   N. Barroca, F. J. Velez and P. Chatzimisios, "Block Acknowledgment Mech-anisms for the optimization of channel use in Wireless Sensor Networks", in *Proc. of the 24th Annual IEEE International Symposium on Personal, Indoor and Mobile Radio Communications (PIMRC 2013)*, London, UK, Sep. 2013, pp. 1565–1570.

[37]   A. Batra, S. Lingam, and J. Balakrishnan, "Multi-band OFDM: A cognitive radio for UWB," in *Proc. IEEE Intl. Symp. Circuits Syst. (ISCAS)*, Island of Kos, Greece, May 21–24, 2006, pp. 4094–4097.

[38]   D. Dash and A. Sabharwal, "Secondary Transmission Profile for a Single-band Cognitive Interference Channel," in *Proc. of 42nd Asilomar Conference on Signals, Systems and Computers*, Pacific Grove, California, USA, Oct. 2008, pp. 1547–1551.

Chapter 8

Two innovative energy efficient IEEE 802.15.4 MAC sub-layer protocols with packet concatenation: employing RTS/CTS and multichannel scheduled channel polling

Fernando J. Velez[1], Luís M. Borges[1], Norberto Barroca[1], and Periklis Chatzimisios[2]

This chapter proposes an IEEE 802.15.4 medium access control (MAC) sub-layer performance enhancement by employing request-to-send/clear-to-send (RTS/CTS) combined with packet concatenation. The results have shown that the use of the RTS/CTS mechanism improves channel efficiency by decreasing the deferral time before transmitting a data packet. In addition, the sensor block acknowledgement (ACK) (BACK) MAC (SBACK-MAC) protocol has been proposed that allows the aggregation of several ACK responses in one special BACK response packet. Two different solutions are briefly considered. The first one considers the SBACK-MAC protocol in the presence of BACK request (concatenation) while the second one considers the SBACK-MAC in the absence of BACK request (piggyback). The throughput and delay performance is mathematically derived under both ideal conditions (a channel environment with no transmission errors) and nonideal conditions (a channel environment with transmission errors). An analytical model is proposed, capable of taking into account the retransmission (RTX) delays and the maximum number of backoff stages. The simulation results successfully validate analytical model. Besides, an innovative efficient multichannel MAC (McMAC) protocol, based on SCP-MAC, has also been proposed, the so-called multichannel scheduled channel polling MAC (MC-SCP-MAC) protocol. The influential range (IR) concept, denial channel list (which considers the degradation metric of each slot channel), extra resolution (ER) phase algorithm and frame capture effect are explored to achieve the maximum performance in terms of delivery ratio and energy consumption. It is shown that MC-SCP-MAC outperforms SCP-MAC and multichannel lightweight medium access control (MC-LMAC) in denser scenarios, with improved throughput fairness. Considering the IR concept reduces the redundancy level in the network facilitating to reduce the energy consumption whilst decreasing the latency. The conclusions from

[1]Faculdade de Engenharia, Instituto de Telecomunicações and Departamento de Engenharia Electromecânica, Universidade da Beira Interior, Portugal
[2]Department of Informatics, CSSN Research Lab, Alexander TEI of Thessaloniki, Greece

this research reveal the importance of an appropriate design for the MAC protocol for the desired wireless body sensor network (WBSN) application.

8.1   Introduction

Wireless sensor network (WSN) is one of the most exciting and important technological innovations in the field of ad-hoc wireless communications of our time. They are responsible for interconnecting several wireless sensor nodes by providing global ad-hoc communication and computational capabilities. This type of networks are capable of linking the physical (PHY) layer with the digital world by sensing, processing and transmitting the real-world phenomena, and by converting these into a form that can be processed, stored and acted upon.

One of the components with the largest power consumption in the sensor nodes is the radio transceiver, as it has a decisive influence in network lifetime. In order to achieve energy efficiency, a MAC protocol must be chosen, to determine and change the operation mode of the radio. Furthermore, recent advances in the field of microelectronic circuits are causing an increase on the interest in the development of WSNs with continuously enhanced capabilities. Nowadays, WSNs can be deployed in many scenarios including healthcare, and medical applications, with the so-called WBSNs or medical body area network (BANs).

Although in the literature most of the proposed communication protocols improve energy efficiency to a certain extent by exploiting the collaborative nature of nodes within WSNs (and in particular WBSNs) and its correlation characteristics, the main commonality of these protocols is the compliance to the traditional layered protocol architecture, allied to the fact they are mainly implemented in commercially available platforms that employs the IEEE 802.15.4 standard.

In the WSNs domain, there is a huge amount of proposals for energy-efficient MAC protocols [1]. However, these MAC protocols have not been capable of having enough success to be set as a real commercial application due to the lack of standardization. To design an optimized WSN MAC protocol, the following main aspects must be considered [2]:

- The first one is the energy efficiency, since there are strong limitations to power supply WSN tiny devices, and one of the main goals is to prolong the network lifetime.
- Other important aspects are scalability and adaptability to changes. Changes in network size, node density and topology should be handled rapidly and effectively for a successful adaptation. Some of the reasons behind these network property changes are limited due to node lifetime, addition of new nodes to the network and varying interference which may alter the connectivity and hence the network topology.
- Finally, quality of service (QoS) attributes, such as latency, throughput and bandwidth efficiency, will also be considered, since this type of metrics is very common when we deal with multimedia applications and real-time applications.

Allied to the aforementioned design issues, IEEE 802.15.4 has been widely accepted as the *de facto* standard for WSNs. Due to its reduced power consumption, it has been used as a basis for ZigBee®, WirelessHart® and MiWi™ applications. Moreover, it represents a significant breakthrough from the "bigger and faster" standards that the IEEE 802 working groups continues to develop and improve. Instead of higher data rates and more functionality, this standard addresses the simple and low-data universe, applied to control and sensor networks, which had existed without global standardization through a series of proprietary methods and protocols.

Actually, annual shipments of IEEE 802.15.4 and ZigBee wireless chipsets are forecasted to reach a cumulative 2.5 billion chipset sales in 2020, as described in [3]. As such, in the context of this work, we propose new innovative MAC sub-layer protocols that can also be considered as a future possible contribution to the standard itself.

Frame concatenation facilitates the aggregation of several consecutive packets by means of channel reservation and different types of ACK and/or network allocation vector (NAV) procedures. In this context, the RTS/CTS mechanism enables to reserve the channel and avoids to repeat the backoff phase for every consecutive transmitted packet, probably reducing overhead. In the presence of RTS/CTS, two solutions are considered, one with DATA/ACK handshake and other without ACKs, simply relying in the establishment of the NAV.

In particular, the SBACK-MAC protocol allows the aggregation of several ACK responses into one special packet *BACK Response* being compliant with the IEEE 802.15.4 standard. Two different solutions are addressed. The first one considers the SBACK-MAC protocol in the presence of *BACK request* (concatenation mechanism), while the second one considers SBACK-MAC in the absence of *BACK request* (the so-called piggyback mechanism).

The following objectives have been identified as the key points for research:

- Evaluate the IEEE 802.15.4 MAC layer performance by using the RTS/CTS combined with packet concatenation.
- Mathematically derive the maximum average throughput and the minimum average delay for the proposed mechanisms, either under ideal conditions (a channel environment without transmission errors) or nonideal conditions (a channel environment with transmission errors), by varying the data payload and the number of transmitted packets. A comparison will be made with IEEE 802.15.4 in the basic access mode.
- Study of the RTX scenarios for both IEEE 802.15.4 and SBACK-MAC, where two nodes simultaneously identify the idle channel during clear channel assessment (CCA) and start transmitting, causing data collisions.

While studying the exchange of frames in the BACK procedures, one assumes a distributed scenario, with a single-destination and single-rate frame aggregation, where there are always frames available for aggregation (saturation conditions – details are given in Section 8.2.1).

This research also involves the proposal of the MC-SCP-MAC protocol. This new MAC protocol explores the advantages of multichannel features in addition to the frame capture effect, present in recent radio transceivers, the IR concept (to mitigate the energy losses due to overhearing) and cognitive-based capabilities, such as the channel degradation sensing jointly with opportunistic channel selection. The denial channel list (which considers the degradation metric of each slot channel), ER phase algorithm and frame capture effect have also been explored to achieve the maximum performance in terms of delivery ratio and energy consumption. Each node maintains a denial channel list which contains the levels of degradation of each channel. This degradation level is a non-negative metric. When a node chooses a channel to transmit a packet, it switches to the chosen channel and performs a first carrier sensing (CS) within the first contention window (CW_1). If the node detects an idle channel, it decreases the degradation level of the channel in one unit (the metric cannot be lower than zero) and passes to the second contention window (CW_2). Otherwise, the node increases the channel degradation in one unit. The nodes that passed to the CW_2 will perform another CS to the channel. In case an idle channel is sensed, the degradation level of the channel is decreased in one unit. Otherwise, the same metric is increased in one unit. Other conditions imposed to channel's degradation level consider that a data packet collision increases the degradation metric by two units in all the nodes that won access to the CW_1 but lost the contention in the CW_2 while detecting a data-packet collision. Besides, if a node sends a data packet and does not receives an MC-ACK, it increases the channel's degradation level by two units.

A simulation approach is considered, where different aspects are addressed, namely the following:

- Energy consumption evaluation and comparison (between single and multi-hop topologies) while considering different traffic generators.
- Evaluation of the expected benefits in terms of packet collision ratio from a multichannel-based protocol with a two-phase contention window mechanism.
- Analysis of the impact of the number of channels considered in the determination of the energy consumption, delay and aggregate throughput.
- Analysis of the energy consumption, delivery ratio and aggregate throughput while varying the sensor nodes density (dense and sparse scenarios).
- Analysis of the impact of enabling the IR concept in the MC-SCP-MAC protocol in terms of energy consumption, delay and delivery ratio.
- Comparison of the performance results for the considered IR thresholds (for the definition of optimal values).

The proposed MAC protocols must be compliant with the IEEE 802.15.4 standard, since we envisage that the proposed solutions can be integrated in the IEEE 802.15.4 standard or serve as the basis for wireless next-generation networks.

The rest of the chapter first presents an overview of IEEE 802.15.4 MAC enhancements, namely the improvements brought by the proposed two-phase contention window mechanism and by RTS/CTS combined with packet concatenation. Then, an analysis in the absence and presence of RTS/CTS, the MC-SCP-MAC and SBACK-MAC protocols is presented. Details are given on the IR concept, extra resolution phase algorithm (concatenation) and node channel rendezvous scheduler, followed by the description of the SBACK-MAC protocol. Throughput and energy consumption are finally evaluated, followed by an overview of the chapter content, conclusions and discussion of topics for future research.

8.2   IEEE 802.15.4 MAC enhancements

8.2.1   Analysis of the overhead in IEEE 802.15.4 MAC sub-layer

In the IEEE 802.15.4 basic access mode, nodes use a nonbeacon-enabled carrier sense multiple access with collision avoidance (CSMA-CA) algorithm for accessing the channel and transmit their packets. The backoff phase (N.B.: This time period is not generally called contention window in IEEE 802.15.4) algorithm is implemented by considering basic units of time called backoff periods. The backoff period duration is equal to $T_{BO} = 20 \times T_{symbol}$ (i.e. 0.32 ms), where $T_{symbol} = 16\,\mu s$ is the symbol time [4]. Before performing CCA, a device shall wait for a random number of backoff periods, determined by the backoff exponent (BE). Then, the transmitter randomly selects a backoff time period uniformly distributed in the range $[0, 2^{BE} - 1]$. Therefore, it is worthwhile to mention that even if there is only one transmitter and one receiver, the transmitter will always choose a random backoff time period within $[0, 2^{BE} - 1]$. Initially, each device sets the BE equal to *macMinBE*, before starting a new transmission and increments it, after every failure to access the channel. In this work, we assume that the BE will not be incremented since we are assuming ideal conditions. Table 8.1 summarizes the key parameters from the IEEE 802.15.4 standard.

The maximum backoff contention window, $CW_{\max}$, is given as follows:

$$CW_{\max} = (2^{BE} - 1) \times T_{BO} \tag{8.1}$$

The time delay, due to CCA, is given by

$$ccaTime = rxSetupTime + T_{CCA} \tag{8.2}$$

The *rxSetupTime* is the time to switch the radio between the different states and must be extracted from the datasheet from the radio transceiver [5–7]. During CCA, which lasts T_{CCA}, the radio transceiver must determine the channel state within the duration of eight symbols (1 symbol period is equal to $16\,\mu s$). Figure 8.1 presents the frame sequence for the IEEE 802.15.4 basic access mode with DATA/ACK.

There is a random deferral period of time before transmitting every data packet, given by

$$D_T = InitialbackoffPeriod + ccaTime + T_{TA} \tag{8.3}$$

In this research work, we only consider the nonbeacon-enabled mode (not the beacon-enabled one, since collisions can occur either between beacons, or between beacons and data or control frames, making a multi-hop beacon-based network difficult to be built and maintained [8]).

Another important attribute is scalability, due to changes in terms of network size, node density and topology. Nodes may die over time. Others may be added later

Table 8.1   Parameters, symbols and values for the IEEE 802.15.4 standard and SBACK-MAC protocol

Description	Symbol	Value
Backoff period duration	T_{BO}	320 µs
CCA detection time	T_{CCA}	128 µs
Setup radio to RX or TX states [6]	$rxSetupTime$	1720 µs
Time delay due to CCA	$ccaTime$	1920 µs
TX/RX or RX/TX switching time	T_{TA}	192 µs
ACK wait duration time	T_{AW}	560 µs
PHY length overhead	L_{H_PHY}	6 bytes
MAC overhead	L_{H_MAC}	9 bytes
DATA payload	L_{DATA}	3 bytes
DATA frame length	L_{FL}	18 bytes
ACK frame length	L_{ACK}	11 bytes
DATA transmission time	T_{DATA}	576 µs
ACK transmission time	T_{ACK}	352 µs
Short interframe spacing (SIFS) time	T_{SIFS}	192 µs
Long interframe spacing (LIFS) time	T_{LIFS}	640 µs
RTS ADDBA transmission time	T_{RTS_ADDBA}	352 µs
CTS ADDBA transmission time	T_{CTS_ADDBA}	352 µs
BACK request transmission time	$T_{BRequest}$	352 µs
BACK response transmission time	$T_{BResponse}$	352 µs
Number of TX frames	n	1 to 112
Data rate	R	250 kb/s

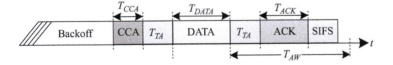

Figure 8.1   Acknowledgement process timing

and some may move to different locations. Therefore, for such kind of networks, the nonbeacon-enabled mode seems to be more adapted to the scalability requirement than the beacon-enabled mode. In the former case, all nodes are independent from the personal area network (PAN) coordinator and the communication is completely decentralized. Moreover, for beacon-enabled networks [4], there is an additional timing requirement for sending two consecutive frames so that the ACK frame transmission should be started between the T_{TA} and $T_{TA} + T_{BO}$ time periods (and there is time remaining in the contention access period (CAP) for the message, appropriate

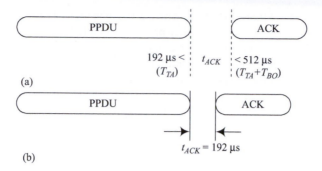

(a)

(b)

Figure 8.2   IEEE 802.15.4 acknowledgement frame timing: (a) beacon and (b) nonbeacon-enabled modes

inter-frame spacing (IFS) and ACK). Figure 8.2 presents the timing requirements for transmitting a packet and receive an ACK for the beacon and nonbeacon-enabled modes, respectively.

By analysing Figures 8.1 and 8.2, we conclude that overhead is one of the fundamental problems of MAC inefficiency. It includes the PHY and MAC headers, backoff duration, IFS (i.e. short interframe spacing – SIFS, and long interframe spacing – LIFS) and ACKs. Moreover, IEEE 802.15.4 radio compliant transceivers also have restricted hardware constraints. The inefficient switching delay time periods between the radio states (i.e. *rxSetupTime*) cannot be neglected, since nodes are continually switching between them, resulting in significant energy spent [9]. The direct sequence spread spectrum (DSSS) mode of IEEE 802.15.4 standard supports a maximum over-the-air data rate of 250 kb/s for the 2.4 GHz band. However, in practice, the effective data rate is lower due to the protocol/hardware timing specifications. This is explained by the various mechanisms that are employed to ensure robust data transmission, including channel access algorithms, data verification and frame ACK. In this work, we analyse the maximum throughput, S_{max}, and the minimum delay, D_{min}, for the IEEE 802.15.4 standard. S_{max} is defined as the number of data bits generated from the MAC layer that can be transmitted per second to its destination including the ACK reception, on average. D_{min} is the time needed to transmit a packet and the successfully reception of the ACK, on average. Although we are considering the 2.4 GHz band, the proposed formulation is also valid for other frequency bands. As explained before, initially, the BE is set to *macMinBE*. By considering the default value $BE = 3$ for *macMinBE*, and assuming the channel is free, the worst case channel access time that corresponds to the maximum backoff window is given by (8.1).

The average backoff window is given by

$$\overline{CW} = \left(\frac{CW_{max}}{2}\right) \times T_{BO} \tag{8.4}$$

S_{max} and D_{min} can be determined for the best case scenario (i.e. an ideal channel with no transmission errors). During one transmission cycle, there is only one active node that has always a frame to be sent, whereas the other neighbouring nodes can only accept frames and provide ACKs. We then propose an analytical model to evaluate S_{max} and D_{min}. Table 8.1 presents the key parameters, symbols and values. Hence, there is no need to redefine every parameter after every equation again. The transmission times, in seconds, for the DATA and ACK frames are given as follows:

$$T_{DATA} = 8 \times (L_{H\_PHY} + L_{H\_MAC} + L_{DATA})/R \qquad (8.5)$$

$$T_{ACK} = 8 \times (L_{H\_PHY} + L_{ACK})/R \qquad (8.6)$$

S_{max}, in bits per second, is given by

$$S_{max} = \frac{(8L_{DATA})}{D_{min}} \qquad (8.7)$$

where $D_{min} = (\overline{CW} + ccaTime + T_{TA} + T_{DATA} + T_{TA} + T_{ACK} + T_{IFS})$, in seconds.

For IFS, SIFS is considered when MAC protocol data unit (i.e. $LH_{PHY} + LH_{MAC} + L_{DATA}$) is less or equal to 18 bytes; otherwise, LIFS is considered. By analysing (8.5) to (8.7), we conclude that, if a short frame is transmitted, the data transmission time is relatively short when compared to the associated overhead time, resulting to relatively low throughput. When a long frame is transmitted (by increasing the payload), data transmission time increases. This way, IEEE 802.15.4 is capable of achieving a much higher throughput. Moreover, we also conclude that the effective data rate of IEEE 802.15.4 in the basic access mode is lower than the maximum over-the-air data rate of 250 kb/s for the 2.4 GHz band. As stated in [10], this lower effective data rate is explained by the fact that IEEE 802.15.4 was built into the frame structure, and various mechanisms have been employed to ensure robust data transmission, including the channel access algorithms, data verification and frame ACK.

A distributed scenario, with single-destination and single-rate frame aggregation, is considered in the modelling approaches considered in this work. In practice, a 2-hop network, with two sources, one relay and two sinks (and interferers), has been considered, as shown in Figure 8.3, not only for determining the throughput and delay but also to evaluate the amount of energy consumed in the nodes. The packets from source node **A** flow, through node **C**, to sink node **D** while the packets originated by source node **B** flow, through node **C**, to reach sink node **E**. In single-destination approaches, frames can be aggregated if they are available (saturation conditions) and have the same source and destination address.

8.2.2 Discovery-addition state

In MC-SCP, the data packet includes the ID of the sender and the time when the packet was sent, along with the data collected from the sensors. Any node that receives this packet can adjust or set its internal clock and know the channel hopping sequence of the sender, based on the sender's ID, which is the seed of the linear congruential generator (LCG).

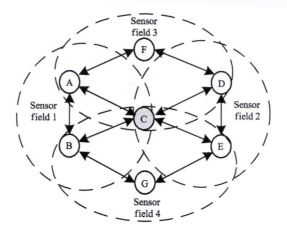

Figure 8.3   Multi-hop star topology simulation scenario with two interferers

8.2.2.1   Choice of channels

In this work, the sensor nodes that try to join the WSN for the first time must follow an algorithm that allows for joining the network and start sensing. There are two algorithms envisaged in MC-SCP-MAC protocol that allow the addition of nodes: first, the node uses the sensing node mechanism, in which it randomly chooses a channel and senses it for an initial period of 10 s. The channel choice is performed by considering a uniform distribution, as shown in (8.8)

$$\varphi_{ch} \in [12; N_{ch} - 1] \tag{8.8}$$

It is preferable the node performs a random channel choice based on uniform distribution than considering the LCG choice based on his ID, since it is less probable that any other node has chosen the same channel with a different seed.

When the new node detects a wake-up tone, a data packet or any packet being transmitted, it waits and receives it. With this procedure, the new node can synchronize his internal clock and assumes that the node that sends the received packet is the temporary parent node. The node is temporary because with the internal clock synchronized, the node can try to listen for the control and synchronization channel to check if it can directly communicate with the PAN coordinator. If so, it will keep the information of the 1-hop neighbour that it considers as a temporary parent node, just in case the PAN coordinator switches off or moves from the initial position (considering mobility). The timestamp stored in the packet must be taken after any queuing or any algorithm delay at the MAC layer.

If the new node does not sense any packet in the chosen channel, it will initiate a second mechanism called hopping sequencing, which hops throughout all the channels. However, it will sense each channel half or 1/4 of the maximum time a node can use each channel to communicate. By decreasing the sensing time in each channel, the number of times a channel is sensed by the node in a frame duration (1 s) is twice (if it senses half of the time of a slot channel) or four times (if it senses a quarter of the time of the slot channel). This mechanism is the last resource of the new node to join the network.

8.2.2.2   Required number of slot channels per frame

In slot contention-based-MAC protocols with single channel, the required number of time slots in the MAC frame is a parameter of paramount importance. Since these protocols rely on the slot choice to perform CS and send a packet when an idle medium is sensed, the number of time slots affects the probability of collision (i.e. characterizing nodes that choose the same slot) and the service time. The majority of single-channel slot contention-based-MAC protocols (e.g. SCP-MAC [11]) consider a fixed number of slots in the MAC frame. However, some MAC protocols already consider a dynamic slot allocation mechanism. The required number of slots in a MAC frame depends on the expected number of nodes that compose the WSN. Moreover, a trade-off must be established in the definition of the number of slots in the MAC frame by relating the number of slots, collision probability and number of nodes. On the one hand, the choice of few slots in a high node density leads to high collision probability. On the other, the choice of a large number of slots in a low node density leads to low collision probability. However, the energy costs are higher due to the energy wasted while waiting for packet transmission.

In the single-channel SCP-MAC protocol, within the 2-hop neighbourhood, a node should not use the same slot from the contention windows 1 or 2. Otherwise, if the sensor nodes sense the channel as idle and decide to transmit packets simultaneously, a packet collision will occur.

Since MC-SCP-MAC is a multichannel protocol, the nodes can choose one channel from the available ones to transmit their packets. As a consequence, in this case, a node can use the same slot as the 2-hop neighbours but on a different channel. In this case, co-channel interference is negligible [12]. The number of required slot channels dedicated to data exchange is calculated in the start-up phase by considering the node density information. At the start-up phase, since only a maximum of 15 channels can be used for a maximum number of nodes, n_{max}, deployed in the WSN, the following relation is considered:

$$
N = \begin{cases} \left\lceil \frac{n_{max}}{2} \right\rceil & 0 < n_{max} \leq 2N_{ch} \\ 15 & n_{max} > 2N_{ch} \end{cases}
\tag{8.9}
$$

Only the default channel (slot channel 11, for synchronization purposes) is known and common to all the nodes. During the network start-up, the node knows the node density and assigns the number of channels to be considered by the channel choice mechanism. It is assumed that a channel supports at most four nodes contending in the same channel. If new nodes are later added to the network, the parent node (PAN coordinator) can decide whether to restart the network or continue with the chosen number of channels. The associated collision probability is decreased by using an enhanced-two phase contention window mechanism, which considerably reduces the number of collisions compared to a single-contention window protocol (e.g. McMAC [12]).

8.2.3   Enhanced-two phase contention window mechanism

After the node starts up and synchronizes with the network while selecting the first slot channel, it enters into the enhanced-two phase contention window mechanism, corresponding to the medium access phase. In the medium access phase, the node follows the state transition diagram presented in Figure 8.4. For every packet the node wants to transmit, the node repeats the choice of the slot channel given by the LCG and the choice of the slot for each enhanced two-phase contention window mechanism phases (N.B.: this is one of the key mechanisms to reduce potential collisions in the same channel).

After the reception and transmission of the packets, the nodes will check if any packet is stored in the queue. If so, the child node uses the remaining time of the slot channel to send the packets without the need to perform the choice of the slot in the two contention windows. This greatly reduces the time needed to transmit all the remaining packets.

It is therefore important to explain the frame and channel assignment structure in detail, i.e. the packet structure and the enhanced-two phase contention window mechanism involved in the message-exchange procedure.

8.2.3.1   Frame and channel structure

A frame corresponds to a hopping sequence of the available channels, in which the parent node senses the data whilst the child nodes choose a channel to transmit their data packets. Each channel is assumed as a slot channel, S_c, as shown in Figure 8.4.

The enhanced-two phase contention window mechanism is employed in each slot channel, in order to facilitate that nodes contend for the channel. This time period is denoted as the node-contention (NC) phase. After the NC phase, the slot channel has an ER phase that can be used by the sensor nodes to transmit more packets that are stored in their queue, after winning the two-phase contention window mechanism while having already sent the first data packet with no collisions. If the node needs to send more than one packet per slot channel, it must enable the 'more' bit and include the number of remaining packets from the queue (to be added in the control information from the first data packet sent to the parent node). This allows for the parent node to wait for the remaining packets, suppressing the need of performing all the two phase contention window mechanism, when the sender needs to transmit consecutive packets to the parent node, in the current slot channel. This mechanism is already used by the WiseMAC [13] protocol, in which a high performance is attained when consecutive transmissions are envisaged. After the ER phase, the node (child or parent node) will enter into the inactive phase, in which nodes go to the sleep mode and save energy. The slot channel structure presented in Figure 8.4 is considered in all the available channels from the IEEE 802.15.4 standard, except for channel 11, which has a different purpose (SYNC and control packets exchange) and, consequently, a different slot channel structure. The slot channel structure for the SYNC and control channel is also presented in Figure 8.4.

The number of slot channels, S_c, is equal to the number of channels (defined in the start-up phase, as it depends on the number of nodes initially deployed), and each

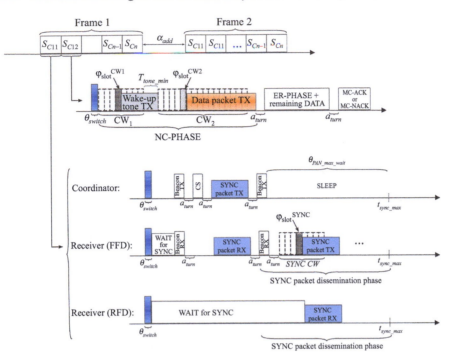

Figure 8.4   Frame structure for the MC-SCP-MAC protocol

slot channel is indexed by a channel number. The slot channel S_{c11} is dedicated just for SYNC and control packets exchange, while the remaining ones ($S_{c12} \cdots S_{c26}$) are dedicated to the NC and data packets exchange. The number of slot channels, on which the radio transceiver can be adjusted to, depends on the nodes density during the deployment, as expressed by (8.9). However, there is the option to disable this dynamic channel adjustment, allowing for considering a fixed number of channels, independently of the nodes' density. This dynamic mechanism is more efficient for networks with low node density, since it uses less channels. The PAN coordinator will be more energy efficient, as the channel hopping sequence is performed for less slot channels, allowing for the nodes to consume less energy. In Figure 8.4, each frame has a duration of $t_F = 1$ s. The duration of each slot channel is $\Delta t_{SC} = t_F / N_{ch}$, with $N_{ch} = 16$. This means that the duration of each slot channel is $\Delta t_{SC} = 62.5$ ms. Figure 8.4 also depicts the different slot channels that are reserved for the senders of the packets whose number, given by the LCG generator, is equal to the slot channel index (the first slot channel is reserved for the SYNC and control packets exchange, the second slot channel is for channel 12, and so on, up to channel 26). In the SYNC and control channel, S_{c11}, during the first time, the child nodes are switched on, they initiate the sensing mode awaiting for the SYNC packet. Hence, the direct (1-hop) communication links between child and parent nodes can be established. The nodes (child and parent nodes) need to switch to the channel 11 before performing the synchronization procedure, as shown in Figure 8.4. This switching channel time in our protocol is 200 µs. However, depending on the radio transceiver, this switching channel time may vary. After switching to the slot channel S_{c11}, the child nodes wait for a beacon packet sent by the parent node, in order to initiate the synchronization of the nodes. As soon as the beacon packet is completely sent, the parent node initiates the transmission of the SYNC packet to the child nodes. This SYNC packet has a length of 18 bytes. The control information in these 18 bytes includes the time of the parent node, the source node ID, destination ID (broadcast), number of considered channels and the next synchronization period. Once the child nodes receive the SYNC packet, they are able to synchronize their internal clocks with the clock of their parent node (the PAN coordinator). The length of the SYNC packet is equal to the one employed in SCP-MAC and is sufficient to the child nodes to receive it on time, even if serious clocks drifts persist in the child nodes. The first synchronization procedure finishes with the transmission of another beacon packet (by the PAN coordinator), to indicate to the child nodes that the synchronization procedure has finished and they could go to sleep mode until the next wake-up to send data packets. This synchronization procedure coincides with the one employed in the beacon-mode from the IEEE 802.15.4 standard.

The maximum allowed packets that can be received by the parent node is limited to the remaining time of the slot channel and must take into account the ACK or NACK packet sent by the parent node. It allows for reducing the queuing delay at the parent node, especially when traffic bursts are considered. If the parent node receives the first data packet with the information that the child node has more packets ready to be sent in the queue and misses receiving the last data packet of the sequence, it has a time out timer, $t_{guard} = 400$ µs. This timeout timer allows for the parent node to be aware of transmission errors in the packet sequence sent by the child node. As soon as the ER phase ends, the parent node sends an ACK or a NACK packet, depending on if it receives more than one packet or node from the child node, respectively. This ACK/NACK packet has a particular structure that allows for RTXs and synchronization of the child nodes. By considering two contention stages (CW_1 and CW_2), the collision probability decreases when compared to the single contention window case with equivalent length (sum of the lengths of CW_1 and CW_2). This is because only the nodes succeeding in CW_1 enter into CW_2. This choice decreases the number of nodes effectively accessing the medium during the second contention stage. After the first phase of the contention procedure, only the successful nodes from CW_1 (the first one(s) that transmit the wake-up tone) contend in CW_2. With less contending nodes, the effective collision probability is considerably reduced. Each slot channel can accommodate the transmission of multiple data frames from the child node that wins the contention, as shown in Figure 8.4. Each slot channel has a fixed length, but the NC and ER phases duration varies according to the number of data packets involved and the choice of slots in CW_1 and CW_2. Since the number of slot channels depends on the number of nodes initially deployed (defined in the start-up phase), the nodes may maintain the same frame time, t_F, and use one of the 15 channels, leading to lower energy consumption.

8.2.4   Adoption of the nonbeacon-enabled mode

Regarding the MAC channel access, the IEEE 802.15.4 standard allows two types of channel access mechanisms: beacon- and nonbeacon-enabled. The latter case uses unslotted carrier-sense multiple access with collision avoidance (CSMA/CA): each time a device needs to access the radio channel, it waits for a random backoff period; at the end of which, it senses the channel. If the channel is found to be idle, then the device transmits the data; otherwise, it waits for another random period before trying to access the channel later on [14].

In this research work, we only consider the nonbeacon-enabled mode (not the beacon-enabled one, since beacon collisions can occur with other beacons, data or control frames, making a multi-hop beacon-based network difficult to be built and maintained [8]). Another important attribute is scalability, due to changes in terms of network size, node density and topology. Nodes may die over time. Others may be added later and some may move to different locations. Therefore, for such kind of networks, the nonbeacon-enabled mode seems to be more adapted to the scalability requirement than the beacon-enabled mode. In the former case, all nodes are independent from the PAN coordinator, and the communication is completely decentralized. Moreover, for beacon-enabled networks [10], there is an additional timing requirement for sending two consecutive frames, so that the ACK frame transmission should be started between the T_{TA} and $T_{TA} + T_{BO}$ time periods (and there is time remaining in the CAP for the message, appropriate interframe space, IFS and ACK).

8.3   IEEE 802.15.4 in the presence and absence of RTS/CTS

The IEEE 802.15.4 standards do not include the use of RTS/CTS mechanism, involving the transmission of short RTS and CTS control packets prior to the transmission of the actual data packets. In this work, we show that IEEE 802.15.4 MAC layer also benefits from the inclusion of RTS/CTS. Although the proposal of employing RTS/CTS is not new and has already been standardized and implemented in legacy Wi-Fi (since it shortens packet collision duration, as shown in [15]), this reservation scheme has not been considered in any of the existing IEEE 802.15.4 standards. In our proposal, we also assume that both the RTS and CTS packets have the structure of an ACK packet, which is assumed to have a limited size of 11 bytes, as shown in Table 8.1.

Figure 8.5 presents the IEEE 802.15.4 employing RTS/CTS with packet concatenation being composed by the following time periods:

- Backoff phase
- CCA mechanism
- Time needed for switching from receiving to transmitting
- RTS transmission time
- Time needed for switching from transmitting to receiving
- CTS reception time.

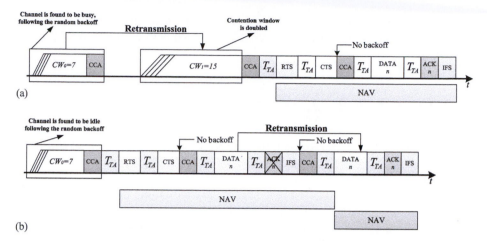

(a)

(b)

Figure 8.5   IEEE 802.15.4 with RTS/CTS with retransmissions: (a) channel is found to be busy and (b) channel is found to be idle

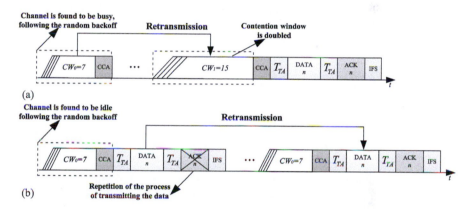

(a)

(b)

Figure 8.6   IEEE 802.15.4 basic access mode with retransmissions: (a) channel is found to be busy and (b) channel is found to be idle

As shown in Figure 8.5, in IEEE 802.15.4 employing RTS/CTS, nodes will use the same backoff procedure as IEEE 802.15.4 basic access mode. However, this process is not repeated for each data packet sent, but only for each RTS/CTS set. Therefore, the channel utilization is maximized by decreasing the deferral time period before transmitting a data packet, as shown in Figure 8.6. The proposed solution shows that, by considering the RTS/CTS mechanism combined with packet aggregation, the network performance is improved in terms of maximum throughput, minimum delay and bandwidth efficiency. This work also introduces an analytical model capable of accounting the RTX delay and the maximum number of backoff stages.

The minimum delay due to CCA, $D_{min\_CCA\_RTS}$, for determining if the channel is found to be busy or idle, after the backoff phase, and before each RTS/CTS set is given by

$$D_{min\_CCA\_RTS} = \sum_{i=1}^{n/N_{agg}} \sum_{k=0}^{k \leq NB} \left(\overline{CW_k} + ccaTime \right) \tag{8.10}$$

By analysing (8.10), we can conclude that nodes only determine the channel state once per RTC/CTS. Therefore, if a node has $n = 100$ data packets to send and the number of aggregated packets is equal to $N_{agg} = 10$, nodes only determine the channel state ten times ($n/N_{agg} = 10$) plus the time needed for transmitting the packets (until the maximum retry limit, $NB_{max} = 4$, is reached).

If the channel is found to be idle during CCA and after sending a data packet an ACK is not received within a duration of T_{AW}, the RTX process will not consider the use of the backoff phase between two consecutive data packets, which allows to decrease the total overhead (as shown in Figure 8.2(b)). Since any other stations can receive both the RTS, CTS, DATA or ACK packets, in the first transmission attempt they will set an internal timer called NAV that is responsible for defining the time period a node will defer the channel access in order to avoid collisions.

In the IEEE 802.15.4 MAC protocol with RTS/CTS, the minimum delay due to packet RTXs, $D_{minDataRetRTS}$, is obtained; when the channel is found to be idle during CCA (after the backoff phase), there is a data transmission, and an ACK is not received within a duration of T_{AW}, as given below:

$$D_{minDataRetRTS} = \begin{cases} H_3, & \text{for } j = 0 \\ H_4, & \text{for } j \in [1, MaxRet] \end{cases} \tag{8.11}$$

where j is the number of packet RTXs and varies between 1 and $MaxRet$ as defined in [16]. The following lessons can be learnt from the analysis of (8.11):

- If, after CCA, a node determines that the channel is found to be idle, and an ACK is correctly received for each packet sent, the minimum delay, $D_{minDataRetRTS}$, is determined by

$$H_3 = T_{TA} + T_{RTS} + T_{TA} + T_{CTS}$$
$$+ \sum_{i=1}^{N_{agg}} (ccaTime + T_{TA} + T_{DATA} + T_{TA} + T_{ACK} + T_{IFS}) \tag{8.12}$$

which means that there are no transmission errors. In this case, the number of RTXs is given by $j = 0$;
- If after CCA, a node found the channel to be idle and an ACK has not been received within the duration of T_{AW}, for one or more packets sent (by considering

aggregation), the minimum delay due to packet RTXs, $D_{minDataRetRTS}$, is determined by

$$H_4 = T_{TA} + T_{RTS} + T_{TA} + T_{CTS} + \sum_{i=1}^{N_{agg}-m} (ccaTime + H_1) + \cdots$$

$$+ \sum_{i=1}^{m} (j_i) \times (ccaTime + H_2) + \sum_{i=1}^{m} (ccaTime + H_1) \qquad (8.13)$$

where H_1 is given by $H_1 = T_{TA} + T_{DATA} + T_{TA} + T_{ACK} + T_{IFS}$ [17] and H_2 is given by $H_2 = T_{TA} + T_{DATA} + T_{AW}$ [17].

The term $\sum_{i=1}^{N_{agg}-m} (ccaTime + H_1)$ represents the $(N_{agg} - m)$ transmitted aggregated packets that have successfully received an ACK response, where m denotes the number of transmitted packets that need RTX. Since each individual packet can be retransmitted more than once due to the fact that an ACK has not been received within a duration of T_{AW} duration, the term j_i represents the number of times a packet has experienced RTX until *MaxRet* has been reached. We then assume that the ACK packet is received, which represents the last case given by $\sum_{i=1}^{m} (ccaTime + H_1)$. By combining (8.10) and (8.11), the minimum average delay, $D_{min\_RTS\_CTS}$, due to the channel state (i.e. busy or idle) and packet RTXs is given by

$$D_{min\_RTS\_CTS} = \frac{D_{min\_CCA\_RTS} + D_{minDataRetRTS}}{n} \qquad (8.14)$$

In IEEE 802.15.4 by employing RTS/CTS combined with packet concatenation, if an erroneous channel is considered, the maximum average throughput, $S_{max\_RTS\_CTS}$, in bits per second, by considering packet RTXs, is given by (8.7) while replacing D_{min} by $D_{min\_RTS\_CTS}$.

8.4   MC-SCP-MAC protocol

MC-SCP is a scheduled and channel polling McMAC protocol. The rationale behind this proposal is based on the single-channel SCP-MAC [11] protocol, which is an energy-efficient medium-access protocol designed for WSNs. The SCP protocol goes one step further in preamble-sampling-based protocols by combining scheduling with channel polling to minimize energy consumption. Compared to [18,19], the main aspects for SCP are the following ones:

- The two-phase collision avoidance mechanism employs a wake-up tone on a single channel between the two contention windows. The wake-up tone is similar to a busy tone and informs the other listening nodes of an ongoing transmission.
- It achieves synchronization by piggybacking the schedules in broadcast data packets.
- By default, nodes wake-up and poll the channel for activity. A sending node reduces the duration of the wake-up tone by starting it right before the receiver starts listening.

- Energy efficiency is also achieved, since channel polling, together with the two-phase contention mechanism, mitigates the problems that may result from collisions while avoiding the waste of energy.

Moreover, time-scheduled communication facilitates the coordination of multichannel communication. Since the nodes (senders and receivers) have to switch their radio interfaces between different channels, coordination of channel switching is required. This channel switching mechanism facilitates that the sender and receiver nodes to simultaneously exchange packets in the same channel. Scheduled and synchronized access provides a way for the nodes to meet in the same channel.

Even though single window slotted contention protocols may suffer collisions when the transmitter node fails to successfully allocate the medium, the SCP protocol MAC contention mechanism employs double-slotted contention whilst minimizing the collisions at the expense of short overhead. In fact, SCP alleviates the strict behaviour of time division multiple access (TDMA) MAC protocols (e.g. LMAC), which may lead to high values for the delay and low values for the throughput, while presenting the advantage of access with reduction of the number of collisions [20], as there is contention in the access to slots for the mote who fails' in the double window contention.

In high scalability scenarios, the proposed MC-SCP protocol is considered as a hybrid MAC protocol. This is due to the employment of a deterministic TDMA-based channel hopping mechanism, a slotted contention basis that relies on an enhanced two-phase contention window mechanism similar to the one from the SCP protocol. Besides the SCP protocol, other new mechanisms, from other recent MAC protocols, are also being considered and adapted here, in order to increase the performance of the proposed MC-SCP MAC protocol, namely the IR concept (concatenation) and the node rendezvous scheduler.

8.4.1 Influential range concept

In IEEE 802.15.4 compliant WSNs, the overhearing problem appears during any normal operation of the network and is widely considered because of the associated energy waste in various MAC sub-layer protocols implementations, as mentioned in [21]. In [21], the authors propose a so-called IR that takes advantage of the overhearing in order to save energy in the context of convergecast applications. IR is not a generic MAC, so its use in the context of this chapter should merely be seen as an application. It is worthwhile to note that IR is relevant for a specific convergecast scenario where packets with information are sent to a unique sink following a tree-shaped WSN.

When a node sends a packet, all the other nodes can overhear this packet. Considering contention-based MAC protocols, the node(s) that loses the channel changes to the receiving mode in order to listen to the transmission. It listens to the full transmission of the packet to know the packet destination. If the packet destination is not meant to it, then it simply drops the packet (overhearing problem). The authors from OB-MAC [21] consider that, based on this overhearing, nodes in the same area probably sense the same information. Hence, all the nodes in a given area sense the same phenomenon, leading to the delivery of the same information to the sink node.

This redundant transmission can be avoided by comparing packets with the ones in the queue, therefore reducing the energy waste.

In the majority of the MAC protocols, when a node receives a packet not intended to it, it simply discards the received packet. However, it spends time to receive and decode it. After receiving and decoding the packet, if it verifies that the packet destination ID does not coincide with the node ID, then it discards the packet.

Based on this technique, even if the packet destination is not itself, the node compares the overheard packet with the stored packets in the queue, to check if there is any packet containing the same information to be sent to the sink node. If so, the node discards the packet from the queue, since the same information has been already sent by neighbouring node(s). If the overheard packet information does not match any of the packets stored in the queue, the node discards the overheard packet and tries to send the packets stored in the queue at the adequate time.

The authors from [21] define the concept of IR applied to OB-MAC, which is a range where the nodes are likely to observe the same information. The IR must always be shorter than the maximum achievable range of the radio transceiver, as shown in Figure 8.7.

In Figure 8.7, the small circle is considered as the IR, while the outer circle is the radio range. The employment of the IR concept by the authors of OB-MAC is restricted to a single channel. As it clearly presents improved energy efficiency due to the reduction of redundant transmissions, we strongly believe that applying the IR concept to a McMAC protocol, such as MC-SCP, will lead to energy efficiency gains.

To achieve this energy efficiency, some conditions must be fulfilled in order to decide when the IR can be applied to the MAC protocol. The IR concept states that a node applies this technique if and only if it is within the IR of the transmitting node. When the phenomenon has taken place, all the nodes that are within the area enable the IR verification. To apply the assertions from this decision, the node that overhears a packet needs to know if it is within the IR of the node that has sent the overheard packet while having the same parent node. A suitable indicator of whether the node is in the IR, is the RSSI from the overheard packet. If the received signal

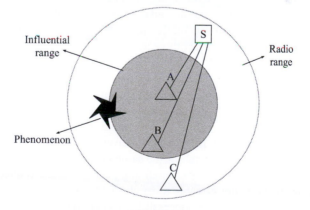

Figure 8.7   Influential range

strength indicator (RSSI) of the overheard packet is higher than a defined sensing range threshold (Π_{irmax}), the node is in the same IR. Otherwise, if the RSSI is lower than Π_{irmax}, the node is not in the same IR. If the node that overhears the packet is in the IR, the overhearing node performs information comparison.

As shown in Figure 8.7, if nodes A, B and C contend for the channel and only node B wins the channel, while the remaining ones continue to listen to the medium, the node B sends its packet to the sink. However, nodes A and C overhear the node's A packet. Both nodes (A and C) verify if node B is in the same IR. If the node A checks that the RSSI of the overheard packet is higher than Π_{irmax}, then it verifies if any packet stored in the queue has the same phenomenon information. If so, it discards the packet stored in the queue. Otherwise, it sends the packet in the next attempt to transmit. For the node C, the comparison of the RSSI from the overheard packet with the value of Π_{irmax} allows to decide whether the packet is to be discarded (or not). Since the RSSI is lower than the threshold, the overheard packet is discarded and the queued packet is sent later. The defined threshold depends much on the density of nodes in the WSN. The denser the WSN is, the more overheard packets the nodes will receive in their IR. Therefore, reducing the value of Π_{irmax} leads to additional energy saving. The values of Π_{irmax} considered by the authors from the OB-MAC protocol vary between -90 and -60 dB m.

8.4.2   *Extra resolution phase decision algorithm (concatenation)*

Each slot channel has an ER phase, which can be used by the sensor nodes (child nodes) to transmit more packets that may exist in their queues after sending the first data packet. Since the channel can be degraded due to interference, or the nodes may move to other positions (mobility is considered), the node may transmit the data packets unnecessarily, since they will not be received by the parent node. Consequently, a decision algorithm is required to decide whether a child node sends all queued packets during the ER phase. The decision is based on the SNIR from the packet, the 'degradation' level of the slot channel and the neighbouring nodes of the child node that intends to transmit the remaining data packets. The first component of the decision algorithm is based on the degradation level from the slot channel, already defined for the denial channel list. While the second component of this algorithm is based on the IR concept, on the last three accesses of the child node to the medium and the corresponding packet delivery results.

A node that has more packets in queue to transmit compares the degradation level of the current slot channel (stored in the denial channel list) with the threshold, D_{ch}. If the channel is considered has 'good' to send packets, then it passes to the second phase of the EC phase decision algorithm. In the second phase, the node checks the last three calls of the IR algorithm. If the node receives at least three positive feedbacks in the IR algorithm for that channel for the same neighbouring nodes, then the node can use the ER phase to transmit the remaining packets stored in queue. In case the IR algorithm feedback is not possible to be checked, the node will check in the last three channel transitions if it has successfully transmitted the packets to the parent node. If so, then the node can utilize the ER phase to transmit the packet in the queue.

8.4.3   Node channel rendezvous scheduler

This section explains the methods for channel choice, the multichannel extension of the two-phase contention window mechanism of the SCP protocol, and for the pseudo-random number generator that defines the choice of the channel which facilitates the delivery of data packets to the parent node (coordinator or full-function device (FFD) node).

8.4.3.1   Predictive wake-up mechanism

After synchronizing the child nodes with the parent node, they should define a way of delivering the data packets within the different channels. Since the parent node (coordinator) is continuously switching the channel, the child nodes need to know which channel the coordinator is using at a given time instant. When the switching time interval coincides with the start of a data packet transmission, this continuous channel switching may cause packet losses. Therefore, the solution to avoid that the parent nodes are continuously switching among the channels, causing high packet losses and energy inefficiency, is to use a deterministic TDMA-based mechanism. This TDMA-based mechanism relies on the number of the channel and the start of each time interval, provided by a pseudorandom number generator. The MC-SCP MAC protocol can use any pseudorandom function to generate the wake-up schedule for a node, based on the channel chosen for the child node. Recent MAC protocols, such as predictive-wakeup MAC (PW-MAC) [22] and efficient multichannel (EM-MAC) [23], utilize this type of pseudo random number generator (RNG) to enable a sender to accurately predict the wake-up schedule rather than waking up on a truly random schedule. If a pseudorandom wake-up schedule is preferred rather than a fixed schedule, the behaviour is also predictable but more inefficient (e.g. WiseMAC [13]). Hence, the possibility of neighbouring nodes to simultaneously wake-up for different channels is avoided. A mechanism is needed that prevents nodes from choosing different channels or the same time instant to initiate packet transmission. As the sensor node has only one radio transceiver, it can only choose one channel at a time. The number of packets lost because the parent node is not in the same channel of the sender node considerably decreases when this pseudorandom predictive mechanism is enabled. In terms of the chances of collision, the possibility of neighbouring nodes to simultaneously wake-up may significantly increase the probability of collisions between the senders. This is verified in the PW-MAC [22] and EM-MAC [23] protocols. However, in this work, the two-phase contention window mechanism is adapted to operate with multichannels. In our case, the probability of collisions may also increase if no predictive mechanism is considered, since the senders will randomly choose the channel for the transmission in each frame. The predictive-channel-based wake-up mechanism can be applied to receiver-initiated duty cycling MAC protocols (e.g. RI-MAC) as well as in sender-initiated approaches, since it provides a high packet success-rate performance for the sender node, when the predictive wake-up mechanism is considered. This mechanism allows for sensor nodes (senders) to deliver packets to the receiver in a rendezvous basis. In the MC-SCP case, there are many suitable pseudorandom number generators. Since the sensor node has limited computing capabilities, the best choice is the LCG [24]. The LCGs are preferred to WSNs as shown in the PW-MAC and EM-MAC protocols in [22,23], since they are computationally and storage efficient. The LCG generates a pseudorandom number, X_{n+1}, as follows:

$$X_{n+1} = (aX_n + c_+)\,mod(m_{lcg}) \tag{8.15}$$

Here, $mod(m_{lcg} > 0)$ is the modulus, a is the multiplier, c_+ is the increment and X_n ($0 \le X_n < m_{lcg}$) is the current seed. Each of the X_{n+1} generated values can be used as a pseudorandom number and becomes the new seed. In our protocol, a suitable LCG can be the one proposed in the work of Park and Miller [25] with the form $X_{n+1} = 16{,}807{\cdot}X_n mod(2^h - 1)$ to generate the different channels for a node with a unique ID in the network. The value proposed for m_{lcg} is equal to $(2^{31} - 1)$. The presented values for a and c_+ are chosen following the suggested premises in [24], in order to facilitate that the LCG has a full cycle for all the seeds. Since the sensor nodes have limited capabilities, there is no need to use the full cycle of the pseudorandom generator for the channels. Therefore, nodes can allow for repeating the channel sequence generation after a fixed number of hops, h. This allows for nodes that are far away from each other to reuse the same channels, with low probability of other node to choose the same channel. The number of hops, h, must be higher than the number of available channels, in order to decrease the value of m_{lcg} in the LCG. Hence, we assume $m_{lcg} = 65{,}536$. Figure 8.8 presents the LCG's cyclic behaviour, which is a characteristic from these generators. After 65,536 calls to the LCG function, it cyclically repeats all the generated numbers (i.e. slot channels choice). Hereafter, since the channels are going to be chosen by a pseudo-RNG, the channel of the PHY layer is defined as a slot channel.

After each node generates the value of X_{n+1}, it must be mapped to 1 of the 16 available channels (N_{ch}) from the IEEE 802.15.4 standard. In this standard, one channel is reserved for controlling and synchronizing messages exchange. Since we consider CC2420 as the main radio transceiver in our scenarios and it utilizes

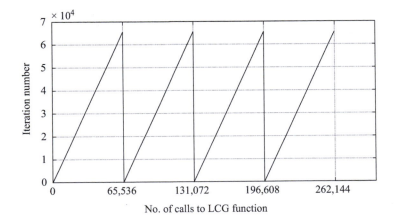

Figure 8.8   LCG cyclic behaviour

the 2.4 GHz frequency band, the associated channels are between 11 and 26 (N.B.: the control and synchronization channel is channel number 11). To map X_{n+1} to a number between 0 and $N_{ch} - 1$, a modular operation is applied to X_{n+1} as presented in the following equation:

$$\lambda_{ch} = X_{n+1} \cdot mod(15) \tag{8.16}$$

To map to the available channel that is assumed by the node, φ_{ch}, the following operation must be performed:

$$\varphi_{ch} = 12 + \lambda_{ch} \tag{8.17}$$

The slot channel 11 is reserved for SYNC packets transmission, while the remaining ones are available for data exchange. This LCG allows for a sender node to know to which channel its potential receiver will change to, when a multi-hop network is considered. If the sender node needs to send a packet to the parent node (1-hop distance), it uses the LCG to choose the channel. However, since the PAN coordinator is hoping all the channels (frequency) during a frame, the sender node must 'exactly' know when it should awake, depending on the chosen channel. If the sender node (after randomly choosing the channel) wants to deliver the packet to the parent node, it has to wait for the adequate time. Since nodes are synchronized, the sender node defines its next wake-up time, $\psi_{wake-up}$, as follows:

$$\psi_{wake-up} = \sigma_{current} + (\varphi_{ch} - 11) \times \Delta t_{SC} + \alpha_{add} - \theta_{switch} \tag{8.18}$$

where $\sigma_{current}$ is the current time of the node (already synchronized), φ_{ch} is the random channel chosen by the sender node, Δt_{SC} is the time interval when the parent node is switched to each channel, α_{add} is the time duration between consecutive frames and θ_{switch} is the time needed by the sensor node to switch to the chosen channel. Figure 8.9 shows the probability of nodes choosing the same slot channel, P_{SC}. As the number of nodes in LoS increases, P_{SC} increases up to the maximum achievable probability

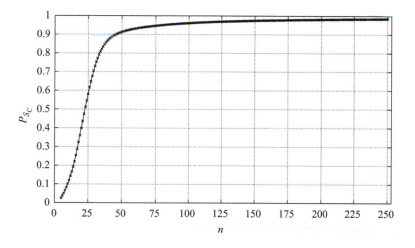

Figure 8.9   Probability of nodes choosing the same slot channel when $N_{ch} = 15$

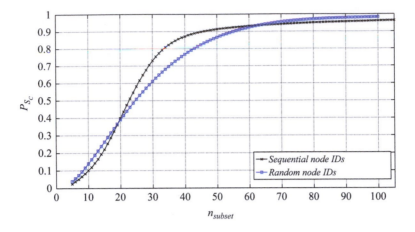

*Figure 8.10   Probability of nodes choosing the same slot channel when $N_{ch} = 15$
for the cases when the IDs of neighbouring nodes are sequential
or random*

(i.e. around 1). P_{SC} increases sharply up to $n = 50$ and continues to increase for
$n > 50$ but at slower rate, as the number of nodes increases.

Figure 8.10 shows the probability of nodes to choose the same slot channel as
the size of the subset of nodes increases for neighbouring IDs of the nodes that
are sequential or random. We averaged 100 MATLAB simulation runs of 100,000 s
(in average) in each run, for each point. The confidence intervals are not presented
because they are negligible. In the simulations, the nodes subset corresponds to a set
of nodes that present a node ID in sequence or randomly by, depending on the chosen
scheme (sequential or random), having impact on P_{SC}. There are some differences in
P_{SC} when the IDs of the neighbouring nodes are sequential or random. If neighbouring
nodes present non-sequential IDs order, P_{SC} is lower when $n \in [20, 60]$, compared
to the case of sequential IDs for the nodes. We can conclude that during deployment
of nodes, if the neighbouring nodes IDs are randomly assigned, the probability of a
node to choose the same slot channel is lower than in the case when the neighbouring
nodes IDs are assigned sequentially.

The sender node will wake-up only at the time when the parent node is switched
to the same channel. By employing a wake-up rendezvous scheme, the sender nodes
will save energy. If more than one node chooses the same channel, an enhanced two-
phase contention window scheme, similar to the one employed in SCP protocol, is
considered, to mitigate collisions.

8.5   SBACK-MAC protocol

8.5.1   Unicast frame concatenation

Following the above introduction of an RTS/CTS mechanism combined with packet
concatenation, this section considers the SBACK-MAC protocol, proposed in

Section 7.7.1, which combines contention-based, scheduling-based and BACK schemes to reduce the energy consumption (due to the protocol overhead) whilst improving the IEEE 802.15.4 MAC sub-layer performance.

Figure 7.10 presents the message sequence chart for the BACK mechanism with *BACK request*, based on [26]. The exchange of *RTS ADDBA* and *CTS ADDBA* special control packets used in the beginning of the BACK mechanism sequence allows for avoiding the hidden-terminal and exposed-terminal problems as in the IEEE 802.11e standard [15,26] (i.e. by using a RTS/CTS handshake). *ADDBA* stands for 'Add BACK'.

The BACK mechanism aims at reducing the power consumption by transmitting less ACK control packets whilst decreasing the time periods the transceivers should switch between different states. By using the BACK, there is no need to receive an ACK for every DATA packet sent, as shown in Figure 7.11. However, during the data transmission, there is no way to know how many packets have successfully reached the destination, except at the end of communication by using the *BACK request/BACK Response*. Like in [27], we set *BE* equal to 0, as if there is no congestion. Therefore, the channel utilization is maximized by decreasing the deferral time before transmitting (i.e. the *InitialbackoffPeriod* will be 0).

As presented in Figure 7.11, to overcome the overhead of the IEEE 802.15.4 MAC, several efficient MAC enhancements are proposed in which the frame concatenation concept is adopted.

The idea is to transmit multiple MAC/PHY frames by using the BACK mechanism (i.e. aggregation), with only one *RTS ADDBA/CTS ADDBA* set between the transmitter/receiver.

8.5.2 *Burst transmissions in the presence of block ACK request*

The IEEE 802.15.4 that employs *BACK request* considers the exchange of two special frames: *RTS ADDBA* and *CTS ADDBA*, where *ADDBA* stands for 'Add BACK'. After this successful exchange, the data packets are transmitted from the transmitter to the receiver (e.g. 10 frames are aggregated). Afterwards, by using the *BACK request* primitive, the transmitter inquires the receiver about the total number of data packets that successfully reach the destination. In response, the receiver sends a special data packet called *BACK Response* identifying the packets that require RTX, and the BACK mechanism finishes.

In our proposed mechanism for every *RTS ADDBA/CTS ADDBA* exchange, we assume that there are always frames available for aggregation (saturation conditions).

As mentioned before, our work addresses a distributed scenario, with single-destination and single-rate frame aggregation. In single-destination approaches, frames can be aggregated if they are available and have the same source and destination address. Moreover, we also assume that the payload of the MAC frames cannot be changed.

In IEEE 802.15.4 with *BACK request*, every time a node has an *RTS ADDBA* to send, the transmission will follow the same backoff procedure like the one presented for IEEE 802.15.4 with RTS/CTS. Therefore, the minimum delay due to CCA,

Figure 8.11   IEEE 802.15.4 with BACK request (concatenation) with retransmissions by using an NAV extra time

$D_{min\_CCA\_BACK}$, in seconds, for determining if the channel state is found to be busy or idle, after the backoff phase, and before each *RTS ADDBA/CTS ADDBA* is given by (7.11) from Section 7.7.1. The associated maximum throughput is given by (7.5).

In SBACK-MAC with *BACK request* by using an NAV extra time, we account the RTX of k lost packets, Figure 8.11. In our proposed mechanism, the value of k will be 20% of the TX aggregated packets. In this case, there is no ACK to confirm that a given packet has successfully reached the destination, and nodes only try to retransmit a packet once, based on the *BACK response*.

In SBACK-MAC with *BACK request*, the minimum delay, $D_{minDataRet\_BACK}$, when the channel is found to be idle during CCA, there is a data transmission (by considering aggregation) and the RTX process is ruled by an NAV extra time as given below:

$$D_{minDataRet\_BACK} = \begin{cases} H_5, & \text{for } j = 0 \\ H_6 & \text{for } j = 1 \end{cases} \tag{8.19}$$

where j is the number of packet RTXs, as defined in [16].

If after CCA, a node determines that the channel is found to be idle, the aggregated packets are sent and a *BACK response* is correctly received, confirming that all the transmitted packets have successfully reach the destination, then, the minimum delay due to packet RTXs, $D_{minDataRet\_BACK}$, is given by

$$H_5 = T_{TA} + T_{RTS\_ADDBA} + T_{TA} + T_{CTS\_ADDBA}$$
$$+ \sum_{i=1}^{N_{agg}} (ccaTime + T_{TA} + T_{DATA} + T_{TA} + T_{IFS})$$
$$+ ccaTime + T_{TA} + T_{BRequest} + T_{TA}$$
$$+ T_{BResponse} + T_{IFS} \tag{8.20}$$

which means that there are no transmission errors. For this case, the number of RTXs is given by $j = 0$.

If after CCA, a node determines that the channel is found to be idle, the aggregated packets are sent and a *BACK Response* is correctly received, indicating that some packets need RTX; the minimum delay due to packet RTXs, $D_{minDataRet\_BACK}$, is given by

$$H_6 = H_5 + \cdots + \sum_{i=1}^{k} (ccaTime + T_{TA} + T_{DATA} + T_{TA} + T_{IFS}) \tag{8.21}$$

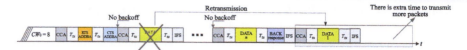

Figure 8.12   *Best-case scenario for IEEE 802.15.4 without BACK request (piggyback)*

where k represents the number of aggregated packets that are allowed to be retransmitted as in [16]. This means that this value must be carefully selected depending on the channel conditions, nodes may only need to retransmit a few data packets, or in an extreme case in which there is the need to retransmit all the aggregated packets.

By combining (7.5) and (8.19), the minimum average delay, $D_{minBACK}$, due to the channel state (i.e. busy or idle) and packet RTXs is given by

$$D_{minBACK} = \frac{D_{min\_CCA\_BACK} + D_{minDataRet\_BACK}}{n}$$

(8.22)

The maximum average throughput (by considering an erroneous channel), S_{max_BACK}, is obtained by replacing D_{min} by $D_{minBACK}$ in (8.7).

8.5.3   Burst transmissions in the absence of block ACK request

The version of the SBACK-MAC protocol without *BACK request* ('piggyback mechanism') also considers the exchange of the *RTS ADDBA* and *CTS ADDBA* packets at the beginning of the communication.

As presented in Figure 7.12, the SBACK-MAC version with piggyback does not consider the use of the *BACK request* primitive. Therefore, the control overhead and the delay is reduced whilst increasing the throughput, by 'piggybacking' the BACK information into the last data fragment; however, with this scheme, the system becomes less robust. If the last aggregated frame (DATA frame n) is lost, the destination does not know that a *BACK response* needs to be sent back.

The version of SBACK-MAC without *BACK request* has the same RTX scheme as in the SBACK-MAC with *BACK request* (Figures 7.12 and 8.12). As a consequence, the minimum delay $D_{min\_CCA\_Piggy}$, in seconds, for determining if the channel state is found to be busy or idle during CCA, following the backoff phase, is given as follows:

$$D_{min\_CCA\_Piggy} = \sum_{i=1}^{n/N_{agg}} \sum_{k=0}^{k \leq NB} \left(\overline{CW_k} + ccaTime \right)$$

(8.23)

In SBACK-MAC with no *BACK request*, the information is piggybacked by using the last data packet transmitted within a burst that includes an M bit bitmap responsible for indicated the packets that successfully reach the destination. Afterwards, if packet RTXs are needed, nodes consider a fixed extra time for retransmitting the packets as in [16] as shown in Figure 8.12. If the last data packet is lost and the *BACK response* is not received within the *BACK response* wait-duration period, $T_{BRW} = T_{AW}$, nodes will try to retransmit the last data packet again once, as shown in Figure 8.12.

Moreover, when a group of data packets is transmitted (i.e. by using aggregation) with only one *RTS ADDBA*/*CTS ADDBA* set between the transmitter/receiver, the receiver confirms the total amount of packets correctly received by using the *BACK response* primitive. All the packets (*RTS ADDBA*, *CTS ADDBA*, DATA and *BACK response*) have a duration field, and the neighbouring nodes are required to set its NAV field accordingly. If packet RTXs are needed, nodes use a longer NAV period, accounting the RTX of k lost packets. In our proposed mechanism, the value of k will be 20% of the TX aggregated packets. In this case, there is confirmation that a given packet has successfully reached the destination, and nodes only try to retransmit a packet once, based on the *BACK response*. The RTX process does not consider the use of the backoff phase between two consecutive data packets, which allows for decreasing the total overhead, as shown in Figures 7.12 and 8.12. This way 'priority' is being created for packet RTXs.

The extra NAV duration due to RTXs include the time period needed for retransmitting the packets plus the *ccaTime* and T_{TA} time periods enabling to avoid packet collisions between neighbouring nodes. This is explained by the fact that when neighbouring nodes wake-up, they will try to access the channel by using backoff phase, defined by the first contention window.

In SBACK-MAC with no *BACK request*, the minimum delay, $D_{minDataRet\_Piggy}$, when the channel is found to be idle during CCA, there is a data transmission (by considering aggregation), and the RTX process ruled by an NAV extra time is given as follows:

$$D_{minDataRet\_Piggy} = \begin{cases} H_7, & \text{for } j = 0 \\ H_8 & \text{for } j = 1 \end{cases} \tag{8.24}$$

where j is the number of packet RTXs and could range between 1 and the maximum number of RTXs, *MaxRet*, as defined in [16].

By analysing (8.25), the following conclusions can be reached:

If after CCA, a node determines that the channel is found to be idle, the aggregated packets are sent and a *BACK Response* is correctly received, confirming that all the transmitted packets have successfully reached the destination, $D_{minDataRet\_Piggy}$ is given by

$$H_7 = T_{TA} + T_{RTS\_ADDBA} + T_{TA} + T_{CTS\_ADDBA}$$
$$+ \sum_{i=1}^{N_{agg}-1} (ccaTime + T_{TA} + T_{DATA} + T_{TA} + T_{IFS})$$
$$+ \cdots + ccaTime + T_{TA} + T_{DATA} + T_{TA}$$
$$+ \cdots + T_{BResponse} + T_{IFS} \tag{8.25}$$

which means that there are no transmission errors. For this case, the number of RTXs is given by $j = 0$.

If after CCA, a node determines that the channel is found to be idle, the aggregated packets are sent and a *BACK Response* is correctly received, indicating that some packets need RTX, $D_{minDataRet\_Piggy}$ is given by

$$H_8 = H_7 + \cdots + \sum_{i=1}^{k} (ccaTime + T_{TA} + T_{DATA} + T_{TA} + T_{IFS}) \qquad (8.26)$$

where k represents the number of aggregated packets that are allowed to be retransmitted as in [16]. This means that this value must be carefully selected depending on the application, since nodes may need to retransmit only a few data packets, or in an extreme case, there is the need to retransmit all the aggregated packets.

By combining (8.23) and (8.24), the minimum average delay, D_{min_Piggy}, due to the channel state (i.e. busy or idle) and packet RTXs is given by

$$D_{min\_Piggy} = \frac{D_{min\_CCA\_Piggy} + D_{minDataRet\_Piggy}}{n} \qquad (8.27)$$

The maximum average throughput (by considering an erroneous channel), S_{max_Piggy}, is obtained replacing D_{min} by D_{min_Piggy} in (8.7).

8.6   Throughput and energy consumption

8.6.1   *Maximum average throughput in the presence and absence of BACK request for the DSSS PHY layer*

Figure 8.13 shows that for the shortest payload sizes (i.e. $L_{DATA} = 3$) it is possible to improve the network performance by using the SBACK-MAC with and without *BACK request* by using an NAV extra time.

When the number of TX packets is higher than seven, SBACK-MAC with and without *BACK request* with and without RTXs achieve shorter delay in comparison to IEEE 802.15.4 in the basic access mode.

Performance results for S_{max} as a function of the number of TX packets show that by considering SBACK-MAC with and without *BACK request* with RTXs, for seven aggregated packets, S_{max} increases by 21% and 35%, respectively, when compared with the IEEE 802.15.4 basic access mode with RTXs.

For 10 aggregated packets, S_{max} increases by 31% and 41%, respectively, when compared with the IEEE 802.15.4 basic access mode with RTXs. For more than 28 aggregated packets, by using SBACK-MAC with and without *BACK request* with RTXs, S_{max} increases by 48% and 53%, respectively, when compared with the IEEE 802.15.4 basic access mode with RTXs. By analysing Figure 8.13, we also conclude that the performance from IEEE 802.15.4 with RTS/CTS and without RTXs, when the number of TX packet is higher than 40, is similar to the one from SBACK-MAC, with and without *BACK request*, with RTXs.

Figure 8.14 presents the analytical results for the maximum average throughput, S_{max}, as a function of the payload size by considering the four different scenarios from Figures 8.5, 8.6, 8.11 and 8.12. The results show that the IEEE 802.15.4 basic

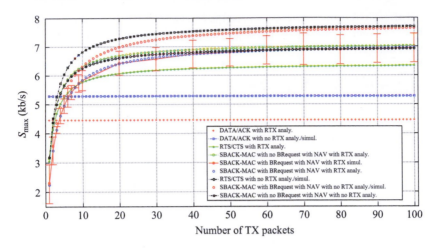

Figure 8.13   *Maximum average throughput as a function of the number of TX packets for a fixed payload size of 3 bytes for IEEE 802.15.4 with and without RTS/CTS and SBACK-MAC with and without BACK request*

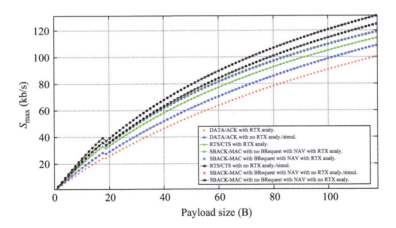

Figure 8.14   *Maximum average throughput as a function of the payload size for a number of TX packets equal to 10 for IEEE 802.15.4 with and without RTS/CTS and SBACK-MAC with and without BACK request*

access mode without RTS/CTS globally presents the worst performance in terms of maximum average throughput, S_{max}, for both the cases, with and without RTXs. The performance from SBACK-MAC with and without *BACK request* with RTXs for small packets sizes (i.e. data payload less than 18 bytes), S_{max}, improves by 41% and 53%, respectively, relatively to IEEE 802.15.4 basic access mode with RTXs. For larger packet sizes, by considering SBACK-MAC with and without *BACK request* with RTXs, the throughput, S_{max}, increases by 18% and 30% when compared with the IEEE 802.15.4 basic access mode with RTXs.

8.6.2.1 High density of nodes

With high density nodes, we have also analysed the aggregate throughput, delivery ratio, energy consumption per node, energy consumption per packet and latency, for the MC-SCP-MAC , MC-LMAC, CSMA and MMSN protocols. We have studied the dependency of these performance metrics on the number of sources nodes, number of slot channels, as well as traffic generation intervals. For this set of tests, we have also considered periodic or exponential traffic patterns. Here, the frame duration is $t_F = 1.57$ s and the time allocated for each slot channel is $\Delta t_{SC} = 0.0625$ s for MC-SCP-MAC. In turn, MC-LMAC, CSMA and MMSN consider a frame duration $t_F = 1.6$ s. The MC-SCP-MAC protocol considers maximum contention window sizes $CW_k^{max} = 80$, $k \in \{1, 2\}$. For this set of results, we consider a maximum transmitter power $P_{tx} = 0.8$ mW for all MAC protocols and a data packet size of $L_{data} = 32$ bytes.

The number of slot channels for data utilized in these tests is $N_{ch} = 8$ for the MC-SCP-MAC, MC-LMAC and MMSN protocols, while a single channel ($N_{ch} = 1$) is considered for CSMA. Since the MC-SCP-MAC protocol achieves enhanced performance in scenarios with higher densities of nodes, for a given number of nodes, the terrain size considered in these tests for the MC-SPC-MAC protocol is $50 \times 50\,\mathrm{m}^2$, whereas for MC-LMAC, CSMA and MMSN protocols an area of 150×150 m^2 is considered. By decreasing the deployment area for the MC-SCP MAC protocol, the nodes will be in the CS range of each other, which, in turn, mitigates the hidden terminals problem allowing for comparison with the MC-LMAC, MMSN and CSMA protocols. In all the MAC protocols, the RTXs are disabled in order to facilitate fair comparison. Figure 8.16 shows the variation of the aggregate throughput with the

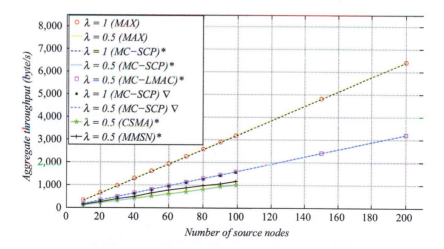

Figure 8.16   Aggregate throughput for the MC-SCP-MAC, MC-LMAC, CSMA and MMSN protocols with periodic (*) and exponential (∇) traffic patterns for different number of sources in the $50 \times 50\,m^2$ deployment scenario

number of source nodes for the MC-SCP, MC-LMAC, MMSN and CSMA, in terms of active source nodes. For **periodic traffic** (uniform distribution) and generation rate $\lambda = 1/2 \text{ s}^{-1}$, the MC-SCP-MAC aggregate throughput is close to the maximum. The same behaviour occurs for the MC-LMAC protocol. We have also investigated the scenarios where nodes generate packets with **exponential** distribution at a rate $\lambda = 1/2 \text{ s}^{-1}$ for the MC-SCP-MAC protocol. In these tests, the aggregate throughput is also near the maximum of the aggregate throughput. With low density of nodes $150 \times 150 \text{ m}^2$ (sparser scenario) for MC-SCP-MAC, under periodic and exponential traffic pattern, the aggregate throughput was much lower than the maximum achievable ones.

The results for the aggregate throughput and delivery ratio, from MC-SCP-MAC, are compared with the ones from MC-LMAC, and CSMA protocols. In order to test the scalability of the MC-SCP-MAC protocol, we have varied the density of the node deployments from $50 \times 50 \text{ m}^2$ to $75 \times 75 \text{ m}^2$, $100 \times 100 \text{ m}^2$, $125 \times 125 \text{ m}^2$, $150 \times 150 \text{ m}^2$ and $200 \times 200 \text{ m}^2$. The authors from MC-LMAC [20] claim that with random deployment beyond a side length (L) of 225 m, unconnected nodes appear. Although we have performed tests with larger deployments, unconnected nodes only start to appear after for side lengths larger than 400 m. However, for the sake of fair comparison with the other MAC protocols, we have only considered side lengths up to 200 m. For this set of tests nodes, we assume periodic or exponential traffic patterns. Here, the frame duration is $t_F = 1.57$ s and the time allocated for each slot channel is $\Delta t_{SC} = 0.17$ s for MC-SCP-MAC, and $t_F = 1.6$ s for MC-LMAC and CSMA consider. For the MC-SCP-MAC protocol, maximum contention window sizes are $\text{CW}_k^{\max} = 80$, $k \in \{1, 2\}$. We consider a maximum transmitter power $P_{tx} = 0.8$ mW for all MAC protocols and data packet size of $L_{data} = 32$ bytes. The number slot channels considered in these tests for data is $N_{ch} = 8$ for the MC-SCP-MAC and MC-LMAC protocols, while a singlechannel ($N_{ch} = 1$) is considered in CSMA. The objective of these tests is to analyse the gains achieved when the IR is enabled for different values of $\Pi_{irmax} \in \{-90; -80; -70; -60\}$ dB m. In all the MAC protocols, the RTXs are disabled for the sake of fairness in comparisons. Figure 8.17 shows the variation of the aggregate throughput with the side length, L, for the MC-SCP-MAC, MC-LMAC and CSMA protocols. **Periodic** traffic with generation rate of $\lambda = 1/2 \text{ s}^{-1}$ is considered. MC-SCP-MAC presents high values of the aggregate throughput if the scenario deployment is denser and IR is enabled. Beyond a side length of 100 m, the aggregate throughput from MC-SCP-MAC starts to decrease, while the aggregate throughput of MC-LMAC continues to increase. However, for a side length of 150 m, the aggregate throughput of MC-LMAC starts to decrease due to the sparsity of the scenario, while the aggregate throughput of MC-SCP-MAC starts to increase. If the IR is disabled, the aggregate throughput of the MC-SCP-MAC achieves its highest values. When the IR is enabled, the values of aggregate throughput are always lower owing to the drop of redundant packets. It reduces the aggregate throughput, but it increases the node lifetime while decreasing the latency, as discussed in earlier sections. For denser scenarios (e.g. $L = 50$ m), the MC-SCP-MAC protocol achieves much higher aggregate throughput than the remaining MAC protocols.

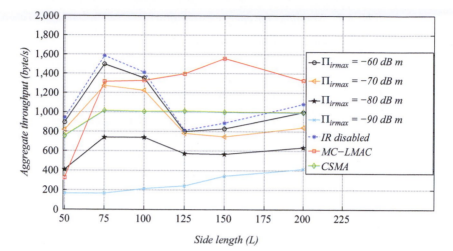

Figure 8.17 *Aggregate throughput of MC-SCP-MAC, MC-LMAC and CSMA protocols with periodic traffic generation ($\lambda = 1/2\,s^{-1}$) for different deployment scenario side lengths when IR enabled or disabled*

8.6.3 Energy consumption for IEEE 802.15.4 in the presence and absence of RTS/CTS and SBACK-MAC

In order to know how much energy is spent by the SBACK-MAC protocol in each state, an analytical model was conceived. A 2-hop network, with two sources, one relay and two sinks (and interferers), has been considered. Figure 8.3 shows the OMNeT++ [30] multi-hop star topology simulation setup. The packets from source node **A** flow, through node **C**, to sink node **D** while the packets originated by source node **B** flow, through node **C**, to reach sink node **E**.

The analysis of the sensor nodes performance is obtained through simulation by considering the CC2420, radio transceiver operating in the 2.4 GHz band. The reason for choosing the CC2420 radio transceiver from Chipcon is related to the fact that it is currently the most popular radio chip on wireless sensor nodes [31].

The energy consumption of a given node over a period of time t is given as follows:

$$E(t) = (t_{tx} \times P_T) + (t_{rx} \times P_R) + (t_{sleep} \times P_S) + (t_{idle} \times P_I) \tag{8.29}$$

The meaning of each variable is presented in Table 8.2.

Figure 8.18 presents the average energy consumption as a function of the packet inter-arrival time for the CC2420 radio transceiver by considering the multi-hop star topology presented in Figure 8.3. Each source node (i.e. Node **A** and **B**) transmits 100 DATA packets with a data payload of 3 bytes (with a data generation interval between 1 and 10 s) to the coordinator node (effective data rate is 250 kb/s) that forwards the packet to the destination (i.e. Node **D** and **E**).

Table 8.2   *Notations for energy estimation*

Notation	Parameter
t_{tx}	Time on TX state
P_T	Power consumption in the transmitting state
t_{rx}	Time on RX state
P_R	Power consumption in the receiving state
t_{sleep}	Time on SLEEP state
P_S	Power consumption in the sleep state
t_{idle}	Time on IDLE state
P_I	Power consumption in the idle state

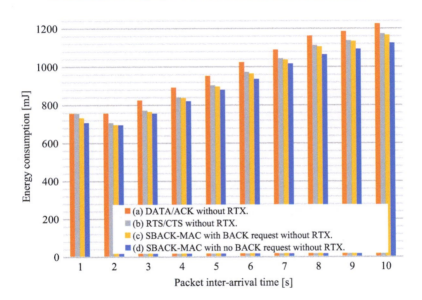

Figure 8.18   *Energy consumption: (a) IEEE 802.15.4 basic access, (b) IEEE 802.15.4 with RTS/CTS, (c) SBACK-MAC with BACK request (concatenation) and (d) SBACK-MAC with no BACK request (piggyback)*

During one transmission cycle, there is only one active node that has always a frame to be sent, whereas the other neighbouring nodes can only accept frames and provide ACKs or *BACK responses*. Therefore, we are considering the best case scenario (i.e. an ideal channel with no transmission errors). Table 8.3 presents the network parameters considered in the simulations.

By analysing the results from Figure 8.18, one concludes that, when the packet inter-arrival time increases, the energy consumption of the radio transceiver also increases. This is explained by the fact that the radio transceiver needs to stay active for longer periods of time, in order to deliver the packets being generated from the sources to the sink nodes. Therefore, the energy consumption is lower in the

Table 8.3   Key simulation parameters

Parameter	Value
Channel bitrate	250 kb/s
Operating frequency	2.4 GHz
Bandwidth	2 MHz
Modulation	O-QPSK
Transmitter power	0 dB m (1 mW)
Channel model	Free-space path loss
Path loss coefficient	2.5
Data Payload	3 bytes
Data packet size	18 bytes
Control packet sizes	11 bytes
(ACK/RTS/CTS/*BACK request*/*BACK Response*)	
Duty cycle	12%
Number of runs	5
Maximum simulation time	100 s
Packet inter-arrival time	From 1 to 10 s

case we have high traffic loads (the lowest packet inter-arrival time), since the nodes are able to deliver packets faster and enter sooner in the sleep mode. In the case of a packet inter-arrival time of 1 s, source nodes are able to deliver all the data packets in the queue to the sink nodes in approximately 10 s, whereas for the case we have a packet inter-arrival time of 10 s, nodes need approximately 100 s to deliver the same amount of data. Moreover, the power spent in the RX/TX states is higher than in the SLEEP state. So, every time a node wakes up and there is no task to perform, there will be an energy waste. This case is more frequent for longer packet inter-arrival times. Moreover, by using the SBACK-MAC protocol with and without *BACK request*, we decrease the total energy consumption of the network for all the inter-arrival periods. By comparing the obtained results for the energy consumption with the ones for the minimum average delay, $D_{\min}$, we conclude that there is an adequate match between the simulations and our analytical model. For the case with no RTXs, SBACK-MAC with and without *BACK request*, as well as IEEE 802.15.4 in the presence of RTS/CTS, presents better performance when compared with IEEE 802.15.4 in the absence of RTS/CTS, with circa 6%–14% decrease in the energy consumption. For all the cases the IEEE 802.15.4 standard in the basic access mode presents the worst performance in terms of energy.

The study of energy consumption for the cases with RTXs is left for further study.

8.6.4   Energy performance on the number of source nodes from MC-SCP-MAC with and without IR

Figure 8.19 presents the dependence of the energy performance on the number of source nodes from MC-SCP-MAC for the traffic-generation patterns considered

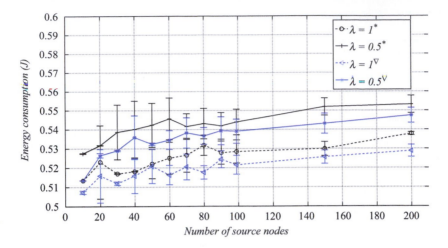

Figure 8.19   Energy consumption per node of MC-SCP-MAC with periodic (*) and exponential (∇) traffic generation for different number of sources in deployment scenario of 50 × 50 m²

within these tests, i.e. the evaluation of the energy consumption per frame successfully delivered by the MC-SCP-MAC protocol, for periodic and exponential traffic patterns with $\lambda \in \{1/2; 1\}$ s^{-1} for a scenario with high density of nodes.

The 95% confidence intervals are also presented in this figure. Here, the energy consumption of the PAN coordinator is not taken into account, since this node has extended capabilities and is not limited by any energy constraint. It is observable in Figure 8.19 that the reduction of the area from the deployment scenario leads to a slight increase of the energy consumption, whilst attaining higher aggregate throughput and delivery ratio. These tests show that higher traffic generation rates lead to the lowest values for the energy consumptions.

In Figure 8.20, it is noticeable that the curves for $\lambda = 1/2$ s^{-1} are the ones that present the highest energy consumption. This is explained by the highest energy consumption observed in Figure 8.19, jointly with the high delivery ratio. This high delivery ratio causes an effective decrease of the energy consumption per delivered frame. Consequently, we conclude that a deployment scenario with higher density of nodes allows for mitigating the hidden terminals problem while achieving less energy waste per successfully delivered packet.

We have evaluated the energy consumption of MC-SCP-MAC as a function of the side length whilst considering periodic traffic profile with a generation rate of $\lambda = 1/2$ s^{-1}, considering the cases where IR is enabled and disabled. These results are presented in Figure 8.21, where the 95% confidence intervals are also shown. As the value of the side length from the deployment scenario increases, the energy consumption is pretty constant up to the value of 125 m for the side length. This is due to the high delivery ratio that MC-SCP-MAC presents in denser scenarios along with the consequent decrease of hidden terminals on the deployment. For side

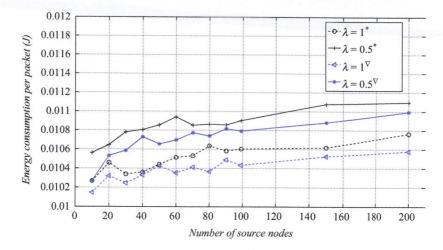

Figure 8.20   *Energy consumption per packet successfully delivered of MC-SCP-MAC with periodic (*) and exponential (∇) traffic generation for different number of sources in deployment scenario of 50 × 50 m²*

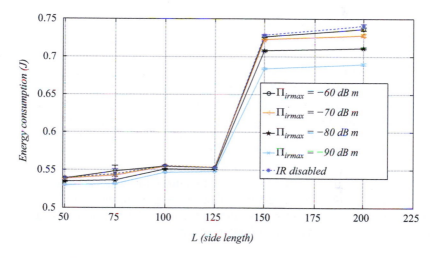

Figure 8.21   *Energy consumption per node of MC-SCP-MAC as a function of the deployment scenarios side lengths with periodic traffic generation ($\lambda = 1/2\,s^{-1}$) when IR enabled or disabled*

lengths larger than 125 m, the MC-SCP-MAC presents high energy consumption. This is explained by the increase of the presence of more hidden terminals, which causes the MC-SCP-MAC to decrease the delivery ratio and consequently suffer from an increase of the energy consumption. MC-SCP-MAC presents the highest values

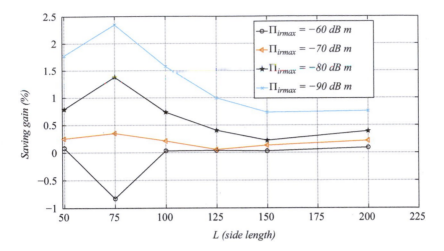

*Figure 8.22   Saving gain percentage per node of MC-SCP-MAC as a function
of the deployment scenarios side lengths when IR is enabled with
periodic traffic generation ($\lambda = 1/2\,s^{-1}$)*

for the energy consumption when the IR is disabled. However, we can observe in Figure 8.21 that, by enabling IR, the energy consumption slightly decreases. This is due to the slight increase of delivery ratio verified when the IR is enabled.

It is now worthwhile to evaluate the energy saving gains that can be achieved with the employment of the IR concept, as shown in Figure 8.22. The respective percentages of saving gain are shown as a function of the side lengths. It considers the **periodic** traffic profile with a generation rate of $\lambda = 1/2\,s^{-1}$. It is observable that the use of IR is only efficient for denser scenarios and for sensing values $\Pi_{irmax} \in \{-90; -80; -70\}$ dB m. For side lengths larger than 100 m, the saving gains are negligible.

We conclude that the employment of IR allows for achieving shorter latency values, higher delivery ratios and considerable energy consumption savings, at the expense of reduced aggregate throughput. Moreover, denser scenarios allow for MC-SCP-MAC to achieve better overall performance.

8.7   Conclusions and discussion

In this chapter, we started by proposing the enhancement of the IEEE 802.15.4 MAC sub-layer by employing **RTS/CTS combined with the packet concatenation** feature for WSNs. The use of the RTS/CTS mechanism improves channel efficiency by decreasing the deferral time before transmitting a data packet. Although the RTXs are addressed in a somehow rigid approach, the proposed solution has shown that, even for the case with RTXs, if the number of TX packets is lower than five (i.e. the number

Table 8.4   *Performance results for the maximum average throughput, S_{max} and minimum average delay, D_{min}, as a function of the number of transmitted packets relatively to IEEE 802.15.4 in the basic access mode by considering a fixed payload size of 3 bytes*

MAC protocol	Number of aggregated Pkts $= 7$		Number of aggregated Pkts $= 10$		Number of aggregated Pkts ≥ 28	
	Increase S_{max} [%]	Decrease D_{min} [%]	Increase S_{max} [%]	Decrease D_{min} [%]	Increase S_{max} [%]	Decrease D_{min} [%]
IEEE 802.15.4 with RTS/CTS	14	14	18	18	30	30
SBACK-MAC with BACK request	21	21	31	31	48	48
SBACK-MAC without BACK request	35	35	41	41	53	53

of aggregated packets), IEEE 802.15.4 employing RTS/CTS combined with concatenation achieves higher values for the throughput in comparison to IEEE 802.15.4 with no RTS/CTS even for shorter packet sizes. The advantage comes from not including the backoff phase into the RTX process like the IEEE 802.15.4 basic access mode (i.e. $BE = 0$). By comparing the analytical and simulation results, we conclude that there is a perfect match. This actually gives hints for future detailed verification of the accuracy of our proposed RTX model. Performance results for the minimum average delay, D_{min}, as a function of the number of TX packets and by assuming a fixed payload size of 3 bytes (i.e. $L_{DATA} = 3$ bytes) show that, by using IEEE 802.15.4 with RTS/CTS with packet concatenation, for 5, 7 and 10 aggregated packets, D_{min}, decreases by 8%, 14% and 18%, respectively. For more than 28 aggregated packets, D_{min} decreases approximately 30%.

Moreover, we observe that by considering IEEE 802.15.4 in the basic access mode, the maximum average throughput, S_{max}, does not depend on the number of TX packets and achieves the maximum value of 5.2 kb/s. For IEEE 802.15.4 with RTS/CTS combined with the packet concatenation feature, the maximum achievable throughput is 6.3 kb/s. In IEEE 802.15.4 employing RTS/CTS with packet concatenation, results for S_{max} as a function of the number of TX packets show that for 5, 7 and 10 aggregated packets S_{max} increases by 8%, 14% and 18%, respectively. For a number of aggregated packets higher than 28, S_{max} increases approximately 30%. Table 8.4 presents a summary of the performance results for maximum average throughput, S_{max}, and minimum average delay, D_{min}, as a function of the number of TX packets by considering a fixed payload size, of 3 bytes.

This chapter also proposes SBACK-MAC, a new innovative MAC protocol that uses a BACK mechanism to achieve channel efficiency for WSNs (**SBACK-MAC**

protocol). The use of a BACK mechanism improves channel efficiency by aggregating several ACK into one special packet, the *BACK response*. Two innovative solutions were proposed to improve the IEEE 802.15.4 performance. The first one considers the SBACK-MAC protocol in the presence of *BACK request* (concatenation mechanism), while the second considers the SBACK-MAC in the absence of *BACK request* (piggyback mechanism). The results showed that, for the shortest payload sizes (i.e. $L_{DATA} = 3$), it is possible to improve the network performance by using the SBACK-MAC with and without *BACK request* by using an NAV extra time. When the number of TX packets is higher than seven, SBACK-MAC with and without *BACK request*, with and without RTXs, achieves lower delay in comparison to IEEE 802.15.4 in the basic access mode. Performance results for D_{min} as a function of the number of TX packets show that, by using SBACK-MAC with and without *BACK request* with RTXs, for seven aggregated packets, D_{min} decreases by 21% and 35%, respectively, when compared with the IEEE 802.15.4 basic access mode with RTXs. For 10 aggregated packets, D_{min} decreases 31% and 41%, respectively, when compared with the IEEE 802.15.4 basic access mode with RTXs. For more than 28 aggregated packets, by using SBACK-MAC in the presence and absence of *BACK request* with RTXs, D_{min} decreases by 48% and 53%, respectively, when compared with the IEEE 802.15.4 basic access mode. Moreover, by considering IEEE 802.15.4 with RTS/CTS with no RTXs when the number of TX packet higher than 40, it presents similar results when compared with SBACK-MAC with and without *BACK request* with RTXs.

Performance results for S_{max} as a function of the number of TX packets show that by using SBACK-MAC with and without *BACK request* with RTXs, for 7 aggregated packets, S_{max} increases by 21% and 35%, respectively, when compared with the IEEE 802.15.4 basic access mode with RTXs. For seven aggregated packets, S_{max} increases by 31% and 45%, respectively, when compared with the IEEE 802.15.4 basic access mode with RTXs. For more than 28 aggregated packets, by using SBACK-MAC with and without *BACK request* with RTXs, S_{max} increases by 48% and 53%, respectively, when compared with the IEEE 802.15.4 basic access mode with RTXs.

We also conclude that by considering IEEE 802.15.4 with RTS/CTS with no RTXs when the number of TX packet exceeds 40, S_{max} presents similar results when compared with SBACK-MAC with and without *BACK request* with RTXs.

The proposed mechanisms also show that for more than 10 TX packets, the bandwidth efficiency and energy consumption will be improved by not considering the backoff phase between two consecutive data packets and by postponing ACKs via the use of the *BACK response* control packet, allowing for the aggregation of several ACK responses into one single packet.

The section on the **MC-SCP-MAC protocol** proposes a new MAC protocol based on the SCP-MAC protocol that envisages multichannel features. In addition, the study of the capture effect in this new protocol is covered, as well as the ER phase algorithm useful to mitigate packet losses. The MC-SCP-MAC performance has been evaluated in terms of collision probability. It is observable that the curves for the MC-SCP-MAC protocol achieve much lower values for the CW_1 and CW_2 collision probabilities (in both regimes) for the MC-SCP-MAC is much lower than SCP-MAC. The improvement is due to the use of multichannel jointly with an enhanced two-phase contention window mechanism to cope with the possible data packet collisions in MC-SCP-MAC. It causes an increase of the overall network performance at the expense of a small increase in the complexity of the protocol algorithm.

In almost all the tests performed, the simulation results consider mostly the CC2420 radio transceiver. However, one set of experiments addresses the comparison between the energy consumption per packet (successfully delivered) between MC-SCP-MAC and SCP-MAC, for the CC2420 and AT86RF231 transceivers. With the AT86RF231 transceiver, the MC-SCP-MAC protocol presents the lowest values for the energy spent per delivered packet, in comparison with the single-channel SCP-MAC protocol. The lowest values for the energy spent per delivered packet are justified by the very high delivery ratio, owing to the use of multiple channels, combined with the ultra-low power consumption features from the AT86RF231 transceiver (with low energy consumption, shorter duration to perform CS, shorter duration to perform channel switching). The same holds for the CC2420 transceiver, in which the MC-SCP-MAC protocol also presents lower values for the energy spent per delivered packet. Since MC-SCP-MAC is a multichannel-based protocol, we have shown how the collision probability varies with the number of slot channels, $N_{ch} \in \{3; 4; 7; 9; 15\}$. Here, SCP-MAC considers a single-channel for data ($N_{ch} = 1$). The network size considered in these tests is $n = 99$ nodes plus 1 PAN coordinator. The increase of the number of slot channels allows for nodes selecting different slot channels. Therefore, it decreases the probability of a node choosing the same slot channel to transmit their data packets.

The difference of the CW_1 collision probability between the MC-SCP and SCP protocols is more noticeable in the unsaturated regime. The curves for the CW_2 collision probability for the MC-SCP-MAC protocol decreases as the number of slot channels increases. In comparison with the SCP protocol, the CW_2 collision probability of the MC-SCP-MAC protocol is always lower. This is due to the low values of CW_1 collision probability that result from the multichannel features and lead to a low number of expected number of nodes that pass from CW_1 to CW_2. There are two main conclusions arising from these tests: the number of slot channels increases along with the proposed enhanced two-phase contention window mechanism, allowing for achieving a much lower collision probability for the data packets.

The difference of the effective collision probability between the MC-SCP and SCP protocols is more notorious in the saturated case. Other tests have been performed to evaluate the energy efficiency with multiple slot channels and contention window sizes. The comparison of the energy efficiency is performed for the MC-LMAC and MC-SCP-MAC whilst considering equal deployment area and transmitter power.

This work also analyses the adaptation of the IR concept to a multichannel-based protocol along with the impact in terms of energy performance of enabling of the **IR concept** in the MC-SCP-MAC. The objective of these tests is to analyse of the gains achieved when the IR concept is enabled, for different values of sensing range, $\Pi_{irmax} \in \{-90; -80; -70; -60\}$ dB m. By enabling the IR, the node that overhears a data packet and verifies that it has redundant data stored in the queue to send to the sink, it discards these redundant packets. The adapted IR concept implies improvements, providing significant reduction of the redundancy of data packets and, consequently, the reduction of the energy consumption per node. As the value of Π_{irmax} increases, the IR of each node decreases which leads to a decrease of the overhearing and the consequent diminishing in the dropping of redundant data packets.

An additional performance metric has also been defined: the energy consumption saving. This metric translates the energy that a node can save when applying the IR concept (for different values of Π_{irmax}). The energy consumption saving is larger if the nodes have to send 600 packets to the PAN (long-term evaluation), compared with the case when nodes have to send 50 packets (short-term evaluation). Moreover, larger network sizes present the largest energy consumption savings when $\Pi_{irmax} = -90$ dB m. For 99 nodes, there is a gain of 21.42%, while for 50 and 20 nodes the gain is 11.70% and 7.07%, respectively. These savings decrease as the value of Π_{irmax} increases. This is due to the reduction of the IR area which, in turn, reduces the overhearing of data packets by the node (as there is a decrease of the level of data packet redundancy), leading to the increase of the energy consumption per node, i.e. there is a reduction of the energy-consumption savings.

Additional tests have been conducted with variable number of slot channels to assess the benefits gained from the use of the IR concept in a McMAC protocol. We conclude that if the number of slot channels used by the MC-SCP-MAC increases, the energy consumption also increases. However, the employment of IR allows for reducing the energy consumption for low values of Π_{irmax}. As the value of sensing range, Π_{irmax}, increases, the number of redundant packets increases. Therefore, more packets are going to be exchanged, leading to an increase of the energy consumption. The energy savings decrease as the value of the sensing range, Π_{irmax}, increases. In turn, these gains are higher as the number of available slot channels increases.

In WSNs, the most usual traffic pattern is the periodic one. In all the tests performed, the periodic traffic pattern is always considered in the simulations. However, the impact of different traffic patterns in the MC-SCP-MAC performance is also addressed. MC-SCP-MAC supports longer traffic generation intervals than other multi-channel based MAC protocols (e.g. MC-LMAC). The results for sparser scenarios have shown that the aggregate throughput does not attain a desirable value. However, tests have been conducted considering a denser scenario for $\lambda = 1$ s^{-1} (under periodic and exponential traffic generation). An aggregate throughput was achieved close to the maximum. With these results, we can conclude that the MC-SCP protocol presents a better performance in high density scenarios, but the presence of hidden terminals may lead to a degradation of the aggregate throughput as the number of source nodes increases. A method that enables to decrease the impact of the hidden terminals problem is therefore imperative.

Another conclusion is that the increase of the traffic generation frequency is well supported by the MC-SCP-MAC in the scenarios that do not present the hidden terminals problem. In terms of latency, as the number of source nodes increases, the values of the latency increase. However, in scenarios with larger density of nodes, the values for the latency are longer than in scenarios with sparser nodes density. Since MC-SCP-MAC is more robust and efficient in denser scenarios (and the envisaged scenario for this protocol is a deployment of sensor nodes in a forest or an agricultural field), we have considered that the performance evaluation in a cluster topology is useful to extract insights for further improvements in the protocol. The objective has been to assess the impact of the IR concept enabling and the number of slot channels available for data exchange in MC-SCP-MAC, whilst considering different traffic patterns and data-generation rates.

In general, when the packet inter-arrival time increases, the **energy consumption** of the radio transceiver also increases because the radio transceiver needs to stay active for longer periods of time, in order to deliver the packets being generated from the sources to the sink nodes.

In MC-SCP-MAC, we have learnt that the employment of IR allows for achieving shorter latency values, higher delivery ratios and considerable energy consumption savings, at the expense of reduced aggregate throughput. Moreover, denser scenarios allow for MC-SCP-MAC to achieve better overall performance.

8.8 Suggestions for future work

A suggestion for further research is to implement the SBACK-MAC without *BACK request* in the OMNeT++ simulator, since in the scope of this work, we only derive the analytical model for the maximum average throughput and minimum average delay. In addition, we plan to compare the IEEE 802.15.4 with and without RTS/CTS with SBACK-MAC with and without *BACK request* by considering the new optional CSS PHY layer at the 2.4 GHz frequency band, which enables to achieve a maximum data rate of 1 Mb/s, against the results obtained in this work where the maximum data rate is 250 kb/s.

Besides, the performance of the innovative MAC sub-layer protocols proposed in this research work will be verified through experimental measurements by using WSN hardware platforms (e.g. Waspmote or IoT OpenMote/Open Mote B platforms), while varying the number of transmitted packets, the payload and the number of nodes.

For the MC-SCP-MAC protocol, we intend to investigate the collision probability model by considering limited RTXs and other queueing models, such as the M/M/1/K Markov chain model, which assumes finite buffers. In addition, we plan to use the model in the routing layer of WSNs, in a real testbed, in order to use the model's output to decide whether to transmit a packet, depending on the number of neighbour nodes.

It can also be suggested to analytically compare the advantages/disadvantages of using one or two contention windows in slotted contention-based MAC protocols. The idea is to apply the two phase contention window mechanism of the SCP-MAC protocol with the specifications of the IEEE 802.11 distributed coordination function (DCF) access method whilst comparing the performance between sequential and random IDs for the neighbouring nodes. The objective is to derive the achieved throughput based on the different slot probabilities and find the optimal values of the contention windows sizes that maximize the throughput for a certain network size. In addition, a throughput comparison between the IEEE 802.11 DCF and the modified IEEE 802.11 DCF access methods is planned.

Our ongoing work also consists of further investigating energy consumption and in particular proposing innovative energy-saving mechanisms that will be combined with the presented MAC mechanisms and will allow further performance enhancement. Although our work considers the realistic case of RTXs (due to an imperfect channel), we also plan to extend the proposed mechanisms in order to take into account the frame error rate process under noisy environments that encounter bursty frame losses. Moreover, we also envision to evaluate the performance of our proposals under specific applications of Internet of Things. Last but not least, we are planning to consider in our research the latest standardization efforts of the IEEE 802.15.4 group and in particular the on-going activities of the High Rate Task Group IEEE 802.15.4t and the Wireless Next Generation Standing Committee (SCwng).

Acknowledgements

This work has been partially supported and funded by CREaTION, UID/EEA/50008/ 2013, COST CA 15104, SFRH/BSAB/113798/2015, 3221/BMOB/16 CMU Portugal Faculty Exchange Programme, CMU/ECE/0030/2017 (CONQUEST) and ORCIP.

References

[1]   Huang P, Xiao L, Soltani S, *et al.* The evolution of MAC protocols in wireless sensor networks: a survey. IEEE Communications Surveys Tutorials. 2013;15(1):101–120.

[2]   Demirkol I, Ersoy C, Alagoz F. MAC protocols for wireless sensor networks: a survey. IEEE Communications Magazine. 2006;44(4):115–121.

[3]   On World. ZigBee and 802.15.4 Shipments to Reach 2.5 Billion by 2020; June 2015 [homepage on the Internet]. PRWeb; [updated 2015 June 17; cited 2018 Jul. 24]. Available from: http://onworld.com/news/ZigBee-and-802.15.4-Shipments-to-Reach-2.5-Billion-by-2020.html.

[4]   IEEE. Approved IEEE Draft Amendment to IEEE Standard for Information Technology-Telecommunications and Information Exchange Between Systems – Part 15.4:Wireless Medium Access Control (MAC) and Physical Layer (PHY) Specifications for Low-Rate Wireless Personal Area Networks (LR-WPANS): Amendment to Add Alternate PHY (Amendment of IEEE Std 802.15.4). IEEE Approved Std P802154a/D7; Jan 2007.

[5]   Texas Instruments. 2.4 GHz IEEE 802.15.4/ZigBee-Ready RF Transceiver (Rev. B). Texas Instrument. CC2420 Datasheet; 2012.

[6]   Rousselot J, Decotignie JD, Aoun M, *et al.* Accurate Timeliness Simulations for Real-Time Wireless Sensor Networks. In: 2009 Third UKSim European Symposium on Computer Modeling and Simulation, Athens, Greece; 2009. p. 476–481.

[7]   Barroca N, Gouveia PTGT, Velez FJ. Impact of Switching Latency Times in Energy Consumption of IEEE 802.15.4 Radio Transceivers. In: Conf. on

Telecommunications – ConfTele., Castelo Branco, Portugal, vol. 1; 2013. p. 45–48.

[8]   Sun M, Sun K, Zou Y. Analysis and Improvement for 802.15.4 Multi-hop Network. In: 2009 WRI International Conference on Communications and Mobile Computing, Kunming, Yunnan, China, vol. 2; 2009. p. 52–56.

[9]   Jurdak R, Ruzzelli AG, O'Hare GMP. Radio sleep mode optimization in wireless sensor networks. IEEE Transactions on Mobile Computing. 2010;9(7):955–968.

[10] Calculating 802.15.4 Data Rates [homepage on the Internet]. [cited 2018 Sep]. Application Note; Available from: https://www.nxp.com/docs/en/application-note/JN-AN-1035.pdf

[11] Ye W, Silva F, Heidemann J. Ultra-low Duty Cycle MAC with Scheduled Channel Polling. In: Proceedings of the 4th International Conference on Embedded Networked Sensor Systems. SenSys '06. New York, NY, USA: ACM; 2006. p. 321–334. Available from: http://doi.acm.org/10.1145/1182807.1182839.

[12] So HSW, Walrand J, Mo J. McMAC: A Parallel Rendezvous Multi-Channel MAC Protocol. In: 2007 IEEE Wireless Communications and Networking Conference, Hong Kong, China; 2007. p. 334–339.

[13] El-Hoiydi A, Decotignie JD. WiseMAC: An Ultra Low Power MAC Protocol for the Downlink of Infrastructure Wireless Sensor Networks. In: Proceedings of the Ninth International Symposium on Computers and Communications, ISCC 2004 (IEEE Cat. No.04TH8769), Alexandria, Egypt, vol. 1; 2004. p. 244–251.

[14] Buratti C, Verdone R. Performance analysis of IEEE 802.15.4 non beacon-enabled mode. IEEE Transactions on Vehicular Technology. 2009;58(7): 3480–3493.

[15] Chatzimisios P, Boucouvalas AC, Vitsas V. Effectiveness of RTS/CTS handshake in IEEE 802.11a wireless LANs. Electronics Letters. 2004;40(14): 915–916.

[16] IEEE. IEEE Draft Standard for Information Technology – Telecommunications and Information Exchange Between Systems – Local and Metropolitan Area Networks - Specific Requirements – Part 15.4: Wireless Medium Access Control (MAC) and Physical Layer (PHY) Specifications for Low Rate Wireless Personal Area Networks (WPANs) (Revision of IEEE Std 802154-2006). IEEE P802154REVi/D07; April 2011. p. 1–310.

[17] Barroca N, Borges LM, Velez FJ, *et al.* IEEE 802.15.4 MAC Layer Performance Enhancement by Employing RTS/CTS Combined with Packet Concatenation. In: 2014 IEEE International Conference on Communications (ICC), Sydney, Australia; 2014. p. 466–471.

[18] Han S, Nam Y, Seok Y, *et al.* WLC29-4: Two-phase Collision Avoidance to Improve Scalability in Wireless LANs. In: IEEE Globecom 2006, San Francisco, CA, 2006. p. 1–5.

[19] Yang X, Vaidya NH. A wireless MAC protocol using implicit pipelining. IEEE Transactions on Mobile Computing. 2006;5(3):258–273.

[20]   Incel OD, van Hoesel L, Jansen P, *et al.* MC-LMAC: a multi-channel MAC protocol for wireless sensor networks. Ad Hoc Networks. 2011;9(1): 73–94. Available from: http://www.sciencedirect.com/science/article/pii/ S1570870510000624.

[21]   Le HC, Guyennet H, Felea V. OBMAC: An Overhearing Based MAC Protocol for Wireless Sensor Networks. In: 2007 International Conference on Sensor Technologies and Applications (SENSORCOMM 2007), Valencia, Spain; 2007. p. 547–553.

[22]   Tang L, Sun Y, Gurewitz O, *et al.* PW-MAC: An Energy-efficient Predictive-wakeup MAC Protocol for Wireless Sensor Networks. In: 2011 Proceedings IEEE INFOCOM, Shanghai, China; 2011. p. 1305–1313.

[23]   Tang L, Sun Y, Gurewitz O, *et al.* EM-MAC: A Dynamic Multichannel Energy-efficient MAC Protocol for Wireless Sensor Networks. In: Proceedings of MobiHoc'11 Proceedings of the Twelfth ACM International Symposium on Mobile Ad Hoc Networking and Computing, Paris, France, May 2011; 2011.

[24]   Knuth DE. The Art of Computer Programming, Volume 2 (3rd Ed.): Seminumerical Algorithms. Boston, MA, USA: Addison-Wesley Longman Publishing Co., Inc.; 1997.

[25]   Park SK, Miller KW. Random number generators: good ones are hard to find. Communications of the ACM. 1988;31(10):1192–1201. Available from: http://doi.acm.org/10.1145/63039.63042.

[26]   IEEE. ISO/IEC Standard for Information Technology – Telecommunications and Information Exchange Between Systems – Local and Metropolitan Area Networks – Specific Requirements Part 11: Wireless LAN Medium Access Control (MAC) and Physical Layer (PHY) Specifications (Includes IEEE Std 802.11, 1999 Edition; IEEE Std 802.11A.-1999; IEEE Std 802.11B.-1999; IEEE Std 802.11B.-1999/Cor 1-2001; and IEEE Std 802.11D.-2001). ISO/IEC 8802-11 IEEE Std 80211 Second edition 2005-08-01 ISO/IEC 8802 11:2005(E) IEEE Std 80211i-2003 Edition; 2005. p. 1–721.

[27]   Woo S, Park W, Young Ahn S, *et al.* Knowledge-based exponential backoff scheme in IEEE 802.15.4 MAC. International Conference on Information Networking, St. Louis, MO. 2008;5200:435–444.

[28]   Cano C, Bellalta B, Cisneros A, *et al.* Quantitative analysis of the hidden terminal problem in preamble sampling WSNs. Ad Hoc Networks. 2012;10(1):19–36. Available from: http://www.sciencedirect.com/ science/article/pii/S1570870511000862.

[29]   Sloane NJA. A Handbook of Integer Sequences. Cambridge, MA: EUA, Academic Press; 1973.

[30]   OMNeT++ Network Simulation Framework [homepage on the Internet]. OMNeT++; 2013 [cited 2013 Sep.]. Available from: http://www.omnetpp.org/.

[31]   Healy M, Newe T, Lewis E. Efficiently securing data on a wireless sensor network. Journal of Physics: Conference Series. 2007;76(1):012063. Available from: http://stacks.iop.org/1742-6596/76/i=1/a=012063.

Chapter 9

A precise low power and hardware-efficient time synchronization method for wearable systems

Fardin Derogarian[1], João Canas Ferreira[2],
Vítor Grade Tavares[2], José Machado da Silva[2],
and Fernando J. Velez[1]

This chapter presents a one-way method for synchronization at the media access control (MAC) layer of nodes and a circuit based on that in a wearable sensor network. The proposed approach minimizes the time skew with an accuracy of half of clock cycle in average. The work is intended to be used in a router integrated circuit (IC) designed for wearable systems. In particular, we address the need for good time synchronization in the simultaneous acquisition of surface electromyographic signals of several muscles. In our main application case, the electrodes are embedded in patient clothes connected to sensor nodes (SNs) equipped with analog-to-digital converters. The SNs are connected together in a network using conducting yarns embedded in the clothes. In the context of such wearable sensor networks, the main contributions of this work are the evaluation of existing protocols for synchronization, the description of a simpler, resource-efficient synchronization protocol, and its analysis, including the determination of the average local and global clock skew and of the synchronization probability in the presence of link failures. Both theoretical analysis and experimental results, in wired wearable networks, show that the proposed protocol has a better performance than precision time protocol (PTP), a standard timing protocol for both single and multihop situations. The proposed approach is simpler, requires no calculations, and exchanges fewer messages. Experimental results obtained with an implementation of the protocol in 0.35 μm complementary metal oxide semiconductor (CMOS) technology show that this approach keeps the one-hop average clock skew around 4.6 ns and peak-to-peak skew around 50 ns for a system clock frequency of 20 MHz.

[1] Instituto de Telecomunicações and Departamento de Engenharia Electromecânica – Universidade da Beira Interior; Faculdade de Engenharia, Portugal
[2] Department of Electrical and Computer Engineering, Faculty of Engineering, University of Porto and INESC TEC, Portugal

9.1   Introduction

A time synchronization method with enough accuracy is a necessary part of many wired or wireless sensors applications. Monitoring of the environment, localization, security, time division multiple access (TDMA)-based protocols, coordinated sleep wake-up scheduling mechanisms, and data fusion are some of applications based on time synchronization [1]. For instance, in gait analysis for evaluation and diagnosis of mobility impairments, it is important to know the relative timing of several different surface electromyography (sEMG) signals and their time relation to the inertial information about the movements of the lower limbs. Otherwise, a healthy condition may be diagnosed as a disorder.

Each individual SN is usually equipped with an independent clock generator. The use of a crystal quartz oscillator ensures good local clock accuracy. Although this kind of oscillators have good stability (with variation around a few parts per million (ppm)), it is not enough to keep synchronization during the whole time of activity: since the clock generators are working independently, small drifts in clock generation lead to synchronization loss. By using a Global Positioning System (GPS) receiver, SNs can achieve clock synchronization in the nanosecond range. However, sensors usually do not need such a high accuracy. On the other hand, using GPS receivers in resource limited systems is not always practical or possible (for indoor use, for instance). Another approach is to use a synchronization mechanism to overcome and compensate the clock drifts in each node relative to a clock reference [2–5].

The time synchronization problem has been studied extensively in all areas of networking [1,2,6]. Many parameters affect the synchronization between the nodes. Limited power supply and unknown message delay in wireless sensor network (WSN) are important challenges that make synchronization difficult in practice [7]. In addition, the latency of message communication also affects synchronization [5]. The delay associated with the process of sending a message has four components:

1. Send time: Delay for assembling a message and handing it over to the MAC layer.
2. Access time: Delay for accessing the channel.
3. Propagation time: Delay due to signal propagation between nodes.
4. Receive time: Delay for receiving and processing message at the receiver.

Except for propagation time, which depends on the communication environment and may be highly deterministic, the other delay sources are nondeterministic in general. In order to reduce the effects of delay on synchronization, in some protocols, such as Reference-Broadcast Synchronization (RBS), send and access delays are shortened by sampling and injecting time information into the message while transmitting [8].

Many synchronization protocols use two-way message exchanges between pairs of nodes to estimate the time skew and offset [9–11]. In this approach, delay and offset are calculated by sending and receiving messages with timing information and measuring the round trip delay. Depending on the protocol, the exchange may be started by the clock source node or by the client node (i.e., the node whose clock is to be adjusted). Each protocol uses different mechanism for calculating skew and offset. A second approach uses only one-way communication [8,12]. In this case, the clock source sends timing information to client nodes and client nodes estimate skew and offset depending on the protocol.

9.2 Related work

A classification of synchronization protocols based on synchronization issues and application-dependent features can be found in [6]. In a master–slave approach, the master node is the time reference and slave nodes synchronize their clock with master, while in peer-to-peer protocols, each node can directly get timing information from the other nodes. The advantages of peer-to-peer protocols are robustness and flexibility, but their implementation is more complex. Because of network size and topology, it is not always possible for two nodes to communicate directly. In many networks, specially when the number of sensors increases, communication is set through multi-hop connections via intermediate nodes. In this case, synchronization must be performed by sender-to-receiver or receiver-to-receiver approaches. In the former, the receiver node synchronizes its clock with the sender and resends timing data to the next node after synchronization; in the latter, the sender broadcasts timing messages and receivers in its coverage area exchange messages among themselves instead of interacting with the sender.

The network time protocol (NTP) is a well-known time synchronization protocol based on two-way message exchange [10]. NTP is one the most used protocols, specially on the Internet, and is known as an effective, secure, and robust protocol.

The NTP clients synchronize their local clocks to the NTP time servers by statistical analysis of the round-trip time. To achieve highly precise time synchronization, the time servers are equipped with or synchronized to atomic clocks or GPS signals. The significant complexity of NTP makes its implementation in sensor networks difficult. In addition, the nondeterminism in transmission time of WSNs can introduce large delays. Therefore, NTP is suitable only for synchronization in WSNs with low precision demands. Authors from [13] have discussed the practical limitations of NTP and addressed the uncertainty of NTP time transfer when network asymmetry is largely eliminated.

If a clock source broadcasts its time, adjacent nodes will receive the same timing message at approximately the same time. In this way, receiver nodes can obtain timing information that is subject to very little variability in its delay. This one-way method has been employed in the RBS protocol [8]. In this approach, a node is selected as a time reference node for all other nodes in the network. Each node records its local time immediately after receiving the message and compares it with the received time. This protocol utilizes a sequence of synchronization messages from clock source node in order to estimate both offset and skew of the local clocks relative to each other.

The timing-sync protocol for sensor networks (TPSN) is a sender–receiver, two-way synchronization protocol [11] and was designed specially for multi-hop networks. TPSN performs synchronization in two phases: level discovery phase and synchronization phase. In the first phase, a root node as clock reference elects and builds a spanning tree of the network. In the subsequent synchronization phase, nodes synchronize to their parent in the tree by exchanging messages at the initiative of the child node. If the root node or the network topology changes, TPSN has to start again with the discovery phase. Redoing the level discovery phase after any changes increases the traffic and the network energy consumption.

The flooding time synchronization (FTSP) is another one-way protocol for multi-hop ad hoc networks [14]. A root node periodically broadcasts a message with global time stamping (TS) information. Each receiver nodes estimates its clock offset and skew by combining both global and local time information. The root node is elected dynamically and periodically based on the smallest node identifier and is responsible for keeping the global time of the network. FTSP is an effective synchronization mechanism in multi-hop ad hoc networks, but [12] shows that clock skew increases exponentially with increasing network diameter.

The PTP, defined by IEEE standard 1588 [15], is a hierarchical master–slave architecture for clock distribution and time synchronization with highly precise synchronization [9]. In this protocol, the clock server periodically sends synchronization messages. A client replies to the server with a message including the reception time of the first message from the server and the replying time. Then the server estimates the delay as well as the offset and sends this information with another message. The client uses the information in the second message for local clock adjustment. To avoid or minimize the effects of clock drift on TS, client nodes periodically set their time to the server time. PTP is a highly precise synchronization protocol, which can ensure clock skew values in the sub-microsecond range.

Although, the aforementioned protocols are widely used to provide synchronization, they are not hardware-aware protocols. In other words, their performance depends on the implementation characteristics of different systems. The main challenge in synchronization is the estimation of the nondeterministic delay of data transmission due to the properties of the hardware. If it is possible to keep variations of the communication and processing delay at the hardware level at a minimum, then achieving synchronization with a simpler and more resource-efficient method is possible.

The time synchronization method presented in this chapter is implemented by a fully digital circuit for one-way master–slave, hardware-aware, highly precise synchronization. The circuit is designed to perform synchronization in the MAC layer, so that the deterministic part of the clock skew between nodes is kept constant and compensated with a single message exchange. In each sensor node, the synchronization circuit provides a programmable clock signal and a real-time counter for TS.

9.3   Motivation

The development of the synchronization protocol described here was motivated by the requirements of the wearable system presented in previous chapters. Each sensor node is capable of acquiring sEMG signals from electrode pairs and kinematic data from 3D inertial sensors.

Since the relative timing of the acquired data is relevant for subsequent processing, a synchronization mechanism for TS of data is required. Data rates in body area network (BAN) applications are significantly higher than in other sensor networks (e.g., 32 kbps for a 16-bit EMG sensor) [16]. Therefore, the synchronization protocol has to ensure timing accuracy in the range of a few microseconds in a wired, multi-hop network. The main aspects that have been considered in the design of the time synchronization protocol are

- Energy efficiency: Due to the portability requirements of the aforementioned system, its sensor nodes use a small battery for power supply. To ensure an adequate operation time with such a limited energy source, the efficient usage of energy in all parts of the system has to be considered. The communication infrastructure, including transmitter and receiver, consumes a significant amount of energy when they are in active mode. Therefore, the number of control messages and transmitted data packets has to be reduced as much as possible. In the proposed protocol, synchronization is based on one-way message exchange, which consumes less energy than two-way methods. Utilizing a one-way method may increase the processing cost or the time to achieve synchronization, but the performance of this work is not affected by aforementioned disadvantages.
- Accuracy: Unknown delay and latency of messages generally effect on time synchronization. Messages experience different delays while passing between network layers. Increasing the number of involved layers produces a cumulative increase of end-to-end delay. This applies specially to the network layer, because the buffering of messages and the variable service time due to changing network traffic make it hard to estimate message delay and reduce the accuracy. Therefore, many synchronization protocols have been implemented in the MAC layer, to achieve high accuracy and fast synchronization. The current protocol has also been designed for MAC layer uses.
- Simplicity: The limited computational resources available in sensor nodes always constitute a key factor, which has to be considered in the design of both hardware and software. Keeping the communication and data processing delays within a fixed range reduces message processing and the size of the hardware needed to implement the protocol in the MAC layer. In comparison to two-way protocols, because of minimizing delay changes, the proposed method requires less calculations.

A one-way master-slave sender-to-receiver mechanism that satisfies these criteria, while enabling a resource-efficient hardware implementation, is described in the next section.

9.4   Description of the synchronization protocol

This section describes the proposed time synchronization protocol and its intended deployment. Figure 9.1 shows an example of a mesh network of SNs connected to each other by wires (conductive yarns in the intended application). All SNs are equipped with a multi-port network module. Such a network module includes

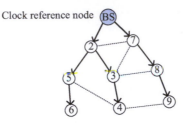

Figure 9.1 A network of sensors with the BS node acting as clock time reference. The bold lines indicate the tree used for synchronization messages

synchronization and packet routing sub-modules working independently of each other. Each device can perform time synchronization with the sender node over one port, while participating in packet routing with the other free ports. Node 1 is the base station (BS) responsible for network management, master clock time reference, and data collection. A routing protocol is needed to route packets from SNs to BS, and to exchange setting, timing and control messages from BS to SNs. Here, the energy-efficient routing protocol from [17] is used, which is based on the combination of source routing and minimum cost forwarding.

Before starting normal activity, the system must go through a setup phase. In this step, the routing protocol finds all minimum-cost paths between BS and SNs, as shown by bold lines in Figure 9.1. The dashed lines are redundancy links that increase the robustness of the system against line breaks caused by wear. Recall that the source routing for minimum cost forwarding (SRMCF) protocol requires that each node determines its neighboring node on the path with minimum cost to the BS (called the *near-node*), and then sends the path information to the BS, who keeps a database of available SNs and the corresponding paths. Synchronization information is transmitted as control messages over the minimum-cost spanning tree built within this phase.

After the network setup, the BS as clock reference synchronizes the network by sending timing information whenever a neighbor SN sends a request. Each SN synchronizes with its *near-node* and rejects timing messages from the other nodes. For instance, node 3 will synchronize with node 2 and reject messages from nodes 7 and 8. Because messages received over different paths experience different delay variations (due to different number of hops, network traffic and packet loss), accepting timing information from various sources would cause message jitter.

End-to-end communication time is usually not deterministic and suffers specially from buffering in the network layer. This protocol is designed for implementation in the MAC layer in order to reduce the effects of delay and the clock time skew. A dedicated hardware transmitter module on each node is responsible for the generation of the timing messages. Also it should be noted, as referred earlier, that to reduce energy consumption, the communication modules are in sleep mode when there is no data to be transferred.

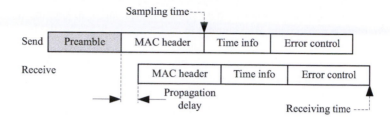

Figure 9.2   Format of the timing information message (MAC layer) and received message at the receiver

Figure 9.2 shows a data frame with a timing message as constructed by the sender and as processed by the receiver. The preamble of a timing message is a string of "1"s, which is used to wake up the receiver before starting the actual data transfer. The MAC header enables the receiver to recognize an incoming timing message. Then the receiver activates a hardware module designed to process the timing information directly without involving the other parts of the receiver. Immediately after sending the MAC header, the master samples the current time, formats the timing information, and sends it. Sampling the time at the sender in this way reduces the delay error due to the message-sending process. Error-control information is put at the end of the message, so that the message integrity can be checked by the receiver.

Delay due to propagation of signals over the link between the nodes is typically very small and negligible (for instance, about 3.3 ns/m for our wearable network). The main delay is due to the difference between the instant when time is sampled at the sender and the message is received. For the data frame shown in Figure 9.2, the transfer takes a fixed number of system clock cycles, because the number of bits inserted in the message after the sampling of time is fixed. Therefore, the receiver always gets messages after a fixed delay (ignoring clock drift and propagation time). If the transmitter adds this fixed number of clocks to the sampled time, then the receiver has all the necessary information to determine exactly the instant when the incoming packet arrived (as measured by the clock of the sender). If the receiver validates the new timing information, then it can set its own time to the received time without further processing.

After synchronizing its time with sender node, each SN sends a timing message to its successors on the clock time tree. For example, in Figure 9.1, node 2 sends timing messages to the nodes 3 and 5. After adjusting their time, node 5 will send timing messages to nodes 5 and 4, and node 3 will send a message to node 4.

We assume that each SN uses the time information to manage a globally synchronized clock signal *Clk-sync*, whose frequency is smaller than the system frequency. A simple way to generate such a clock signal is to use an auto-reload down-counter: when the value of the counter reaches zero, an active transition of *Clk-sync* is generated and the counter is reloaded with a predefined value (equal for all nodes).

In this case, synchronizing *Clk-sync* signals is equivalent to keeping the values of all counters synchronized. The counter in each node is driven by the local system clock, which exhibits some clock skew relative to the system clock of other nodes. To limit the skew between nodes, each SN periodically updates its counter with the value specified in the timing message coming from its reference node.

An SN may also manage a time stamp to annotate any acquired data. We assume that the current time stamp at each SN is determined by an up-counter driven by the globally synchronized *Clk-sync* signal. The time resolution depends on the frequency of *Clk-sync*, which is application dependent and must be selected according to characteristics of the sensed phenomenon. For example, the sampling rate of EMG signals is 1–2 kHz, which is much smaller than the system clock frequency (usually above 1 MHz).

The use of *Clk-sync* ensures that the time stamp counter operates at the same frequency in all nodes. However, it is still necessary to ensure that its contents are the same by using a dedicated message. After starting up, each SN first synchronizes the *Clk-sync* down counter. Then, the time stamp counter contents is synchronized once. Afterwards, the *clock-sync* counter must be synchronized periodically to compensate for the drift of the local system clock.

Under the proposed protocol, the processing of timing messages at the transmitter side is limited to adding a fixed number to the sampled time, and the receiver does no processing to estimate clock time skew. It is important to reduce any processing as much as possible, because the protocol is intended for hardware implementation in an energy-constrained environment.

Elimination of the effects of sending and receiving delays on timing message as described above minimizes the clock time skew between SNs and leads to a high precision synchronization mechanism for wired wearable networks with a simple hardware implementation.

9.5   Analytic characterization of the protocol

This section provides an analytic characterization of the proposed protocol's behavior for use in wired wearable sensor networks. A comparison with PTP is done where appropriate. The aspects addressed are delay and offset bounds, average local and global time skew, as well as probability of synchronization as a function of packet loss.

9.5.1   Instantaneous delay and skew

We assume that all nodes have the same nominal system clock period τ. However, the actual clock period of node i will exhibit a small difference due to random local clock drift p_i, so that, in general, the clock period τ_i at node i is given by

$$\tau_i = \frac{1}{f_i} = \frac{\tau}{1 + p_i} \tag{9.1}$$

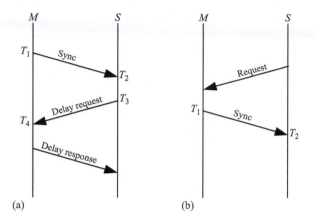

Figure 9.3   Timing diagram: (a) PTP protocol, (b) proposed protocol

The p_i depends on the oscillator that it can be in range of few ppm for a stabilized oscillator with crystal quartz. Consider the multi-hop network of Figure 9.1 with the minimum cost paths forming a spanning tree, and assume that node 1, the BS, is the time reference, which periodically broadcasts its time to other nodes. Figure 9.3(a) shows the synchronization timing diagram for PTP. The time reference node (node 1) starts by sending a message to the client node (node 2) at time $T_M(t_1)$ (master clock time). Node 2 receives it at $T_S(t_2)$ (slave clock time) and replies at $T_S(t_3)$ with a message that is received by node 1 at $T_M(t_4)$. Node 1 calculates the delay and offset according to (9.2) and (9.3) [9,18]:

$$\text{delay} = \frac{(T_S(t_2) - T_M(t_1)) + (T_M(t_4) - T_S(t_3))}{2} \tag{9.2}$$

$$\text{offset} = \frac{(T_S(t_2) - T_M(t_1)) - (T_M(t_4) - T_S(t_3))}{2} \tag{9.3}$$

If the time value included in the *Sync* request is $T_M(t_1)$, then $T_S(t_2)$ is given by

$$T_S(t_2) = T_M(t_1) + s\,\tau_1 + c\,\tau_d + d_\ell + r\,\tau_2 + S_r(t), \tag{9.4}$$

where s is the number of clock cycles used in internal processing after $T_M(t_1)$, c is the number of bits transmitted at data rate period τ_d, d_ℓ is the propagation delay between nodes 1 and 2 (the sum of delays of line and I/O ports), r is the number of clock cycles for message reception at the receiver (with clock period τ_2), and $S_r(t)$ determines the sampling range and is the time difference between signal appearance at the input of the sampling module of receiver and the actual sampling instant. It should be noted that the data rate is not necessarily equal to the system clock frequency of the transmitter. In most of base band communication systems, we have

$$\tau_d = m\,\tau_1, \qquad m \in \mathbb{N}^+. \tag{9.5}$$

The propagation delay $d_{\ell_{1,2}}$ is usually very small in comparison with the clock speed. We will also assume that $d_{\ell_{1,2}} = d_{\ell_{2,1}} = d_\ell$. Parameters c and r are usually fixed, depending on the packet size. Because we assume that the implementation is in the MAC layer, the receiver node processes the message almost immediately after receiving it, i.e., without putting it in a queue, which decreases the uncertainty delay at the receiver. Under these conditions we have

$$
\begin{aligned}
\text{delay} &= \frac{((s+mc)\tau_1 + d_\ell + r\tau_2 + S_r(t))}{2} \\
&\quad + \frac{((s+mc)\tau_2 + d_\ell + r\tau_1 + S_r(t))}{2} \\
&= d_\ell + S_r(t) + \frac{(s+mc+r)\tau}{2}\left(\frac{2+p_1+p_2}{(1+p_1)(1+p_2)}\right).
\end{aligned}
$$
(9.6)

Under the same assumptions, the offset is given by

$$
(s+mc+r)\tau\left(\frac{p_2-p_1}{(1+p_1)(1+p_2)}\right).
$$
(9.7)

Since p_1 and p_2 are very small values, (9.6) simplifies to

$$
\text{delay} \approx d_\ell + S_r(t) + (s+mc+r)\tau.
$$
(9.8)

Since PTP measures time only in multiple units of τ and, for this application, $d_\ell < \tau$ and $S_r(t) < \tau$, calculation of the delay with a round trip message exchange would always produce the constant value $(s+mc+r)\tau$. If there is always a fixed transmission delay, then there is no need to recalculate it. On the other hand, the offset value calculated by (9.7) is very small and returns zero value by a processor or microcontroller. In other words, there is no need to have a round trip message, when the delay is fixed. In this case, a single message (as in Figure 9.3(b)) is enough to send timing information, while achieving the same precision. Therefore a simpler protocol with fewer message exchanges can be used. In the remainder of this subsection, the one-way method used by the proposed approach to achieve minimum clock time skew is analyzed. Here the clock time skew refers to the total time deviation of each node from the time reference.

Although SNs communicate in asynchronous mode, they are implemented as synchronous logic circuits, where nodes process data at the edge of the clock signal. Since the hardware clock generators of each node are independent, two nodes do not have exactly the same clock signal, even without any error or failure. The situation is illustrated by the timing diagram of Figure 9.4, which shows the events at two communicating nodes. The transmitter sends data at the positive edge of the clock and the receiver detects incoming data at the negative edge. In this example, bit b_{n-1} is sent at time t_1, and the receiver will acquire it after $d_\ell + S_r$. The ideal value of $S_r(t)$ is half of the data rate period, so that sampling occurs exactly in the middle of the incoming signal. In practice, due to receiver clock variation and jitter, the sampling point changes around the middle of incoming signal, as shown with gray area in Figure 9.4.

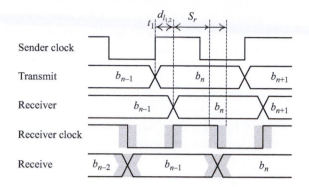

Figure 9.4   Clocks and signals in sender and receiver nodes

Equation (9.4) still applies for a single message (Figure 9.3(b)). The proposed approach determines that the transmitter add n clock cycles to T_M in order to minimize the time skew or error. Assuming $\tau_1 \approx \tau_2$, we have

$$T_S(t_2) \approx T_M(t_1) + n\tau_1, \quad n \in \mathbb{N} \tag{9.9}$$

The integer value of n calculated from (9.4) and (9.9) is

$$n = \left[\frac{(s+mc)\tau_1 + d_\ell + r\tau_2 + S_r(t)}{\tau_1}\right] \approx (s+mc+r) + \left[\frac{d_\ell + S_r(t)}{\tau_1}\right]. \tag{9.10}$$

By adding n to the time included in the packet, the receiver sets its own time to be approximately equal to the time of the reference node:

$$T_S(t_2) \approx T_M(t_2). \tag{9.11}$$

Since t_1 and t_2 are arbitrary times, this approach ensures that (9.11) is always valid.

Using (9.4) and (9.9), the instantaneous time skew between transmitter and receiver is found to be

$$C_S(t) = T_M(t) - T_S(t)$$

$$= (s+mc)\tau_1 + d_\ell + r\tau_2 + S_r(t) - n\tau_1 \tag{9.12}$$

$$= r(\tau_2 - \tau_1) + d_\ell + S_r(t) - \tau_1\left[\frac{d_\ell + S_r(t)}{\tau_1}\right].$$

In comparison with other terms, the value of $(\tau_2 - \tau_1)$ is very small (e.g. 30 ps at 20 MHz with a crystal quartz oscillator) and may be ignored. Therefore, time skew mainly depends on propagation time and sampling range $S_r(t)$.

This previous analysis applies to all nodes that are *one hop* away from the reference node but it should be noted that regardless of the master clock variation or position, the slave node always follows the master clock. So, the above calculation can be applied to all nodes with one hop distance from each other. In this way, the time skew accumulates as the hop count increases. In other words, for a node which is h hops away from the time reference node, time skew under the proposed approach is given by

$$C_S^h(t) = h\left(d_\ell + S_r(t) - \tau\left[\frac{d_\ell + S_r(t)}{\tau}\right]\right). \tag{9.13}$$

where the upper index of C_S^h refers to the number of hops.

The variation of the time skew can be estimated from (9.13) as

$$\frac{\partial C_S^h}{\partial t} = h\left(\frac{\partial d_\ell}{\partial t} + \frac{\partial S_r}{\partial t}\right). \tag{9.14}$$

Signal propagation is almost a constant in a wired wearable network, so the previous expression simplifies to

$$\frac{\partial C_S^h}{\partial t} \approx h\frac{\partial S_r}{\partial t}. \tag{9.15}$$

This means that the time skew deviation is a linear function with $S_r(t)$. So, it is expected that the any variation on the recovered clock at the receiver appeases linearly on the synchronized time. Here, no restriction is made on $S_r(t)$ and it can be any arbitrary synchronization method.

9.5.2 Average time skew

The value of n given by (9.10) depends mostly on constant parameters but varies with the instantaneous value $S_r(t)$. The use of a fixed value of n implies that $S_r(t)$ has to be constrained to a certain range. For baseband communications, the optimal average value of $S_r(t)$ is the middle of the incoming data bit. This condition can be used to calculate a fixed value of n for use in the proposed approach. From (9.5), we have

$$\langle S_r(t)\rangle = \frac{\tau_d}{2} = \frac{m}{2}\tau, \tag{9.16}$$

where $\langle S_r\rangle$ is the average of $S_r(t)$. Using $\langle S_r\rangle$ in (9.10) results in

$$n = (s + mc + r) + \left[\frac{d_\ell + \frac{m}{2}\tau}{\tau}\right]. \tag{9.17}$$

Since $d_\ell \ll \tau$ in wired wearable sensor networks, (9.17) simplifies to

$$n = (s + mc + r) + \left\lceil\frac{m}{2}\right\rceil \tag{9.18}$$

and the average skew is

$$\langle C_S^h\rangle = h\left(d_\ell + \frac{m}{2}\tau - \left\lceil\frac{m}{2}\right\rceil\tau\right). \tag{9.19}$$

This equation gives different results for even and odd values of m:

$$\langle C_S^h\rangle = \begin{cases} h\left(d_\ell + \frac{1}{2}\tau\right), & m \text{ odd} \\ hd_\ell, & m \text{ even} \end{cases} \tag{9.20}$$

For even m, $\langle C_S^h \rangle$ grows as the number of hops and only depends on the signal propagation delay between the nodes; when m is odd, $\langle C_S^h \rangle$ also depends on the system clock period.

The behavior of $\langle C_S^h \rangle$ for odd values of m can be controlled by a modification of the adding strategy. For such a system, the average time skew at nodes that are one hop away from the reference node is

$$\langle C_S^1 \rangle = d_\ell + \frac{1}{2}\tau. \tag{9.21}$$

Now for the second hop, add the value $n - 1$ instead of n to the reference time in the message. The average time skew for nodes that are two hops away becomes

$$\langle C_S^2 \rangle = d_\ell + \frac{1}{2}\tau + d_\ell - \frac{1}{2}\tau = 2d_\ell. \tag{9.22}$$

As a consequence, the average time skew decreases and the value becomes the same as the one for even m. By continuing this procedure of adding n for nodes with odd hop counts and $n - 1$ for nodes with even hop counts, $\langle C_S^h \rangle$ for nodes h hops away from the reference is

$$\langle C_S^h \rangle = d_\ell \sum_{i=1}^{h} i + \sum_{i=1}^{h} \frac{1}{2}\tau(-1)^{1+i}, \quad m \text{ odd}. \tag{9.23}$$

In general, for any $m \in \mathbb{N}^+$, (9.20) and (9.23) give

$$\langle C_S^h \rangle = d_\ell \sum_{i=1}^{h} i + (m \bmod 2) \sum_{i=1}^{h} \frac{1}{2}\tau(-1)^{1+i}. \tag{9.24}$$

It can be concluded that by alternatively changing the additive value, the average time skew for networks with odd m decreases significantly. It should be noted that in any case and independently of m, the instant time skew of (9.15) is valid. Figure 9.5 depicts the $\langle C_S^h \rangle$ values calculated from (9.20) and (9.23) assuming that $d_\ell = 0.1\,\tau$. Using the adjustment included in (9.23), the value of $\langle C_S^h \rangle$ for odd m is near to the values for even m and is much smaller than the unadjusted value obtained from (9.20).

9.5.3   Impact of clock drift and update interval

The validity of the calculated bounds for the clock time skew between nodes requires a process of updating and synchronizing the nodes periodically, since the accumulation of skew produces a clock offset that needs to be accounted for. For that, the appropriate updating moment has to be determined, as the time between updates depends on the expected clock drift.

Immediately after a successful synchronization, (9.15) and (9.24) are valid. To estimate the effects of clock drift, suppose that the local system clocks have a normally distributed frequency with mean value $f_c = 1/\tau$, and that nodes 1 and 2 initially have the same clock offset $T(0)$. The time difference due to clock drift after k clock cycles is

$$\Delta t_{2,1} = k(\tau_2 - \tau_1), \tag{9.25}$$

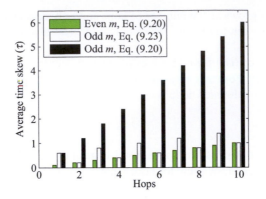

Figure 9.5   Average skew $\langle C_S^h \rangle$ as a function of hop count h

where $\tau_1 = 1/f_1$ and $\tau_2 = 1/f_2$ are the system clock periods of nodes 1 and 2, respectively. Adjusting (9.24) to account for $\Delta t_{2,1}$ results in

$$\langle C_S^h \rangle = d_\ell \sum_{i=1}^{h} i + (m \bmod 2) \sum_{i=1}^{h} \tfrac{1}{2} \tau \, (-1)^{1+i} + \langle \sum_{i=1}^{h} \Delta t_{i+1,i} \rangle. \qquad (9.26)$$

Minimizing the effect of Δt requires reducing the value of k. The best case occurs when the *Sync* message arrives at the receiver just before the active edge of *Clk-sync* is generated.

The necessary synchronization accuracy depends on the application. For a signal sampling rate of 1 kHz, a synchronization accuracy around 100 μs (10% of the sampling rate) may be chosen. In many cases, it will not be necessary to update the time in each period of *Clk-sync*. In general, the update interval can be determined by considering the required accuracy, (9.26) and the clock drift of the oscillators.

9.5.4   Probability of synchronization

The entire process of synchronization is based on the successful reception of the timing messages. Consider the network of Figure 9.1, where SN2 receives its synchronization messages from SN1. The probability of having both nodes synchronized is

$$p_{2,1} = P(t_2 = t_1) = p_{request} \times p_{sync}, \qquad (9.27)$$

where $p_{2,1}$ is the probability of successful message exchange, which is the product of the probabilities for successful delivery of *Request* and *Sync* messages. The probability of message delivery depends on network traffic, topology, and node or link failures.

For successful synchronization of a given node, all previous nodes on the path to the reference node must be synchronized.

Equation (9.27) can then be extended to a node located h hops away from the reference node:

$$p_{h,1} = \prod_{i=2}^{h} p_{i,i-1} \tag{9.28}$$

If the probabilities $p_{request} = p_{sync} = p$ are equal for all links, (9.28) simplifies to

$$p_{h,1} = p^{2(h-1)}. \tag{9.29}$$

Now consider the situation when PTP is used. This protocol uses three packets to complete the synchronization. So, the (9.30) for SN2 will be

$$p_{2,1} = p_{sync} \times p_{request} \times p_{response}, \tag{9.30}$$

where $p_{response}$ denotes the probability of a successful *Response* packet message transmission. Assuming that all messages have the same probability p of being successfully transmitted, (9.29) can be rewritten for PTP as

$$p_{h,1} = p^{3(h-1)}. \tag{9.31}$$

A comparison of (9.29) and (9.31) shows that, for the application in wearable networks, PTP is more prone to failure in the synchronization process than the protocol proposed in this work, because PTP requires more messages to be transmitted.

9.5.5   General protocol operation

The aim of the proposed synchronization protocol is to ensure that a globally synchronized clock signal *Clk-sync* is available at every sensor node. The main purpose of this signal is to be used for synchronized sampling. In addition, each node maintains a counter that can be used to time stamp the acquired signals.

The simplest way to generate *Clk-sync* is to use an auto-reload counter: when the value of a down counter (called *TC* in the following description) reaches zero, an active transition of *Clk-sync* is generated and the counter is reloaded with a predefined value (called *TCPR*). The value of *TCPR* in all SNs must be the same.

Having all *Clk-sync* signals synchronized is equivalent to say that all *TC* have to be synchronized. The counter *TC* is driven by the local system clock, which exhibits some clock skew relatively to the system clock of the other nodes. The procedure described in the previous subsection can be used to keep the average value of the skew within bounds. For this, each SN periodically updates its *TC* value with the *TC* value received from the timing message.

The current time stamp at each SN is maintained by an up counter *TS* (TS-counter), which is driven by the synchronized clock signal *Clk-sync*. The time resolution of *TS* depends on the frequency of *Clk-sync*, which must be chosen according to the sensed phenomena: e.g., the electromyography (EMG) signals of [19] have frequencies below 500 Hz, so a sampling rate of at least 1 kHz is necessary.

It should be noted that using *Clk-sync* for the *TS*-counter only ensures that it counts synchronously. It is still necessary to synchronize the contents of *TS* at the start of the session with a timing message that includes the global current value of *TS*.

Figure 9.6   Sequence of MAC messages for clock synchronization generated by an SN

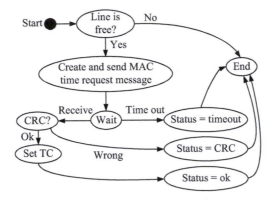

Figure 9.7   Sending the Request message and receiving Sync messages (slave node)

Because each SN synchronizes its *Clk-sync* periodically, it is not necessary to do the same for *TS*: after a one-time *TS* synchronization, TS will stay synchronized while the system is running.

The process is summarized in Figure 9.6. After starting, an SN first updates the *TC*-counter for synchronization of the local *Clk-sync* signal; next, the *TS*-counter is synchronized only once. Counter *TC* must then be synchronized periodically to compensate for the drift of the local system clock.

The protocol defines that synchronization starts when the slave node sends a timing *Request* message to the master node. Note that the same node can act as a slave (to synchronize its own clock signal) and as a master for several other nodes. The synchronization process executed in slave mode is shown in Figure 9.7. The node starts by checking the availability of the connection to the master. If it is free, the slave node creates a *Request* in the MAC layer and sends it to the master. The request message specifies whether the request is for *TC* or *TS* synchronization. The slave node will then update its counters with the information received in the *Sync* message from the master node. The status of the received *Sync* message is saved in the *Status* register, so that it can be used in higher level processing by a microcontroller or microprocessor. For example, the value *Status = CRC* indicates that the *Sync* message was received with a cyclic redundancy check (CRC) error. So, the timing information in the message is not valid and the receiver has to try again.

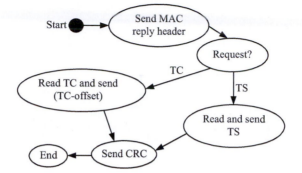

Figure 9.8   Replying to a Request message with a Sync message (in master node)

Master mode operation depends on the type of the *Request* message, as shown in Figure 9.8. When a time stamp value is requested, the master simply replies with a message containing its own *TS* value. For a clock counter value request (*TC*), the master replies with the value of the difference *TC-Offset*, where *Offset* is the fixed value used to compensate the deterministic part of delay, which is fixed and equal for all sensor nodes. For the prototype described in Section 9.7, this value is set by software as part of the node's initialization sequence.

9.6   The synchronization circuit

This section describes a synchronization circuit that is based on the method just described. The circuit is part of the fully digital sensor node communication system, which has been implemented both in an Actel low-power field programmable gate array (FPGA) and in a CMOS application-specific integrated circuit (ASIC).

Figure 9.9 shows the block diagram of the circuit including all necessary modules to send and receive timing information. The three modules in gray (*Signal Detector, TX Line SW and Line Driver*) do not belong to the synchronization circuit, but they are involved in the communication process. An 8-bit internal system bus connects the module to the system core, which controls the time synchronization circuit and reads or writes to the registers (e.g., the real-time time stamp). A 16-bit bus is used to connect the internal modules and to access the two counters *TC* and *TS*. The signal *Clk-sync* is available at one of the output pins. The synchronization circuit also manages a real-time time stamp that is available via the internal bus. The operation of each module is described next.

9.6.1   Control module

The *Control* module manages all the activities related to time synchronization. It is connected to the system core by the internal 8-bit bus and also to the *Signal Detector*.

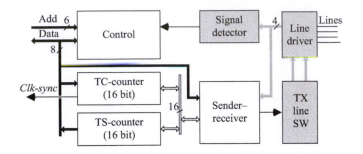

Figure 9.9   Block diagram of the circuit

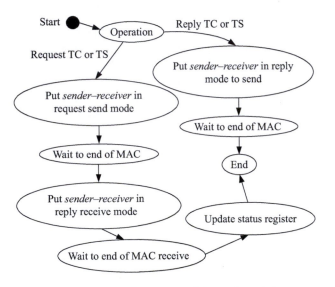

Figure 9.10   Operation of Control module

All the register values can be read or written via this module. The main operation of *Control* is shown in Figure 9.10.

Normally, this module is in sleep mode unless one of the following events occurs:

1. Incoming timing message: The SNs use several kinds of MAC messages, with timing synchronization being just one of them. The detection of a timing request message from an adjacent SN is carried out by the *Signal Detector* module. This module is responsible for determining the type of the MAC message by decapsulating and processing the MAC header, and forwarding the rest of incoming data to the destination module (which is the synchronization circuit for timing

messages). For timing messages, the MAC header also determines whether the request is for updating the *TC*- or the *TS*-counter. The *Control* module receives the timing message directly from the *Signal Detector* module and creates the appropriate reply message.

2. Local command: The request procedure for either *TC* or *TS* can be started by the node's microcontroller with a command sent via the SPI port to the *Control* module. The latter sets up the synchronization module for sending and receiving timing information and starts the transmission. After sending the message, the *Control* module sets up the *Sender–Receiver* module to directly receive the MAC reply message.

9.6.2   TC-counter and TS-counter modules

The *TC-counter* module shown in Figure 9.11(a) generates the clock signal *Clk-sync* for TS and is used as a synchronized clock by other parts of the sensors. Whenever the 16-bit down counter *TC* reaches zero, it is reloaded with the value of the 16-bit *TCPR* register. The periodic *Reload* signal is used to drive the clock generation circuit *Clock_gen*, which produces the reference clock signal *Clk-sync*. As mentioned before, this clock signal is used as input clock for the *TS-counter* module and is also available for use as clock reference in other parts of the sensor node. The output of this module is a pulse signal with duration of 1 or 32 system clock cycles. The wider pulses are necessary for some circuits. For example, when *Clk-sync* is used as an external interrupt of a microcontroller, a short pulse width may not be sufficient to generate a valid interrupt.

The wearable system in [19] is designed to capture EMG signals using a 1 kHz sampling frequency. To generate the *Clk-sync* reference for the EMG signal sensing part, from a 20 MHz system clock, the *TC*-counter must have at least a 16-bit, which is the length chosen for the current implementation. Therefore, the value of *TCPR* must be in range 1–65,535 (33–65,535 for the wide pulse mode). The frequency f_{sync} of *Clk-sync* is given by

$$f_{sync} = \frac{f_s}{1 + \text{TCPR}}, \tag{9.32}$$

where f_s is the frequency of system clock. For $f_s = 20$ MHz, the value of f_{sync} is within the range 300 Hz to 10 MHz.

Figure 9.11(b) shows the *TS-counter* module. This module includes *TS* and generates a time stamp that can be read by the system, or even written, via a 16-bit internal bus. As for *TC*, other SNs are able to obtain the value of *TS* for updates of their own *TS*-counter. The time interval between *TS* overflows is

$$T_{ov} = \frac{65536}{f_{sync}}. \tag{9.33}$$

In fact, *TS* does not count the time ticks continuously, instead it is periodically reset every time interval T_{ov} (when overflow occurs). However, this will happen simultaneously in all nodes, which means that the information on relative event occurrence,

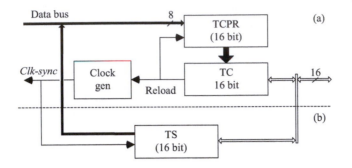

Figure 9.11   Counters: (a) TC-counter, (b) TS-counter

between nodes, is always preserved. The BS can keep the absolute time count for all events.

9.6.3   Sender–Receiver module

In our context, synchronizing two SNs is equivalent to ensuring that their *TC*-counters have identical values at the same time. This ideal situation cannot be granted at all times, but a small bound on the difference between the counters may be enough for practical purposes. This is achieved by an exchange of special messages between the nodes. All processing of timing messages is performed by the *Sender–Receiver* module. It may operate in one of three different modes:

- Request for timing information (Figure 9.12): Start the synchronization process by sending a request message.
- Reply to a request (Figure 9.13): Reply to a timing request message.
- Reception of timing message (Figure 9.14): Receive timing information in response to a request message.

9.6.4   Timing request message

In order to illustrate the operation of *Sender–Receiver* module, we will consider the situation where node SN3 in Figure 9.1 needs to synchronize its *TC* value with its *near-node* SN2.

The process is started by the upper layers of the control software of node SN3 by sending a command (via the internal bus) instructing the *Control* module to activate the request mode. The *Control* modules instructs the *Sender–Receiver* module to start communication with SN2. The configuration of the module in this mode is shown in Figure 9.12.

The *Control* module generates both of the *EN_req* and *Mac_sel* signals: the first enables the request mode and the second specifies that the request concerns the value

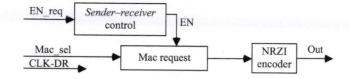

Figure 9.12   *Configuration of Sender–Receiver to send a request*

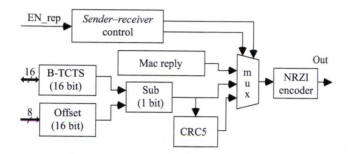

Figure 9.13   *Configuration of sender–receiver for reply to a timing information request*

of *TC*. Based on the request type (which could be *TC* or *TS*), the *Mac_Request* module generates the appropriate MAC message, which must then be encoded for transmission. The line encoding scheme used in our implementation is non-return to zero inverted (NRZI). Therefore, the output of *mac request* is encoded before driving the line. The module *TX SW* (Figure 9.9) forwards the MAC message to the line connected to SN2. At the end of the process, the *Control* module changes the configuration of the *Sender–Receiver* module to reception mode and waits for a reply from SN2.

9.6.5   *Timing reply message*

When the *Signal Detector* module of SN2 detects the timing message request, it forwards the incoming data to the synchronization circuit. The *Control* module enables the reply mode by asserting the *EN_rep* (Enable-reply) signal. In this mode, node SN2 must send the value of its *TC*-counter to SN3, after performing an adjustment to account for the offset between the nodes. The configuration of the *Sender–Receiver* module is the one shown in Figure 9.13. In that figure, *B-TCTS* is a 16-bit register for buffering the sampled value of *TC* in node SN2 at the start of reply processing. The 16-bit register *Offset* is used to memorize the offset value, which will be used to compensate for the skew introduced by the communication delay. The value of *Offset* can be managed by the upper software layers through the *Control* module. Module *Sub* is a serial subtracter that is used to subtract *Offset* from the *TC* value buffered in *B-TCTS*. According to (9.18), the *Offset* value must be added to the *TC* value that is sent back to SN3. However, *TC* is a down-counter, so the *Offset* value must be subtracted instead. It should be noted that a 1-bit serial subtracter can be used to perform the subtraction of two 16-bit values, since communication is serial and the subtraction can be performed during reception.

To protect and check the validity of the data at the receiver node, the *CRC5* module generates a 5-bit CRC, which is concatenated to the outgoing data. A multiplexor controlled by *Send–Receive Control* selects, in order, the *Mac reply*, *Sub* and *CRC5* modules to build the full reply message at the output, which goes through the NRZI encoder before being sent to node SN3.

9.6.6   Reception of timing messages

As mentioned previously, node SN3 changes the configuration of the *Sender–Receiver* module after sending the timing request message. The configuration for reception is shown in Figure 9.14. In this case, register *B-TCTS* acts as a Serial-In-Parallel-Out buffer to receive the *TC* value. After the *CRC5* module confirms the validity of the received data, the value buffered in *B-TCTS* is loaded to the *TC*.

The processor for updating *TS* is the same as the one used for *TC*, with the only difference that, in the reply step, the value of *TS* will be sent without any adjustment. As mentioned in Section 9.5.5, *TS*-counters inherit synchronization from *Clk-sync* signals; exchanging *TS* values is necessary only for time stamp alignment.

9.6.7   Implementation characteristics

The time synchronization module in the fabricated IC is plotted in Figure 9.15. The box labeled *TS* indicates the approximate area occupied by the synchronization circuit. It uses 971 cells (765 gates and 206 flip flops) and occupies 0.1388 mm^2 (19% of IC), a number which is significantly smaller than the 15,115 cells required by the hardware implementation of the more complex IEEE-1588 protocol reported in [20].

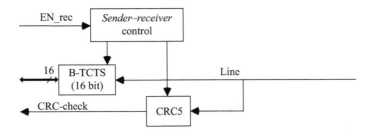

Figure 9.14   Configuration of sender–receiver to receive timing message

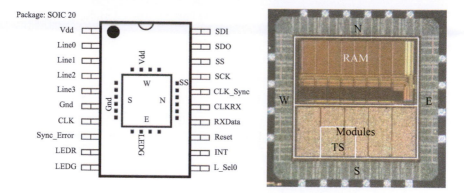

Figure 9.15   *CMOS ASIC 0.35 μm microphotograph with indication of the synchronization circuit (TS)*

9.7   Experimental results

This section presents experimental results obtained with the IC version of the circuit. The system clock frequency of each node is 20 MHz, which results in a data rate of 10 Mbps. With this system clock frequency, the synchronized clock can be set in the range 305 Hz–10 MHz. The network of SNs used for the measurements is composed by the nodes BS, SN2, SN3, SN4, SN5, and SN6 of Figure 9.1.

9.7.1   *Physical layer signaling*

Signals generated in the physical layer during message exchange (request and reply) for synchronization of the *TC* are shown in Figure 9.16. As mentioned before, the line between the nodes is bidirectional, so the reply message appears on the same line immediately after the request message. The least significant bit (LSB) of the adjusted *TC* value is sent first; the message trailer consists of the CRC check bits.

For the evaluated setup, the entire process, from sending the request message to the end of the reply message, takes 5.2 μs. To keep network nodes synchronized, it is necessary to exchange timing messages periodically, which requires the utilization of system resources, such as channel time, and should be taken into account. For an interval between timing messages of 1 ms only 0.52% of the channel time will be used for synchronization, a very small percentage of the total traffic on the network.

Figure 9.17 presents part of a logic simulation showing the *TS* synchronization of SN3 and SN4 for a scenario in which *Clk-sync* is configured to operate at 81 kHz. The simulation shows precisely when the changes of the *TS*-counter occur. This internal information cannot be obtained by observing the external IC signals. The final *TS* value can be read via the SPI port, but its instantaneous value is not externally accessible. In the figure, *L* indicates the state of the communication line between the

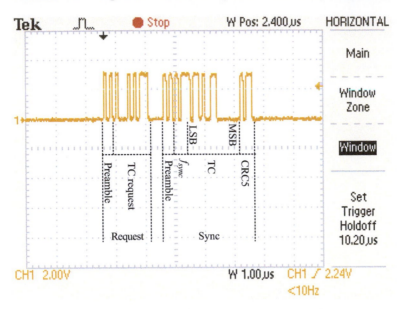

Figure 9.16   *Physical layer signals during timing message exchange between SN2 and BS*

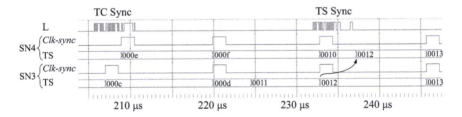

Figure 9.17   *Simulation showing synchronization of Clk-sync signal and TS-counter*

two nodes. For each of the nodes, the simulation shows the *Clk-sync* signal and the contents of the *TS*-counter.

At the beginning of the simulation, both *Clk-sync* (generated by *TC*) and *TS* are out of synchronization. The synchronization of *TC* starts at $t = 205.7$ μs, when node SN4 sends a *Request* message to SN3. The round trip message takes 5.2 μs to complete (ending at $t = 210.7$ μs). The receiver takes 50 ns (one system clock cycle) to check the validity of the message and loads the receiver's *TC*-counter at $t = 210.7$ μs, resulting in a synchronized *Clk-sync* at $t = 220$ μs.

To synchronize the time stamp *TS*, node SN4 sends a request at $t = 232$ ns. Node SN4 replaces its *TS* value with the value received from SN3 (which is 0x0012) at time $t = 237.2$ μs. So, after 237.2 μs, both nodes are synchronized. A further

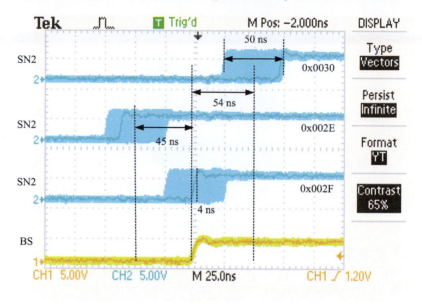

Figure 9.18   Measured one-hop clock skew for offset = 0x002E, 0x002F, and 0x0030

noteworthy aspect of the simulation is that during the previous process node, SN3 has synchronized its *TS* value with node SN2 at $t = 224.5$ µs (new value 0x0011).

9.7.2   One-hop and multi-hop clock skew

The one-hop clock skew between SN2 and BS for various values of the parameter *Offset* can be seen in Figure 9.18. The oscilloscope signals shown in this figure are generated by using the "infinite time persist" display mode. The system clock of the BS is used as reference in all skew measurements. The offset value for compensating the constant delay in the proposed circuit was experimentally measured to be $s = 0x0030$. The observed clock skew variation range is 50 ns, that is one period of the system clock, as described by (9.12) and (9.13). The measured average one-hop clock skew is 54 ns, i.e., the sum of signal propagation delay and $S_r(t)$. The measurements confirm that the synchronization circuit is able to keep the clock skew in the bounds defined by theoretical analysis.

By setting the *Offset* parameter to $s' = s - 1 = 0x002F$, the *Clk-sync* shifts one clock cycle to the left, causing to decrease the skew. In the setup used for the measurements, the average clock skew is 4.6 ns, which is much smaller than the total clock skew variation of 54.6 ns. This value is the minimum skew achievable by the proposed method and shows the feasibility of minimizing the average clock skew by selecting appropriate values for the offset. The effect of using the offset value 0x002E is also shown in Figure 9.18, with one more clock period shift as expected. So, by changing the offset value, it is possible to shift the *clock-sync* in both directions by an

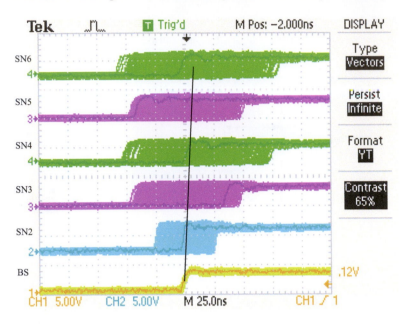

Figure 9.19   Measured multi-hop clock skew with Offset = 0x002F

arbitrary value. Regardless of the offset value and its effect, the peak-to-peak clock jitter is 50 ns.

The clock skew increases with the number of hops. Figure 9.19 depicts the clock signal at different nodes (the offset value, 0x002F, is the same for all the nodes). As can be seen, and as expected, the average clock skew increases by almost 4.6 ns as the number of hops increases. In agreement with (9.13), the clock skew variation range also increases: 50 ns at SN2, 100 ns at SN3 and SN5, and 150 ns at nodes SN4 and SN6. The comparison between the measured and calculated values is shown in Figure 9.20, confirming the correct operation of the circuit.

9.7.3   Power consumption

Figure 9.21 depicts the measured current consumption of the circuit as a function of the *Clk-sync* frequency from 305 Hz to 5 MHz. The current increases linearly from 0.18 to 0.64 mA as the frequency increases. With *Clk-sync* = 1 kHz, the total current is almost 0.18 mA. Most of the current is used to drive the *Clk-sync* pin that is available as an output of the ASIC.

9.7.4   Effects of timing message interval and failure on synchronization

In many applications, message failure is inevitable and it may happen for several reasons such as collisions and high traffic load. To evaluate the effect of message failure on synchronization, the clock skew between BS and SN2 has been measured

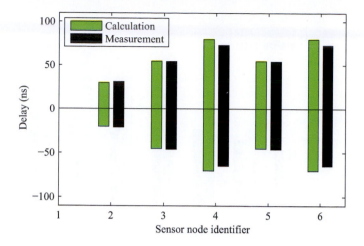

Figure 9.20   *Comparison between the measured and calculated values for multi-hop clock skew with offset = 0x002F*

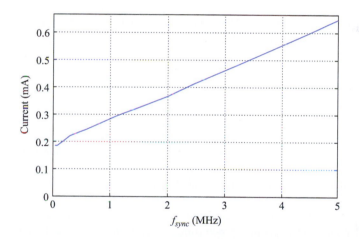

Figure 9.21   *The circuit current consumption for f_{sync} up to 5 MHz*

as a function of the number of consecutive missing messages. The results are shown in Figure 9.22 for two different *Clk-sync* frequencies. In both cases, the timing update messages are sent just before the rising edge of *Clk-sync*. Therefore, the interval between updates used for the 2 kHz clock signal is half the interval used for the 1 kHz case. Therefore, the clock skew for the latter case is larger than the skew for the 2 kHz case.

According to (9.25), the linear growth of clock skew depends on the clock frequency difference between the nodes. In the absence of failures, the synchronization

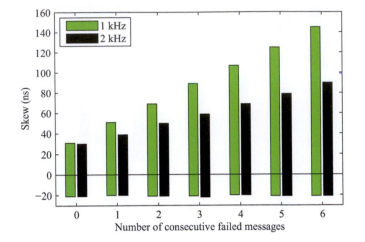

Figure 9.22   One-hop skew after consecutive timing message failures

accuracy is within the range [−21 ns; 29 ns], as expected. The observed increase of the clock skew in the presence of failures means that, in this case, $\Delta t > 0 \Rightarrow \tau_S > \tau_M$. This means that the upper (positive) bound of the skew grows with the number of missing messages, while the lower (negative) bound remains the same. For situations where $\Delta t < 0$, it is the lower bound that becomes increasingly more negative. Regardless of the clock skew, a successful synchronization message puts the skew of the receiver back in the range [−21 ns; 29 ns].

If a line between two nodes disconnects or breaks, we expect the skew of the disconnected segment to grow without limit, but the synchronization between nodes in each segment to be maintained. To confirm this behavior, nodes SN2 and SN3 were disconnected at $t = 37$ s in order to obtain the results for average clock skew shown in Figure 9.23. Until the disconnection occurred, all nodes remained synchronized. At $t = 37$ s, SN3 starts running free and its skew starts to increase as expected. The nodes on the disconnected segment get their time reference (directly or indirectly) from SN3 and stay synchronized with it. The clock skew after the disconnection depends on the clock difference between BS and SN3. In the present case, the positive slope of the skew curve is due to the fact that the clock of SN3 has a somewhat higher clock frequency than the clock of BS. If the clock of the disconnected node has a lower frequency, the slope of the skew curve will be negative.

The periodic exchange of timing messages generates additional traffic. Measurements show that with a 1 ms interval between synchronization messages, 0.52% of the channel bandwidth is used for synchronization, which is almost negligible. This allows the system to achieve high precision with an average clock skew below 5 ns. Some BAN applications may not need such a high precision. If precisions in the range of a few microseconds is enough, the time interval between messages can be larger (in the range of a second), consuming even less energy and channel bandwidth.

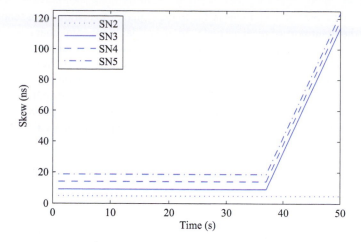

Figure 9.23   Average clock skew after disconnection of the line between SN2 and SN3 @37 s

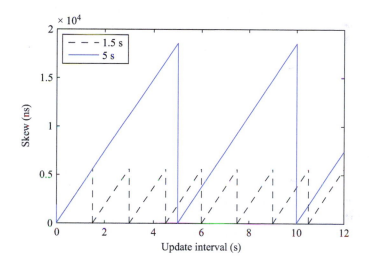

Figure 9.24   Clock skew for update interval of 1.5 and 5 s, Clk-sync = 1 kHz

Figure 9.24 shows the measured skew between BS and SN as a function of time for two update intervals: 1.5 and 5 s. The skew depends on the clock frequency difference between the nodes, which has been measured to be 3.7 ppm. Since the clock frequency of each node stays constant, the accumulation of the small difference grows linearly, reaching 18.5 μs (for a 5 s interval) and 5.58 μs (for a 1.5 s interval). In both cases, the absence of synchronization message for a long time results in significantly higher skew. Nevertheless, a message synchronizes *Clk-sync* immediately.

The measured value of 4.6 ns is due to the signal propagation from sender output node to the input port of receiver node, an unavoidable inherent characteristic of the communication path (I/O pads and connecting yarn). In resource-limited systems, such as sensor networks, achieving a smaller clock skew is difficult in practice, even if more sophisticated two-way methods are used, as it would require more processing and a more complex implementation. Therefore, the low-overhead approach described here makes it practical for many wearable systems to achieve a very-low skew in range of sub-microsecond (4.6 ns for the wired, yarn-based application used for evaluation) with a one-way method and without any further processing.

9.8 Conclusion

The circuit described in this chapter was designed for establishing time synchronization between sensor nodes of the wearable part of the system. The synchronization is based on one-way master-to-slave message exchange implemented in the MAC layer, in order to avoid the nondeterministic delays caused by data processing and buffering in the higher levels of the protocol stack. By directly sending and processing the timing information without buffering, the proposed approach leads to an average clock skew of a few nanoseconds. The circuit generates two synchronized values: a programmable clock signal and a real-time counter for TS purposes.

Experimental evaluation with IC implementation obtained an average one-hop clock skew of 4.6 ns which is the time required for signal propagation from sender output to the receiver input. Based on theoretical calculations, in a multi-hop network, the global average time skew grows linearly with hop count; this is supported by the experimental results.

The low skew values provided by this approach satisfy the requirements of many BAN applications. Even for networks whose nodes are 10 hops away from the time reference node, the average global skew will typically be under 50 ns. A value of ten hops exceeds the largest internode distance of many, if not all, existing wearable systems. The proposed circuit achieves the maximum synchronization performance that could be achieved by PTP, but with fewer timing messages and calculations, less complexity, and better energy efficiency.

References

[1] Ranganathan P., Nygard K. 'Time Synchronization in Wireless Sensor Networks: A Survey'. *International Journal of on Pervasive and Ubiquitous Computing (UbiComp)*. 2010;1(2):92–102.

[2] Sivrikaya F., Yener B. 'Time Synchronization in Sensor Networks: A Survey'. *IEEE Journal of Network*. 2004;18(4):45–50.

[3] Lasassmeh S., Conrad J. 'Time synchronization in wireless sensor networks: A survey'. *IEEE Conf. on SoutheastCon*; 2010. pp. 242–245.

[4]   Wu Y.C., Chaudhari Q., Serpedin E. 'Clock Synchronization of Wireless Sensor Networks'. *IEEE Magazine on Signal Processing*. 2011;28(1):124–138.

[5]   Rhee I.K., Lee J., Kim J., Serpedin E., Wu Y.C. 'Clock Synchronization in Wireless Sensor Networks: An Overview'. *Journal of Sensors*. 2009;9(1):56–85. Available from: http://www.mdpi.com/1424-8220/9/1/56.

[6]   Sundararaman B., Buy U., Kshemkalyani A.D. 'Clock Synchronization for Wireless Sensor Networks: A Survey'. *Journal of Ad Hoc Networks*. 2005;3:281–323.

[7]   Leng M., Wu Y.C. 'Low-Complexity Maximum-Likelihood Estimator for Clock Synchronization of Wireless Sensor Nodes Under Exponential Delays'. *IEEE Journal of Signal Processing*. 2011;59(10):4860–4870.

[8]   Elson J., Girod L., Estrin D. 'Fine-grained network time synchronization using reference broadcasts'. *5th Symp. on Operating Systems Design and Implementation (SIGOPS)*; 2002. pp. 147–163. Available from: http://doi.acm.org/10.1145/1060289.1060304.

[9]   Lee K., Eidson J. 'IEEE-1588 standard for a precision clock synchronization protocol for networked measurement and control systems'. *34th Annual Precise Time and Time Interval (PTTI) Meeting*; 2002. pp. 98–105.

[10]   Mills D. 'Internet Time Synchronization: The Network Time Protocol'. *IEEE Journal of Communications*. 1991;39(10):1482–1493.

[11]   Ganeriwal S., Kumar R., Srivastava M.B. 'Timing-sync protocol for sensor networks'. *1st Intl. Conf. on Embedded Networked Sensor Systems (SenSys)*; 2003. pp. 138–149. Available from: http://doi.acm.org/10.1145/958491.958508.

[12]   Lenzen C., Sommer P., Wattenhofer R. 'Optimal clock synchronization in networks'. *7th ACM Conf. on Embedded Networked Sensor Systems (SenSys)*; 2009. pp. 225–238. Available from: http://doi.acm.org/10.1145/1644038.1644061.

[13]   Novick A.N., Lombardi M.A. 'Practical limitations of NTP time transfer'. *Frequency Control Symposium & the European Frequency and Time Forum (FCS), 2015 Joint Conference of the IEEE International*; IEEE; 2015. pp. 570–574.

[14]   Maróti M., Kusy B., Simon G., Lédeczi A. 'The flooding time synchronization protocol'. *2nd Intl. Conf. on Embedded Networked Sensor Systems (SenSys)*; 2004. pp. 39–49. Available from: http://doi.acm.org/10.1145/1031495.1031501.

[15]   IEEE. *IEEE Standard for a Precision Clock Synchronization Protocol for Networked Measurement and Control Systems*. IEEE Std 1588-2008 (Revision of IEEE Std 1588-2002); July 2008, pp. 1–300.

[16]   Latré B., Braem B., Moerman I., Blondia C., Demeester P. 'A Survey on Wireless Body Area Networks'. *Journal of Wireless Networks*. 2011;17(1):1–18.

[17]   Derogarian F., Ferreira J.C., Tavares V.M.G. 'A routing protocol for WSN based on the implementation of source routing for minimum cost forwarding method'. *5th Intl. Conf. Sensor Technologies and Applications (SENSORCOMM)*; ThinkMind; 2010. pp. 85–90.

[18]   IEEE. *IEEE Standard for a Precision Clock Synchronization Protocol for Networked Measurement and Control Systems*. IEEE; 2008.

[19]   Zambrano A., Derogarian F., Dias R., *et al.* 'A wearable sensor network for human locomotion data capture'. *9th Intl. Conf. on Wearable Micro and Nano Technologies for Personalized Health (pHealth)*; 2012.

[20]   Park J.W., Hwang J.H., Chung W.Y., Lee S.W., Lee Y.S. 'Design time stamp hardware unit supporting IEEE 1588 standard'. *Intl. Conf. on SoC Design (ISOCC)*; 2011. pp. 345–348.

Chapter 10
Wearable sensor networks for human gait

José Machado da Silva[1], Fardin Derogarian[2],
João Canas Ferreira[1], and Vítor Grade Tavares[1]

A new wearable data capture system for gait analysis is being developed. It consists of a pantyhose with embedded conductive yarns interconnecting customized sensing electronic devices that capture inertial and electromyographic signals and send aggregated information to a personal computer through a wireless link. The use of conductive yarns to build the myoelectric electrodes and the interconnections of the wired sensors network as well as the topology and functionality of the sensor modules are presented.

10.1 Introduction

The average life expectancy at birth across the OECD (The Organisation for Economic Co-operation and Development) countries has reached 80 years [1]. In order that this longer life can be lived with adequate quality, it is imperative to find economic and efficient healthcare procedures that allow for greater patients' autonomy in outpatient care. The conventional health system centred in the hospital, focused on reaction to illness and malaise, is changing to a more customized care, with a focus on prevention, detection of risk factors and early treatment [2,3]. This paradigm shift has been raising the interest for wearable electronics and wearable health systems, often comprising body sensor networks (BSNs).

However, for home-care solutions being widely adopted, it is necessary to increase their acceptability by the patients who, as final users, need to interact with them for long times. Under this perspective, wearable systems built on textile platforms represent a convenient solution for outpatient care, provided these are comfortable, non-invasive and require no technological skills [4]. These can be used as a clinical tool applied in the rehabilitation and diagnosis of medical conditions, as well as in the monitoring of sport activities.

[1] Departamento de Engenharia Eletrotécnica e de Computadores, Faculdade de Engenharia da Universidade do Porto, INESC TEC, Portugal
[2] Instituto de Telecomunicações and Departamento de Engenharia Eletromecânica – Universidade da Beira Interior; Faculdade de Engenharia, Portugal

Both wireless and wired networks can be found in BSN applications [5]. Solutions based on wireless communications are often preferred for their flexibility in terms of sensor nodes (SNs) deployment and scalability, but these imply higher power consumption, which is a critical aspect on a system where the number of batteries and bulky components is to be kept to a minimum [6]. Wired networks provide higher speed, better reliability and allow for lower power consumption communications [7].

One type of information that commonly needs to be monitored is patients' physical activity correlated with bio-signals that provide feedback on the respective health condition, namely electromyogram (EMG) and electrocardiogram (ECG) signals. In fact, gait impairments are among the most common limitations in the elderly and, in general, reduced mobility markedly impairs the quality of humans' lives at any age.

This chapter addresses the development, implementation and evaluation of a wearable system for gait analysis and monitoring of human physical activity. It consists of a wearable textile substrate (leggings) with embedded conductive yarns interconnecting custom electronic devices in a mesh network that acquire EMG and inertial signals. All data are aggregated in a central processing module (CPM), where are then sent via a wireless link to a smartphone or personal computer for final processing.

Next section provides a brief overview on human gait, modelling and characterization. Section 10.3 presents gait observation systems commonly used today and reports different research efforts that have been done to develop wearable gait observation systems. These have been made possible with the availability of conductive yarns (Section 10.4) that allow implementing electrical conductors in textile substrates, as well as of electrodes for bio-signals. The ProLimb BSN is then presented in Section 10.5. The network topology, SNs architecture and results obtained with first prototypes are presented. This BSN is meant to be implemented on a fabric substrate with the purpose of capturing relevant data in a practical and non-invasive way, even for people with strong impairments or disabilities.

10.2   Human gait

Walking (bipedal gait) is the natural exercise most commonly performed by humans to move from one position to another. It is an activity consisting of a sequence of steps with a period frequently defined from the heel contact of one leg to the next heel contact of the same leg. The normal gait cycle, accomplished by a healthy person in a stride interval being all strides very similar among them, comprises two main phases, the stance phase (performed over about 60% of the gait cycle) and the swing phase (over the remaining 40%) [8].

For characterization and evaluation purposes, the gait cycle can be further subdivided (Figure 10.1). The following subdivision in eight subsequent phases is commonly accepted.

- Stance phase (60%) – When a foot strikes the ground and ends when the foot is lifted
    Weight acceptance
    – Initial contact/heel strike (0%–2% of gait cycle);
    – Loading response/foot flat (2%–10% of gait cycle);

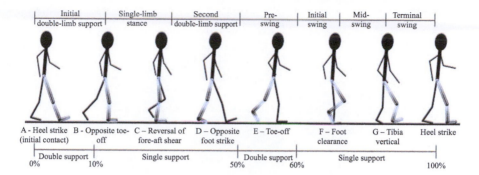

| Initial double-limb support | Single-limb stance | Second double-limb support | Pre-swing | Initial swing | Mid-swing | Terminal swing |

A - Heel strike (initial contact)   B - Opposite toe-off   C – Reversal of fore-aft shear   D – Opposite foot strike   E – Toe-off   F – Foot clearance   G – Tibia vertical   Heel strike

Double support 0%–10%   Single support   Double support 50%–60%   Single support   100%

Figure 10.1   Gait cycle of a healthy subject Reference [9]

Single limb support
– Mid-stance (10%–30% of gait cycle);
– Terminal stance/push-off (30%–50% of gait cycle);
• Swing phase (40%) – initiated when the foot is in the air and advances to the next-heel contact position
Limb advancement
– Pre-swing (50%–60% of gait cycle);
– Initial swing (60%–70% of gait cycle);
– Mid-swing (70%–85% of gait cycle);
– Terminal swing (85%–100% of gait cycle).

Gait is a complex and multifactorial phenomenon of the nervous and musculoskeletal systems, involving the coordination of different joints, muscles, ligaments, neuronal activity and afferent modulation, with the purpose of obtaining a balanced walking (Figure 10.2). It is commanded by spinal pattern generators, under regulation by supraspinal control. Evidence has been found that the basic neural networks, or central pattern generators (CPGs), involved in the generation of locomotion evokes are placed in the spinal cord [10,11]. In spite the CPG networks realize simple functions, still they are adaptable to changes of environmental demands. The analysis of evoked cerebral potentials has allowed to find that during gait group I muscle afferents (muscle fibre spindles that monitor how fast a muscle stretch changes) are blocked at both segmental and supraspinal levels, being polysynaptic reflexes involved in the compensation of perturbations introduced [12,13].

The gait activity is defined with space-time and force (inner and outer body) parameters that characterize the kinematics, kinetics and energetics of the movement [14,15].

Kinematics of locomotion are characterized by positions, angles, velocities and accelerations of body segments and joints during motion. Stride length, stride time, cadence and speed are commonly used parameters. Stride length is the distance between one heel-strike to the next of the right foot in the walking sense; stride time is the time required to complete the stride length, i.e. the duration of the gait

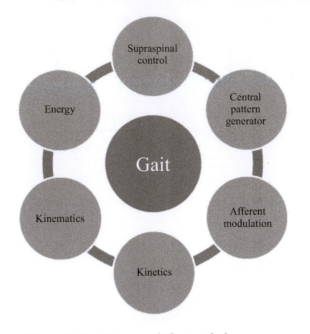

Figure 10.2   Gait, a multifactorial phenomenon

cycle; cadence is given by the number of strides per minute (strides/min); speed measures the speed (m/s) with which the subject moves when walking. Step length and step time give a subdivision of the stride length and time and are, respectively, the distance covered from the contact of one foot to the contact of the contra-lateral foot, and the time difference between the two contact instants.

Different indexes and reference time-varying curves have been proposed for mean values of kinematics data, such as those proposed by Winter, Stoquart, Schutte (the Gillette Gait index), Schwartz (the Gait Deviation Index), Beynon (the Gait Profile Score) and Baker (the Movement Analysis Profile derived from the Gait Profile Score) [16–21].

The lower limb kinetics during walking is formulated from quantities such as, three-dimensional forces (or torques) at the hip, knee and ankle joints; net powers (product of the torque and angular velocity about a joint) and work (area under the net power vs. time curve), as well as ground reaction force (GRF) vectors and gravity [8,22–24].

Since the walking activity results from the combined action of joints and muscle activities, ageing, pathologies, fatigue, carrying of weights, walking surface conditions, and even clothing characteristics can affect an individual's walking pattern on the fly or gradually during its lifetime. In [25], an additional parameter, the constraint force, is used to characterize how gait is affected by tissues.

Each individual presents specific values of these parameters in his/her normal walking cycle, making it repeatable, what allows gait to be used as a flexible biometric quantity given its unobtrusiveness and the fact that it is based on information observable at a distance [26].

Walking speed varies with age, gender and physical condition. It is accepted that the average walking speed of a normal person is in the range of 4–6 km/h [27–29]. Still it has been considered a reliable vital sign of humans for it reflects both functional and physiological changes, providing information that determine potential for rehabilitation and allows predicting falls or fear of falling.

The detection and analysis of gait abnormalities (deviation from normal) requires the capture of quantities related to strength, sensation and coordination, i.e. forces and moments in the joints, muscle activity, velocity and acceleration of each segment of the limb, etc., and their processing to evaluate whether the respective values and synchronization are maintained within normal ranges.

10.2.1 Gait modelling

Gait analysis constitutes a diagnosis method that relies on records of joint spatio-temporal features captured during a certain gait exercise, providing quantitative data to analyse and diagnose movement disorders.

A biomechanical model is required to infer the movements of body segments from the measured quantities. When combined with EMG records, as well as with other behavioural features, such as oxygen consumption and foot pressures, flexible diagnostic and inspection tools can be obtained. The availability of proper gait models allows to accurately characterize and simulate gait activity and to evaluate potential interventions.

Gait patterns are commonly described by quantifying the observed changes in angular rotation of hip, knee and ankle joints during the gait cycle. Each joint presents a specific behaviour and gait pattern anomalies can be detected and characterized after the measure of displacements from the expected normal range of rotation curves.

For that purpose, clinicians resort to evaluating a set of spatio-temporal features that can be extracted from the raw captured data, particularly inertial (body segment motion and GRF) and surface EMG (sEMG) signals, to predict how performance (movement) is affected by, e.g., a pathology or intervention. Different muscle strength, coordination of muscle activity and body orientation can affect performance. An affection in one forces changes in the other.

Two approaches are possible on analysing the locomotion dynamics, forward and inverse dynamics. With the former one estimates segment motions from joint torques and applied forces. The second seeks estimating joint torques from applied forces and segment motions. It is called inverse dynamics because the kinematics are worked back to derive the kinetics responsible for the motion.

The biomechanics equations that govern locomotion can be formulated as [30]

$$[\boldsymbol{H}(\theta)]\ddot{\theta} = \boldsymbol{C}(\theta,\dot{\theta}) + \boldsymbol{G}(\theta) + M(\theta) \tag{10.1}$$

where for an n degree of freedom model with $\theta, \dot{\theta}, \ddot{\theta}$ as $n \times 1$ vectors of angular displacement, velocity and acceleration, $[\boldsymbol{H}(\theta)]$ is an $n \times n$ inertia matrix, $\boldsymbol{C}(\theta,\dot{\theta})$

is an $n \times 1$ vector of Coriolis and centrifugal terms, $G(\theta)$ is an $n \times 1$ vector of gravitational terms and $M(\theta)$ is an $n \times 1$ vector of applied moments.

For a forward dynamics simulation, the different angular accelerations can be obtained from (10.2), whose integration in time in dynamic simulations provides the motion trajectories responses corresponding to inputs, i.e. joint moments.

$$\ddot{\theta} = [H(\theta)]^{-1}C(\theta,\dot{\theta}) + G(\theta) + M(\theta) \qquad (10.2)$$

For an inverse dynamics solution, the knowledge of the forces applied to the different segments allows calculating the joint torques.

$$M(\theta) = [H(\theta)]\ddot{\theta} - C(\theta,\dot{\theta}) - G(\theta) \qquad (10.3)$$

The solution of these differential equations is computationally intensive, and the availability of accurate simulation tools has evolved with the development of computers' hardware and software technology.

The neuromusculoskeletal model employed in the human gait animation system presented in [31] comprises 14 rigid bodies, 19 degrees of freedom, 60 muscular models, 16 pairs of neural oscillators and other neuronal systems. With this model, it is possible to generate different locomotion patterns, such as normal gait, pathological gait, running and ape-like walking. It takes as inputs kinematic data, as well as *in vivo* dynamic data such as energy consumption.

A musculoskeletal model for muscle-driven simulation of human gait was developed by Rajagopal *et al.* [32], which comprises 22 rigid bodies. The lower body is governed by seven degrees of freedom in each lower limb plus three rotational and three translational degrees of freedom in the pelvis. The actuators of the lower limbs are 80 muscle-tendon units modelled with Hill-type models. Hill-type muscle actuators provide a good trade-off in complexity vs. computation time terms, comparing to simple torque-driven models and the more complex finite-element simulations capable of representing well the muscles geometry [33,34].

The recruitment and activation dynamics that determine the force inputs for forward dynamics simulation can be obtained from the analysis of EMG signal intensities (EMG-driven modelling). Transfer functions can be derived that enable the estimation of the active state of each muscle [35,36]. The neural activation or muscle recruitment amplitude can be obtained from $u(t) = \alpha \times emg(t - \delta - \beta_1 u(t-1) - \beta_2(ut - 2))$, where $emg()$ maps the EMG values, δ is the electromechanical delay, and α and $\beta_{1,2}$ are coefficients that define the second-order dynamics [37,38].

The majority of commercially available systems used to derive joint kinematics and kinetics are based on the capture, by means of video cameras, of the position of retroreflective markers placed on the subject body (anatomical landmarks) in relation to bony landmarks. The International Society of Biomechanics standard and the so-called conventional gait model (Figure 10.3) for which many variations exist, widely used within video-capturing-based gait observation systems, define the minimum number of reflective markers that should be used to determine inverse dynamics three-dimensional kinematics and kinetics of the lower limb [39–41].

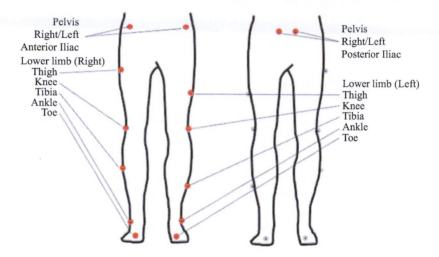

Figure 10.3   Placement of reflective markers according to the conventional gait model [42]

10.2.2   Gait data characterization

Typically, the maximum frequency of the human body motion is around 12 Hz and gait is a relatively lengthy phenomenon. The mean comfortable gait speed is 1.36 m/s. At fastest walking speeds (2.14 m/s in average), a stride period is about 1 s [28,43] and thus, in general, the kinematic and kinetic quantities that govern human locomotion are low frequency signals [8].

The GRF of a walking subject shows a spectrum with frequencies mostly in the 0 and 4 Hz range, but some components up to 15 Hz (99% of the power) can also be observed [44,45]. The peak velocity of the ankle during the stance period can reach 3.6 ± 0.2 rad/s, and its torque bandwidth is approximately 3.5 Hz [46]. Research on the development of bipedal walking robots has shown that joint torques control bandwidths go from 6 to 38 Hz [47]. The gait kinematic quantities of the whole human body show frequencies lower than 10–12 Hz. This provides useful information to detect whether one is in presence of abnormal subject performance, namely due to disease [44,48], to track the rehabilitation [17] as well as for the process of selecting the appropriate filter cutoff frequency of the data conditioning circuitry [49].

Sampling frequencies (f_s) used to capture these signals range typically between 50 Hz and 1 kHz. Gabriel [50] determined that the best cut-off frequency (f_c) of the low-pass digital filter (zero lag, fourth-order Butterworth) used to filter random noise in human body motion data is given by $f_{c1} = 71 \times 10^{-3} f_s - 3 \times 10^{-5} f_s^2$. A refinement of the f_{c1} calculation is obtained using $f_{c2} = 6 \times 10^{-2} f_s - 2.2 \times 10^{-5} f_s^2 + 5.95/\varepsilon$, where ε is the relative mean residual of the difference between the unfiltered (x_{nu}) and the filtered (x_{nf}) data obtained with the first approach of the cut-off frequency f_{c1}, i.e. $\varepsilon = \sqrt{\left(\sum_{n=0}^{N}(x_{nu} - x_{nf})^2\right)/\left(\sum_{n=0}^{N}(x_{nu} - \bar{x}_{nu})^2\right)}$, being $\bar{x}_{nu}$ the mean of x_{nu}. Van

Table 10.1   Characteristics of EMG signals

Classification	Bandwidth (Hz)	Dynamic range
Intramuscular (single fibre)	500–10,000	1–10 µV
Intramuscular (motor unit)	5–10,000	100 µV–2 mV
Surface (skeletal muscle)	2–500	50 µV–5 mV

den Bogert and de Koning also propose different cutoff frequencies in the 3–94 Hz range, for different body segments and analysis purposes [51].

Electromyographic (EMG) signals are the electrical manifestation of neuro-muscular activity that occurs when muscles are contracted and comprise a linear summation of the action potentials of each activated muscle fibre [52]. Each motor unit action potential can be seen as a stream of Dirac impulses associated to the time of occurrence of the individual motor nerve fibre action potentials. In humans, the amplitude of these action potentials depends on of the muscle fibre radius (r) as $V_{ap} = kr^{1.7}$, where k is a constant. The criteria used to assume that a recording electrode is close to an active muscle fibre are a fast rise time and an amplitude higher than 150 µV [53].

The EMG signal can be captured either using (planar) electrodes placed on the skin surface or subcutaneously (in-dwelling) with needle or wire electrodes inserted in the muscle fibres (intramuscular). Surface EMG (sEMG) signals are weighted by the distance and the electrical characteristics of the tissues interposed between the active muscle fibres and the skin, as well as the detecting electrode itself. These introduce a low-pass filtering effect whose bandwidth and gain decrease as that distance increases. The EMG signal bandwidth can be reduced from more than 1,000 Hz, seen for intramuscular signals, to less than 400 Hz. Table 10.1 summarizes the bandwidth and amplitude of intramuscular and surface captured EMG signals [54].

For intramuscular recordings, different types of several mm long needles have been used, which provide different recording volumes covering different portions of a motor unit volume [55]. Christensen *et al.* [56] report the use of 0.56 mm diameter concentric needles inserted 2–3 cm into the biceps brachial muscle. The signal was conditioned using a bandpass (20 Hz to 10 kHz) filter sampled at 51.2 kHz. In these conditions, the rms amplitude of the captured EMG signals was 180 ± 55 µV range. When using surface electrodes that amplitude decreased to 55 ± 19 µV.

Contrary to needles, the electrodes used in sEMG recordings are easy to apply, practically painless, provide more reproducible signals, and are usable in applications involving movement. Their disadvantage resides in a larger pick-up area and are thus more susceptible to crosstalk from signals from adjacent muscles. Surface electrodes can be passive or active. The passive ones are commonly based on a simple silver/silver-chloride detection surface. Active electrodes are named so for already comprising a preamplification unit, either in the electrode itself or nearby.

Typically, the significant sEMG activity is seen between 5 and 450 Hz and shows peak-to-peak amplitudes up to 10 mV. The recommended sampling frequency should be at least four to five times that of the highest frequency present in the signal to be captured, i.e. about 2,000–2,500 Hz for sEMG and 4,000–5,000 Hz for intramuscular signals. Regarding filtering, 500 and 1,000 Hz are recommended for the cutoff frequencies of the respective loss-pass filters, to eliminate artefacts [57–59].

Electromyographic (EMG) signals can originate from different voluntary or involuntary causes and the sensing contact area of an electrode will almost always captures information activity from more than one motor unit. Thus, a typical EMG represents the activation of multiple motor units, but still these signals contain two useful types of information, timing of muscle activity and its relative intensity, that are used in clinical and kinesiological EMG. The former is performed by psychiatrists and neurologists for diagnostics purposes and addresses the study of the characteristics of the motor unit action potential for duration and amplitude. Kinesiological EMG concerns movement analysis after the study of the relationship of muscular activity with the kinetics of the body segments, i.e. the correlation between the timing, strength and force produced by the muscles with the kinetics and kinematics of gait, for the diagnostics of neuromuscular disorders [60], as well as to develop biologically command-based gait rehabilitation robot controllers [33,61].

10.3   Gait observation systems

Gait analysis is usually carried out in dedicated laboratories or clinics. Direct observation, image acquisition and sensor-based data acquisition systems have been used, eventually complemented with the opinion of the patient or athlete concerning pain. Simple direct observations are rarely sufficient to provide a thorough and insightful analysis and are mainly subjective.

Image acquisition and processing systems, mostly resorting to digital cameras (video-based systems) and marker tracking based systems, are the most commonly employed and least obtrusive human motion tracking methods. Video-based systems resort to the movement-recorded data to determine the relative position of the body segments. In marker tracking systems, the markers can be active (like light emitting diodes) or passive (light reflectors) and are placed on specific body places in relation to reference landmarks. Opto-reflective markers are illuminated and detected with video cameras. The spatial positions of the markers provide general time-distance parameters, like rotation angles, angles velocities and accelerations of the hip, knee, and ankle joints. The three-dimensional position of each marker is obtained after crossing the information provided by two or more cameras, once their position and orientation are known.

Optical marker-based tracking systems have been considered one of the best-suited motion data acquisition methods. It has been reported that these systems provide joint angle measurement errors as low as 0.6° (with millimetre resolutions), while video-based systems yield errors four times higher (2.3°) [62]. In the first motion capture systems, low resolution and a reduced number of cameras were used and consequently also a few spaced markers. Currently, clinical systems with more than eight cameras operating with sampling frequencies higher than 100 Hz, detecting tens of markers, can be found [40,41,62]. Nevertheless, a wrong placement of the markers can lead to wrong data acquisition. Another source of error is the movement of soft tissues in relation to the bones that occurs during walking [41].

Among the markerless systems, one can also find time-of-flight-based cameras [63,64]. In this case, the sensor camera emits, e.g. infra-red light or ultrasounds, towards the scene where the person is walking. The reflected wave is, for each pixel p, detected by the sensor as a signal $I(p) = a(p) \exp^{j\phi(p)}$, whose amplitude is proportional to the reflectivity of the object and its phase is proportional to the round trip time $\phi(p) = t\pi ft(p)$. Here, f is the modulating frequency, and the round trip time is proportional to the round trip distance $t(p) = d/v$. For infrared-based systems, this corresponds to a maximum action range of 14.99 m, considering v the velocity of light and a modulating frequency of 10 MHz. For ultrasound-based systems that range is 17.16 m, considering $v = 343.2$ m/s and a modulating frequency of 10 Hz.

For kinematics detection and analysis purposes, GRF or pressure mapping platforms for dynamics measurement can be used together with sEMG detection to determine muscle activity. Different GRF observation platforms such as AccuGait (Advanced Mechanical Technologies), PhysioSensing (SensingFuture), Platform Pro (Mediologic), FDM-T system (Zebris), Footwalk (AM cube), WIN-POD (Medicapteurs), Sports Balance Analyzer (Tekscan), footscan (RSscan) and *emed* (Novel Electronics) are available [30,65], which show pressure measurements in the 10–1,270 (kPa) range. These systems are often combined and synchronized with image acquisition systems to obtain comprehensive data analysis systems [66].

In general, these systems tend to be expensive and difficult to use. They require high levels of expertise, and thus the presence of technicians are time consuming, uncomfortable for the patient and their use must be confined to a lab. Brand and Crowninshield reported that, for gait analysis, the measurement technique should not affect the function it is measuring [67]. The observation with floor platforms has the advantage of being unobtrusive, but the parameters that can be measured are limited. Baker stated that being walking performed in a dedicated analysis laboratory, with the patient concentrated on what he is doing, the obtained results might provide results that are not representative of his/her normal walking [41].

Gait analysis systems should preferably be inexpensive, easy to use, no limited in the walking distance range and provide continuous time monitoring under real-life conditions. Nevertheless, inevitably some invasiveness is necessary to capture sEMG, measure muscle tension or joint contact forces [68,69].

10.3.1   *Wearable body sensor networks*

In the last decade, a number of wired and wireless networks for wearable applications have been proposed for both general [70–73] and specific medical and healthcare purposes [74–78]. The majority of these systems comprise a network of SNs interconnected with copper wires. XSENS and Tec Gihan Co. are two of the manufacturers who are already commercializing wearable systems, the MVN BIOMECH and WS M3D systems, respectively [63]. These are made of wireless or cabled motion trackers fastened to the body with Velcro straps or inserted in a Lycra suite, always with the support of a technician.

Solutions to incorporate the sensors in fabric have also been explored. An electronic e-textile pants is presented in [79] which comprises SNs in the form of e-tags with microcontrollers, sensors (accelerometers and gyroscopes), and communication devices attached near the ankles and the knees. A piezoelectric pressure sensor is used to detect heel contact during gait. Copper conductors are used to interconnect the e-tag modules. All information is transferred via an I2C bus to a final e-tag module that transmits data through a Bluetooth link.

Post and Orth proposed in [80] an electronic system that resorts to conductive yarns as a data bus and described some primary techniques for building circuits from commercially available fabrics, yarns, fasteners and components. It is also shown that all of the input devices can be made by seamstresses or clothing factories, entirely from fabric.

Based on theoretical analysis and experimental results, the ability of conductive yarn to carry both data and power supply was demonstrated by Wade and Asada [81] who introduced a method for the analysis and design of these conductive-fabric, sensor-network garments, applying DC-PLC (DC power line communication) to a wearable garment, a method for analysing the garment characteristics and the details of the physical implementation.

A wearable personal network (WPN) system based on a serial bus implemented with conductive yarns was introduced by Lee *et al.* [7]. This system implements a four-layer WPN using a bus topology controlled by a host node. The system is implemented on a FPGA running at 50 MHz with a data rate of up to 10 Mbps. The profile layer is provided to make the application development process easy. The data link layer exchanges frames in a master–slave manner in either the reliable or best-effort mode. The lower part of the data link layer and the physical layer of the WPN are made of a fabric serial-bus interface, which is capable of measuring bus signal properties and adapting to medium variation. The WPN communication modules were implemented on small flexible printed circuit boards (PCBs) and used together with conductive yarns to build a WPN shirt prototype.

An inductive coupling transceiver with a fault-tolerant network switch for multi-layer wearable BSN applications was proposed by Yoo and Lee [82]. The inductive coupling is made with conductive yarns and employs a resonance compensator (RC) with a digitally controlled on-chip capacitor bank and a variable hysteresis Schmitt trigger comparator to compensate dynamic and static deviations of the woven inductor. The communication is carried out at a data rate of 10 Mbps. Each node consists of two chips: transceiver and switches and RC, both fabricated with a 0.25 μm CMOS technology and occupying areas of 2 and 0.8 mm^2, respectively.

An interference resilient 60 kbps to 10 Mbps body channel transceiver, for multimedia and medical data transactions resorting to body channel communication (BCC), is presented by Cho *et al.* [83]. The transceiver uses the human body as a signal transmission medium in the 30–120 MHz frequency range for energy-efficient and scalable data communication around the body. In the frequency range of BCC, the human body may operate as a receiving antenna and pick up large interferences, degrading the signal-to-noise ratio (SNR) to 22 dB. In order to avoid body-induced interferences, a four-channel adaptive frequency hopping (AFH) scheme that monitors channel status continuously and selects only the clean channels is introduced. To support a 10 Mbps data rate, wideband frequency-shift keying signalling with fast AFH, a direct-switching modulator and a delay locked loop (DLL)-based demodulator are incorporated in the transceiver design. The dual frequency synthesizers in the modulator reduce the frequency hopping overhead to 4.2 μs without degradation of the phase noise and the reference spur. The transceiver was fabricated with a 0.18 μm CMOS technology and withstands 28 dB SNR, with an operating distance of over 1.8 m with a SNR degradation of 3 dB.

An impulse radio type transceiver for BCC applications has been presented by Shikada and Wang [84], in which the information data is modulated by pulses. First, the information with a date rate of 1.2 Mbps is modulated with eight pulses for bit "1" and nothing for bit "0". The pulses with a width of 10 ns are produced by a 4.9 MHz oscillator and a XOR gate. The modulated pulses are then spectrum-formed by a band-pass filter to have their main spectrum components within the 30–50 MHz range. The output signal drives an electrode placed on the body. The received signal is first filtered and amplified, being then adjusted to an adequate level by an automatic gain controller, prior being formatted with an envelope-detection scheme (diode and a low-pass filter). The detected signal is then converted to a 1-bit stream with a comparator.

A human body communication technology employing a magnetoquasistatic field has been developed by Ogasawara *et al.* [85]. The system is designed to mitigate the influence of the surroundings on the communication channel. It uses the human body as one component of a loop that creates magnetic coupling between two transceivers. A circuit model was devised and its validity was experimentally demonstrated. The obtained results demonstrate the validity of the magnetoquasistatic human body channel (HBC) model and its robustness regarding the surrounding environment. It is also shown that HBC based on magnetic coupling is a feasible technology. The available data rate of the system is not reported.

The wearable wireless ECG sensor presented by Nemati *et al.* [86] combines the ANT wireless protocol for data communication with capacitive ECG signal sensing and processing. The ANT™ protocol was used as a low-data-rate wireless module to reduce the power consumption and size of the sensor. Small capacitive electrodes were integrated into a cotton T-shirt together with a signal processing and transmitting board on a two-layer standard PCB design technology. Signal conditioning and processing were implemented to remove motion artefacts.

10.4 Conductive yarns

Conductive fibres and yarns have drawn considerable attention during the last decade. There are different types of these conductors, which have been selectively used according to the requirements of the particular applications. Generally, textile materials made of organic polymers are perfect insulators. Methods of producing conductive yarns can be summarized as follows [87–90]:

1. Addition of carbon or metals such as silver, steel, nickel to the structure, in the form of wires, fibres and micro or nanoparticles. Carbon fibres and carbon-filled fibres exhibit good conductive properties, but on the other hand, they have some aesthetic problems. Metal fibres and wires which can be incorporated into textile structures have high conductivity, but they have also some disadvantages like weight, cost and difficulty of use within textile machinery.
2. Use of inherently conductive polymers such as polyanyline, polyvinyl alcohol, polypyrrole and polyamide 11 (PA 11). Amongst these, polyanyline has attracted higher attention due to its good environmental, thermal and chemical stability. Currently conductive polymers are gaining more and more importance due to their advantages, but these polymers are still rather costly. They can be used in applications where flexibility, low weight and conductivity are required.
3. Coating with conductive substances. Highly conductive fibres can be produced by metallic or galvanic coating, but these methods have some limitations with adhesion and corrosion resistance and suitability of the substrate. Metallic salt coatings have some limitations in conductivity.

Textiles with conductivity properties have been used in many technical applications such as protection of people and electronic devices from electromagnetic interference and electrostatic discharge, heating, wearable electronics, data storage and transmission, sensors and actuators [91–94].

Conductive yarns have also been explored to implement electrodes resorting to different fabrication processes, such as screen printing, weaving or knitting [90,95–97]. The electrical characteristics of these electrodes allow for a reliable capture of bio-signals without significant signal quality degradation, compared to the conventional silver/silver chloride (Ag/AgCl) electrodes [98]. Furthermore, textile electrodes are flexible, unobtrusive, do not cause skin irritation due to the gel and are a good candidate for chronic applications. Moreover, unlike Ag/AgCl electrodes, textile electrodes do not need to be changed in long-term applications and can be found in different shapes, materials and formats.

The removable embroidery electrodes presented in Figure 10.4 were designed to be washable and long-lasting, comfortable to wear in contact to the skin and inexpensive [99]. A male snap fastener, electrically connected to the electrode in the front-side of the sensor, is included on their back-side. They are made of electrically conductive embroidery patterns, realized with commercial conductive threads, having a circular shape with a diameter of 16 mm. Table 10.2 shows the characteristics of the upper and lower commercial threads used in the embroidery patterns, respectively. The reported electrode-skin impedance is 0.1 Ω.

The biopotentials recorded with these new electrodes might not be comparable in terms of signal quality with the gold standard of gel electrodes, but it has been shown that still good results can be obtained. In fact, due to the lack of hydrogel, textile electrodes present different electrical contact characteristics. The skin-electrode

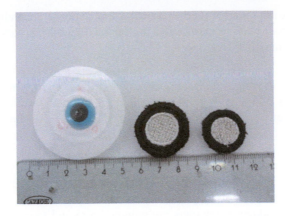

Figure 10.4   Detachable embroidery electrodes

Table 10.2   Characteristics of thread materials

Embroidery thread	R (Ωm^{-1})	Manufacturer
Upper	40,000	Less EMF Inc.
Upper	200	TibTech
Lower	70	Imbut Gmbh

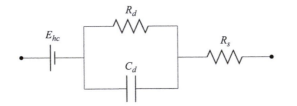

Figure 10.5   Electrical model of the electrode-skin impedance

impedance is an important feature since it affects the captured signal quality. Although a low impedance is desired, a comfortable wearable system should not require the electrodes to be covered by the hydrogel or be moistened.

The electrode-skin impedance is often modelled with the equivalent circuit shown in Figure 10.5 [100]. The voltage source E_{hc} models the half-cell potential, capacitance C_d is related with the electrical charge accumulated between the electrode and the skin, R_d is the resistance between the electrode and the skin during the charge transfer and R_s is related with the electrolyte gel, sweat and the underlying skin tissue resistances.

$$Z(j\omega) = R_s + \frac{R_d}{1 + j\omega R_d C_d} \tag{10.4}$$

Besides the electrode-skin contact impedance dependency with the signal frequency, it also depends on the exerted pressure which can be considered in the impedance (10.4) after the use of pressure dependent R_d and C_d elements. Albulbul *et al.* [101] studied the effect of an externally applied pressure on Ag/AgCl electrodes. It was concluded that a light to moderate applied pressure can decrease the electrode-skin impedance, but a large one may result in an impedance increase. According to the authors, this increase may be caused by a redistribution of the electrolyte used with nonpolarizable electrodes.

The electrode-skin impedance is known to affect the measured biopotentials, namely the noise level. Puurtinen *et al.* [102] measured the impedance and noise of sEMG signals captured with dry textile electrodes, textile electrodes moistened with water and textile electrodes covered with hydrogel, for five different electrode sizes. The authors noted that the noise level increases as the electrode size decreases. The noise level obtained with dry textile electrodes was high, but that obtained with wet textile electrodes was low and similar to the noise level obtained with textile electrodes covered with hydrogel. The authors also found that hydrogel did not seem to improve noise properties; however, it may have effects on movement artefacts. In this study, they concluded that it is feasible to use textile-embedded sensors in physiological monitoring applications when moistening or hydrogel is applied.

Experiments were carried out to study the influence of the pressure applied to the electrode-skin interface of embroidered textile electrodes with no hydrogel or moistened water on a long-term basis, i.e. after observing the variation of the electrode-skin impedance components over time. Measurements of EMG were also performed simultaneously with the purpose of studying the influence of the impedance in the recorded biopotentials.

A forearm sleeve made of plain weave fabric, integrating embroidered textile electrodes and elastics to conform to human forearms (see Figure 10.6) developed at the University of Beira Interior [99] was used to acquire sEMG signals and measure the electrode-skin interface impedance from four volunteers. The forearm sleeve has elastic properties that allow us to vary the pressure the electrodes apply against the skin. It also has a strip with four tightening positions to control the applied pressure. The rectangle tissue seen around the electrodes comprises a pressure sensor, which allows monitoring the applied pressure.

The skin impedance is affected by temperature and total body water content [103]. In order to avoid detecting changes in the measured impedance that were not related to the electrode-skin impedance variations over time, all the measurements were performed in a room kept at a constant temperature. During the impedance measurements, the test subject did not ingest any food or beverage. Prior to the measurements, the volunteer also went through skin preparation (skin was cleaned with alcohol).

The textile electrodes pair was placed on the surface of the *flexor carpi radialis* muscle to acquire the EMG signal, and a standard Ag/AgCl electrode with 1 cm diameter (Dormo, Telic S.A., Spain) was placed on the elbow to establish the reference.

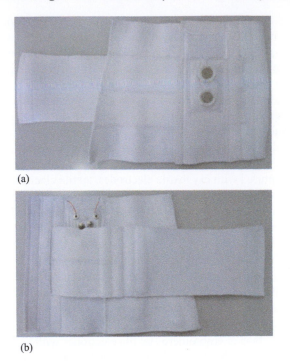

(a)

(b)

Figure 10.6   Partial views of the forearm sleeve: (a) inner part with textile
electrodes (b) outer part with a pair of snap fasteners connected to
the data acquisition terminals and a pair of wires connected to the
pressure sensor

10.4.1   Variation of electrode-skin impedance and SNR during long-term monitoring

To evaluate long-term effects, impedance and sEMG measurements were performed over time in four volunteers. The following measurements were performed with the sleeve tightened with the highest level, in order to accomplish a lower electrode-skin impedance, and therefore a better signal quality. Due to hardware restrictions, the impedance and the sEMG signals were not measured simultaneously, but both measurements were taken within less than 1 min interval. All the volunteers performed their daily activities in the period during which measurements were made.

The impedances (magnitude and phase) obtained over time for all the volunteers are illustrated in Figures 10.7–10.10.

The parameters for the electrode-skin impedance equivalent circuit (see Figure 10.5) were extracted, and the SNR was calculated for each measurement. The variation of the impedance parameters R_s, R_d and C_d are plotted for all the measurement sessions with the volunteers (Figures 10.11–10.14). The evolution of the SNR is also included in the plots to provide a better insight of the parameters evolution and their influence on the signal quality.

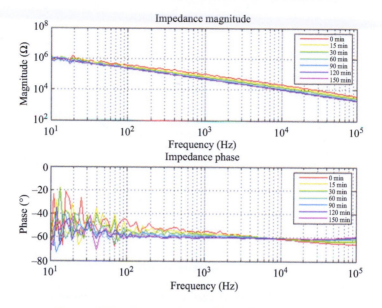

Figure 10.7 *Measured impedance for volunteer 1*

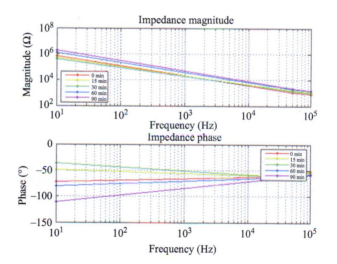

Figure 10.8 *Measured impedance for volunteer 2*

After observing the values of the electrode-skin impedance over time, one can notice that the impedance variation is very different among the volunteers. Although all of them were asked not to drink or eat anything during the measurements, they were free to perform their daily activities in the measurement period. During that

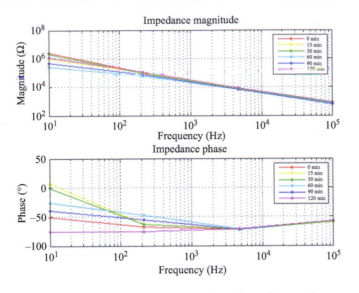

Figure 10.9   Measured impedance for volunteer 3

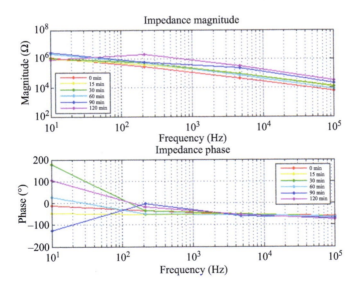

Figure 10.10   Measured impedance for volunteer 4

time, the hand sleeve and the electrodes might suffer small displacements that could cause a shift in the impedance of the electrode-skin interface. From the observation of the impedance plots, one could not see a relation with the SNR of the sEMG signal. As the impedance measurements are influenced by the properties of the

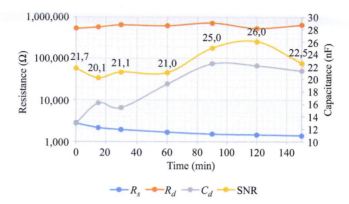

Figure 10.11   Variation of R_s, R_d, C_d and SNR for volunteer 1

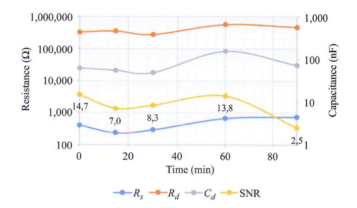

Figure 10.12   Variation of R_s, R_d, C_d and SNR for volunteer 2

electrode/electrolyte and skin, the values of R_s, R_d and C_d were extracted to study their relation with the SNR individually.

If we hypothesize that the main factor for the SNR variation is the contact between the electrodes and the skin, the contact area would define how the impedance parameters would relate with the SNR. When the contact is better (higher effective area), the SNR and the C_d should increase and the resistances R_d and R_s should decrease. Conversely, if the contact between the electrodes and the skin is loose, SNR and C_d values decrease, and R_d and R_s values increase.

From Figures 10.11–10.14, we notice that SNR is correlated with C_d and R_s. When the SNR increases, both C_d and R_s also increase, likewise for the reverse behaviour. From our observations, the R_d has no relation with the SNR. Since these measurements were performed during an average of 2 h, the electrode-skin interface changes, namely due to sweet, and the electrode surface adapts to the skin.

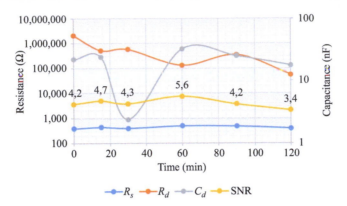

Figure 10.13   Variation of R_s, R_d, C_d and SNR for volunteer 3

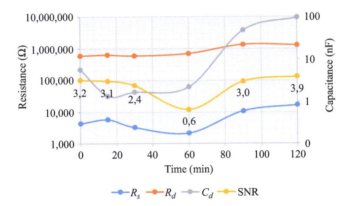

Figure 10.14   Variation of R_s, R_d, C_d and SNR for volunteer 4

Also, the human body impedance of each person varies differently during the day. The R_d could be more influenced by these parameters than by the contact area itself. This study shows that an impedance magnitude measurement is not enough to monitor the quality of biosignals, since the resistance if affected by the person's own impedance variation. The extraction of the parameters of the electrode-skin impedance electrical model, after performing phase and magnitude measurements, is more reliable for the purpose of the signal-quality monitoring.

10.5   A BSN for gait analysis – the ProLimb system

The most comfortable and easiest way to monitor physiologic signals consists in using garments with conductive yarns as wire lines [104,105]. A new instrument

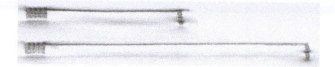

Figure 10.15   Detail of sEMG electrodes and interconnection line, both embedded in the knitted fabric

infrastructure, ProLimb, specifically dedicated to capturing locomotion data is being developed. It is a wired network with conductive yarns, meant to be a wearable BSN for gait analysis and monitoring. This BSN includes, in a single infrastructure, the means to capture inertial and sEMG signals of the lower limbs.

10.5.1   e-Leggings

The system is based on a leggings with elastic properties, which allows the correct positioning of the textile electrodes (Figure 10.15) and electronics. The e-legging is meant to be comfortable, which implies an adequate combination of materials, compression effect with electronic components incorporated in the textile and interconnected with data and power tracks made with embedded conductive yarns. The conjugation of the BSN with a textile support makes the infrastructure portable, minimally invasive and allows performing the monitoring of typical movement activities, in and outside of a laboratory environment, for prolonged periods of time and under everyday living conditions. This piece of garment is meant to be easy to dress, reusable and can be cleaned and maintained with traditional methods (Figure 10.16(a)).

It is presented as a network of eight SNs, four SNs per leg (Figure 10.16(b)), interconnected through textile-conductive yarns. The SNs provide the measurement of kinematic variables, namely linear and angular movement, as well as of sEMG signals in the muscles that are the most important for locomotion. Each SN monitors a different target muscle, namely the *Biceps Femoris*, the *Rectus Femoris*, the *Tibialis Anterior* and the *Gastrocnemius*, and comprises a sEMG sensor, an accelerometer and a gyroscope.

The acquired data is transmitted via the conductive yarns to a CPM attached to the patients' belt. Aggregated information is sent to a personal computer, or other processing device, through a Bluetooth wireless link from an on-body CPM, as seen in Figure 10.17.

10.5.2   Interconnections in a mesh network

In order to successfully produce fabrics with conductive yarns, and particularly to sew the sEMG electrodes and the interconnections among SN on the leggings, yarns made with twisted filaments, each one a polymeric filament covered with silver, were used. This material is widely used in textile industry for being very comfortable, easy to knit into cloth and for its elastic properties. One of the inherent properties of most

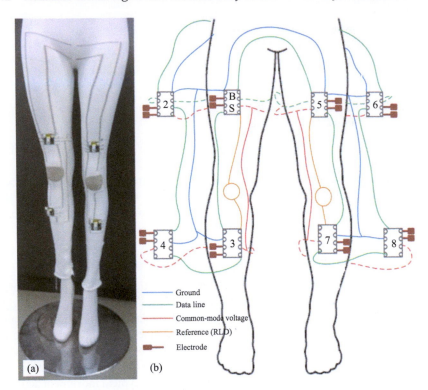

Figure 10.16   e-Legging: (a) photograph of the prototype (b) diagram of interconnections

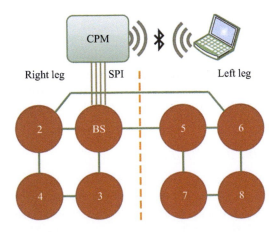

Figure 10.17   System data connections

of the existing conductive yarns is their ability to stretch. Other technologies were available, namely silver covered pure polyamide or polyester yarns, stainless steel covered yarns or even pure stainless steel yarn.

The yarns are embedded in the leggings during the fabrication process carried out at the Textile Engineering Department of the University of Minho and using a technique to embody the textile electrodes, interconnections and the reference electrode in the stockings. A suitable but complex routing (Figure 10.16) of the conducting tracks is necessary to knit the yarns without them crossing each other and in order to avoid more production stages with the inclusion of isolation layers. This is where the capabilities of the production machines assume a critical importance, justifying the seamless technology that was adopted. The machine is capable of drawing a pattern where the conductive yarn is carefully selected to be inserted in specific places, thus ensuring that the tracks will not intersect.

The use of conductive yarns in the interconnections raises two issues. On the one hand, the higher impedance (compared with copper conductors) limits communication frequency and, on the other hand, requires that testing facilities are available to evaluate whether the frequency bandwidth has degraded. The conductive yarn can be modelled as a series R–L impedance. Figure 10.18 shows the impedance profile of a 60 cm long yarn. It can be seen that both resistance and inductance vary with frequency, i.e. $Z(f) = r(f) + j2\pi\omega L(f)$. Table 10.3 shows the measured values of these parameters at 10 MHz for two different conductive yarns. These show that the values are different

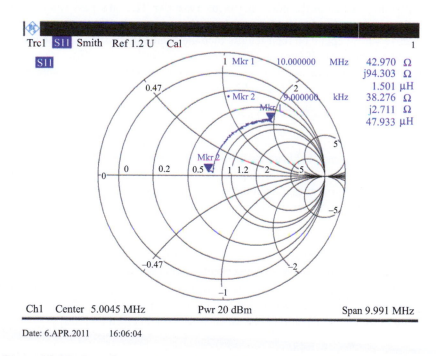

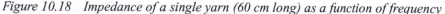

Figure 10.18   Impedance of a single yarn (60 cm long) as a function of frequency

Table 10.3   Yarn impedance per unit length

Yarn	Relaxed (natural length)		Stretched (1.5 natural length)	
	Ω/cm	nH/cm	Ω/cm	nH/cm
1	0.65	16.4	0.47	6.68
2	6.76	35	4.43	33

for relaxed and stretched modes and both resistivity and inductance decrease with stretching. This electrical behaviour requires using several yarns in parallel in order to obtain a suitable communication frequency.

An intra-network of SNs in wearable systems comprises a set of SNs in an interior network, all connected to a collector node. The topology of the network depends on the number of SNs and data rate. Star and serial bus topologies are useful for cases with a small number of SNs. With an increased number, the performance degrades: the number of ports increases in a star network and also the bus load in a serial network. In networks with a high number of SNs, the mesh topology is a better option to overcome these limitations while keeping high data-rates.

The combination of the electrical properties of the conductive yarns, with the capacitive impedance of the ports, performs a n-order RLC low-pass filter in serial topology and second-order in mesh topology. Considering that the parameters of the conductive yarns depend on their length, then the behaviour of such a low-pass filter will be also a function of yarn length. Increasing the order of the filter, the signal's attenuation and phase delay increase, as well. The Fourier series of a square wave with levels between $(0, A)$ and period T is given by

$$f(t) = A\left(\frac{1}{2}\right) + \sum_{n=1}^{\infty} \frac{2}{\pi n}\sin\left(\frac{\pi n}{2}\right)\cos\left(\frac{2\pi n}{T}t\right) \tag{10.5}$$

therefore, the low-pass behaviour of the line generates a malformed signal at the receiver nodes, which is hard or impossible to detect because the attenuation and phase change the harmonics. Given the vulnerability of conductive textile yarns and to ensure a high data rate given the number of sensors used and the issues introduced by variable interconnection impedances, a mesh topology was adopted to organize the SN. With this topology, a more reliable communication is obtained due to the presence of redundant links.

The yarn impedance together with circuits' output and input capacitances create a second-order RLC-network with a transfer function $G(s) = \omega_n^2 / \left(s^2 + 2\zeta\omega_n s + \omega_n^2\right)$, where $\omega_n = 1/\sqrt{LC}$ is the undamped natural frequency and $\zeta = (R/2)\sqrt{C/L}$, the damping ratio. The propagation delay of such network is given by [106]

$$t_{pd} = \frac{e^{-2.9\zeta^{1.35}} + 1.48\zeta}{\omega_n} \tag{10.6}$$

and its step response shows a rise time given by $t_r \approx 2.2RC + 0.6\sqrt{LC}$. Considering an interconnection length of 60 cm ($L \approx 980$ nH), a typical average output resistance of 300 Ω and a total capacitance of 20 pF, to obtain a critically damped response ($\zeta = 0.7$), the interconnection should present about 16 Ω, i.e. the equivalent of three yarns in parallel. This would provide rise and propagations times of, respectively, $t_r \approx 3.4$ ns and $t_{pd} \approx 5.5$ ns. These values ensure a bandwidth higher than 40 MHz.

The mesh topology facilitates a fully distributed network configuration. However, as it is not possible to cross conductive yarns using a single fabrics layer, the number of possible interconnection that can be knit in the legging is limited. Taking this into consideration, the solution reached for the data connections is shown in Figure 10.17.

One critical aspect with BSN is power management. The SNs may be powered by small batteries or by energy distributed from the CPM to the whole network through the conductive yarns. In a first prototype, each sensor is provided with its own power battery, but the final objective is to have each SN harvesting power from the mesh network, being PLC used to convey data. For that purpose, a specific transceiver has been developed [107].

10.5.3   Sensor nodes and central processing module

Figure 10.16(b) shows all interconnections implemented in the e-legging prototype. Each knee acts as a common-mode reference point (right leg driver, RLD) for the sEMG circuits placed in that leg. The common-mode voltage of the four circuits present in one leg is averaged in the RLD bias circuit of one of these SNs. In total, each leg shows a textile conductor to interconnect all common-mode voltages, a ground connection, the data lines and one RLD (with a second redundant option).

Each SN contains all necessary communication modules, as well as an operation managing microcontroller (μC) responsible also for routing data in the established mesh network, and counts with at least two paths with its neighbor SN by which collected data can be routed. In fact, each SN acts as a router device to handover packets from the source SN to the BS (Base Station), which is responsible for collecting all data captured by the SNs. A low-power microprocessor-based system (the CPM, Figure 10.19(a)) gathers the information from the wearable network via a communication link to the BS module. The CPM accesses the different SNs dispersed throughout the leggings, using the wire mesh network embedded in the technical fabric. To transfer data to a computer or saving them on a memory, the CPM has been equipped with a wireless Bluetooth module, USB port and a micro secure digital (MicroSD) card to record data whenever wireless communication is unavailable.

Each SN comprises two functional modules: *SensorV2* (Figure 10.19(b)) and *EMGV3* (Figure 10.19(c)), fitted on top of one another. *SensorV2* includes a 16-bit μC (PIC24FJ64GA104), a FPGA (Actel AGLN250), an inertial measurement unit (IMU – Accelerometer and Gyroscope) and a three-channel shunt voltage monitor. The *EMGV3* module is the sEMG acquisition circuit based on the Texas Instruments ADS1291 chip. This board includes contact pads to allow it to be stitched to the leggings to interconnect with the embroidered conductive yarns and the textile electrodes.

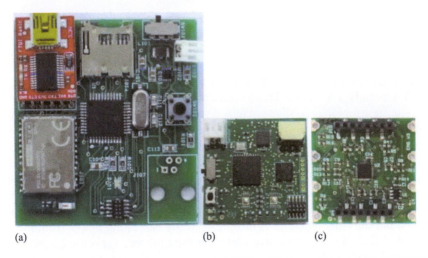

Figure 10.19   Hardware prototypes: (a) CPM, (b) SensorV2 and (c) EMGV3

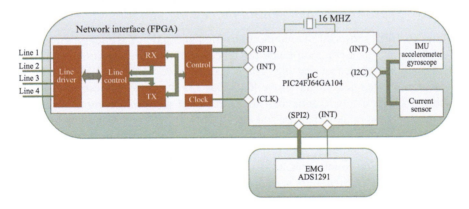

Figure 10.20   Sensor node architecture

Vertical pin connectors are used to fasten the *SensorV2* board on top of the *EMGV3* board. Both boards have a $30 \times 30\,mm^2$ area.

The oscillator of the μC (16 MHz) is the clock source for both the μC and the FPGA. The μC implements the networking and application layers and acquires the sEMG and inertial data. The physical and MAC layers of the data routing protocol are implemented in the FPGA, as shown in the block diagram of Figure 10.20.

The power supply is a small ($5.7 \times 28\,mm^2$) Li-Po battery, with a nominal voltage of 3.7 V at 110 mAh. A DC-to-DC converter generates 3.3 V power supply voltage for the μC, IMU and *EMGV3* module. To supply the FPGA and line driver circuits, a 1.5 V linear regulator is used.

10.5.4 Data acquisition and network capability

Surface EMG (sEMG) and kinematic data are captured using the ADS1291, a low power 24-bit analogue front-end, and a triaxial IMU whit a 16-bit resolution.

By default, the system configuration is 1 ksps for sEMG acquisition in eight SNs and 50 sps for the IMU in four SNs (the IMU units in the four SNs that are on the back of the legs can be off, due to information redundancy). However, several configurations are possible, being the data communication restriction limit (1 Mbps) set by the serial communication with the Bluetooth link. The baud rate between SNs is 4 Mbps but can work at a maximum of 9 Mbps, even with the higher impedance textile conductors, and the SPI communication between the BS and CPM can be performed at 8 Mbps.

10.5.5 Routing protocol

Energy efficiency and integration of systems are considered fundamental milestones for the proposed BSN. The system should work for long periods of time, especially during prolonged monitoring. Thus, an energy-efficient source routing for minimum cost forwarding (SRMCF) routing protocol, described in [108], was specifically developed and adopted for the network data layer.

The SRMCF routing protocol is established on the basis of fundamental concepts in source-based routing for ad-hoc networks and minimum cost forwarding methods. In this protocol, the BS collects all the captured data from the SNs. These use the minimum cost forwarding method as the routing algorithm for sending acquired data to the BS. Communication from the BS to a specific SN over a minimum cost path typically requires intermediate nodes to have stored information about the minimum cost routing paths. With the SRMCF protocol, the routing path is carried in the packets header, avoiding the necessity of the routing information to be stored in intermediate nodes.

The present version of the CPM does not include the hardware needed to perform the SRMCF routing protocol, leading to the need of the BS functions to be executed in a SN. In the final version that capability will be available also in the CPM, allowing to replace the SPI connection with the same type of SN-to-SN interconnection, improving thus the network reliability.

10.5.6 Sensors time synchronization

In order to properly synchronize the data of different SNs, a time-stamping technique is used. Time stamping consists in recording the event time using a marker synchronized with a central clock source. The method used here is based on getting clock information from the node which is in the minimum cost path to the BS node. That is, each node sends a request to the aforementioned neighbour node and receives a reply with a message including the real-time clock. The main clock source is in the BS node. When an SN wants to synchronize with the network, it sends a query message to the BS and this replies with its time [109].

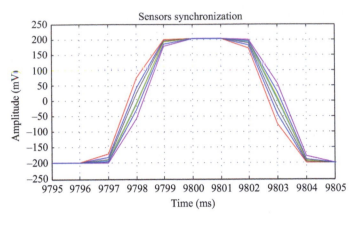

Figure 10.21   Sensor nodes synchronization

Each data packet generated by an SN includes the time stamping information of the first sample. In the final data processing, this information is then used for time synchronization, extrapolating the time of the remaining captured samples according to the known sampling frequency.

The synchronization among sensors is crucial in this kind of applications, where information from several sensors has to be correlated. The system is completely asynchronous and has three types of sensors: sEMG, accelerometer and gyroscope. The accelerometer and the gyroscope are internally synchronized as they are integrated into the same chip. To synchronize the sEMG information with the information from the IMU, as well as the information among SNs, the time stamping from the routing protocol is used as a reference. The time stamping resolution is 1 ms, but can be increased if necessary.

To test the data synchronization process, a square wave with a frequency of 100 Hz was simultaneously captured with the eight sensors. As shown in Figure 10.21, the delay between the BS (clock reference) and the remaining SNs is shorter than 1 ms, which is less that the time stamping and the sampling period.

10.6   Lab experiments

In order to validate the BSN functionality, a series of experiments with the sensor network was carried out to capture the signals in two types of exercises, gait and isometric. The sensor network was used without the integration in the leggings and with conventional electrodes, so as to obtain a better comparison with the signals captured using a commercial system comprising several Trigno (DELSYS [110]) wireless sensors.

The SNs were placed at intermediate points of the thigh and leg as shown in Figure 10.22. The observed muscles are *quadriceps femoris, rectus femoris, biceps femoris, long head, short head, tibialis anterior* and *gastrocnemius medialis*.

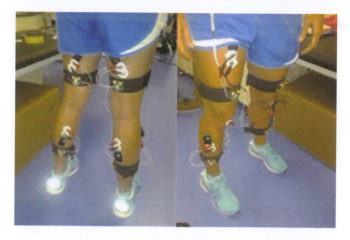

Figure 10.22   Systems placement for the experiments

The experiments followed a preparation protocol including skin preparation and correct electrodes placement by a physiotherapist. For the sEMG electrodes, cut Ag/AgCl electrodes were used to obtain a separation between the respective centres of 2.5 cm.

The Trigno system consists of 16 wireless modules with a dimension of 37 mm × 26 mm × 15 mm. Each module comprises an EMG channel and a triaxial accelerometer. Each channel has 16-bit resolution and sampling frequency of 4 kHz for EMG and 300 Hz for the accelerometer. The modules include their own sEMG electrodes with a separation of 1 cm but need adhesive to fix the modules to the person's body. In the experiments, eight Trigno modules were used, placed as near as possible to the conventional electrodes in order to have the two systems monitoring the activity of the same muscles.

In the gait exercises, 36 steps were performed in a straight line, 18 steps one way and 18 round back. Figure 10.23 shows the kinematic data and EMG signals captured by the SN placed on the *Tibialis Anterior* muscle of the right leg. It can be seen from Figures 10.23 and 10.24 that similar information is provided by the two systems.

The isometric exercise measurements were made for the left leg *Tibialis Anterior* muscle. In this exercise, a Biodex system [111] was used, with the configuration shown in Figure 10.25. The performed exercise corresponds to a unilateral maximum voluntary contraction with the foot in a neutral position with zero degrees flexion. Figure 10.26 shows the sEMG signals captured by both systems. Knowing that one cannot directly compare the two sEMG signals, since the systems are completely independent, have different configurations and are not placed exactly in the same place, it can still be seen that muscle activity patterns can be correctly detected in both cases and that the two signals present similar quality.

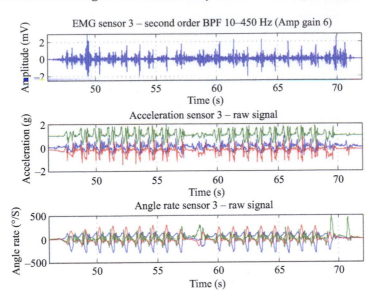

*Figure 10.23   sEMG, acceleration and angular rate signals captured with the
ProLimb BSN during gait exercise*

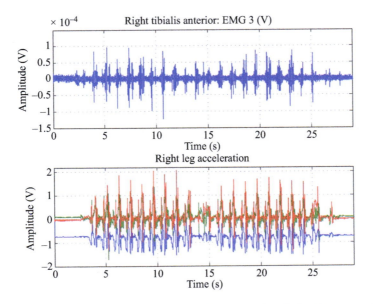

*Figure 10.24   sEMG and acceleration signals captured with the Trigno modules
during gait exercise*

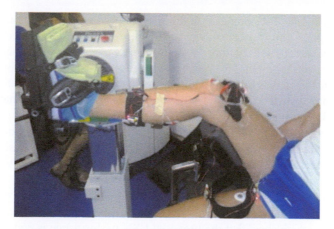

Figure 10.25   Isometric exercise configuration

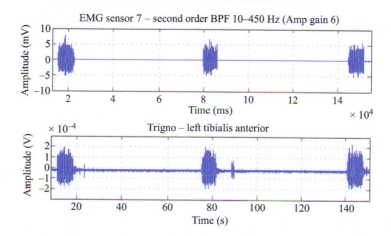

Figure 10.26   sEMG data captured during the isometric exercise with the ProLimb system (top) and with the Trigno system (bottom)

10.7   Conclusion

The recent technological advances in sensors, low-power miniaturized microelectronics, wireless communications and information technologies, enabled the design and proliferation of sensor networks that are promoting the rapid development of various remote wellness or disease monitoring systems and assisted healthcare environments. Smart textiles are also contributing to this new paradigm as a technology that allows developing devices and systems that can be worn by patients and elderly people, as well as to realize dry electrodes to monitor vital signs. The development of wearable sensor networks implanted within smart textiles and using conductive yarns as powering and intercommunication links is an on-going research domain towards achieving high level of performance in smart garments for future m-health contexts.

Gait analysis and fall monitoring are two fields where this combination of technologies has been explored. That requires acquiring knowledge on the biomechanics of gait and human movement and on the characteristics and acquisition of electromyographic signals, which are required to develop the kinematics models that govern the analysis and detection tools.

This chapter describes the development, operation and characteristics of a wearable BSN for human locomotion data capture for gait analysis. The system explores technical fabrics to build a more comfortable and easier to manipulate e-leggings, being conductive yarns used to wire-up SNs, as well as sEMG signals. The architecture and main characteristics of the SNs are described. Experimental results obtained in a biomechanics laboratory, for a real-time human locomotion data acquisition, are presented which show that the proposed BSN performance is equivalent to that of a commercial system with the advantage of consuming lower wireless transmission power and providing also the measurement of an additional quantity – angular velocity.

References

[1]   OECD. OECD Better Life Index: Health.   OECD Publishing. [Online;   accessed   10-November-2016].   Available from: http://www. oecdbetterlifeindex.org/topics/health/.

[2]   Health Research Institute. Healthcare Delivery of the Future: How Digital Technology Can Bridge Time and Distance Between Clinicians and Consumers. Delaware: PricewaterhouseCoopers LLP; 2014. Available from: https://www.pwc.com/us/en/health-industries/top-health-industry-issues/assets/pwc-healthcare-delivery-of-the-future.pdf.

[3]   Mountford N., Kessie T., Quinlan M., *et al.* Connected Health in Europe: Where Are We Today? University College Dublin: COST Action (ENJECT TD1405), supported by COST (European Cooperation in Science and Technology); 2016.

[4]   Piccini L., Ciani O., Andreoni G. New Emerging Biomedical Technologies for Home-care and Telemedicine Applications: The Sensorwear Project. In: Campolo D, editor. New Developments in Biomedical Engineering. London: InTech; 2010.

[5]   Tao W., Liu T., Zheng R., Feng H. 'Gait analysis using wearable sensors'. *Sensors.* 2012;12(2):2255–2283.

[6]   Xing J., Zhu Y. 'A survey on body area network'. *IEEE 5th International Conference on Wireless Communications, Networking and Mobile Computing;* 2009. pp. 1–4.

[7]   Lee H.S. 'Wearable personal network based on fabric serial bus using electrically conductive yarn'. *Electronics and Telecommunications Research Institute.* 2010;32(5):713–721.

[8] Winter D.A. Biomechanics and Motor Control of Human Movement. 4th ed. Hoboken NJ: John Wiley & Sons, Inc.; 2009.

[9] Gait in prosthetic rehabilitation. (2015, June 15). Physiopedia. Retrieved on September 27, 2016 from https://www.physio-pedia.com/index.php?title=Gait_in_prosthetic_rehabilitation&oldid=115494.

[10] MacKay-Lyons M. 'Central pattern generation of locomotion: a review of the evidence'. *Physical Therapy*. 2002;82(1):69–83. Available from: http://ptjournal.apta.org/content/82/1/69.

[11] Guertin P.A. 'Central pattern generator for locomotion: anatomical, physiological, and pathophysiological considerations'. *Frontiers in Neurology*. 2012;3(183). Available from: https://www.ncbi.nlm.nih.gov/pmc/articles/PMC3567435/.

[12] Dietz V., Quintern J., Berger W. 'Afferent control of human stance and gait: evidence for blocking of group I afferents during gait'. *Experimental Brain Research*. 1985;61(1):153–163. Available from: http://dx.doi.org/10.1007/BF00235630.

[13] Schieppati M., Nardone A., Siliotto R., Grasso M. 'Early and late stretch responses of human foot muscles induced by perturbation of stance'. *Experimental Brain Research*. 1990;105(3):411–422. Available from: http://dx.doi.org/10.1007/BF00233041.

[14] J.D. 'Kinematics and kinetics of gait: from lab to clinic'. *Clinical Sports Medicine*. 2010;29(3):347–364.

[15] Farris D.J., Sawicki G.S. 'The mechanics and energetics of human walking and running: a joint level perspective'. *Interface, Journal of the Royal Society*. 2012;9(66):110–118. Available from: https://www.ncbi.nlm.nih.gov/pubmed/21613286.

[16] Brand R.A. 'The biomechanics and motor control of human gait: normal, elderly, and pathological'. *Journal of Biomechanics*. 1991;25(8):949.

[17] Stoquart G., Detrembleur C., Lejeune T. 'Effect of speed on kinematic, kinetic, electromyographic and energetic reference values during treadmill walking'. *Neurophysiologie Clinique/Clinical Neurophysiology*. 2008;38(2): 105–116. Available from: http://www.sciencedirect.com/science/article/pii/S0987705308000257.

[18] Schutte L., Narayanan U., Stout J., Selber P., Gage J., Schwartz M. 'An index for quantifying deviations from normal gait'. *Gait & Posture*. 2000;11(1):25–31. Available from: http://www.sciencedirect.com/science/article/pii/S0966636299000478.

[19] Schwartz M.H., Rozumalski A. 'The gait deviation index: a new comprehensive index of gait pathology'. *Gait & Posture*. 2008;28(3): 351–357. Available from: http://www.sciencedirect.com/science/article/pii/S0966636208001136.

[20] Beynon S., McGinley J.L., Dobson F., Baker R. 'Correlations of the gait profile score and the movement analysis profile relative to clinical judgments'. *Gait & Posture*. 2010;32(1):129–132. Available from: http://www.sciencedirect.com/science/article/pii/S0966636210000238.

[21] Baker R., McGinley J.L., Schwartz M.H., *et al.* 'The gait profile score and movement analysis profile'. *Gait & Posture*. 2009;30(3): 265–269. Available from: http://www.sciencedirect.com/science/article/pii/ S0966636209001489.

[22] Keenan G.S., Franz J.R., Dicharry J., Croce U.D., Kerrigan D.C. 'Lower limb joint kinetics in walking: the role of industry recommended footwear'. *Gait and Posture*. 2011;33(3):350–355.

[23] Schache A.G., Blanch P.D., Dorn T.W., Brown N.A., Rosemond D., Pandy M.G. 'Effect of running speed on lower limb joint kinetics'. *Medicine & Science in Sports & Exercise*. 2011;43(7):1260–1271.

[24] Oh S.E., Choi A., Munand J.H. 'Prediction of ground reaction forces during gait based on kinematics and a neural network model'. *Journal of Biomechanics*. 2012;46(14):2372–2380.

[25] Luengas L.A., Camargo E., Sanchez G. 'Modeling and simulation of normal and hemiparetic gait'. *Frontiers of Mechanical Engineering*. 2015;10(3): 233–241. Available from: http://dx.doi.org/10.1007/s11465-015-0343-0.

[26] Boyd J.E., Little J.J. Biometric Gait Recognition. Berlin, Heidelberg: Springer Berlin Heidelberg; 2005. pp. 19–42. Available from: http://dx.doi.org/10.1007/11493648_2.

[27] Borghese N.A., Bianchi L., Lacquaniti F. 'Kinematic determinants of human locomotion'. *Journal of Physiology*. 1996;494(3):863–879.

[28] Fritz S., Lusardi M. 'White paper: "walking speed: the sixth vital sign"'. *Journal of Geriatric Physical Therapy*. 2009;32(2):2–5.

[29] Lelas J.L., Merriman G.J., Riley P.O., Kerrigan D. 'Predicting peak kinematic and kinetic parameters from gait speed'. *Gait & Posture*. 2002;17(2): 106–112.

[30] DeLisa J.A. Gait Analysis in the Science of Rehabilitation. U.S. Department of Veterans Affairs, Veterans Health Administration, Rehabilitation Research and Development Service. Baltimore, MD: American Printing House for the Blind, Inc., M.C. Migel Library; 1998.

[31] Hase K., Miyashita K., Ok S., Arakawa Y. 'Human gait simulation with a neuromusculoskeletal model and evolutionary computation'. *The Journal of Visualization and Computer Animation*. 2003;14(2):73–92. Available from: http://dx.doi.org/10.1002/vis.306.

[32] Rajagopal A., Dembia C.L., DeMers M.S., Delp D.D., Hicks J.L., Delp S.L. 'Full-body musculoskeletal model for muscle-driven simulation of human gait'. *IEEE Transactions on Biomedical Engineering*. 2016;63(10): 2068–2079.

[33] Lee S.S.M., Arnold A.S., de Boef Miara M., Biewener A.A, Wakeling J.M. 'Accuracy of gastrocnemius muscles forces in walking and running goats predicted by one-element and two-element Hill-type models'. *Journal of Biomechanics*. 2013;46(13):2288–2295.

[34] Seo J.W., Kang D.W., Kim J.Y., *et al.* 'Finite element analysis of the femur during stance phase of gait based on musculoskeletal model simulation'. *Bio-Medical Materials and Engineering*. 2014;24(6):2485–2493.

[35] Lee S.S., de Boef Miara M., Arnold A.S., Biewener A.A., Wakeling J.M. 'EMG analysis tuned for determining the timing and level of activation in different motor units'. *Journal of Electromyography and Kinesiology*. 2011;21(4):557–565. Available from: http://www.sciencedirect.com/science/article/pii/S1050641111000575.

[36] Sartori M., Reggiani M., Farina D., Lloyd DG. EMG-driven forward-dynamic estimation of muscle force and joint moment about multiple degrees of freedom in the human lower extremity. *PLoS ONE*. 2012;7(12): e52618. Available from: https://doi.org/10.1371/journal.pone.0052618

[37] Lloyd D.G., Besier T.F. 'An EMG-driven musculoskeletal model to estimate muscle forces and knee joint moments in vivo'. *Journal of Biomechanics*. 2003;36(6):765–776. Available from: http://www.sciencedirect.com/science/article/pii/S0021929003000101.

[38] Shao Q., Bassett D.N., Manal K., Buchanan T.S. 'An EMG-driven model to estimate muscle forces and joint moments in stroke patients'. *Computers in Biology and Medicine*. 2009;39(12):1083–1088. Available from: http://www.sciencedirect.com/science/article/pii/S0010482509001644.

[39] Wu G., Siegler S., Allard P., *et al.* 'ISB recommendation on definitions of joint coordinate system of various joints for the reporting of human joint motion – part I: ankle, hip, and spine'. *Journal of Biomechanics*. 2002;35(4): 543–548. Available from: http://www.sciencedirect.com/science/article/pii/S0021929001002226.

[40] Żuk M., Pezowicz C. 'Kinematic analysis of a six-degrees-of-freedom model based on ISB recommendation: a repeatability analysis and comparison with conventional gait model'. *Applied Bionics and Biomechanics*. 2015;2015:9. Available from: https://www.hindawi.com/journals/abb/2015/503713/#B8.

[41] Baker R. 'Gait analysis methods in rehabilitation'. *Journal of NeuroEngineering and Rehabilitation*. 2006;3(1):4. Available from: http://dx.doi.org/10.1186/1743-0003-3-4.

[42] Baudet A., Morisset C., d'Athis P., *et al.* 'Cross-talk correction method for knee kinematics in gait analysis using principal component analysis (PCA): a new proposal'. *PLoS One*. 2014;9(7). Available from: http://dx.doi.org/10.1371/journal.pone.0102098.

[43] Bohannon R.W. 'Comfortable and maximum walking speed of adults aged 20–79 years: reference values and determinants'. *Age and Ageing*. 1997;26(1):15–19.

[44] McGrath D., Judkins T.N., Pipinos I., Johanning J. 'Peripheral arterial disease affects the frequency response of ground reaction forces during walking'. *University of Nebraska Omaha, Journal Articles*. 2012;12(132):12. Available from: http://digitalcommons.unomaha.edu/biomechanicsarticles/132.

[45] Antonsson E.K., Mann R.W. 'The frequency content of gait'. *Journal of Biomechanics*. 1985;18(1):39–47. Available from: http://www.sciencedirect.com/science/article/pii/0021929085900430.

[46]   Au S.K., Herr H.M. 'Powered ankle-foot prosthesis'. *IEEE Robotics Automation Magazine*. 2008;15(3):52–59.

[47]   Wang S., Meijneke C., van der Kooij H. 'Modeling, design, and optimization of Mindwalker series elastic joint'. *2013 IEEE 13th International Conference on Rehabilitation Robotics (ICORR)*; 2013. pp. 1–8.

[48]   Zhang S., Qian J., Shen L., Wu X., Hu X. 'Gait complexity and frequency content analyses of patients with Parkinson's disease'. *2015 International Symposium on Bioelectronics and Bioinformatics (ISBB)*; 2015. pp. 87–90.

[49]   Fink C., Preiß R., Schöllhorn W.I. 'How to find the optimal cutoff frequency for filtering kinematic data?' *16 International Symposium on Biomechanics in Sports (ISBS)*; 1998. p. 3.

[50]   Gabriel D. 'Estimate of the optimum cutoff frequency for the Butterworth low-pass digital filter'. *Journal of Applied Biomechanics*. 1999;15(3): 318–329.

[51]   Van den Bogert A., de Koning J. 'On optimal filtering for inverse dynamics analysis'. *Proceedings of the IXth Biennial Conference of the Canadian Society for Biomechanics*; 1996. pp. 214–215.

[52]   Luca C.J.D. Myoelectrical Manifestations of Localized Muscular Fatigue in Humans. Bethesda, MD: CRC Critical Reviews in Biomedical Engineering; 1984.

[53]   Rosenfalck P. Intra- and Extracellular Potential Fields of Active Nerve and Muscle Fibers. Denmark: University of Copenhagen; 1969.

[54]   Booker G.M. Introduction to Biomechatronics (Materials, Circuits and Devices). Raleigh, NC: IET – Institution of Engineering and Technology; 2012.

[55]   Merletti R., Farina D. 'Analysis of intramuscular electromyogram signals'. *Philosophical Transactions of the Royal Society of London A: Mathematical, Physical and Engineering Sciences*. 2009;367(1887):357–368. Available from: http://rsta.royalsocietypublishing.org/content/367/1887/357.

[56]   Christensen H., Sögaard K., Jensen B.R., Finsen L., Sjögaard G. 'Intramuscular and surface EMG power spectrum from dynamic and static contractions'. *Journal of Electromyography and Kinesiology*. 1995;5(1): 27–36. Available from: http://www.sciencedirect.com/science/article/pii/S1050641199800030.

[57]   Luca C.J.D. 'The use of surface electromyography in biomechanics'. *Journal of Applied Biomechanics*. 1997;13(2):135–163. Available from: http://dx.doi.org/10.1123/jab.13.2.135.

[58]   'Standards for reporting EMG data'. *Journal of Electromyography and Kinesiology*. 2016;27:I – II. Available from: http://www.sciencedirect.com/science/article/pii/S1050641116300116.

[59]   van Boxtel A. 'Optimal signal bandwidth for the recording of surface EMG activity of facial, jaw, oral, and neck muscles'. *Psychophysiology*. 2001;38(1): 22–34. Available from: http://dx.doi.org/10.1111/1469-8986.3810022.

[60]   Bovi G., Rabuffetti M., Mazzoleni P., Ferrarin M. 'A multiple-task gait analysis approach: kinematic, kinetic and EMG reference data for healthy

young and adult subjects'. *Gait & Posture*. 2011;33(1):6–13. Available from: http://www.sciencedirect.com/science/article/pii/S0966636210002468.

[61]   Ma Y., Xie S., Zhang Y. 'A patient-specific EMG-driven neuromuscular model for the potential use of human-inspired gait rehabilitation robots'. *Computers in Biology and Medicine*. 2016;70:88–98. Available from: http://dx.doi.org/10.1016/j.compbiomed.2016.01.001.

[62]   Song M.H., Godøy R.I. 'How fast is your body motion? Determining a sufficient frame rate for an optical motion tracking system using passive markers'. *PLoS One*. 2016;11(3):14. Available from: http://doi.org/10.1371/journal.pone.0150993.

[63]   Muro-de-la Herran A., Garcia-Zapirain B., Mendez-Zorrilla A. 'Gait analysis methods: an overview of wearable and non-wearable systems, highlighting clinical applications'. *Sensors (Basel, Switzerland)*. 2014;14(2):3362–3394.

[64]   Altuntas C., Turkmen F., Ucar A., Akgul Y.A. 'Measurement and analysis of gait by using a time-of-flight camera'. *ISPRS – International Archives of the Photogrammetry, Remote Sensing and Spatial Information Sciences*. 2016;XLI-B3:459–464. Available from: http://www.int-arch-photogramm-remote-sens-spatial-inf-sci.net/XLI-B3/459/2016/.

[65]   Pau M., Coghe G., Atzeni C., *et al.* 'Novel characterization of gait impairments in people with multiple sclerosis by means of the gait profile score'. *Journal of the Neurological Sciences*. 2014;345(1–2): 159–163. Available from: http://www.sciencedirect.com/science/article/pii/S0022510X1400478X.

[66]   O'Connor B.J., Yack H.J., White S.C. 'Reducing errors in kinetic calculations: improved synchronization of video and ground reaction force records'. *Journal of Applied Biomechanics*. 1995;11:216–223.

[67]   Brand R.A., Crowninshield R.D. 'Comment on criteria for patient evaluation tools'. *Journal of Biomechanics*. 1981;14(9):655. Available from: http://dx.doi.org/10.1016/0021-9290(81)90093-2.

[68]   Zizoua C. 'Detecting muscle contractions using strain gauges'. *Electronics Letters*. 2016;52(2):1836–1838. Available from: http://digital-library.theiet.org/content/journals/10.1049/el.2016.2986.

[69]   Jung Y., Phan C., Koo S. 'Intra-articular knee contact force estimation during walking using force-reaction elements and subject-specific joint model'. *Journal of Biomechanical Engineering, ASME*. 2016;138(2):9. Available from: http://dx.doi.org/10.1016/0021-9290(81)90093-2.

[70]   Chen M., Gonzalez S., Vasilakos A., Cao H., Leung V.C.M. 'Body area networks: a survey'. *Springer Journal of Mobile Networks and Applications*. 2010;16(2):171–193.

[71]   Reddy G.P., Reddy P.B., Reddy R. 'Body area networks'. *Journal of Telematics and Informatics*. 2013; 1(1):36–42.

[72]   Toh W.Y., Tan Y.K., Koh W.S., Siek L. 'Autonomous wearable sensor nodes with flexible energy harvesting'. *IEEE Journal of Sensors Journal*. 2014;14(7):2299–2306.

[73] Movassaghi S., Abolhasan M., Lipman J., Smith D., Jamalipour A. 'Wireless body area networks: A Survey'. *IEEE Communications Surveys & Tutorials*. Third Quarter 2014;16(3):1658–1686. Available from: doi: 10.1109/SURV.2013.121313.00064

[74] Akita J., Shinmura T., Sakurazawa S., *et al.* 'Wearable electromyography measurement system using cable-free network system on conductive fabric'. *Journal of Artificial Intelligence in Medicine*. 2008;42(2):99–108. Available from: http://dx.doi.org/10.1016/j.artmed.2007.11.003.

[75] Ullah S., Khan P., Ullah N., Saleem S., Higgins H., Kwak K.S. 'A review of wireless body area networks for medical applications'. *International Journal of Communications, Network and System Sciences*. 2009;2(8):797–803.

[76] Pantelopoulos A., Bourbakis N.G. 'A survey on wearable sensor-based systems for health monitoring and prognosis'. *IEEE Transactions on Systems, Man, and Cybernetics, Part C: Applications and Reviews*. 2010;40(1):1–12.

[77] Crosby G.V., Ghosh T., Murimi R., Chin C.A. 'Wireless body area networks for healthcare: a survey'. *International Journal of Ad hoc, Sensor & Ubiquitous Computing*. 2012;3(3):1–26.

[78] Boulis A., Smith D., Miniutti D., Libman L., Tselishchev Y. 'Challenges in body area networks for healthcare: the MAC'. *IEEE Communications Magazine*. 2012;50(5):100–106.

[79] Liu J., Lockhart T.E., Jones M., Martin T. 'Local dynamic stability assessment of motion impaired elderly using electronic textile pants'. *IEEE Transactions on Automation Science and Engineering*. 2008;5(4):696–702.

[80] Post E.R., Orth M. 'Smart fabric, or wearable clothing'. *1st Intl. Symp. on Wearable Computers*, IEEE; 1997. pp. 167–168.

[81] Wade E., Asada H. 'Conductive fabric garment for a cable-free body area network'. *IEEE Journal of Pervasive Computing*. 2007;6(1):52–58.

[82] Yoo J., Lee S., Yoo H. 'A 1.12 pJ/b inductive transceiver with a fault-tolerant network switch for multi-layer wearable body area network applications'. *IEEE Journal of Solid-State Circuits*. 2009;44(11):2999–3010.

[83] Cho N., Yan L., Bae J., Yoo H.J. 'A 60 kb/s to 10 Mb/s adaptive frequency hopping transceiver for interference-resilient body channel communication'. *IEEE Journal of Solid-State Circuits*. 2009;44(3):708–717.

[84] Shikada K., Wang J. 'Development of human body communication transceiver based on impulse radio scheme'. *2nd IEEE CPMT Symp. Japan*; 2012. pp. 1–4.

[85] Ogasawara T., Sasaki A.I., Fujii K., Morimura H. 'Human body communication based on magnetic coupling'. *IEEE Transactions on Antennas and Propagation*. 2014;62(2):804–813.

[86] Nemati E., Deen M., Mondal T. 'A wireless wearable ECG sensor for long-term applications'. *IEEE Communications Magazine*. 2012;50(1):36–43.

[87] Inoue M., Itabashi Y., Tada Y. 'Development of bimodal electrically conductive pastes with Ag micro- and nano-fillers for printing stretchable E-textile systems'. *European Microelectronics Packaging Conference (EMPC)*; 2015. pp. 1–5.

[88] Quarterly T. 'Woven electronics: An uncommon thread'. *The Economist.* 2014/03/08. Available from: http://www.economist.com/news/technology-quarterly/21598328-conductive-fibres-lighter-aircraft-electric-knickers-flexible-filaments.

[89] Rattfalt L., Chedid M., Hult P., Linden M., Ask P. 'Electrical properties of textile electrodes'. *29th Annual International Conference of the IEEE Engineering in Medicine and Biology Society*; 2007. pp. 5735–5738.

[90] Stoppa M., Chiolerio A. 'Wearable electronics and smart textiles: a critical review'. *Sensors.* 2014;14(7):11957. Available from: http://www.mdpi.com/1424-8220/14/7/11957.

[91] Patel P.C., Vasavada D.A., Mankodi H.R. 'Applications of electrically conductive yarns in technical textiles'. *IEEE International Conference on Power System Technology (POWERCON)*; 2012. pp. 1–6.

[92] Sandrolini L., Reggiani U. 'Assessment of electrically conductive textiles for use in EMC applications'. *International Symposium on Electromagnetic Compatibility – EMC Europe*; 2009. pp. 1–4.

[93] Šafářová V., Malachová K., Militký J. 'Electromechanical analysis of textile structures designed for wearable sensors'. *Proceedings of the 16th International Conference on Mechatronics – Mechatronika*; 2014. pp. 416–422.

[94] Uzun M., Sancak E., Usta I. 'The use of conductive wires for smart and protective textiles'. *E-Health and Bioengineering Conference (EHB)*; 2015. pp. 1–4.

[95] Tao D., Zhang H., Wu Z., Li G. 'Real-time performance of textile electrodes in electromyogram pattern-recognition based prosthesis control'. *Proceedings of IEEE-EMBS International Conference on Biomedical and Health Informatics*; 2012. pp. 487–490.

[96] Paiva A., Carvalho H., Catarino A., Postolache O., Postolache G. 'Development of dry textile electrodes for electromyography a comparison between knitted structures and conductive yarns'. *9th International Conference on Sensing Technology (ICST)*; 2015. pp. 447–451.

[97] Pylatiuk C., Muller-Riederer M., Kargov A., *et al.* 'Comparison of surface EMG monitoring electrodes for long-term use in rehabilitation device control'. *Proceedings of IEEE 11th Internat. Conf. on Rehabilitation Robotics (ICORR 2009)*; 2009. pp. 300–304.

[98] Meziane N., Webster J.G., Attari M., Nimunkar A.J. 'Dry electrodes for electrocardiography'. *Physiological Measurement.* 2013;34(9):R47. Available from: http://stacks.iop.org/0967-3334/34/i=9/a=R47.

[99] Trindade I.G., Machado da Silva J., Miguel R., *et al.* 'Design and evaluation of novel textile wearable systems for the surveillance of vital signals'. *Sensors.* 2016;16(10):1573. Available from: http://www.mdpi.com/1424-8220/16/10/1573.

[100] Neuman M.R. Biopotential Electrodes. In: Bronzino J, editor. *The Biomedical Engineering Handbook*, 2nd ed. 2 Volume Set. Boca Raton, FL: CRC Press; 2000.

[101]   Albulbul A., Chan A. 'Electrode-skin impedance changes due to an externally applied force'. *2012 IEEE International Symposium on Medical Measurements and Applications Proceedings (MeMeA)*; 2012. pp. 1–4.

[102]   Puurtinen M., Komulainen S., Kauppinen P., Malmivuo J., Hyttinen J.A.K. 'Measurement of noise and impedance of dry and wet textile electrodes, and textile electrodes with hydrogel'. *28th Annual International Conference of the IEEE Engineering in Medicine and Biology Society. EMBS'06*, 2006, pp. 6012–6015.

[103]   Taji B., Shirmohammadi S., Groza V., Batkin I. 'Impact of skin–electrode interface on electrocardiogram measurements using conductive textile electrodes'. *IEEE Transactions on Instrumentation and Measurement.* 2014;63(6):1412–1422.

[104]   A. Lymberis and L. Gatzoulis, "Wearable health systems: From smart technologies to real applications," *2006 International Conference of the IEEE Engineering in Medicine and Biology Society*, New York, NY, 2006, pp. 6789–6792. Available from: doi: 10.1109/IEMBS.2006.260948

[105]   Gatzoulis L., Iakovidis I. 'Wearable and portable eHealth systems'. *IEEE Engineering in Medicine and Biology Magazine.* 2007;26(5):51–56.

[106]   Ismail Y.I., Friedman E.G. 'Repeater insertion in RLC lines for minimum propagation delay'. *ISCAS'99. Proceedings of the IEEE International Symposium on*; vol. 6; 1999. pp. 404–407.

[107]   Carvalho J., Machado da Silva J. 'A transceiver for E-textile body-area-networks'. *MeMeA 2014, 9th IEEE International Symposium on Medical Measurement*; 2014.

[108]   Derogarian F., Canas Ferreira J., Grade Tavares V. 'A routing protocol for WSN based on the implementation of source routing for minimum cost forwarding method'. *SENSORCOMM 2011, The Fifth International Conference on Sensor Technologies and Applications*; 2011. pp. 85–90.

[109]   Derogarian F., Dias R., Ferreira J.C., Tavares V.G., da Silva J.M. 'Using a wired body area network for locomotion data acquisition'. *27th Conference on Design of Circuits and Integrated Systems (DCIS 2012)*; Avignon, France; 2012.

[110]   Trigno Wireless System; 2018. Available from: https://www.delsys.com/products/wireless-emg/.

[111]   System 4 Pro – Dynamometers – Physical Medicine | Biodex; 2017. Available from: http://www.biodex.com/physical-medicine/products/dynamometers/system-4-pro.

Chapter 11

Integration of sensing devices and the cloud for innovative e-Health applications

Albena Mihovska[1], Aristodemos Pnevmatikakis[2], Sofoklis Kyriazakos[1], Krasimir Tonchev[3], Razvan Craciunescu[4], Vladimir Poulkov[3], Harm op den Akker[5,6], and Hermie Hermens[5,6]

E-health environments are extremely complex and challenging to manage, as they are required to cope with an assortment of patient conditions under various circumstances with a number of resource constraints. E-health devices usually indicate a piece of equipment with the mandatory capabilities of communication and some optional capabilities of sensing, actuation, data capture, data storage and data processing. The devices would collect various kinds of information and provide it to the information and communication networks for further processing and would operate in a dynamic environment, which would require unobtrusive monitoring and interacting with the inhabitants (i.e., patients) while they perform their activities of daily living (ADL). The devices should be able to recognise abnormal events as well as slowly emerging shifts in behaviour, and able to inform associated users (caregivers, healthcare professionals, family) appropriately and timely, in order to provide a feeling of safety and comfort for all involved parties. This chapter describes an e-Health Home Caring Environment designed to cater to patients with Mild Cognitive Impairments (MCI), Chronic Obstructive Pulmonary Disease (COPD) and seniors with frailty conditions. It details the required software and hardware technological innovations to design an affordable, easy-to-install smart "caring home" cognitive environment, which "senses" intuitively the wishes and "learns" the needs of the person living in the home. As a result, the environment provides unobtrusive daily support, notifying informal and formal caregivers when necessary and serving as a bridge to supportive services offered by the outside world.

[1] Department of Business Development and Technology (BTECH), Aarhus University, Herning, Denmark
[2] Athens Information Technology, Multimodal Signal Analytics Group, Greece
[3] Teleinfrastructure R&D Laboratory, Faculty of Telecommunications, Technical University of Sofia, Bulgaria
[4] Telecommunication Department, University Politehnica of Bucharest, Romania
[5] Roessingh Research and Development, Telemedicine Group, The Netherlands
[6] Biomedical Signals and Systems Group, Faculty of Electrical Engineering, Mathematics and Computer Science, University of Twente, The Netherlands

11.1   Introduction

The design of e-health environments has the objective to provide personalised services and applications to their primary users (i.e., the patients) by breaking the barrier of technology acceptance and addressing their daily needs, under strict regulation and security constraints. A typical scenario would employ wireless and wired sensors and local or cloud-based processing units to collect, process, store and communicate data related to the patients' needs and condition. E-health devices can be located on the patients' bodies or immediate environments to monitor and interact with the patients, while they perform their activities of daily living (ADL). The system should be able to recognise abnormal events, as well as slowly emerging shifts in behaviour of the primary users, and to inform secondary users (caregivers, healthcare professionals, family) appropriately and timely, hence providing a feeling of safety and comfort for all involved parties.

This chapter describes an e-Health Home Caring Environment designed to cater for patients with MCIs, COPD and seniors with frailty conditions [1–4]. It details the required software and hardware technological innovations to design an affordable, easy-to-install smart "caring home" cognitive environment, which "senses" intuitively the wishes and "learns" the needs of the person living in the home. As a result, the environment provides unobtrusive daily support, notifying informal and formal caregivers when necessary and serving as a bridge to supportive services offered by the outside world. This chapter is organized as follows: Section 11.2 overviews current e-Health networks, the required enabling technologies and devices and the challenges to their real-life deployment. Section 11.3 describes a case study of an e-Health system based on the eWALL platform [1–3], which integrates a sensing environment and a cloud environment. Section 11.4 analyses the eWALL home environment, addressing the devices, their networking and the processing of their signals. Section 11.5 discusses the eWALL cloud environment, analysing reasoning techniques and the applications offered to the user. Section 11.6 analyses the innovation potential of the eWALL system, and lists some of the outstanding challenges in the area of devices, e-Health policies and so forth for unleashing it fully. Section 11.7 concludes the chapter.

11.2   Requirements for e-health technical solutions

The average life expectation worldwide and the expected quality of life of the ageing or chronically ill population have drastically increased over the past years. This has been also the main goal to look for viable and user-friendly ambient assisted living (AAL) solutions. The form of elderly care provided varies greatly among countries and there are specific requirements related to the personal lifestyle of the primary users (e.g., the patients), which need to be incorporated in the offered services. Traditionally, elderly care has been the responsibility of family members and was provided within the extended family home.

Impaired mobility is a major health concern for older adults, affecting 50% of people over 85 and at least a quarter of those over 75. As adults lose the ability to walk, to climb stairs and to rise from a chair, they become completely disabled. The problem should be addressed because people over 65 constitute the fastest growing segment of the western world population. The therapies that are designed to improve the mobility in elderly patients would usually be built around diagnosing and treating specific impairments, such as reduced strength or poor balance. It is appropriate to compare older adults seeking to improve their mobility to athletes seeking to improve their split times. People in both groups perform best when they measure their progress and work towards specific goals related to strength, aerobic capacity and other physical qualities. Someone attempting to improve an older adult's mobility must decide what impairments to focus on, and in many cases, there is little scientific evidence to justify any of the options.

Depending on the patient's condition and lifestyle, various therapies may be offered. What the primary target users of an e-Health system would have in common, regardless of whether they belong to a cognitive or to a physical impairment group, is that they may lose trust in their cognitive or physical abilities and they would then gradually abandon their former activities. As a result, their confidence and mood are affected and they often succumb in isolation. This has an impact on their overall health and social life. Moreover, rehabilitation therapies are expensive and have certain duration; even if they are proven to be effective, patients often lack the motivation to continue the training on their own.

This cycle is shown in Figure 11.1.

In Figure 11.1, the main points of the cycle described in the attitude of various types of patients can be identified; almost all personas, without exception, experience loss of self-trust and subsequently, loss of motivation. The result is isolation, lack

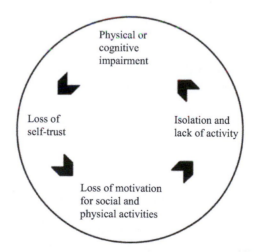

Figure 11.1   Common characteristics of a primary user profile in an ambient environment

of activity and low or no adherence to therapy, which can potentially deteriorate the physical and/or cognitive state of the patient even further. Therefore, we should strive that these two emotional parameters, self-trust and motivation, are placed at the core of an e-Health system.

Most of the existing AAL systems address a broad audience; from cognitively healthy adults to older people with diagnosed MCI and even people with dementia. Thus, they provide a wide range of functionalities, most of which have a common core, the monitoring of the user and the user environment. Further, health has many different forms of correlation/association with the following factors, which are location- and person-dependent:

- Lifestyle (eating, physical activities, working environment, use of hazardous materials);
- Weather conditions (rainy, cloudy, snowing);
- Outdoor air quality (temperature, humidity, pressure);
- Indoor air quality (temperature, humidity, pressure);
- Stress factors (work environment, car traffic, noise pollution);
- Electromagnetic pollution.

Information about the user in the immediate environment can be collectively obtained from various devices integrated into a network [5,6], and which, generally can be considered as uncertified information sources:

- Sensors integrated within the smartphone (accelerometers, gyroscopes, microphone, camera, GPS receiver, luxmeter);
- Devices connected with the smartphone (bluetooth cardio bracelet, heart rate monitor, pedometer, digital thermometer, weighing scale, etc.);
- Weather databases from organisations which distribute current and historical weather information;
- Air pollution monitoring stations;
- Outdoor air quality stations;
- Indoor air quality stations.

The information provided by the cloud environment is generally data from certified information sources, and originates from:

- Patient record data (from hospitals);
- Databases of medical laboratories;
- Portable medical devices such as blood glucose meter, pulse oximeter.

A typical patient's portfolio will include motion-type of exercises, dietary advice, cognitive games as a mind stimulator and so forth. The supporting system should be highly secured, responsive and controlled and one, where the users' privacy and the protection of their personal data, remains intact. Because a large amount of personal information is required to generate personalised care services, the providing systems demand integrity, accessibility and availability along with interoperability, which is even more important with the colossal pool of data defining the infrastructure of today. In general, e-Health services and applications rely on the collection, storage, processing and analysis of data, which to a large extent is user-related and often made available to different sectors irrespective of boundaries ranging from machine learning and engineering to economics and medicine [3]. Therefore, an effective e-Health system will rely on collectively acquired up-to-date knowledge regarding all these domains, from medical symptoms of certain diseases to data collected by in-home monitoring technology. This can be achieved by the integration of various devices and software into a complete platform. The collected data will then be exchanged and transformed into actionable intelligence by means of a data mining computing platform (i.e., the cloud environment). For enabling high level of personalisation, full knowledge on the profiles of the end users needs to be implemented into the latter [4].

In-home monitoring refers to the process of acquiring sensor and audiovisual data focused on the user and the respective home environment and applying machine learning, fuzzy logic and other artificial intelligence algorithms to make sense out of it. In-home monitoring primarily aims to give insights into the user's life; particularly his/her behaviour and performance of ADL activities, the user's adherence to coaching instructions and the condition of the home, in terms of safety. Moreover, data collected over longer periods of time can result in identifying patterns and warning signs in the user's life. The insights gathered by in-home monitoring are utilised in the following domains:

- Feedback to the user his/herself and communication with his/her family and/or his/her caregivers. The content of the communicated message can be ADL support reminders, home automation notifications, alerts for dangerous situations, progress reports.
- *Physical training*: in-home monitoring is used to determine the adherence of the user during physical training and/or the health status of the user in order to adjust it accordingly. Motion sensing provides the system with the necessary input to offer an interactive training to the user, according to his/her abilities and pace.
- *Cognitive training*: in-home monitoring can derive insights regarding the user's cognitive state and consequently, propose therapy, action to be taken or training. An example of monitoring in cognitive coaching is in dem@care, where the user is asked to repeat a meaningful sentence after the system has pronounced it. According to his/her memory of the sentence, the system can produce a diagnosis of the cognitive functioning of the user.
- Home automation: functionalities of this cluster include remote control of windows, doors, cameras and other devices or control with badges.

An example of the various devices that may be integrated into the in-home environment to provide input about the patient's state is shown in Figure 11.2.

The requirements for an e-Health system, therefore, should include the user's point of view and the provider's point of view. The following two groups of requirements can be summarised:

(1) System requirements:

- Flexibility – the ability to support a variety of market available or specifically developed user and network devices;

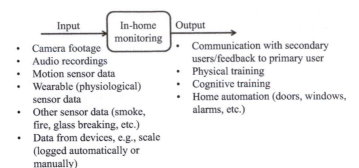

Figure 11.2   Input and output of in-home monitoring

- Scalability – the ability to easily scale the AAL platform to all envisioned use-cases;
- Traceability – the ability to log and track the taken actions throughout the platform operation;
- Extensibility – the ability to easily integrate novel devices in the platform (transparent protocol formats and protocol messages);
- Reliability – the ability to provide reliable communication within the platform and always ontime reaction;
- Compatibility – the ability to integrate various information from various devices in a user-transparent manner;
- Responsiveness –the ability to dynamically react and/or reconfigure platform elements;
- Multiuser capability – the ability to support multiple users with guaranteed profiling;
- User mobility detection – the ability to track the position and the movement of users in-house (important for various AAL services);
- Emergency reaction capability – the ability to extract contextual information and respond accordingly;
- Interoperability – the ability to interconnect with other systems (e.g., epSOS);
- Security – the ability to secure the eWALL users' data from obtrusive and accidental eavesdropping;
- Social networking – the ability to interact with family and friends in order to ease the potential social isolation experienced;
- Audio and video (A/V) interaction with the user – the ability to supply appropriate and user relevant information in audio and visual formats;
- Privacy – the ability to keep personal information from being disclosed and shared with unauthorised parties;
- Context information – the ability to provide context information that is useful for services to adapt themselves according to the needs, preferences and situation of the user;

- Service orientation – the ability of a system to ensure reusability and composability of services and service components;
- Semantic interoperability – the ability to enable semantic interoperability between applications and services for ensuring the highest degree of decoupling (enables an open system and facilitates reuse of existing services and applications);
- Maintainability and configurability – the ability to easily maintain and configure the system after deployment;
- Multi-modal user interaction – the ability to support multi-modal user interaction;
- User data separation – the ability to create pseudoidentifiers for privacy protection supported by the system user and network devices;
- Distributed decision making – the ability to make decisions in the patient's home system in addition to the AAL cloud environment;
- Anonymity – the ability to switch off sensors and devices and manage the deletion of raw data from these sensors.

Some specific system requirements may include the following aspects:

- User pattern recognition and detection – the ability to extract repetitive user behaviours;
- Interference detection and localisation – the ability to detect potential interference sources (important for the wireless communication part within the AAL platform);
- Reconfiguration – the ability to reconfigure the system and technical parameters according to the environmental context;
- Synchronisation – the ability to provision synchronous operation of user devices and sensors;
- Spectrum sensing – the ability to support opportunistic spectrum access and interference minimisation;
- Learning – the ability to learn and store user behaviours;
- Reasoning – the ability to reason upon stored user behaviours and choose appropriate actions (e.g., reminders for taking prescribed medicine);
- Adaptive A/V formats – the ability to support A/V bit rate adaptation to accommodate various possible display devices;
- Remote accessibility – the ability to provide remote access to the AAL platform;
- Priorities management – the ability to handle different simultaneous requests and messages with different priority levels;
- A/V-based user tracking – the ability to process audio and visual data from in-house sensors in order to track humans, eye movements, etc.;
- Monitoring of household appliances – the ability to handle potential hazardous situations when household appliances are turned on;
- Memory training games – the ability to interact with the users and aid their potential memory loss problems;
- Ambiental parameters monitoring – the ability to monitor critical ambiental parameters such as temperature, air quality, smoke, etc.;
- Monitoring users' vital parameters – the ability to monitor (using wearable sensors) body temperature, blood pressure, etc.;

- Monitoring users' associated vital parameters – the ability to monitor vital parameters that affect the users' status such as sleep length, sleep quality, amount of conducted physical activity, etc.;
- Identification, authentication and authorisation – the ability of system components to identify, authenticate and authorise an entity (human users and other system components) that wants to use them before allowing them access to resources;
- Confidentiality – the ability to maintain confidentiality (the way in which the information disclosed or managed by the system is treated) of identifiable data, including controls on storage, handling and sharing of data;
- Integrity – the ability to detect data modifications and prevent unauthorised modifications, especially related to service user data, sensor data and commands sent to actuators.
- Non-repudiation – the ability to trace back every action on sensitive assets to the person or system component that performed it;
- Auditing – the ability of a system to log all actions on sensitive assets, including failed access attempts;
- Consent specification – the ability to provide a usable interface to capture the consent of the end user about sharing data with services;
- Communication – the ability to enable inter-component message-based (or event-based) and call-based communication between distributed components;
- Context history – the ability to provide access to past/historic context information (the history length will depend on the actual context information);
- Conflicting context information – the ability to provide means for resolving conflicting context information coming from different context sources;

(2) User requirements:

- Technological transparency and unobtrusiveness;
- Safety;
- Security;
- Trust that private data will not be breached.

The acceptability of a sensor-based e-Health system depends on the risks felt by older adults, caregivers, hospitals, public authorities and policymakers. Unobtrusiveness is an important feature that can enhance the user experience. One way to achieve it is a simple approach to the implementation of the human–computer interaction, for example, by means of hand gestures, thus avoiding A/V recognition. This can be realised by use of infrared sensors for gestures detection and an Arduino Leonardo board to process them as part of the hardware components, integrated with a method based on the hand position detection to recognise the gestures (also known as, the software component). Different gestures can be detected based on the position and the moving direction of the hand. A physical implementation of a prototype non-A/V gesture control interface is shown in Figure 11.3. With a good calibration of the sensors and good software components, a low-cost non-A/V system can recognise

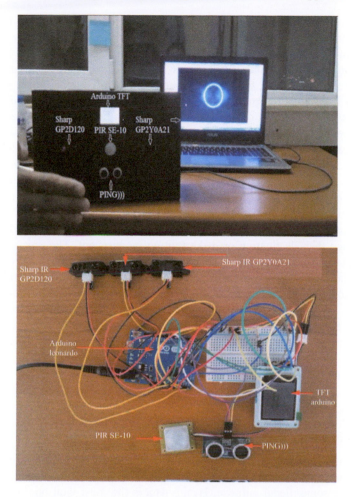

Figure 11.3   Prototype of a gesture control interface

a hand gesture in at least 95% of the cases with a low CPU usage of the microcontroller and a time of execution (from the gesture is made to the time the PC executes the command) of minimum 60 ms per gesture. The prototype has been made user-friendly by developing a PCB to integrate the Arduino board and the sensors and designing a case. It integrates on an open source Arduino development board based on the Atmega32u4 microprocessor the following hardware components: an Arduino TFT LCD Screen with a 1.77 in. diagonal, and 160 × 128 pixel resolution; calibrated sensors comprising ultrasonic-PING sensors (active sensors that emit an ultrasonic wave at a frequency greater than 20 kHz and receive the signal reflected on the target object); passive infrared (PIR) sensors and calibrated SHAPR infrared sensors for accurate distance measurement. If it is possible to detect a gesture based on the output of one sensor, then this sensor is kept active and the other ones are disabled.

The output of the active sensor is read and interpreted (translated into a command that is sent to computer). If no more gestures can be detected, the other sensors are enabled and the reading cycle restarts. Arduino emulates the mouse and keyboard to send commands to computer. For example, if the user moves his/her hand from left to right/from right to left, then the command sent is "press right arrow key"/"press left arrow key". In order to do this, Arduino has to emulate the keyboard. If the user moves his/her hand forward/backward, then Arduino has to emulate the mouse and sends the command for scroll wheel movement.

In healthcare applications, the use of the sensor devices for monitoring poses similar security and privacy risks as the ones reported for sensor networks. Since healthcare applications of sensor networks and conventional wireless sensor networks (WSNs) applications are similar, most of their security threats are equivalent.

Security threats and attacks can be classified as passive and active [7]. For example, a passive attack may occur while routing data packets in the network. Attackers may change their destination or make routing inconsistent. They may also "steal" health data by eavesdropping to the wireless communication media. Active attacks cause greater damage than their passive counterparts. For example, attackers may find the location of the user by eavesdropping which could lead to life-threatening situations. The common design of sensor devices incorporates limited external security features and, therefore, makes them prone to physical tampering. This increases the vulnerability of the devices and poses more complex security challenges. Most commonly, the attacks in health monitoring are related to eavesdropping and modification of medical data, forging of alarms on medical data, denial of service, location and activity tracking of users, physical tampering with devices and jamming attacks. Also, people with depraved intent may use the information for harmful activities. The generic attacks, which can occur in a WSN-based healthcare system, are classified as follows:

- Data modification – The attacker can delete and/or replace part or all of the information and send the modified information back to the original receiver to achieve some illegal purpose. Health data is the most vital one in this case of attacks. Modifying them may result in system failure and cause severe problems regarding the persons' health.
- Impersonation attack – If an attacker eavesdrops a wireless sensor node's identity information, it can be used to deceive the other nodes by impersonating as a valid sensor node.
- Eavesdropping – In case of open (unsecured) wireless channels, any opponent can intercept radio communications between the wireless sensor nodes.
- Replaying – The attacker can eavesdrop a piece of valid information and resend it to the original receiver after a while to achieve the same purpose in a totally different case.

The increasing use of big data analytics brings about additional challenges to consider when designing an e-Health system [8,10]. There are numerous security and privacy concerns in moving the health data under the big data approach. The privacy of the patients and the safeguarding of their personal data are major issues in applying big data analytics to e-Health. It is important that an e-Health Data Privacy Model

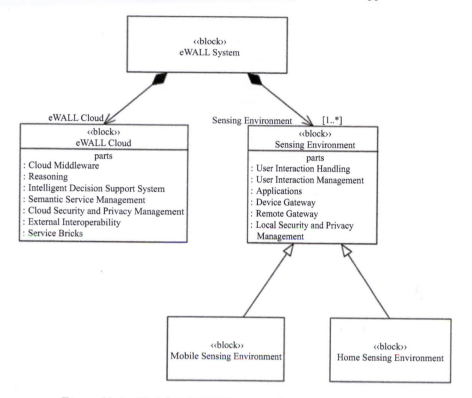

Figure 11.4   High-level eWALL system decomposition model [8]

for enabling transparency in the flow of data over the network and that only the data relevant to a particular health service provider would be delivered [9].

11.3   eWALL system

eWALL [1–4] is an AAL caring environment with a high degree of personalisation and interaction. The system was designed to comply with all the requirements specified in Section 11.2.

The eWALL system is composed of two main subsystems: (i) the eWALL Sensing Environment and (ii) the eWALL Cloud. These systems are detailed further in this Chapter. The eWALL Sensing Environment is a logical environment, deployed over a physical space, which is mainly responsible for explicit and implicit interaction with the primary user. Implicit interaction is referring to the collection of various data about the user and its environment from medical and environmental sensors and control of environment through actuators. Explicit interaction is referring to the direct interaction with the user using A/V devices and user interaction sensors. The eWALL Cloud is a central processing and data storage subsystem. A high-level system decomposition model is shown in Figure 11.4.

Figure 11.4 is a SysML Block Definition Diagram (BDD) of the high-level components of the eWALL system. There are two types of eWALL Sensing Environment, namely, a *stationary* one that is referred to in the rest of this chapter as the Home Sensing Environment (HSE) and is related to the physical surrounding of the user in its home, and the Mobile Sensing Environment (MSE) centralised around the mobile device of the user when the user is outside of the home environment.

11.4   eWALL home environment

The home providing the e-Health services needs to be aware of the care recipient. This can only be done by taking measurements, either of the environment or of the body of the care recipient. Hence such a home is equipped with sensors, fixed at the home, used occasionally or worn by the care recipient. All these sensors are networked to transfer their data to the cloud.

Some of the data though can be quite sensitive or simply too bulky to send to the cloud. Video data are examples of both categories: Transferring them requires excessive bandwidth, poses security risks and raises ethical questions. To accommodate for such cases, the data are processed locally, sending away only anonymised metadata to the cloud.

Hence the data from the sensors are gathered from the sensor boards into a home processing unit usually utilising wireless radios. There they are distributed to the different processing algorithms via the device gateway. The resulting metadata are indexed in a local database, from which they are streamed via the remote proxy to the cloud. This is shown in Figure 11.5.

The choice of the devices is not trivial. They can be broadly categorised as experimental, commercial or consumer. They differ in many aspects ranging from

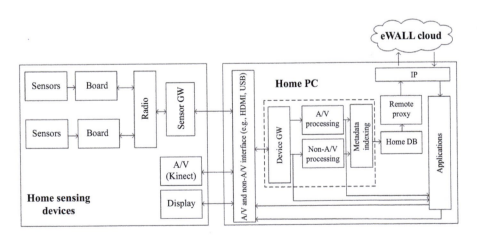

Figure 11.5   Home environment, comprising sensors and the home processing unit

ease of use, robustness and regulations compliance to cost, battery life and integration potential.

Experimental devices are typically used in research laboratories for developing proof-of-concept solutions of theoretical research ideas. They usually provide flexibility for researchers to deploy various algorithms and protocols for processing and networking purposes. However, the experimental devices almost always come without organised support and may be unstable under different circumstances. Their robustness, battery life and regulations compliance are problematic for practical deployments.

Nevertheless, they are the easiest and the most convenient enabler of rapid prototyping in laboratory conditions.

Commercial devices are more reliable than the experimental ones for practical deployments, but they usually only apply to niche markets satisfying specific market needs. The level of transparency of these devices for potential integration into end-to-end e-Health solutions strongly depends on their type and their manufacturer. These devices usually provide a high level of robustness in exchange for a steep price tag.

Consumer devices are widely used by the public and are cheaper than the commercial devices. These devices are not very flexible for experimentations as the resulting measurements are controlled by their manufacturers. Web-based APIs are provided, but in many cases exposing a limited range of the measurements. Real-time data provision is out of the question since they require synchronisation with the manufacturer's server. Nevertheless, these devices are sometimes preferred as they can provide the best user experience in terms of robustness and battery life.

The eWALL home environment adopts a cooperative approach to the choice of its networked devices in the caring home. The caring home's needs and the devices' capabilities led to an integration of various types from all three device categories. The resulting home environment is shown in Figure 11.6.

All the devices are connected to the home PC for collecting and processing the home signals to extract the care recipient's context. The primary user interface (UI) device is a large touch screen. An optional Android smartphone is the secondary UI device, also facilitating the collection of data from consumer devices like the wearable Fitbit Charge HR activity/sleep monitor and the Beddit sleep monitor. Two other consumer devices are the Plugwise Stick and Circles that form a ZigBee mesh network for socket sensing and controlling, and the Philips Hue gateway and lamps for lighting controlling. Audiovisual sensing in the living room is facilitated by the Kinect for Xbox One and the required PC adapter. Three sensors for monitoring vital signs are connected via Bluetooth: (i) the Nonin pulse oximeter, (ii) the Omron blood pressure monitor and (iii) the ThinkLabs One stethoscope. These are meant to be used by the care recipient at given times every day, or during exercising at home. The environment monitoring sensors are experimental; the Arduino explorer communicates via ZigBee radio with custom packaged Arduino sensing modules, one per monitored room.

The non-audiovisual processing is 2-fold. The trivial part is handling the environmental signals to model the expected sensor values and understand deviations from those. Much more elaborate processing is needed to distinguish the heart and

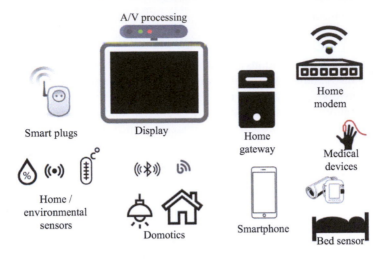

Figure 11.6   The eWALL home environment

lung signals from the stethoscope and measure metadata related to these two organs. There, denoising is followed by blind source separation. The latter employs non-negative matrix factorisation and an unsupervised basis spectral learning technique, detailed in [1].

The audiovisual processing is about voice commands (handled by the Kinect software) and face tracking and analytics. A real-time face tracker for pose and 3D position that utilises a single camera and face detection as the only visual measurement on the frames is built. This is made possible by the enhanced persistence offered by the selected face detector based on the histograms of oriented gradients. Facial landmarks' detection facilitates pose and 3D position estimation of the tracked faces. The detector together with elaborate initialisation and target handling modules are integrated in a face tracking system based on Kalman filters, offering enhanced face tracking and processing speed performance, no matter the visual complexity of the scene. The algorithm is detailed in [2].

11.5   eWALL cloud environment

The eWALL cloud environment consists of the reasoning, applications and data persistence parts.

11.5.1   *Reasoning*

The reasoning within the eWALL is implemented through different types of Lifestyle Reasoners and the virtual coach [4,11]. Lifestyle Reasoners within eWALL are components that process and store long-term data that follows certain patterns or routines that define the lifestyle of the user. The aim of these components is to predict behaviour and to detect (slow or fast) changes that might indicate a change in the user's health status. To do so, the Lifestyle Reasoners consume medium-level data (e.g., processed data from service bricks) from multiple sources and derive semantically meaningful patterns. The data is processed, stored and compared with medium- and long-term data stored in the cloud and the reasoner determines, e.g., whether a variation falls within the expected thresholds, or employs more complex methods to determine deviation. These results can then be exposed through an API for use in other applications or other processing components. The output of a reasoner can also be an input for a combined Lifestyle Reasoner. The main distinction between Lifestyle Reasoners and the Intelligent Decision Support System (IDSS) is that the Lifestyle Reasoners make decisions about the short-, medium- or long-term past, whereas the IDSS reasoners reason about the *now*.

Given this distinction between IDSS components and Lifestyle Reasoners, the Lifestyle Reasoners are for the most part "passive" components in the eWALL cloud, in that they tend not to initiate immediate engagement with the user. The only exception to this is the Vital Signs Lifestyle Reasoner where detection of a deviation of a lifestyle pattern could be considered so important that the reasoner itself sends notifications to the primary user and/or caregivers. In general though, the reasoning over long-term does not necessitate alerting and as such the components are designed and set up as information sources. Examples of some of the reasoners implemented in the eWALL are given in Sections 11.5.1.1–11.5.1.6.

11.5.1.1 Vital Signs Lifestyle Reasoner

The Vital Signs (VS) Lifestyle Reasoner (LR) gathers information about the primary user's vital signs with respect to their *heart rate, blood pressure, heart rate variability* and *oxygen saturation*. Based on the collected data, the reasoner analyses and tracks the medical condition of the primary user. The focus of the reasoner is to monitor the primary user's health status and to identify possible changes that develop in the user's health such as occurrence of hypertension, atrial tachycardia, decreased heart rate variability, etc.

11.5.1.2 Daily Physical Activity Lifestyle Reasoner

The Daily Physical Activity Lifestyle Reasoner [11] has as main goal to calculate and keep track of a personalised (tailored) daily physical activity pattern for each individual user, which is subsequently used in the goal-setting IDSS component to set a suitable goal for each user and for each specific day of the week. The Daily Physical Activity Lifestyle Reasoner is part of the suite of services and applications in eWALL that support everyday physical activity, as shown in the overview in Figure 11.7. The acronyms used in the figure are for Physical Activity (PA) and Main Screen (MS), which is the primary UI on the eWALL touch screen device.

As can be seen from the figure, the Lifestyle Reasoner is only a small part in a complex set of modules that provide the full daily activity service. Nevertheless, it provides a crucial level of intelligence that is required to provide appropriate motivation for our users. The core functionality of this reasoner is to categorise each day

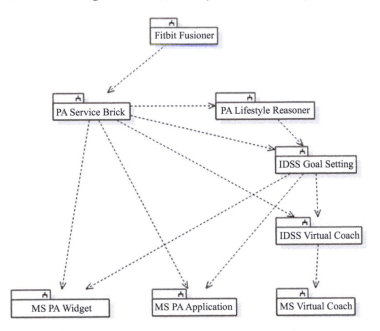

Figure 11.7   High-level overview of the software modules involved in the daily physical activity service suite for eWALL

of the week according to the amount of physical activity performed on those days. In other words, each day of the week is labelled according to the averaged amount of physical activity performed on that day.

11.5.1.3   Mood lifestyle reasoner

Mood and the emotional state of the user are valuable characteristics of well-being. This reasoner is devoted to monitoring the mood of the user across time and register when the user is in a low or bad mood. Based on this, recommendations that aim to improve the user's mood are issued. The main challenge in machine mood classification is that a specific user may express fairly complex and dynamic mood changes. These changes are usually conditional on a large set of possible states that the user can experience which imposes difficulties. The reasoner is activated at the end of the day and analysis is done based on the collected data through the day. If there is not enough data for reliable estimate of the mood a history data is requested for no more than 5 days and the mood is estimated based on this sample.

Daily functions are important cues to the well-being of a care recipient. eWALL monitors them automatically, adding to the system the ability to offer help, both in the form of motivational messages to the care recipient and in the form of notifications and alarms to the care providers.

11.5.1.4   Well-being Advertisement Reasoner

In order to provide notifications to older adults regarding eating, drinking, showering, going for a walk, etc., we created a concept that we name "wellbeing advertisements". These are graphic advertisements of different well-being activities that are shown for a little while on the eWALL screen with impressive graphics and motivational scope. Since eWALL is of quite impressive dimensions, the well-being ads make eWALL appear like user's personal home billboard. This is a completely different paradigm from the notifications used so far for such kind of activities, which asked for feedback and/or interrupted the user in the middle of another action. This reasoner is the intelligent component in the eWALL cloud behind the well-being advertisements that are shown on the eWALL Main Screen. These well-being advertisements are aimed to provide subtle reminders that aim to nudge people towards a certain behaviour. The Well-being Advertisement Reasoner decides which advertisement is most relevant to show to the user, based on his/her historical behaviour and his/her current context. The well-being ads provide a non-intrusive way to stimulate user behaviour.

11.5.1.5   Virtual Coach

The Virtual Coach reasoner is a combination two reasoners – the "Physical Activity Message Timing" and "Physical Activity Message Content". These two reasoners are merged for technical reasons and for their combined functionality but their aims remain the same. The initial goal of this reasoner is to provide motivational messages to the user regarding his/her daily levels of physical activity, but actually the core principle of the system is to gather as much information as possible about the user and his/her context, and use this information in deciding on what to say and how to say.

11.5.1.6   Automatic Goal Setting

The Automatic Goal Setting IDSS Reasoner sets personalised goals with minor or no involvement from the healthcare professional. The goals are set based on the user's previous behaviour (user's routine) and healthcare recommendations (national guidelines or direct involvement of healthcare professional through the eWALL web portal). The Goal Setting IDSS Reasoner runs once every week and defines the goals for the upcoming week. When a new goal needs to be set, the user's **pattern of daily physical activity** is averaged and the **average "end points"** compared to the **ultimate goal**. Days that are considered *outliers*, given the lifestyle pattern of the user (this task is performed by the Daily Physical Activity Lifestyle Reasoner) are not considered in this calculation. If the average daily point is close to the ultimate goal, the new goal set corresponds to the ultimate goal. Otherwise, the new goal will be the average daily end point weighted by a **deviation allowance factor**.

11.5.2   Applications

The applications are one of the core components of eWALL, as they provide the interface between the user and the system. Due to this, it was important to design them in a way that provides good quality of service and user-friendly interfaces. The

Daily Functioning Monitoring (DFM) utilises the sensing environment in t
e-care home to understand what is happening in the daily life of the care recipien
From going to bed, to cooking and receiving visitors, all the activities have thei
importance in understanding the well-being of the care recipient.

DFM in eWALL is achieved utilising two probabilistic models: (i) an intermediate
one gives the probability of the location of the care recipient in any room in the home
or outdoors and (ii) feeding the main one for the probability of certain activities at
any moment in time.

While the location model is fed by the signals from the infrared presence sensors
and the main home door sensor, activity probabilities are updated based on a decision
tree, outlier detection in all the non-binary measurements in the home and physical
activity levels.

The activities recognised by DFM and the measurement cues used by the decision
tree in order to do so are summarised in Table 11.1.

This reasoner is intended to reason upon user's daily functioning activities to be
used as an automated diary. When analysing these activities, patterns are formed based
on which the reasoner estimates user behaviour and assists. The reasoner implements
two primary functionalities: (a) habits building – the reasoner searches for repeating
activities based on which forms the habits by taking into account the natural variations
introduced by user's behaviour, and (b) reason upon the duration or frequency of
different functionalities. The primary usage of this reasoner is to support the building
of the automatic diary for the user. This diary is further utilised by the eWALL Personal
Daily Support Service (PeDaS) to remind the user about upcoming activities which
are recognised as habits. The secondary usage is to collect statistics about duration
on each activity and display these collections to the primary or secondary user. This
can anticipate user action which can improve overall well-being.

Table 11.1 *Activities recognised by the DFM and the measurement cues supporting
the decisions*

Activity	Measurement cues
Cooking	Kitchen presence, high humidity, high temperature
Sleeping	Bedroom presence, physical inactivity, bedroom TV not on, or Living room presence, physical inactivity, living room TV not on
Entertaining	Living room presence, low physical activity, up to one face being tracked by the face tracker, living room TV on, or Bedroom presence, low physical activity, bedroom TV on
Showering	Bathroom presence, high humidity, high temperature
Sanitary visits	Bathroom presence, normal humidity, normal temperature, low physical activity
Visited	Living room presence, multiple faces tracked by the face tracker
Housework	Presence in the home, high physical activity, not any other indoors activity recognised by the decision tree
Outdoors	Presence outdoors
Unknown	No activity recognised by the decision tree

eWALL applications are web-based, meaning that they are designed to be executed and rendered in a web browser. The main reason for choosing this approach is that, being based on central hosting, it facilitates the applications maintenance, as it eliminates the need to distribute software to the end users. Another well-known reason for using web applications is their cross-platform compatibility. Since eWALL will support mobile platforms such as smartphones or tablets, the web application approach will provide universality in the design approach. From the developers' point of view, the readily available software from open source communities in the web applications development area makes them excellent candidates for rapid development.

11.5.2.1   Daily Physical Activity Monitoring

One of the most important factors for the healthy living of elderly people is their regular physical activity. The Daily Physical Activity Monitoring (DPAM) application is a web-based application intended to provide the user information about their daily activities such as walking, jogging and exercising. It is a monitor in the sense that it displays different metrics, how active the user was for a selected day, week or month. This feedback can be a reason or a motivation for the user to take necessary actions in case there was not enough activity during some period of time. Using this application, the user has the possibility to follow his/her progress, how much he/she walks in a given time period, how many calories he/she burned, or if he/she reached the designated goal for the day. Additional statistics on his/her different types of activities are provided. Based on these, the user can decide upon the necessity of increasing different physical activities such as running or exercising. The application can also serve the caregivers where they can use it to track the physical activity to their beloveds.

11.5.2.2   Daily Functioning Monitoring

The maintenance of daily functioning can be a valuable factor for the independent living of elderly people. The monitoring of daily functioning can help improve the quality of life of the elderly people because immediate actions can be taken in case a deviation is detected from specific daily patterns. The Daily Functional Monitoring application is intended to provide to the user information about their daily functioning such as cooking, bathing, sleeping, entertaining, etc. It displays how much time the user spent on a specific activity or in a specific location. Daily Functioning Monitoring could improve eldercare by providing a sense of safety and quality of life by detecting key events and by enabling timely intervention. Long-term monitoring can allow caregivers to make informed decisions or to assess response to a treatment.

11.5.2.3   My Sleep application

The "My Sleep" application has the purpose of presenting quantitative and qualitative interpretation of the sleep behaviour of the user based on the home sensing devices. The application consumes service brick data describing sleep events such as the time the user went to sleep, the time the user woke up, the duration of sleep, the sleep efficiency, etc. This data is processed and semantically represented within the interface. Complementary, the Sleep Lifestyle Reasoner is requested to provide patterns of

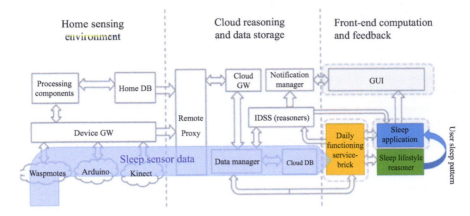

Figure 11.8 Connection between the sleep application and the dependent eWALL components

sleep behaviour for the user and the information is displayed according to each sleep period described in the interface. Additionally, the application computes the quality of sleep parameters taking into account objective parameters according to medical norms in measuring the quality of sleep. The sleep data can be visualised for one day at the time, or one week at the time. These connections are depicted in Figure 11.8.

11.5.2.4 Healthcare monitor

"My Health" application is a key component of the eWALL system and it brings major value for COPD and ARI patients. It supports taking vital signs measurements (pulse, oxygen saturation in blood and blood pressure) by the primary users with the medical devices that are compliable with the eWALL platform. Knowing the health measurements will help both the eWALL primary users and caregivers to understand how the primary users' health is progressing. My Health application is fed up with vital signs data coming from measurements that the eWALL primary users are expected to take on their own. For this application to be realised, the eWALL is made currently compliant with the Omron blood pressure monitor and the Nonin Pulse oximeter which the primary user uses to take measurements of his/her heart rate, oxygen saturation in blood (SPO_2) levels and blood pressure.

11.5.2.5 Domotics

Important information for elderly people living independently is the real-time temperature in different rooms in the home. This information is provided by the Domotics application. The current humidity level is also shown in the Domotics application which is especially important for the elderly people suffering from COPD. In addition to current temperature, humidity and illumination level and control, the Domotics application provides historical overview for the current day, last week and last month of the data for temperature and humidity in different rooms (e.g., living room, kitchen, bedroom). The available overview could be very useful for the user and/or caregiver in order to track these parameters and its possible influence to other behaviour or health condition of the user. The Domotics application is developed as an interactive web application in order to be compatible with targeted eWALL devices.

11.5.2.6 Calendar application

The Calendar is a service that registers the external activities of the user as well as indoor activities supported by eWALL, such as medical measurements and indoor workouts. The calendar is connected to the PeDaS, contributing to the intelligence and the personalisation of the eWALL system by providing information on user's schedule. In the calendar are included periodic and irregular activities/events such as periodic medical checkups and appointments with doctors; medical measurements; periodic activities (e.g., once a month meeting of his/her hometown's cultural association); appointments with friends; shower and grooming; physical activities; cognitive exercise and other irregular events and activities.

11.5.2.7 Personal Daily Support Service

The eWALL PeDaS provides support for elderly users with MCI in structuring, timing (when) and content (how, what) of their daily activities. These activities can include, but are not limited to, sleeping (waking up, going to sleep, quality of sleep and wandering); preparing and consuming meals; scheduled meetings/activities; taking medicine; doing physical and cognitive training and personal hygiene. The functional architecture of the PeDaS is depicted in Figure 11.9.

The PeDaS system includes four different components: (i) calendar, (ii) home sensing system, (iii) reasoner and (iv) user interaction. The structuring of the day is done by filling the calendar. This can be done by the informal caregiver and/or the user. The user can automatically fill in the calendar through dialogues with the

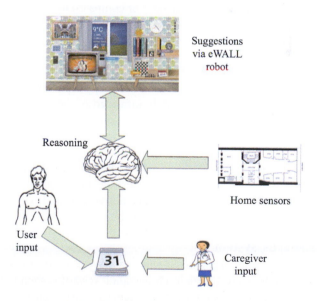

Figure 11.9 Functional architecture of the PeDaS

notification robot: e.g., "At what time do you usually have lunch?" The response of the user to such questions is automatically inserted in the calendar. In this way, a calendar can be created for every day of the week and automatically generated for next weeks. Additionally, persons that have permission are allowed to insert additional calendar items, like for a visit, calls, etc. The home sensing system provides information on the activities that the user is performing in terms of the time during which the user is carrying out a predefined activity and a label of the activity (e.g., sleep). It is a modular system in the sense that it is able to use multiple sensors to distinguish different activities.

The reasoner component of the PeDaS is based on the connection to several eWALL services for sensing information (e.g., sleep) and services that are used to promote healthy habits, like the well-being advertisements. As a basic requirement, the reasoner is able to link the information in the calendar, like lunch, to an activity observed by the sensors, e.g., the user is cooking. Each item in the calendar has certain boundaries indicating the timeframe in which the activity should be done. Outside these boundaries, a notification/alarm is generated. Different levels of notifications/alarms are possible, depending on the urgency of staying within the timeframe, which is different for the different activities, e.g., medication is more restricted than lunch. At a higher level, the intelligence is able to adapt the timing of these activities gradually from an initial – healthy – setting by, e.g., the (informal) caregiver to a personal setting, while keeping an eye on healthy boundaries. When eWALL detects (slow) changes in certain daily activities or habits (e.g., the user wakes up in the middle of the night, and sleeps in the afternoon), notifications are sent to the caregivers. The results of the reasoner are given back to the user via the eWALL robot. At the basic level, the robot will provide feedback on, e.g., deviation between the planned lunchtime and the actual lunchtime, observed by the sensor system. Certain activities will be automatically entered to the calendar or compared to items therein. The user should confirm the calendar items. For the rest, the user or his/her informal caregiver will be prompted by the eWALL robot.

11.5.2.8 Falls Prevention Program

Falls pose a physical and psychological risk for elderly people. Falls can be prevented if the elderly user of eWALL is adequately informed of the factors involved in the risk of falling and follows advice on mitigating those risks.

The Falls Prevention Program of eWALL focuses on eight risk factors and evaluates their constituents, either utilising the user's profile, questionnaires or measurements from the home environment, originating from the user's home or body. These factors are (i) home hazards, (ii) medication intake, (iii) vital signals, (iv) vision, (v) feet and (vi) leg conditions, (viii) cognitive state and (viii) social life. The UI of the application is demonstrated in Figure 11.10.

11.5.2.9 Rewards System

In order to increase overall engagement with the eWALL platform as well as tie the various eWALL applications together, an engaging reward system is implemented. Every activity of the user is rewarded by the system with virtual currency (i.e., eWALL Coins). The eWALL Coin represents a virtual credit unit for performing any kind of

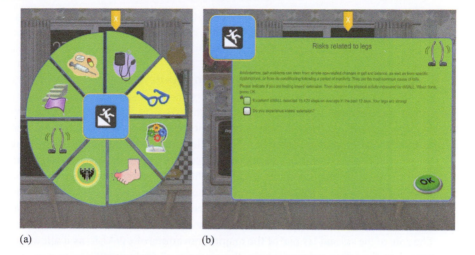

(a)   (b)

Figure 11.10   *Falls Prevention Program UI: The eight risk factors (a) and the two constituents of the legs condition (b), one measured automatically by the home environment (related to the number of steps) and another asking the opinion of the user*

Figure 11.11   *Rewards System UI showing the total coins currently owned and their source*

"good behaviour" (like taking steps, performing a video exercise or even opening one of the health overview books). The coins can then be used as a currency throughout the system to purchase rewards (i.e., UI customisations).

The UI of the Rewards System shown in Figure 11.11 displays the amount of eWALL Coins earned, specified per "source of income". These sources are related to user physical activity and sleep quality, as well as to the user's interaction with any of the apps, usage of the video trainer and scores in the games.

11.6   eWALL innovation potential and challenges

Innovation management these days is challenging, as it is a term widely used by almost every research- and market-oriented activity. Innovation in eWALL is defined by those technical breakthroughs that advance the state-of-the-art and create value for the user. The issue of value creation by e-Health systems has been explored in a number of research projects and it has been shown to identify the economic and financial benefits of the e-Health solutions. Value creation is essential for assessing the innovation potential of a product, taking into consideration the overall operational context, within which these applications and services lie.

The role of the "cloud" is one of the major innovations of eWALL, as it allows on the same time the advanced processing, the easiness of deployment, the modularity and the openness of the system. The owner of the cloud infrastructure can be the service provider or enabler of the e-Health services deriving from eWALL. The users can therefore easily connect their home environment through a home gateway, seamlessly integrated with various things; thus creating a Home Caring Environment of advanced sensing and interaction. Part of the reasoning and data processing, including the possibility of Health Big Data Analytics, will be made securely on the cloud and that results in a highly available and reliable environment with sufficient performance. On the same time, developers and hardware vendors can develop their "service bricks" and applications, thanks to the openness of eWALL and these services and applications can be made available to the users by applying several business models, e.g., freemium services, revenue-share, etc.

Another major innovation of eWALL is the persuasion of non-obtrusiveness, despite the fact that the system brings numerous devices in their home environment. The User Experience was in the centre of focus during the prototype design and deviates significantly from the typical interaction approaches. Users can customise their interaction environment in a way to create a familiar environment that will increase their engagement. The lack of the latter is the major reason of failure for most systems of that kind.

Finally, with regard to the value creation, as mentioned earlier, it is linked with the financial and economic benefits of a product. eWALL's cloud architecture and software open-source licensing are aligned to the direction of promoting crowdsourcing by users that develop services and applications; thus enabling the usage of more disruptive business models that can make eWALL affordable and sustainable.

11.7   Conclusions

Despite the existence of numerous e-health solutions, there are only a few with an integrated approach that managed to penetrate the market of chronic diseases and Silver

Economy. That is mainly because of the entry barriers that need to be addressed, among them the feeling of non-obtrusiveness, the technology acceptance, the affordability and the personalisation.

The EC-funded FP7 project eWALL performed an innovative and highly user-oriented research and development project that successfully integrated sensing and the cloud towards a smart ambient application platform. The eWALL platform integrates sensors, actuators, reasoners, supported by a state-of-the-art cloud infrastructure to enable the patient-relevant personalisation. The proposed gesture recognition for the human–computer interaction can yield fast response times and make the system more user-friendly. Based on its features described in this chapter, eWALL is a promising solution that can break the entry barriers in the e-Health sector and provide impactful applications to its users.

A major role for making any e-Health platform, like eWALL, successful and to ensure that the technology is deployed and accepted by the end users, is to define strict user and usage requirements and also, make the technology compatible with standards that would guarantee the safety of the end users and the ease of deployment. eWALL is engineered with the continuous feedback of the end users following strict requirements and addresses all critical needs related to reliability, interoperability, security, privacy and trust, therefore aims to be a leading solution of the future.

References

[1] Kyriazakos, S., Prasad, R., Mihovska, A., *et al.* "eWALL: An open-source cloud-based eHealth platform for creating home caring environments for older adults living with chronic diseases or frailty". *Wireless Pers. Commun.* 2017; **97**: 1835. Available from https://doi.org/10.1007/s11277-017-4656-7.

[2] Zaric, N., Pejanovic-Djurisic, M. and Mihovska, A. "Ambient assisted living systems in the context of human centric sensing and IoT concept: eWall case study", in *Proc. of IEEE 2016 Personal, Indoor, and Mobile Radio Communications (PIMRC)*, London, UK, September 2016.

[3] Kyriazakos, S., Mihaylov, M., Anggorojati, B., *et al.* " eWALL: An intelligent caring home environment offering personalized context-aware applications based on advanced sensing". *Wireless Pers. Commun.*, (April 2016); **87**(3): 1093–1111. Available from: https://doi.org/10.1007/s11277-015-2779-2.

[4] Tonchev, K., Tsenov, G., Mladenov, V., Manolova, A. and Poulkov, V. "Personalized and intelligent sleep lifestyle reasoner with web application for improving quality of sleep part of AAL architecture". *FABULOUS 2016, MindCare 2016, IIOT 2015: Pervasive Computing Paradigms for Mental Health*, pp. 107–112. In Lecture Notes of the ICST, Vol. 207 (March 2018). Available from: https://doi.org/10.1007/978-3-319-74935-8_15.

[5] Shah, G. and Papadias, C. "On the blind recovery of cardiac and respiratory sounds". *IEEE J Biomed. Health Inform.* (January 2015); **19**(1): 151–157.

[6]   G. Bardas and A. Pnevmatikakis. "Real-time face tracker yielding 3D pose and position", in *Proc. of Global Wireless Summit 2016.* Gistrup: River Publishers, December 2016, e-ISBN: 9788793609297.

[7]   ICT Project eWALL, Deliverable 2.4. "Ethics, Privacy and Security", April 2014, Available from: ewallproject.eu.

[8]   ICT Project eWALL, Deliverable 2.3. "Preliminary User and System Requirements", April 2014, Available from: ewallproject.eu.

[9]   Nidhi, A. Mihovska and R. Prasad. "User privacy in big data analytics for eHealth: A data privacy model", in *Proc. of GWS 2016*, November 27–30, 2016, Aarhus, Denmark.

[10] Wang, W. and E. Krishnan. "Big data and clinicians: A review on the state of the science". *J Med. Inform.*, 2014; **2**(1): e1.

[11] Tonchev, K., Velchev, Y., Koleva, P., Manolova, A., Balabanov, G. and Poulkov, V. "Implementation of daily functioning and habits building reasoner part of AAL architecture". *FABULOUS 2016, MindCare 2016, IIOT 2015: Pervasive Computing Paradigms for Mental Health*, pp 113–118. In Lecture Notes of the ICST, Vol. 207 (March 2018). Available from: https://doi.org/10.1007/978-3-319-74935-8_15.

Chapter 12

VitalResponder®: wearable wireless platform for vitals and body-area environment monitoring of first response teams

João Paulo Silva Cunha[1,2], Susana Rodrigues[1],
Duarte Dias[1], Pedro Brandão[3,4], Ana Aguiar[2,4],
Ilídio Oliveira[5], José Maria Fernandes[5], Catarina Maia[1],
Ana Rita Tedim[1], Ana Barros[1], Orangel Azuaje[2,4],
Eduardo Soares[3,4], and Fernando de La Torre[6]

Under the VitalResponder® (VR) line of research, mostly funded by the Carnegie Mellon University (CMU)-Portugal program, we have been developing, in partnership with colleagues from CMU, novel wearable monitoring solutions for hazardous professionals such as first responders (FR). We are exploring the synergy between innovative wearable technologies, scattered sensor network and precise localization to provide secure, reliable and effective first-response information services in emergency scenarios. This enables a thorough teams' management, namely on FR exposure to different hazardous elements, effort levels and critical situations that contribute to team members' stress and fatigue levels.

Wearable is a keyword in this project, aiming to create on and around each subject a wireless body and body-area sensing network capable of acquiring both the vitals and individual surrounding ambient measurements. Furthermore, we developed a second wireless connectivity level where each team member is a node of a disconnect-tolerant ad hoc wireless network, so that a multi-hop protocol may carry the sensing information to a data collector unit usually housed at an FR vehicle nearby. This collector constitutes the team's data buffer to a cloud-based information technology (IT) system that constitutes the backbone of the storage and analytics components of the VR platform.

[1]INESC TEC, Portugal
[2]Faculdade de Engenharia, Universidade do Porto, Portugal
[3]Faculdade de Ciências, Universidade do Porto, Portugal
[4]Instituto de Telecomunicações, Portugal
[5]IEETA e DETI, Universidade de Aveiro, Portugal
[6]Robotics Institute, Carnegie Mellon University, USA

VR developments led to new wearable technology wireless devices that are connected to a body-area aggregator and are capable of acquiring medical quality electrocardiogram (ECG), body temperature, actigraphy, and several environmental measures such as luminosity, temperature, humidity, pressure, altitude, toxic gases and Global Positioning System (GPS) coordinates. The combination of all the data gathered by the wearable sensors is being used to create personalized methods for each subject to evaluate stress and fatigue factors both in (near) real time and in a retrospective off-line evaluation. This aims to provide decision-makers reliable indicators (such as levels of carbon monoxide exposure, stress levels, etc.) for better team management.

Several studies with different FR professionals were already performed with the VR platform. In this chapter, we present two examples; one performed with firefighters (FF) and another with police officers (POs) totaling 100 h of real data collected with the system in two different configurations. Results showed the adaptability and utility of the VR platform for different FR scenarios and paved the way for more complex and close to routine FR scenario studies that we are presently conducting.

Finally, we have already started the process to study the potential markets, possible business models, and future product(s) supply chain requirements so that we guide the following phases of the project to result on a product prototype that will try to go into the market by the end of the project in the summer of 2018.

12.1   Introduction

First responders (FRs) are considered a high-risk occupational group that is constantly exposed to different hazardous environments, severe stress and fatigue, with a negative impact on their health and performance [1]. Nevertheless, FR's gear and management procedures do not routinely include any quantitative information on these aspects and exposures. This urgent need of assessment of these professionals' occupational health led us to the development of a technological platform for real-time monitoring of both physiology and body-area environment data. From this platform, we are able to extract stress and fatigue quantified indicators, providing more information for better personnel management by FR commanders, and therefore introducing the implementation of a *quantified Occupational Health* (qOHealth) platform. Other hazardous professions (pilots, public transportation drivers, miners, air traffic controllers, etc.) also face demanding work scenarios and may benefit from our novel *qOHealth* approach to promote better levels of health and well-being among these professionals.

Given this background, we had the vision to explore the synergies between innovative wearable technologies, scattered sensor networks, and precise localization in order to provide more secure, reliable and effective first-response information services.

This vision is sketched in Figure 12.1 where a team of FR, in this use case firefighters (FFs), is wearing physiology and surrounding environment sensors, which have wireless communications among them and with a team commander that is located near a vehicle. This vehicle has other more sophisticated environment sensors and a data collector small server that connects to the cloud for data analysis. Finally, the events' data, such as FR geo-referenced vitals and body-area environment data, wind

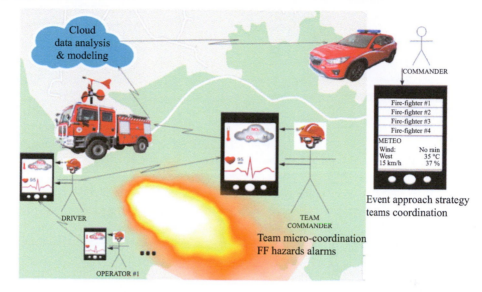

Figure 12.1   Vital Responder® vision, for the FF teams scenarios

speed, temperature, etc. are relayed to the back-office system where event coordinators access all the information to take decisions so as to optimize critical event approach and teams management, in a secure and effective way.

This work has been developed within the VitalResponder® (VR) research line and it is now part of a Carnegie Mellon University (CMU)-Portugal program-funded project named "VR2Market: Towards a Mobile Wearable Health Surveillance Monitoring Product for First Response and other Hazardous Professions." This project is hosted in Portugal, with a consortium of Portuguese research centers, a medical device company, end-users organizations, consulting services of psychology and CMU partners from both Pittsburgh and Silicon Valley campus.

In this chapter, we report the R&D efforts and achievements of this large project by describing its vision and the different components of the resulting VR platform. Given the book theme, we focus this report on the wearable wireless body-area network (WBAN) and on the multi-hop-enabled Mobile Ad hoc NETwork (MANET) built on top of 802.11 technology developed to collect FR team data, using each team member as a network node. We also report results already achieved and on-going studies that used/are using the described technologies in different hazardous professionals, demonstrating the scalability and adaptability of the developed solution.

12.2   System architecture

Within the VR platform, we developed an integrated computational system (VR-System) to cover the qOHealth use cases, including distributed data gathering, on-line monitoring and post-mission review [2,3].

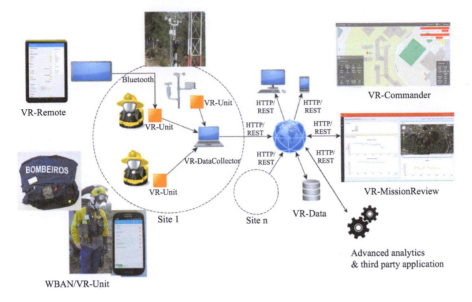

Figure 12.2   VR platform architecture

The VR-System comprises several layers that can be mapped to four main contexts: local sensing of psychophysiological and body-area environment data (within the WBAN); site-wide transport of data in the operation theater (using a MANET); cloud-enabled data management and visualization tools and data analytics. Figure 12.2 illustrates these contexts showing the components developed.

In the VR scenarios, each team member carries a personal aggregator, the VR-Unit application, running on a smartphone or a Raspberry Pi (RPi) device. The VR-Unit acts as a body-area data aggregator, gathering both physiological and environmental information from wearable sensors, connected to the device via Bluetooth. The VR-Unit also acts as the personal-area gateway to the network, sending the sensor data from the WBAN to the site's VR-DataCollector node. This node acts as a data sink for a team operating at some site and provides a local repository for the WBAN information. Data from each site is forwarded by the VR-DataCollector to the remote, long-term storage, VR-Data services.

The VR-Data services store the missions' data and provide public services to manage and review team information for online and offline mission review. For this, different dashboards were developed (VR-Commander, VR-Remote and VR-MissionReview) and an application programming interface (API) is provided so that third-party applications can be developed, for example for advanced analysis of the collected data.

The next sub-sections describe the sensing devices that compose the WBAN, both the physiological as well as the environment. Following this, we discuss the network infrastructure (MANET) used to convey the data to the VR-DataCollector. At the end

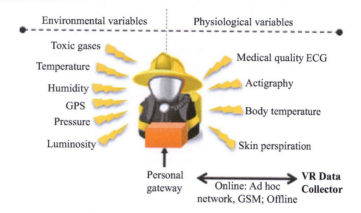

Figure 12.3 WBAN with two types of variables

of this section, we describe the ICT platform developed to collect and analyze the gathered data.

12.2.1 Wearable on-body and body-area sensing devices

The correlation between events and human body physiological changes are leading to a better understanding of how the human body reacts to specific events. These can be related to stress or fatigue, aspects that are nowadays measured qualitatively. In the present project, we aim to obtain a quantified perspective into these aspects and related body-area environment "stressors," such as high ambient temperature, dangerous levels of toxic gases and extreme dry environments. This is achieved by sensing body vital signals and body-area environment parameters. The VR wearable devices developed to address both aspects, through a physiological monitoring unit, in two different form-factors, where the electrocardiogram (ECG), actigraphy, body temperature and skin perspiration are monitored. In many hazardous professionals teams, understanding the subject's physiological changes is not enough given that their surroundings can change drastically affecting their physiological status. For example, during fire combat, an FF should not be exposed to high temperatures for long-periods nor to certain levels of some gases. Furthermore, exposure to these and other environmental elements can increase the stress level. To measure these factors, several prototypes for environmental variables monitors have been developed. Figure 12.3 presents a scheme with all the variables that can be gathered in the current version, creating the wireless sensor network (WSN) of each professional.

Each BSN node communicates with the personal gateway using Bluetooth, which enables the use of several devices all connected and synchronized with the personal gateway clock. This important feature allows sending wirelessly reliable real-time data during monitoring sessions between each WSN and the Data Collector. This gateway also has a data logger capacity to store locally all the data that is being gathered in real time.

This new approach, to monitor each subject individually, is only possible due to the continuous progress of microelectronics that is enabling the reduction of electronic components and devices. This allows developing new, smaller and more "body friendly" wearable gadgets, contributing for a better quality and more reliable individual monitoring in ambulatory scenarios, thus creating a WBAN. The VR wearable devices took advantage of this state-of-the-art microelectronics and novel textile embedding technologies to monitor both physiologic and body-area environment variables in the same integrated platform. To better understand each body sensor network (BSN) node from the VR WBAN, the following sub-sections are going to discuss the solutions and prototypes developed in this research project.

12.2.1.1 Physiology monitoring devices

Physiological monitoring is nowadays in the center of wearable devices and an effort is being done to try to acquire the most reliable data to obtain physiological measures with health and medical viability. In this project, there is a special attention to ECG acquisition and its quality, even during hazardous situations and physically demanding events. Due to this reason, we partnered with Biodevices, S.A., a medical devices' company that "develops, commercializes and exports biomedical engineering solutions for medical diagnose acquisition" [4].

One of the products of this company is the VitalJacket® t-shirt (Figure 12.4(a)) [5,6]. This is a wearable medical device with European medical certification (MDD-93/42/EEC), capable of acquiring medical quality multi-lead ECG with a sampling rate of 500 Hz and actigraphy data in ambulatory scenarios, transmitting all the data through Bluetooth and saving it on an SD Card. Normally, for ambulatory scenarios a simple heart rate (HR) monitor is used. In this case, we acquire medical grade ECG, not only to measure HR with a higher reliability, but also to extract features from the ECG morphology and use them to understand physiological changes better, as we will present further down. VitalJacket®'s characteristics make it a promising physiological wearable device to use during daily activities of hazardous professionals and it was the first physiological monitor device used in this project as a BSN node for each subject.

After some field tests and data acquisition, mainly with FFs, we determined more requirements to adapt the commercial version of VitalJacket® to be used with hazardous professionals. A prototype of a new version of the VitalJacket® t-shirt (Figure 12.4(b)) was developed, changing the textile to a more breathable one, with a higher burning temperature to tolerate the higher heat on the FF's surroundings. A body temperature sensor was also added. This sensor is a very small probe that is in contact with the skin, measuring its temperature each minute. This sensor was totally incorporated in VitalJacket®'s electronic technology. Its measurements are sent at the same time as the other physiological variables. We used this development for two years among FR with remarkable success (details in Sections 12.3.1 and 12.3.2).

Nevertheless, we did not stop the wearable devices development with the t-shirt form-factor. Following our "macro-to-micro" line of research, we further reduced our wearable physiological device size, transforming it into a skin patch form-factor, making the device even more user-friendly and maintaining all the data reliability and robustness of the t-shirt version. This new device technology is a patent

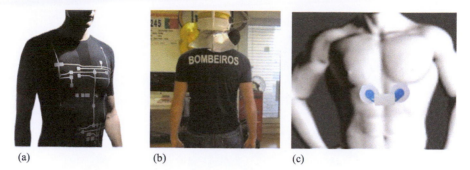

(a)   (b)   (c)

Figure 12.4   *(a) VitalJacket® commercial version from Biodevices, S.A.;*
              (b) Extended version of VitalJacket® designed for hazardous
              professionals (BOMBEIROS is a Portuguese word that means FFs);
              (c) New patent-pending version with a skin patch form-factor

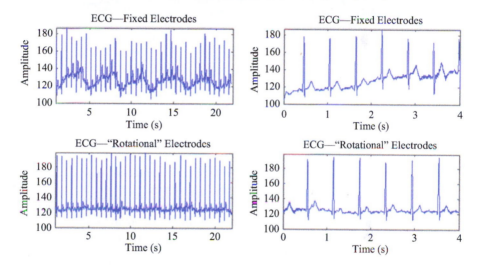

Figure 12.5   *Difference in raw ECG waveform (no filter) during chest motion*
              actions from a patch with (lower tracings) and without (upper
              tracings) our patent-pending technology

pending technology [7] with the form-factor of a "snap to skin" patch device that is able to acquire medical grade quality ECG, highly robust to motion artifacts (major problem in skin surface devices) using only two electrodes (Figure 12.4(c)). Named *VitalSticker*, it also sensors actigraphy and body temperature, with Bluetooth data transmission and SD Card data storage. This technology uses a unique rotational flexible connection between the electrodes and the device that prevents skin stretching, reduces chest movement noise and daily/sports activities noise, acting as a natural filter in the ECG waveform as can be seen in Figure 12.5. It is more comfortable to use and easier to wear than the previous t-shirt version, just needing to snap it over the skin.

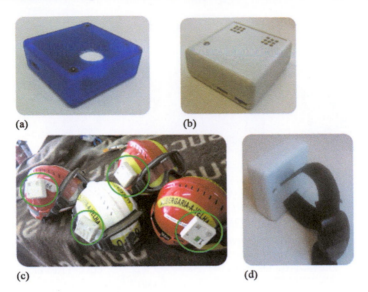

Figure 12.6 *(a) FREMU prototype (5 × 5 × 2 cm); (b) AmbiUnit prototype (7 × 7 × 3 cm); (c) AmbiUnit devices (green circles in the image) attached to FF helmets—already used for over 500 h in real fire combat situation and (d) AmbiUnit device with a strap*

A set of *VitalSticker* prototypes was already used by FFs in real scenarios as part of their WBAN. One of the main advantages of *VitalSticker* reported by these hazardous professionals is the absence of a textile that needs to be washed after every shift. They also conveyed the fact that it is much easier to "dress" and reported that is more comfortable to use, mainly during high intensity moving events. The next steps to improve this prototype are to further reduce the form-factor and its weight and improve its hardware in several aspects.

12.2.1.2 Body-area environment monitoring devices

In some situations, the environmental changes can be extreme, leading to a health status change, which then reflects at the physiological level. These ambient conditions near the subject are, as such, very important to collect in some hazardous professionals, such as FFs that are subjected to toxic gases and high temperatures. For this purpose, the VitalResponder project has also made some developments in body-area ambient monitoring devices.

The first developed device (Figure 12.6(a)) was able to acquire carbon monoxide values, air temperature, atmospheric pressure and relative altitude. This first prototype was able to send all the data via Bluetooth, with a 10 s period. However, more sensors were needed and the device needed to be adaptable to different environments. As such, we improved the prototype into a new version named *AmbiUnit* (Figure 12.6(b)). This prototype can measure two gases, air temperature, air humidity, luminosity, atmospheric pressure and relative altitude. The adaptable property of this device is the fact that you can choose what gases you want to monitor—it has a snap system that allows the user to open the device and snap two sensors to measure different gases, for example, carbon monoxide, nitrogen dioxide, oxygen, ozone and carbon dioxide. This way, it is possible to measure gas levels according to the environment where each hazardous professional normally acts (e.g., a forest fire combat has different types of toxic gases when compared to a fire in a building). The *AmbiUnit* can be attached to a helmet (Figure 12.6(c)) or it can be tied up, to a belt or a backpack, using a strap (Figure 12.6(d)). *AmbiUnit* also adds the ability to change the sampling rate and to request individual sensor measures remotely using API commands through Bluetooth. These characteristics aim to save battery (slowing sampling rate) and allow the team manager to request a new sensor measurement if needed.

12.2.2   *MANET for WBAN to data collector connectivity*

In the offline mode, the data logged in the WBAN personal gateway (VR-Unit) can be uploaded to the data collector and back office whenever connectivity is established, via LTE, Wi-Fi or other. Following the project vision (Figure 12.1), the challenging task in the targeted scenarios is collecting the data from each professional's WBAN in real time for the online mode when there is a lack or unavailability of network infrastructure, e.g., during forest fires. In this situation, the network connectivity between the gateway in each FF and the data collector (usually located in the firefighters' truck) can be intermittent.

We opted for using a multi-hop-enabled MANET built on top of 802.11 technology to support the expected worst-case aggregate traffic. The 802.11 technology is well established, has decent energy consumption and commercial off-the-shelf devices are widely available at affordable prices. As such, integration into a portable system is feasible.

The communication patterns in the FR scenarios are not limited to data collection. Through observation and interviews with FR teams, we identified the following communication models of interest:

* Many sources to one sink: for data collection from the VR-Units to the VR-DataCollector;
* Dissemination: to send a command to all team members or request a team check;
* Unicast communication: to communicate with a specific team member and for explicit confirmation from team members to a team check;

Reliability for message delivery will also be needed in some of the cases, namely for team checks and their responses. A relevant point for reliability is the connectivity between each FR node and the data collector.

For evaluating the network connectivity, we built two prototypes, one based on Android smartphones and another on RPi.[1] Based on the developed prototypes we measured the coverage ranges of 70 m [8] and 120 m [9] for the Android and RPi

---

[1] https://www.raspberrypi.org/.

solutions, respectively. In the RPi study, several different wireless USB cards were tested, comparing the range and battery expenditure. One of the conclusions was that the use of USB on the RPi B+ amounted for the bulk of the energy cost and the increased energy fee of the card with the greatest range (a TP-Link TL-WN722N) was minor. We measured a surge of ~6.4% in 335 mA for an increase of range from 30 m to over 150 m at 10 Mbps. We measured the consumption of the RPi without USB dongles at ~191 mA.

Using the communication ranges assessed and the Global Positioning System (GPS) traces of a four-person FF team during an 8 h fire exercise, we analyzed the expected connectivity of such a network. Although we cannot be sure about the generality of these traces, they are the first traces to provide some insight into the mobility of FFs on the terrain. The results show that they get as far as 800 m away from the fire truck, which corroborates our decision for a technology that supports multi-hop to provide connectivity. Figure 12.7 shows the movement of FF1 in the fire drill. The diamond markes where the meteorological station and the VR-DataCollector were located. Additionally, we characterized the disconnection times to any other FF and disconnections to the sink. In Figure 12.8, we show the disconnection times for FF1 to other FFs and the sink. The x-axis shows the duration of the disconnection and the y-axis indicates its frequency. We were able to conclude that over 90% of the disconnection times lie below 10 s, but they can sporadically reach 1,000 s [9].

The possibility of disconnections led to the design of a Disconnection-Tolerant Network (DisToNet). This framework enables a network-level buffering (thus in kernel mode) which we implemented as a prototype for RPi [10].

Figure 12.7 *Map of the movement of FF1 (red marker) and the meteorological station (marked with a diamond). (The rough map marked measured maximum distance is 823 m. Image rendered using http://www.gpsvisualizer.com/)*

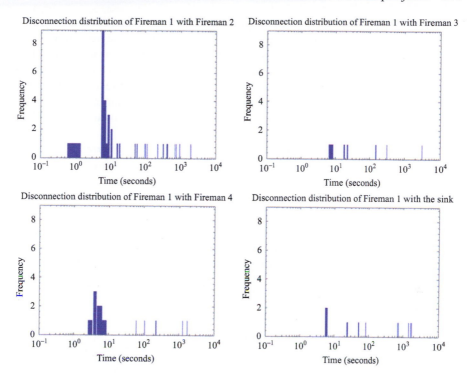

Figure 12.8   Disconnection times for FF1 extrapolated from GPS traces of a fire exercise

As mentioned, we built two prototypes for enabling MANETs in Android and RPi devices. The first one was *AdHocDroid* [3] that enables creating an IP MANET of Android smartphones where one of the smartphones can act as a gateway. The framework enables easy configuration of a MANET of smartphones, with added functionality for experimenting different routing protocols. The associated software has been made available to the community on GitHub.[2] We successfully integrated the *AdHocDroid* MANET as part of the system and deployed it on FFs team during exercises and for 4 weeks in the summer of 2015.

However, *AdHocDroid* suffered from the absence, incompleteness or *buggyness* of the 802.11 ad hoc mode driver implementations from the mobile chipset manufacturers. For this reason, we moved on to our second prototype on RPi, which is a more stable and supported platform and where the USB WiFi network cards have better support for the ad hoc mode (such as the TP-Link TL-WN722N chosen).

In this platform, we addressed with more detail the support for autonomic configuration of the nodes (IP address, address of the data collector, name resolution for other nodes in the network). A fully working prototype with the autonomic configuration

---

[2]https://github.com/eSoares/Android-Ad-hoc/.

based on RPi has been tested and deployed, providing a measured coverage of 120 m. Some implementation details are now being made to enable name resolution. A Uniform Resource Identifier (URI) will identify nodes. The VR-Unit uses this URI to address different nodes. Our network layer will use this string to create a unique IP address on the network (using a defined function, such as a hash function). This simple solution (with some nuances to enable multicast addresses) will allow for configuring nodes with unique IP addresses and enable a unique name resolution from URI to IP address.

The issue of traffic differentiation is also relevant in an intermittent network. We may need to prioritize an ECG data stream over carbon monoxide measures. We have been experimenting with applying policies for different types of traffic based on requirements for delay and throughput. Options for using different paths to the data collector when available are also being considered.

For future work, we are designing a reliable protocol for the MANET and studying multicast on these networks. With these two points, we are able to provide a network API for applications to use the underlying MANET supporting the communication models we identified.

12.2.3 VR ICT platform integration

In the VR Information and communication technology (ICT) solution (Figure 12.2), the VR-DataCollector node is deployed at each operation site and acts as a site data logger. It can accept messages in multiple channels, from plain sockets to higher-level message passing paradigms with common brokering protocols such as AMQP (Advanced Message Queuing Protocol) or MQTT (Message Queue Telemetry Transport—ISO/IEC PRF 20922) [2,3]. This allows additional decoupling between the VR-Unit implementations and the site aggregator.

All data received by VR-DataCollector is relayed to the long-term missions' data repository, the VR-Data component that comprises a repository and a services' interface. The repository is structured in a relational database with JSON extensions (PostgreSQL[3]). The VR-Data provides an API that enables third-party clients to manage and review team information during missions (online scenario) and after (offline scenario). This is possible by making HTTP requests to transfer JSON-encapsulated data, or, alternatively, use the export endpoint to retrieve a compressed file with text-based extracts of the long-term repository. While the former is better suited to integrate third-party applications, the latter is useful for research pipelines, using proprietary toolboxes.

The VR-Data API follows the representational state transfer architectural style (REST), for consistent and intuitive access and is implemented over the Java EE framework (JSR339[4]). It can be used to manage the resources stored in the long-term

---

[3]https://www.postgresql.org/.
[4]JSR 339: JAX-RS 2.0: The Java API for RESTful Web Services, Java Community process (https://jcp.org/en/jsr/detail?id=339).

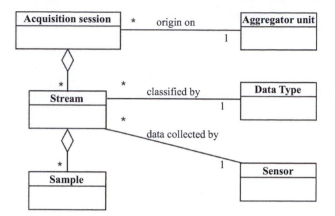

Figure 12.9   The VR-Data model used by the API: the information for each VR operational is captured by VR-Unit (Aggregator unit) during specific operation (Acquisition session). Each Sensor acquiring data will produce a data Stream from a specific Data Type that will have Samples associated

storage, the VR-Data repository. Figure 12.9 illustrates the key resources exposed in the API.

The Aggregator is the wearable device collecting data from the nearby sensors. Each person would wear an aggregator (e.g., the RPi unit). The Acquisition represents a data recording session for an Aggregator in the context of a specific mission, for a specific team. It is expected that an Aggregator would produce many Acquisitions along its use.

The Stream represents a segment of data, in the context of an Acquisition, for a given sensor. If a sensor can provide multiple data types (e.g., GPS, altitude and time) then each type is modeled as a specific Stream.

The Sensor represents a sensing unit, either a single sensor device or a device with an array of related sensors (sensor box, e.g., *AmbiUnit*), identified by the associated Bluetooth device.

The Sample is the value of an observation collected by a sensor, either as a single value or as a structure. Every sample has a timestamp at the origin (added by the VR-Unit) and a timestamp added by the site aggregator. The sample concept can be extended to accommodate specificities of the particular sensors.

Following the REST-style to access resource collections, the following sample request would retrieve all collected samples for the data stream identified by the key 6030, for acquisition session number 107; "v1" denotes the version of the API.

GET http://vr_server/services/v1/acquisitions/107/streams/6030/samples

To avoid retrieving segments that can be very large, the results to be included can be controlled using parameters in the request ("**limit**" and "**offset**" for paging the results; "**after**" and "**before**" timestamps to constrain the time).

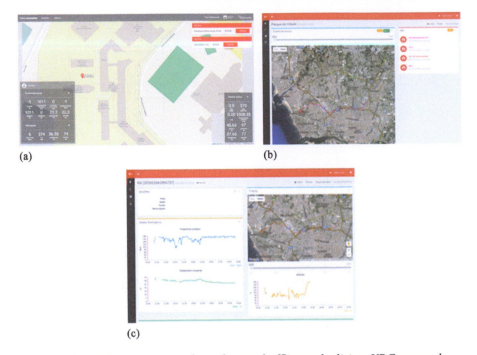

(a)   (b)

(c)

*Figure 12.10   VR-System provides online and offline web clients. VRCommander
(a) is the main online user interface for tracking the location of each
member of a team as well as the gathered values and eventual
alarms. For offline scenario, Mission Review allows an overall team
operational visualization (b) and accessing personal physiological,
environmental and GPS data visualization (c)*

The VR-Data API has already been used to integrate the VR-System with external visualization tools, available from other research groups, allowing the representation of the actions and location of the FRs in the field in a virtual world (avatars).

The VR-System already provides web dashboards for monitoring and data review (Figure 12.10). The monitoring solution (VR-Commander) provides live visualization of data streams, especially location, key physiological variables and alarms. The mission commander or other qualified professional can access all the incoming data of each professional in the field. This allows better management of hazards in teams in real scenarios, namely to support decision making by providing FRs' health status and environmental conditions of the operational scenario (e.g., high HR or high concentration of toxic gases).

The mission review environment (VR-MissionReview) allows recalling saved missions and extract indicators on the professionals' activities. It is possible, for example, to track positions or obtain the levels of exposure to pollutants in the field for each professional. The review process can help to better understand what was observed and help answer operational related questions (e.g., how did each professional performed and reacted under certain circumstance; helping to better know the team and how they can be more efficient). For close online monitoring, we developed a VR-Remote Android App that allows connecting directly with the VR-Unit (WBAN) for in situ assessment.

In addition, the VR-System provides data import and export services for individual acquisitions. The import feature can use the information saved in the VR-Unit devices (packaged as a compressed file) and merge it into the long-term repository. This is useful in some cases for which the information collected in operational scenarios did not reach the VR-DataCollector during a mission. Complementary, the export services allow the extraction of segments of the missions' data (to a compressed file) and their use in external tools (e.g., analyze the data with signal processing toolboxes).

12.3 Results

The VR platform and related developments have been used for different FR studies where the system was adapted with specific analytics modules for each studied scenarios. Successive studies with real scenarios have enabled the identification of new requirements that we incorporated into the platform in iterative agile cycles of innovation.

We present here two different studies, with different FR professional teams—FFs and POs—to better illustrate the usage of the platform, its scalability and adaptability in fulfilling the vision and objectives we have been pursuing in the Vital Responder line of R&D.

12.3.1 Studies with FFs

The objective of this study was to perform a psychophysiological analysis of stress and fatigue of an FF team during a fire drill. Drills are part of the training of new recruits and considered good "ecological" situations, i.e., close to real-work conditions of these professionals. Ecological designs benefit from high ecological validity and less bias [11].

Accordingly, in cooperation with a Portuguese FF brigade, we designed a pilot ecological study involving four FFs in order to assess their psychophysiological stress and fatigue levels under real-world conditions. By using the VR platform, we could monitor ECG, body temperature, activity and self-reported data, such as events, assessments of their "views & feelings" on those events and other observations. This and other studies have been revised and obtained authorization from the Ethics Review Committee of the University of Porto. All participants signed an informed consent prior to being enrolled in the study.

12.3.1.1 Participants
Four FFs aged between 17 and 20 years ($M = 18.25 \pm 1.26$) participated in this study. A total of 23 h of clinical grade ECG, body temperature and actigraphy was collected (Table 12.1).

Table 12.1   Dataset obtained from FF controlled drills

Number of FFs	Dataset#hours (ECG, actigraphy, body temperature, location)	
4	On duty	Off duty
	10 h	13 h

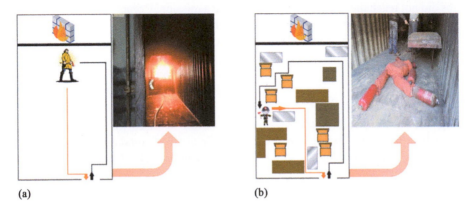

(a)   (b)

Figure 12.11   Fire drill workflows. (a) Event 1 exercise description and (b) Event 2 exercise description

12.3.1.2   Procedure

A controlled fire drill was developed by the Fire brigade in order to provide a training scenario for the FFs trainees. All participants accepted to participate in this study by filling in an informed consent form. At the beginning of the protocol, the participants were equipped with a VitalJacket® for ECG and temperature data collection. The protocol workflow is explained in Figure 12.11.

Participants were required to enter in a container on fire. Then, they have to perform two situations. In the first one, Event 1, participants had to withstand heat, by remaining exposed to high temperatures (>100 °C/212 °F) and smoke. This event lasted about 20 min (Figure 12.11(a)). In Event 2, participants had to enter the same container on fire and perform a rescue situation, by removing an object from a low-visibility dense smoke filled area, in a very high temperature environment and with different obstacles to surpass (chairs, tables, etc.). This event lasted about 5 min (Figure 12.11(b)). Participants were requested to fill in stress and fatigue Visual Analogue Scales—VAS [12,13] at the beginning and at the end of the drill, and after each task. Furthermore, all participants were required to use a VitalJacket® on a day off in order to collect physiological baseline data.

Using one of the features of the mobile app of the VR platform connected via Bluetooth to the wearable physiologic monitor, we were able to perform online exact annotation of events in all connected devices. This was done using

"Radio-buttons," labeling the different events on the wearable monitors' data-stream which was recorded into the SD card memory. Furthermore, it also allowed the association of these "labels" with an internal timestamp and other media (photos, audio and video) to better characterize the reported event. All event data are stored in an SQLite Database, from which a report of the event data can be generated and exported. All data are afterward uploaded to the VR-Data component so that we can perform the needed data analysis.

12.3.1.3 Data analysis

An advanced analytics tool with the needed processing pipelines was implemented for the present study. In order to extract heartbeat information from the ECG recordings, an algorithm based on the one developed by Pan and Tompkins was used [14]. Further improvements were made taking into consideration the work by Hamilton [15]. For stress analysis, pNN20 was considered based on previous recommendations from the literature [16]. HR measures were also considered. We programmed all these parameters in Matlab and integrated them into a graphical user interface for easy analytic procedures of the collected data available from the VR platform.

12.3.1.4 Statistical analysis

Non-parametric tests were applied considering the reduced population sample, and the fact that some of the variables analyzed failed to be normal distributed (Shapiro–Wilk Normality Test). Statistical evaluation was conducted using IBM SPSS AMOS (v.24) software.

12.3.1.5 Results and discussion

As can be depicted in Figure 12.12(a), the collected data analysis showed different changes in physiological behavior, particularly an increase in body temperature and in HR during the two tasks. pNN20 (stress indicator-shown in brown) considerable decreases suggesting an increase in cardiac strain and stress (Figure 12.12(b)).

Self-reports collected through the VAS scales suggested the presence of an anticipatory stress state at the beginning of the protocol and fatigue levels were higher at the end of the drill (Table 12.2).

Findings suggested that FFs perceived more stress at the beginning of the protocol ($M = 3.5 \pm 1.0$) when compared to the end of it ($M = 2.3 \pm 1.5$). These results could be explained by the anticipatory stress state, typical in unknown situations. The physiological analysis also confirms this assumption (Table 12.3). The events were not psychological appraised as being particularly stressful. However, when analyzing physiological data, we can observe an increase in the mean cardiac frequency (bpm) of all FFs, from Baseline 1—briefing to Event 1 and from Baseline 2—briefing to Event 2 (Table 12.3). The highest cardiac activation for all the FFs was obtained during Event 2, which was expectable considering that it was the most intense and demanding of all moments. Physiological data suggested that the events caused higher physiological activation when compared to baseline periods (Table 12.3).

Regarding fatigue levels, results suggested that participants felt more fatigued at the end of the protocol ($M = 3.5 \pm 2.7$) when compared to the beginning

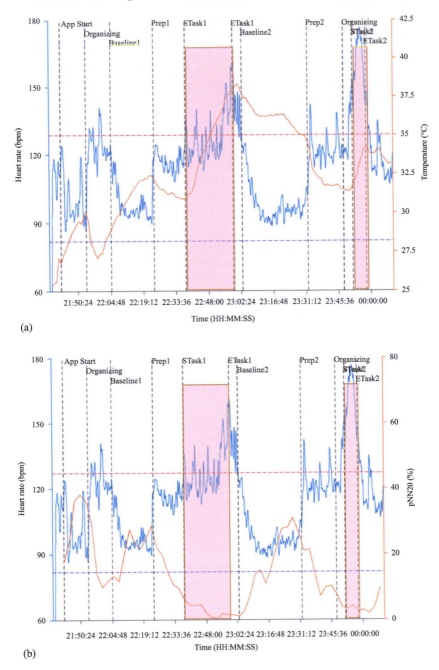

(a)

(b)

Figure 12.12   (a) HR and temperature variation of an FF during a controlled fire
drill. (b) HR and pNN20 variation of an FF during a controlled fire
drill. Shadow areas indicate the two tasks of the drill

Table 12.2   *Mean results obtained from the psychological assessment*
for each of the three assessment moments

	Beginning of the protocol	After Event 1	After Event 2
Vas Stress	3.5 ± 1.0	1.8 ± 1.3	2.3 ± 1.5
Vas Fatigue	2.5 ± 1.9	2.5 ± 1.9	3.5 ± 2.7

Table 12.3   *FFs mean cardiac frequency (bpm) according*
to the different assessment moments

Assessment moments	FF1	FF2	FF3	FF4
Baseline (day off)	66	82	82	63
Baseline 1—Briefing 1	108	117	100	98
Event 1	138	152	127	143
Baseline 2—Briefing 2	108	124	98	101
Event 2	144	166	159	176

($M = 2.5 \pm 1.9$), which was expected. It is important to note that these are only preliminary analysis, no statistically significant results were found, which was expected considering the reduced sample size ($N = 4$). It can be concluded that controlled fired drills had the potential to evoke a psychophysiological stress response. Hence, more studies should be conducted considering that the understanding of psychophysiological stress changes on FFs would help on the design of controlled and more realistic emergency training scenarios.

12.3.2   Studies with POs

In this second study, we developed a protocol to analyze stress among POs working in real scenarios. Considering that stress perceptions are likely to trigger a stress response, psychological and physiological data were analyzed, particularly Heart Rate Variability (HRV) [17]. Specifically, frequency HRV parameters, such as LF power [18] and the ratio between LF/HF power [17] were used in this study. The literature provides normative values for this parameter in healthy populations: for LF power $= 519 \pm 291$ and for LF/HF $= 2.8 \pm 2.6$ [19]. This novel and original interdisciplinary method took advantage of the VR platform to include physiological and psychological measures of stress, combining geo-referenced events with ECG data, using user-friendly and non-intrusive technology. The aim of this study was to investigate PO's stress and its impact in terms of stress reactivity and psychological appraisal during real workplace conditions.

12.3.2.1   Participants

POs ($N = 6$) from a national Police force in the second largest city in Portugal volunteered for this study. Participants mainly performed emergency police duties, since they were part of a rapid intervention team that is called to intervene in critical situations, on-call 24/7. A total of 77 h of clinical grade ECG was collected. See Table 12.4 for a summary of the sample characteristics. The exclusion criteria for the study were participants having a history of cardiovascular disease and/or taking prescription drugs known to affect cardiovascular function. The study was approved by the University of Porto ethics committee and the Portuguese National Police Force Board. All participants signed an informed consent.

Data were collected during one workday and a non-workday. On the non-work days' participants were required to only use the VitalJacket® in order to collect baseline physiological data. During these days, participants were requested to rest and not to perform any physical activity or to go through a stressful situation.

As can be depicted in Figure 12.13, on working days a specific data collection protocol was implemented. At the beginning of each shift, participants put on the VitalJacket® and switched the smartphone on. Following the experience of a stressful situation, POs were required to fill in the event questionnaire, including a description of the event and ratings of stress appraisal. Data were collected and uploaded to

Table 12.4   Dataset description in terms of individuals mean age, gender and total of ECG acquisitions

N	Mean age	Gender	On-duty ECG acquisition	Off-duty ECG acquisition	Total ECG acquisition
6	34.57 ± 4.32	Male	47 h	30 h	77 h

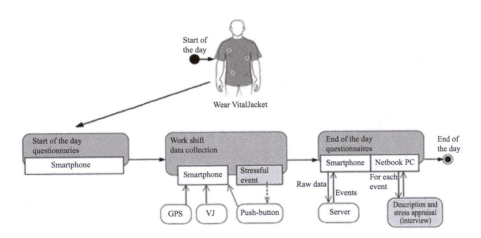

Figure 12.13   Diagram of the followed protocol

Figure 12.14   *Visualization of events in Google Earth. This image represent one PO shift. Ellipses in blue represent the highest LF power metrics obtained for the PO during the shift*

the server for processing. We used the processed ECG data, together with the GPS information and the Google Earth platform, to display the location history for the full workday of a participant (Figure 12.14), with reported events and other moments detected through an automated algorithm, as potentially stressful identified in the map. At the end of the shift, for each of the displayed events and potentially stressful moments, the PO visualized the exact location and trajectory. In addition, a brief end-of-shift interview was conducted at the police station to complete any missing information.

12.3.2.2   Data analysis

The ECG data were divided into blocks of 100 s and processed using the HRV Toolkit from Physionet [20] that was integrated into an advanced analytics pipeline developed for this study. The system then analyzed the HRV metrics and selected the places (using GPS) where a potentially stressful event occurred. These events were selected from all the moments the officers reported an event combined with the blocks having the highest HRV's LF spectral power, but separated at least 5 min between each other. At an end-of-shift interview, we mapped and presented to the PO these events, as illustrated in Figure 12.14. Responses from the software application and the end-of-shift interviews were transcribed verbatim and subjected to an inductive content analysis procedure [21]. Following the geo-referenced event system analysis, an extra procedure for ECG signal analysis pipeline was developed. For this purpose, we conducted an ECG motionless analysis based on the LF/HF parameters in order to understand the cardiac variability of the PO during workdays compared with non-workdays. This analysis also allowed discarding autonomic changes induced by movement. For that purpose, independent 5 min windows (no overlap) where no

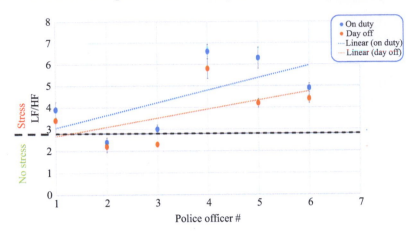

Figure 12.15   *POs LF/HF during shifts and days off. The black dashed line*
represents the normative values for healthy populations with
similar age and gender [19]

VitalJacket® activity was present, i.e., the participants were still (e.g., seated), were
also analyzed for extraction of HRV parameters.

12.3.2.3   Results and discussion

We collected and analyzed a total of 47 h of annotated ECG recordings during workdays and 30 h of data during non-workdays, resulting in a total of 77 h of clinical grade
ECG signal. Results from LF/HF metrics were higher for all POs (with the exception
of Participants 2 and 3) both on duty and off duty when compared to normative values
(Figure 12.15).

In what concerns to self-report measures, overall stress intensity reports for all
events ranged from 0 to 5, mean was 2.1 ± 0.99 and the perceived control range was
0 to 5 and the mean was 4.2 ± 0.42.

Overall results suggested that POs experience high levels of physiological stress
(based on LF/HF), indicating the presence of a chronic stress state when compared to
healthy individuals [19]. However, when questioned about the stressors experienced
during working days, no significant events were self-rated as stressful. If only self-
reported information was contemplated in this study, probably it would reach the
premature conclusion that POs do not experience stress.

These findings provide important information regarding the potential of using a
multi-method approach with both self-report psychological questionnaires and physiology HRV measures for the analysis of stress. Hence, the monitoring of stress
during real world events might be the key for controlling stress-related health events
in this type of FR professionals and might be the essential information to consider
when managing participants' duties. Furthermore, on the basis of these findings it is
recommended that practitioners should implement healthy routines (e.g., increased

fitness exercise levels); monthly medical and physical monitoring, group sessions for verbalization and stress symptoms detection in order to help professionals to better cope with stress. Finally, the current method has a potential impact for working as a reference model for future stress diagnosis methods. Since it allows the understanding of the real impact of stress on health, by overcoming the potential limitations of self-report measures used to investigate stress. The current study is not without limitations. Particularly, the reduced sample size, with specific duties and short time duration of data collection procedures are also limitations. Hence, further research is required, using larger samples, different occupational groups, for longer periods of time and including longitudinal designs.

12.3.3 Business concept validation

Parallel to the technology development, field trials and studies with professionals described in the previous sections, we identified the potential market segments for the application of its technologies. The complete list of these market segments includes: FR professionals (FFs, PO, security professionals, etc.), oil drilling industry, air traffic controllers, military professionals, manufacturing and construction, sports and games, bus and truck drivers, surgeons and miners. The user needs of the professionals in the first three markets were identified by means of face-to-face interviews as shown in Table 12.5.

Furthermore, the first product statement for the FFs market, in terms of features and benefits, was developed from interviews with the project researchers and is presented in the tables below. In sum, the following tables are a summary of the features (Table 12.6) and respective benefits (Table 12.7) of the technology described in the first sections of this chapter.

Table 12.7 lists the benefits of the VR. The main purpose is to provide possible answers to the question: How, and in what degree, do the product features help address the problem and satisfy the needs of the customer?

12.3.4 Supply chain strategy for wearable products

The diversity of potential markets for the VR platform identified in the previous section, increases the need to build flexibility and adaptability in its early design in order to adapt the product and supply chain design to the unfolding market needs during the customer discovery process [22]. For software products, the iterations to create product-market fit are much faster, while for physical products, iterations are expensive, often involve longer processes and components selection may severely impact schedules. Therefore, the VR2Market project identified, on the one hand, the best practices that may be used during the product design phase toward building supply chain flexibility and, on the other hand, the supply chain configuration decisions of companies commercializing wearable products [23]. The results are based on literature review and interviews with seven CEOs of wearable technologies start-up companies (medical and sports) with presence in EU, USA and Asia.

Table 12.5 Customer problems and needs

Market segment	# Interviews	Problems and needs
FFs	20 FFs of four fire stations	Stressful moments before, during and after the fire Difficulties in breathing during the fire Lack of communications Lack of medical follow-up Tiredness due to shifts Lack of temperature sensors Lack of infra-red sensors
Oil drilling industry	Two workers of an oil drilling platform and one worker of a refinery	Problems with high pressure in the platform High stress levels 14 days of stay in the platform causes anxiety problems Problems with the communications Need to track vital signs and positioning of workers Difficulties in monitoring the gases level during the complete work shift Employers provide additional equipment only in case of insurance advantages
Air traffic controllers	10 air traffic controllers	Stressful situations due to: – responsibility for the work itself; – weather conditions; – technical problems (e.g., radar); – personal life episodes; – lack of visibility on the landing strip Tiredness due to: – work shift; – personal life episodes Need for health condition monitoring

12.3.4.1 Designing for supply chain

Designing products for supply chain may create additional competitive advantages for companies in terms of operations efficiency and responsiveness to market needs [24,25]. In fact, the way products are designed to contribute substantially to the total cost of product delivery [26,27]. Therefore, companies that design for supply chain, i.e., design the product considering the impact on the supply chain, are more agile and will be able to have a smoother scale-up in their commercialization process [24,28]. Three practices enable the design for the supply chain:

Modularity: Designing a product in modules of components not only reduces the product development time to keep costs down and accelerate time-to-market but also reduces the complexity of the product, thus reducing the complexity of the supply chain. In fact, the use of product modularity reduces the number of interfaces among parts and components, which consequently potentiate an easier supplier integration [24,25,29]. As wearable technologies consist on electronic components integrated

Table 12.6   General description of the VR and its features

	Feature designation	Feature description
General description	VR is a monitoring system for FFs. It is composed of individual garment (VitalJacket® and helmet) and a climate station that sits on the FFs vehicle. At the individual level (I), it allows individual monitoring of vital and toxic exposure signs, as well as individual positioning on the ground and in relation to other team members. At the team level, the system also allows the monitorization of the relative positioning of the team members to each other and the vehicle.	
Feature 1	(Individual) Vital signs monitoring	Individual sensors on the VitalJacket® (intimate contact with skin) monitor energy consumption (actigraphy obtained by PGA), O2 saturation, ECG, HR, HRV.
Feature 2	(I) Toxic gases and temperature exposure	Sensors on the helmet to measure toxic gases (CO_2 and NO_2) and outside temperature exposure.
Feature 3	(I) Critical event monitoring and alarm	Data mining that identifies patterns (standardization of signals) and allows to infer about critical events like falls, tachycardia, body temperature peaks and loss of cardiovascular capacity
Feature 4	(I) Photo frames	A camera in the helmet collects photo frames from the fire and other field conditions.
Feature 5	(I) Location	A GPS locating device is embedded in the helmet.
Feature 6	(Team) Mesh network	Network set among team members and supporting vehicle, with a 70 m range in a forest setting. Allows communication to a coordination center (every 15 min).
Feature 7	(Vehicle) Environmental station	Gathers environmental temperature, humidity and wind data.
Feature 8	Environmental conditions measurement	Sensors that measure atmospheric pressure and luminosity

in clothes or accessories, the products have naturally these two main modules and start-ups are the orchestrators of the supply network that consists of fabric suppliers, electronic system suppliers and module assembly manufacturers.

Standardization: Another way to reduce product complexity is by using standardized components [24]. Wearable technology companies that use standard electronic components will have lower supplier lead times and less risk of stock disruptions, though increasing their scale-up capability.

Design for postponement: Wearable products have the strong requirement of being adapted to the human body, which relies on the need to customize some parts of the product. Designing the product to enable late customization in the supply chain process increases flexibility and reduces inventory costs [30].

Table 12.7 Benefits of VR

	Benefit designation	Benefit description
Benefit 1	Individual monitorization—critical events	As an alarm is triggered in case of a critical event, the team and the center can act upon it immediately, providing help to the team member and to the team itself.
Benefit 2	Individual monitorization—non-critical events	The continuous monitoring and acquisition of vital signs in real time (toxic gases, temperature exposure, fatigue evolution and stress) allows better management of team members—organization of rest periods, work scheduling, etc. This could avoid adverse events.
Benefit 3	Location	Allows the coordination center to know where each individual and each team is on the field, and to evaluate the relative positioning of the individuals/teams, better management firefighting resources.
Benefit 4	Fire forecasting	Using the photo frames and environmental data gathered by the individual sensors and vehicle station, it is possible for the coordinator center to forecast how the fire will progress.
Benefit 5	Individual monitorization—historical register, extraction of personal recommendations	The continuous monitoring of vital signs allows to record a baseline data and create a set of individual recommendations (like changes of lifestyle) in order to control, for example, cardiovascular capacity and avoid events like cardiac arrest
Benefit 6	Operation without support networks	There is no need of networks 3G or 4G, the ad hoc network allows the transmission of information without the presence of towers or satellites. This is useful for fireworks in remote locations.
Benefit 7	User-friendly	The process of acquisition is automatic: no need of record information in sd cards and download it to a computer. Easy access to the acquired data by showing it in a user-friendly way.

12.3.4.2 Supply chain configuration for the commercialization of wearable products

After developing a supply chain friendly product, companies still face some challenges in order to configure a scalable, reliable and responsive supply chain: identify the right partners, build trust in supply chain relationships, engage suppliers, conduct certification processes, define expansion strategy and fund the production scale-up. Three main decisions are involved which are as follows:

Make or buy: Managers decide on whether to make or buy their components [31]. Start-ups in the wearable technologies industry are mainly integrators of the various modules of the new product they are bringing to the market. Therefore, most of the operations are outsourced and thus the start-up's daily operations are essentially supply chain orchestration and sometimes some final product integration and customization.

Sourcing decisions: Managers define the member firms of their supply chain [32]. Although wearable technologies start-ups source some of their components from the Far East, usually they choose a close-by supplier for the integration of the electronic components with garments or accessories. Key sourcing decisions include: shall the start-up have more than one supplier for each component? From which locations shall the start-up source? Who should be the strategic suppliers?

Supply chain relationships: Managers decide on the type of relationships they want to establish with the other supply chain members [33]. Start-ups tend to bridge their lack of resources and specific competences by leveraging their supply chain partner's expertise. Therefore, the capability to build a strategic partnership with suppliers and distributors may dictate the firm survival.

12.4   Conclusion

The VitalResponder® platform here presented in different perspectives of its current implementation status has been an evolving system that is capable of fulfilling requirements of different first-response scenarios to improve the security and well-being of professionals, such as FF and POs. It relies on the synergies between innovative wearable sensing, geo-location, WBANs and ICT to be able to respond to those scenarios.

The wearable sensing component has been evolving in two directions (1) to accommodate ambient and physiological measures and (2) from a t-shirt form factor to a patch for the physiological variables collection. This component has already resulted in two patents that may be explored in the coming years.

Team data communication has been a major topic of R&D within the project and a MANET built on 802.11 technology approach has been developed to provide the needed connectivity in the target scenarios. This solution is disconnection-tolerant and has shown excellent results.

The ICT integration has been evolving to be more and more complete both at the data collection and at the different data visualization components and has shown to be highly adaptable to the different requirements our first responder partners have been imposing to the VR team.

The platform has been used for many hours of data collection in drills, testing protocols and in real firefighting operations during the summer season. These data have been used for both stress and fatigue studies and to obtain new requirements to iteratively develop the different components of the system so that it reached its current state. We will continue this path, and together with the design and implementation of a business model, we hope to be able to bring to the market the results of this project, taking into account a previous study on supply chain strategy for the specific product targets of these results.

Acknowledgments

This work has been financed by National Funds through the FCT—Fundação para a Ciência e Tecnologia (Portuguese Foundation for Science and Technology) within the project "VR2Market" Grant CMUP-ERI/FIA/0031/2013 and project NanoSTIMA funded through the North Portugal Regional Operational Program (NORTE 2020), under the PORTUGAL 2020 Partnership Agreement, and through the European Regional Development Fund (ERDF). The authors would like to thank all the previous and current members of the project team that involves also end-users from firefighters from Albergaria-a-Velha and Vila Real-Cruz Verde, Polícia de Segurança Pública (PSP) from Porto Metropolitan Command, consultants and researchers. Regarding contributions and suggestions to the work developed, we would like to thank Emanuel Lima and Pedro Santos.

References

[1]   Skogstad M, Skorstad M, Lie A, Conradi H, Heir T, Weisæth L. Work-related post-traumatic stress disorder. *Occupational Medicine.* 2013;63(3):175–82.

[2]   Magalhães T, Oliveira I, Fernandes J. Message based integration in cyber-physical system: firefighters in the field. *Proceedings of the 12th EAI International Conference on Mobile and Ubiquitous Systems: Computing, Networking and Services on 12th EAI International Conference on Mobile and Ubiquitous Systems: Computing, Networking and Services*; Coimbra, Portugal: ICST (Institute for Computer Sciences, Social-Informatics and Telecommunications Engineering); 2015. p. 285–6.

[3]   Aguiar A, Soares E, Brandão P, Magalhães T, Fernandes JM, Oliveira I. Demo: wireless IP mesh on android for fire-fighters monitoring. *Proceedings of the 9th ACM MobiCom Workshop on Challenged Networks*; Maui, Hawaii, USA: ACM; 2014. p. 89–92.

[4]   Biodevices SA. VitalJacket®; 2017. Available at: http://www.vitaljacket.com/pt/empresa/certificacao/ [Accessed on February 2017].

[5]   Cunha JP. pHealth and wearable technologies: a permanent challenge. Studies in Health Technology and Informatics. 2012;177:185–95.

[6]   Cunha JPS, Cunha B, Pereira AS, Xavier W, Ferreira N, Meireles L, Vital-Jacket®: A wearable wireless vital signs monitor for patients' mobility in cardiology and sports. In: *Proceedings of the 4th International Conference on Pervasive computing technologies for healthcare (PervasiveHealth) 2010.* Munchen, 22–25th March, 2010.

[7]   Cunha JPS, inventor; INESC TEC, assignee. Medical Device With Rotational Flexible Electrodes. Portugal patent 20161000064607. 2016.

[8]   Santos PM, Abrudan TE, Aguiar A, Barros J. Impact of position errors on path loss model estimation for device-to-device channels. *IEEE Transactions on Wireless Communications.* 2014;13(5):2353–61.

[9]   Contreras OA. Performance evaluation of an IEEE 802.11 mobile ad-hoc network on the Raspberry Pi. Porto, Portugal: University of Porto; 2015.

[10]   Lima E, Brandão P, Azuaje O, Aguiar A. Demo: DisToNet: disconnection tolerant mobile ad-hoc networks. *Proceedings of the 10th ACM MobiCom Workshop on Challenged Networks*; Paris, France: ACM; 2015. p. 63–4.

[11]   Rodrigues S, Kaiseler M, Queirós C. Psychophysiological assessment of stress under ecological settings. *European Psychologist*. 2015;20(3):204–26.

[12]   Lesage FX, Berjot S, Deschamps F. Clinical stress assessment using a visual analogue scale. Occupational Medicine (Lond). 2012;62(8):600–5.

[13]   Sobel-Fox RM, McSorley AM, Roesch SC, Malcarne VL, Hawes SM, Sadler GR. Assessment of daily and weekly fatigue among African American cancer survivors. *Journal of Psychosocial Oncology*. 2013;31(4):413–29.

[14]   Pan J, Tompkins WJ. A real-time QRS detection algorithm. *IEEE Transactions on Biomedical Engineering*. 1985;(3):230–6.

[15]   Hamilton P. Open source ECG analysis. In: *Proceedings of Computers in Cardiology Conference, IEEE*, pp. 101–104. Memphis, 22–25th September, 2002. doi: 10.1109/CIC.2002.1166717

[16]   Schaaff K, Adam MT. Measuring emotional arousal for online applications: Evaluation of ultra-short term heart rate variability measures. In: *Proceedings of the 2013 Humaine Association Conference on Affective Computing and Intelligent Interaction, ACII 2013*, pp. 362–368. Geneva, 2–5th September, 2013. doi: 10.1109/ACII.2013.66

[17]   Camm AJ, Malik M. Heart rate variability. Standards of measurement, physiological interpretation, and clinical use. Task Force of the European Society of Cardiology and the North American Society of Pacing and Electrophysiology. *European Heart Journal* 1996;17(3):354–81.

[18]   Rodrigues JGP, Kaiseler M, Aguiar A, Cunha JPS, Barros J. A mobile sensing approach to stress detection and memory activation for public bus drivers. *IEEE Transactions on Intelligent Transportation Systems*. 2015;16(6): 3294–303.

[19]   Nunan D, Sandercock GR, Brodie DA. A quantitative systematic review of normal values for short-term heart rate variability in healthy adults. *Pacing and Clinical Electrophysiology*. 2010;33(11):1407–17.

[20]   Goldberger AL, Amaral LA, Glass L, *et al.* PhysioBank, PhysioToolkit, and PhysioNet: components of a new research resource for complex physiologic signals. *Circulation*. 2000;101(23):E215–20.

[21]   Maykut P, Morehouse R. *Beginning Qualitative Research: A Philosophic and Practical Guide*. London: Falmer Press; 1994.

[22]   Bahrami H, Evans S. Super-flexibility for real-time adaptation: perspectives from Silicon Valley. *California Management Review*. 2011;53(3): 21–39.

[23]   Tedim AR, Barros AC, Maia C, Godsell J. Start-ups of wearable technologies: Challenges in supply chain strategic decisions. In: *Proceedings of the 22nd International Annual EurOMA Conference*, pp. 1–8 Neuchâtel, Switzerland, 2015.

[24]   Khan O, Christopher M, Creazza A. Aligning product design with the supply chain: a case study. *Supply Chain Management: An International Journal.* 2012;17(3):323–36.

[25]   Marsillac E, Roh JJ. Connecting product design, process and supply chain decisions to strengthen global supply chain capabilities. *International Journal of Production Economics.* 2014;147:317–29.

[26]   Hong P, Roh J. Internationalization, product development and performance outcomes: a comparative study of 10 countries. *Research in International Business and Finance.* 2009;23(2):169–80.

[27]   Hong P, Doll WJ, Revilla E, Nahm AY. Knowledge sharing and strategic fit in integrated product development proejcts: an empirical study. *International Journal of Production Economics.* 2011;132(2):186–96.

[28]   Dekkers R, Sharifi H, Ismail H, Reid I. Achieving agility in supply chain through simultaneous "design of" and "design for" supply chain. *Journal of Manufacturing Technology Management.* 2006;17(8):1078–98.

[29]   Park Y, Ogawa K, Tatsumoto H, Hong P. The impact of product architecture on supply chain integration: a case study of Nokia and Texas Instruments. *International Journal of Services and Operations Management.* 2009;5(6):787–98.

[30]   Fixson SK. Product architecture assessment: a tool to link product, process, and supply chain design decisions. *Journal of Operations Management.* 2005;23(3–4):345–69.

[31]   Fine C. Value chain design and three-dimensional concurrent engineering. in Kim K, (ed.). *Business Eco-Systems: Relationships between Large and Small Firms.* Seoul: Federation of Korean Industries; 2006.

[32]   Petersen KJ, Handfield RB, Ragatz GL. Supplier integration into new product development: coordinating product, process and supply chain design. *Journal of Operations Management.* 2005;23(3):371–88.

[33]   Ellram LM, Tate WL, Carter CR. Product-process-supply chain: an integrative approach to three-dimensional concurrent engineering. *International Journal of Physical Distribution & Logistics Management.* 2007;37(4):305–30.

Chapter 13

Wearable sensors for foetal movement monitoring in low risk pregnancies

Luís M. Borges[1], Norberto Barroca[1], Fernando J. Velez[1], J. Martinez-de-Oliveira[2], and António S. Lebres[3]

There is no doubt about the interest of evaluating foetal health through its movements' profile, mainly by counting them at determined intervals. Presently, this is only available through mother's intervention, which is, naturally, extremely sensitive to subjective interference, like anxiety. To turn this method more efficient and comfortable a wearable autonomous device is desirable, both for its expected accuracy in identifying their number and also their type and allowing a prolonged observation.

The aim of our research project "Smart-Clothing" has been to find adequate sensors to identify the various foetal movement types (full body, head or limb extension and flexion, respiratory) and to apply them to abdominal belt, tape or girdle, to be continuously used by pregnant women, mainly during the day, but even at night. Although the resulting prototypes have not been considered to fulfil the requirements previously defined, the work done and the performed experiences had given useful information for further developments.

A few years after the end of that project, the description of the trials and results obtained seem to be useful data to serve as a basis for the design of future projects.

13.1 Introduction

The need to monitor the foetal health in the periods between the medical sessions in low risk pregnancies are based in traditional protocols for counting the foetal movements felt by the mother [1]. Although the maternal perception is a relevant characteristic for the evaluation of the foetal health, the monitoring is hard to accomplish and can induce errors caused by the mother's anxiety and concentration. Between the medical sessions that occur weekly during the last five weeks of pregnancy, the foetal

[1] Instituto de Telecomunicações and Departamento de Engenharia Electromecânica, Universidade da Beira Interior, Portugal

[2] Department of Child and Mother Health, School of Health Sciences, University of Beira Interior, Centro Hospitalar Cova da Beira, Portugal

[3] Departamento de Física, Universidade da Beira Interior, Portugal

health can change suddenly. Consequently, the majority of foetus fatalities at the end of pregnancy occur in the low risk group. Therefore, it is important to obtain an obstetric tracing, allowing for the identification of sudden changes in the foetus health, by monitoring the foetus movements and the foetal heart rate (FHR). One example of how this theme of pregnancy monitoring was gaining importance in the research community in the last 15 years is mentioned in [2], which describes an innovative remote monitoring decision support system, used in the early diagnosis of pregnancy complications, through the effective and non-invasive monitoring of maternal and foetal electrocardiograms. Some patents on foetal monitoring are presented in [3–5].

According to [6], in the hospital, the foetal monitoring is done by using TocoCardiographie (CTG), whose equipment records the FHR and the uterus contractions. The FHR is determined by using an ultrasound Doppler sensor ($f = 1$–3 MHz), while the uterine contractions are roughly detected with a pressure sensor. Signalling foetal movements in a CTG record is made by pregnant women, by pressing an event marker device. The pregnant woman can monitor the foetal movements at home, by counting and annotating the number of those she felt, which should be registered in a form for posterior analysis of the physician. It has been shown that an average pregnant women do feel 80% of foetal movements under these conditions. It has been introduced in the market a low cost portable equipment, based on the Doppler technology, which allows for hearing foetal cardiac sounds by the pregnant woman that can be used beyond the 12th week and allows for recording the cardiac sounds. The use of these devices based on long exposition to ultrasound technique can pose possible risks to the foetus. However, their effects are not well known yet.

E-health environments should be designed in order to provide personalised services and applications to their primary users (i.e., the patients), under strict resource, regulation and security constraints. A typical scenario would employ wireless and wired sensors and local or cloud-based processing units to collect, process, store and communicate data related to the patients' needs and condition. E-health devices can be located on the patients' bodies or immediate environments to monitor and interact with the patients, while they perform their activities of daily living (ADL). The system should be able to recognise abnormal events, as well as slowly emerging shifts in the behaviour of the primary users, and to inform secondary users (caregivers, healthcare professionals and family) appropriately and timely, hence providing a feeling of safety and comfort for all involved parties.

Electronic devices for the monitoring of the foetal movement in the last weeks of pregnancy require the integration of sensors in the garment and/or the inclusion of sensors embedded into textiles, along with a hierarchical communication system that allows for the delivery of the collected data to the health professional remotely based. In this work, we will review those tests that have been carried out with different types of sensors integrated in a belt, with the aim of choosing those that are more reliable for the detection of foetal movements.

Wireless body area network (WBANs), wireless body sensor networks (WBSNs) and wireless sensor networks (WSNs) have opened a new world of opportunities in the context of healthcare monitoring, the developments in the microelectronics integration have made possible the creation of small wireless nodes that operate together creating a well-defined multi-hop network, so the future will bring new systems capable to be integrated in specialised medical technology with pervasive, wireless sensor networks. The WSNs in the context of medical applications could be one of the many infrastructures that can be used. The main areas that can benefit from it include investigation in functional textile materials and wireless communication networks in the context of human body and statistical methods for the data analysis and treatment as described by the authors from [7].

One of the primary objectives of the "Smart-Clothing for Health Monitoring and Sports Applications" (Smart-Clothing) project [8] was to monitor foetal movement in the last weeks of pregnancy during a significant part of the day, the longer the better.

As the main supporting device, we have chosen to adopt a basic pregnancy belt. The weight from the pregnant woman augments, on average, 10–14 kg during pregnancy, but there are also pregnant women whose weight increases upto 30 kg. This extra weight pulls abdomen forward and has a negative effect on the wall of the abdomen, hip and spinal column. To avoid extra pain due to the resulting bad posture, supporting belts are very useful for future mothers. In fact, these belts have been conceived aiming at user comfort. They adapt to the shape of the bodies and their evolution during pregnancy. These belts provide maximum protection for the abdominal region and allow the belly growing without pressing the baby. Very positive results are achieved in the decrease of lumbar pain and reduction of stretching in the abdomen and hip zones.

In order to achieve the monitoring of foetal movement in the last weeks of pregnancy goal, when the foetus has gained a significant degree of development in terms of stature, it is important to have a full integration of sensors and a hierarchical communication system. As such, and for the purpose of prolonged monitoring, it is important to use wireless solutions either at home or in the hospital, particularly with the perspective of giving the pregnant woman the chance of expanding her movements beyond a single location. So, with this research, we have proposed some of the options available for these interconnections, and the terminology used to discuss them.

Hybrid communications can be a solution to obtain a network of networks, e.g., by using Internet Protocol (IP). A bottom-up architecture formed by (i) WSNs (LR-WPAN) and (ii) WiFi has been explored to allow for healthcare monitoring anyway, anywhere and anytime.

It is important not only to create a WBAN, WBSNs and WSN platform that should accommodate a wide range of sensors, depending on needs, expandability and cost but also to create a reliable mean to share the information across the network. By using this infrastructure, remote agents can be either collectors of information (by storing the data in a database that could be accessed later on) or being active in a real-time monitoring process. For example, a nurse or a doctor who monitors the foetus in the pregnant women has all this information available by using a Mote Interface Board (Gateway) that is attached to a Personal Computer (PC).

The structure of the rest of the chapter is as follows. Section 13.2 addresses the Flex Sensor technology, whilst analysing the variation of the resistance as a function of the deflection of the sensor. Section 13.3 describes the schematics for different circuits with the Flex Sensors, namely the bend sensor voltage divider, adjustable buffers, variable deflection threshold switch, resistance to voltage converter and Flex Sensor circuit with a Light Emitting Diode (LED). Section 13.4 addresses a voltage divider as Input of the Analogue-to-Digital Converter (ADC), while Section 13.5 analyses the computation of the Flex Sensor angle. Section 13.6 presents the stand-alone Smart-Clothing Flex Sensor belt, including details on the acquisition module and its firmware, while describing the first, second and third versions of the Flex Sensor view software in detail. Section 13.7 describes the developed packet protocol interface developed for the communication with the belt and includes the description of the acquisition module to PC and the PC to acquisition module packets. Section 13.8 presents the experimental part of the work and discusses the results obtained with the Flex Sensor belt. Section 13.9 discusses the efforts on developing other types of belts, e.g., the on-off belt and the belt with piezoelectric sensors. Section 13.10 briefly presents the underlying wireless network architecture, while Section 13.11 draws the Conclusions and presents a suggestion for future work.

13.2   The Flex Sensor technology

The Flex Sensor patented technology is based on variable resistive carbon elements. As a variable printed resistor, the Flex Sensor achieves large form-factor on a thin flexible substrate. When the substrate is bent, the sensor produces a resistance output correlated to the bend radius—the smaller the radius, the higher the resistance value, as shown in Figure 13.1.

An application is its attachment to a door. As the door is opened, it is possible to measure how far the door has opened and how fast it is moving, in angular terms. The sensor is lightweight small, easily packaged and very reliable.

In fact, flex sensors are sensors that change in resistance depending on the extension of bending on the sensor. They convert the change in the bend to electrical resistance variation. As mentioned before, the more the bend is, the more the resistance value is, as shown in Figure 13.2. They are usually in the form of a thin strip, from 1″ to 5″ long that vary in resistance from approximately 10 to 50 kΩ. They are often used in gloves to sense finger movement.

There are two main brands of flex sensors that are affordable and easily available in the market: SpectraSymbol Flex Sensors and the Gentile-Abrams sensor. *SpectraSymbol* manufactured the ones used in Smart Clothing.

Flex Sensors are analogue resistors. They work as variable analogue voltage dividers. Inside the Flex Sensor, there are carbon resistive elements within a thin flexible substrate. More carbon means less resistance. When the substrate is bent, the

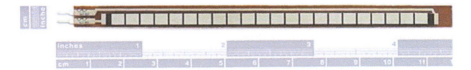

Figure 13.1   Flex Sensor potentiometer

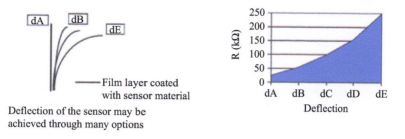

dA   dB
          dE

——Film layer coated
    with sensor material

Deflection of the sensor may be
achieved through many options

Figure 13.2   Resistance versus deflection of the Flex Sensor (adapted from [9])

sensor produces a resistance output relative to the bend radius. With a typical flex sensor, a flexion of 0° corresponds to 10 kΩ, while a flexion of 90° will originate 30–40 kΩ.

The characteristics of the Spectra Symbol sensor are the following ones:

- Size: approximately 0.28″ wide and 1″/3″/5″ long;
- Resistance range: 1.5–40 kΩ, depending on the sensor;
- Lifetime: larger than 1 million life cycles;
- Temperature range: −35°C to +80°C;
- Hysteresis: 7%;
- Voltage: 5–12 V.

13.3   Schematic for the circuit with the Flex Sensor

13.3.1   Bend sensor voltage divider

As a first and simple schematic for experimental deflection-to-voltage conversion, the flex sensor is connected to a measuring resistor whose resistance is R_M in a voltage divider configuration, as shown in Figure 13.3.

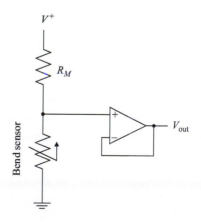

Figure 13.3   Bend sensor voltage divider (adapted from [9])

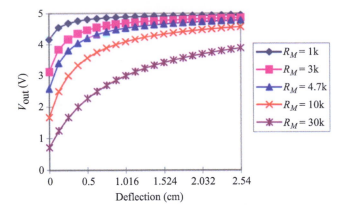

Figure 13.4   Deflection versus voltage for the Flex Sensor (extracted from [9])

The output is very accurately described by the following equation ($I_B \cong 0$, $V_{OS} \cong 0$):

$$V_{\text{out}} = \frac{(V^+)}{[1 + R_M/R_{\text{Flex}}]}. \tag{13.1}$$

In this configuration, the output will increase with the increase of the deflection. If the R_{Flex} and R_M are swapped, the output will decrease and the deflection increases. The value of the measuring resistance, R_M, is chosen to maximise the desired deflection sensitivity, the output voltage range of measurement and the chosen limit to the value of the resistors' current. In this case, if the chosen option is to use op-amps with single supply the most recommended options are LM358 and LM324. Other op-amps like LF355 and TL082 are also appropriate. The first ones are chosen for low offset voltage and the latter for low bias currents. In both cases, they allow for reducing the error due to the source impedance of the voltage divider, being op-amps selectable criterion how near zero we need the output should be, when single supply is used. Selected results for the deflection values as a function of the measuring resistance are presented in Figure 13.4, where $V^+ = 5$, V is considered.

13.3.2 Adjustable buffers

Adjustable buffers are interfaces that isolate the output from the high source impedance of the Flex Sensor. These allow for adjusting the output offset and gain. In Figure 13.5, the ratio of the resistance between resistors R_2 and R_1 sets the gain of the output. Offsets resulting from the non-infinite Flex Sensor resistance at zero deflection are trimmed out with the potentiometer, R_3. According to [9], for the best results, R_3 should be about one-twentieth of R_1 or R_2. Adding an additional pot at R_2 makes the gain easily adjustable. Broad range gain adjustment is made by replacing R_2 and R_1 with a single pot.

The schematic presented in Figure 13.6 gives similar results to the previous one but the offset trim is isolated from the adjustable gain. According to [9], with this

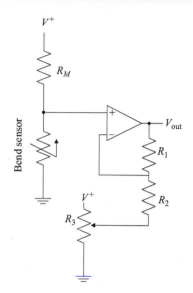

Figure 13.5 *Adjusting the output offset voltage for the flex sensor (adapted from [9])*

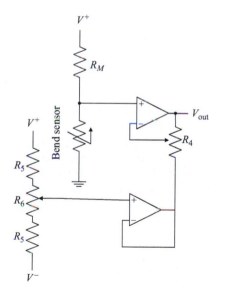

Figure 13.6 *Adjustable gain with buffered offset trimming (The offset trim is isolated from the adjustable gain, adapted from [9])*

separation, there is no constraint on values for R_6. Furthermore, it is possible to reach zero or even negative values at the op-amp output, if needed, for a particular application.

Typical values for R_5 and R_6 are around 10 kΩ.

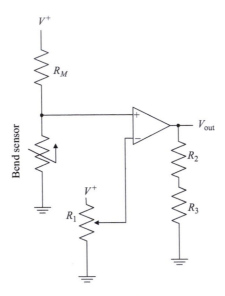

Figure 13.7 Variable deflection threshold switch (adapted from [9])

13.3.3 Variable deflection threshold switch

The simple circuit is ideal for applications that require on-off switching at a specific deflection value, using a touch-sensitive membrane, cut-off and limit switches. The Flex Sensor is arranged in a voltage divider with R_M and an op-amp, as shown in Figure 13.7. The op-amp $U1$ is used as a comparator.

The output from the $U1$ is either high or low. The non-inverting input from the op-amp is driven by the output of the divider (R_M/R_{Flex}), which is a voltage that increases with the deflection. Hence, with zero deflection, the output of the op-amp will be low. According to [9], if the voltage at the non-inverting input of the op-amp exceeds the voltage of the inverting input, the output of the op-amp will toggle high. Therefore, in Figure 13.7, the deflection threshold is set at the inverting input by the potentiometer R_1 and the resistance R_2 acts like a "de-bouncer", eliminating any multiple triggering of the output that can occur. For op-amps selection, op-amps like the LM358 and LM324 are suggested, and the combination of R_2 and R_3 is chosen to maximise the desired deflection output sensitivity range. A typical value for this combination is about 47 kΩ.

13.3.4 Resistance-to-voltage converter

In this type of circuit, the Flex Sensor is the input of a resistance-to-voltage converter. With the positive reference voltage, the output of the op-amp must be able to swing below ground, from 0 V down to $-V_{ref}$, therefore dual sided supplies are necessary. A negative reference voltage will yield a positive output swing, from 0 V up to $+V_{ref}$.

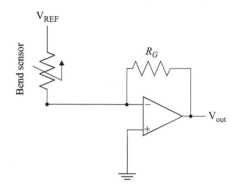

Figure 13.8   Resistance-to-voltage converter (adapted from [9])

Since this is a simple inverse relation between V_{out} and V_{Flex}, the equation for the output in Figure 13.8 (13.2) can be written as

$$V_{out} = \frac{(-R_G * V_{ref})}{R_{Flex}},\qquad(13.2)$$

V_{out} is inversely proportional to R_{Flex}. Changing R_G and/or V_{ref} changes the response slope.

Application Example: For a human-to-machine variable control device, like a joystick, the maximum deflection applied to the Flex Sensor is about 2″. The testing of a typical Flex Sensor shows that the corresponding R_{Flex} at 2 inches is about 4.6 kΩ. If V_{ref} is −5 V, and an output swing of 0 V to +5 V is desired, then the R_G should be approximately equal to this minimum R_{Flex}. R_G is set to 4.7 kΩ. A full swing of 0 V to +5 V is thus achieved. A set of deflection versus V_{out} curves is shown in Figure 13.4 for a standard Flex Sensor using this interface with a variety of R_G values.

The current through the Flex Sensor should be limited to less than 1 mA/cm^2 of applied deflection. As the voltage divider circuit, adding a resistor in parallel with R_{Flex} will give a definite rest voltage, which is essentially a zero-deflection intercept value. This can be useful when enough resolution is desired at low deflection (Figure 13.8).

13.3.5   Flex Sensor LED brightness

The circuit presented in Figure 13.9 is useful for applications where some visual feedback is needed. Basically, the electrical circuit is just a voltage divider plus a LED whose brightness decreases with the increase of the deflection angle of the Flex Sensor. If the flex angle increases the resistance from the Flex Sensor increases; hence, the current of the base from the transistor decreases and the brightness of the transistor will be reduced. The resistance R_L limits the maximum current through the LED. Besides, the transistor controls the current through the LED. Since the circuit depends on h_{fe} (of the transistor), sensitivity may need to be tuned to accommodate the h_{fe} span for common transistors.

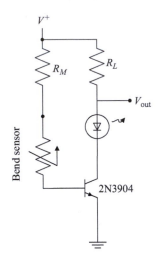

Figure 13.9   Schematic for the Flex Sensor circuit with a LED (adapted from [9])

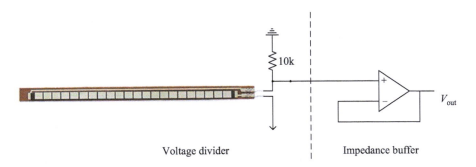

Figure 13.10   Schematic for the application of the Flex Sensor

13.4   Voltage divider as input of the ADC

After being exposed to different types of interfacing and reading from the Flex Sensor the one that better fits to interface with a microcontroller is the one presented in Figure 13.10.

Initially, an impedance buffer has been used, as presented in the schematic from Figure 13.10, but in the microcontroller used in this project, the MSP430F449, there already exists a high impedance port. Hence, there is no need to include this part into the design of each flex sensor, using instead a simple voltage divider that is connected to the ADC pins.

The voltage divider obtained is given by the following equation:

$$V_{\text{out}} = \frac{R_1}{R_1 + R_{\text{Flex}} \times V_{\text{in}}}. \tag{13.3}$$

Developing the equation and solving it in order to find R_{Flex}, one obtains

$$V_{\text{out}} \times R_1 + V_{\text{out}} \times R_{\text{Flex}} = V_{\text{in}} \times R_1,$$

$$V_{\text{out}} \times R_{\text{Flex}} = V_{\text{in}} \times R_1 - V_{\text{out}} \times R_1,$$

$$R_{\text{Flex}} = \frac{R_1(V_{\text{in}} - V_{\text{out}})}{V_{\text{out}}} \tag{13.4}$$

13.5   Computation of the Flex Sensor angle

The manufacturer gives data about the angle and the corresponding resistance from the Flex Sensor. The data given by the manufacturer are presented in Table 13.1.

In order to interpolate the deformation angles for the corresponding value of the resistance, two equations have been proposed, one for the R_{T1} and another for R_{T2}, as there are two distinct regions for the resistance from the Flex Sensor (Figure 13.11).

Table 13.1   Manufacturer data for resistance and angle correspondence

Resistance value [Ω]	Angle [°]
10,000	0
14,000	90
22,000	180

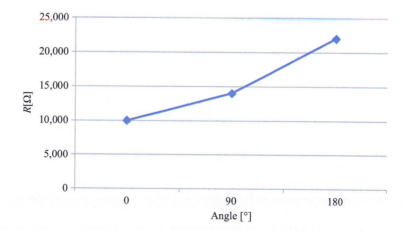

Figure 13.11   Curve for the resistance versus angle (from the manufacturer)

$\boldsymbol{R_{T1}}$:

$$R_{T1} = m_1 * \theta_1 + b_1,$$

$$m_1 = \frac{R_2 - R_1}{\theta_2 - \theta_1} = \frac{14000 - 10000}{90 - 0} = 44.44.$$

For $R_{T1} = 10000\,\Omega$ and $\theta_1 = 0$

$$b_1 = 10000 - 44.44 * 0 = 10000,$$

$$R_{T1} = 44.44 * \theta_1 + 10000,$$

$$\theta_1 = \frac{R_{T1} - 10000}{44.44}. \tag{13.5}$$

$\boldsymbol{R_{T2}}$:

$$R_{T2} = m_2 * \theta_2 + b_2,$$

$$m_2 = \frac{R_3 - R_2}{\theta_3 - \theta_2} = \frac{22000 - 14000}{180 - 90} = 88.88.$$

For $R_{T2} = 14000\,\Omega$ and $\theta_2 = 90$

$$b_2 = 14000 - 88.88 * 90 = 6001,$$

$$R_{T2} = 88.88 * \theta_2 + 6001,$$

$$\theta_2 = \frac{R_{T2} - 6001}{88.88}. \tag{13.6}$$

13.6   Standalone Smart-Clothing Flex Sensor belt

The standalone Smart-Clothing Flex Sensor belt is based on the Flex Sensor described above. Initially, only one sensor was used. Then, a system with the support of up to eight Flex Sensors has been designed. Later a new version of this belt has been integrated with wireless technologies such as ZigBee, WiFi and GSM in order to transmit the data to the doctor without using cables. The system is composed of three essential blocks: the belt, the acquisition module and the PC, as shown in Figure 13.12.

The first block is composed of the Smart-Clothing Flex Sensor belt which has eight Flex Sensors attached to the belt. The second block is composed of electronics circuitry in a Printed Circuit Board (PCB) with connectors to the eight Flex Sensors. Besides, it has the eight resistors so it can be possible to do the voltage dividers an

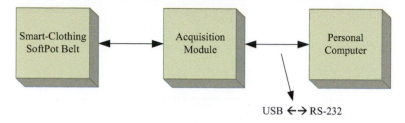

Figure 13.12   Block diagram for the Smart-Clothing Flex Sensor belt

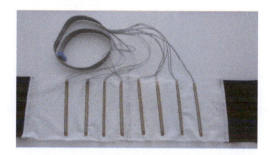

Figure 13.13   Flex Sensor belt with eight flex sensors

appropriate connection to MSP430-449STK2 module, which is responsible to read the voltage from the voltage dividers of each sensor, and then send it via USB$\leftarrow\rightarrow$RS-232 to the PC. The third block is the PC that runs the proprietary software (also developed in Smart Clothing), which allows us to view the deformation angles of each Flex Sensor and count automatically the movements from the foetus.

13.6.1   Smart-Clothing Flex Sensor belt

The first version of the Flex Sensors belt incorporates eight Flex Sensors, as shown in Figure 13.13.

A simple voltage divider is enough to operate this sensor as the manufacturer proposes a correspondence of standard values of the flexion angle to a certain value of the resistance. According to [10], the length of the belt is 20 cm. It has Velcro to facilitate the adjustment to the user's belly. The details of the construction of the belt are given in [10].

13.6.2   Acquisition module

The system diagram is presented in Figure 13.14, where just one Flex Sensor is presented. However, all eight sensors were connected to a unique voltage divider. Besides the Flex Sensor, a button was incorporated to be pressed by the pregnant woman when she feels or detects the foetus moving. These events will be very useful for comparison purposes, as they enable a comparison with the movements detected automatically by the belt.

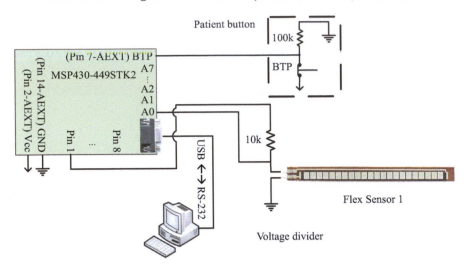

Figure 13.14   Flex Sensor belt acquisition module diagram

A simple voltage divider is enough to operate this sensor as the manufacturer proposes a correspondence of standard values of the flexion angle to a certain value of resistance. The Flex Sensor manufacturer states that a resistance value of 10 kΩ matches an angle of 0°, while a value of 14 kΩ matches an angle of 90° and a value of 22 kΩ matches an angle of 180°. An initial drawing of this system had the power supply connected to each voltage divider, but when the system was turned on the system failed because of not supporting enough current to hold on all the voltage dividers as well as the microcontroller. To overcome this limitation, the solution was to supply the V_{cc} voltage to the voltage divider by using the external pin from the microcontroller and read the voltage from the voltage divider using the ADC from the microcontroller. To compute the value of the resistance from the Flex Sensor the following voltage divider equation is used:

$$R_{\text{Flex}} = \frac{R_1 * (V_{\text{in}} - V_{\text{out}})}{V_{\text{out}}}, \tag{13.7}$$

where R_{Flex} is the resistance value, R_1 is equal to 10 kΩ, V_{in} is the V_{cc} value supplied to the voltage divider, and V_{out} is the voltage value from the voltage divider. The V_{cc} voltage supplied to the voltage divider is measured periodically by a routine that runs in the microcontroller in order to compensate the battery losses during the system operation and to achieve a better accuracy of the values from the Flex Sensors.

For the conversion of the resistance value to the angle value, a linear approximation behaviour of Flex Sensors by segments is considered, i.e., two different equations have been used to extrapolate the angle values, as follows:

$$\theta_1[°] = \frac{R_{\text{Flex}} - 10 * 10^3}{44.44}, \tag{13.8}$$

*Figure 13.15   Theoretical and calibration curve (experimentally measured values)
of the Flex Sensor*

$$\theta_2[°] = \frac{R_{Flex} - 6001}{88.88},$$
(13.9)

where θ_1 and θ_2 are the angles for the corresponding Flex Sensor resistance value.
Equation (13.8) is used when the resistance value is between 10 kΩ and 14 kΩ
while (13.9) is used when the resistance value is between 14 kΩ and 22 kΩ. These
equations are based on the theoretical and calibration curve of the Flex Sensor, as
shown in Figure 13.15. The theoretical line is based on the resistance values, and
corresponding deflection angle, supplied by the Flex Sensor manufacturer and the
calibration curve is the result of an experiment where the flex sensor was bent from
0° to 90° and the resistance value measured was registered at each 10° of bending
increment.

13.6.3   Acquisition module firmware

The *firmware* developed by our team for microcontroller hosted by the MSP430-
449STK2 is based on the algorithm shown in Figure 13.16. The variable "*Ler_canal*"
(read channel) is used in the firmware in order to control the reading of the channels
from the ADC, avoiding erroneous readings caused by unsynchronised readings of
the channels.

   The firmware developed for the Flex Sensor belt system begins with the reset
option, which in fact can be performed starting from all the states. The system initiates
the UART function that is used to send the data from the acquisition module to the
computer. Then, the value of voltage of the power supply is used by the function
"*ler_power_supply()*" from the system. This value is used later in the calculation
of the measured values by the microcontroller ADC. The next step function is the
"*Configura_TIMER_B_200()*" function. It is used to start the microcontroller TIMER
B as well as the interval between each interruption of the timer, which is set to be

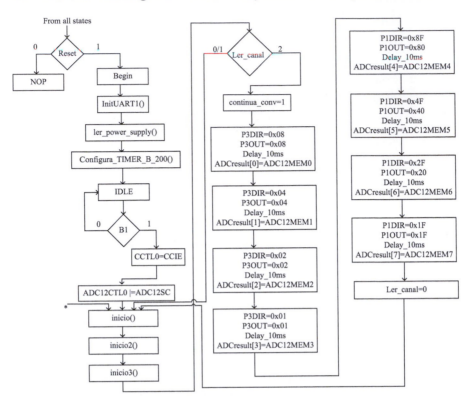

Figure 13.16   Algorithm for the acquisition module firmware (Part 1)

200 ms. All the interruptions in the system, namely the TIMER B ones, are enabled by pressing the button B1 from the acquisition module. Consequently, the interrupt vector from the ADC will be enabled and the reading of the eight ADC channels will start.

The function "*inicio()*" is called to check if the button B1 is pressed. This function is useful when the process of calibration is initiated and then it is needed whenever the interruptions are started again. Then, the function "*inicio2()*" is called to check if the patient button was pressed. If so a unit is added to the global variable "*Patient_counter*". The system interruptions are disabled if the button B2 is pressed. The function "*inicio3()*" is used to start the reception of the data for the calibration of the flex sensors. This occurs when B2 is pressed, disabling the interruptions of the system. The interruptions are enabled again just pressing B1 and the interrupt vector of the ADC converter is initiated. Each time the system is triggered if the global variable "*Ler_canal*" is checked if it is equal to one and if it so the global variable "continua_conv" is set to one. After this variable is set to one, the output P3.3 is set to one, too.

This way, it acts as the positive point for the voltage divider of the Flex Sensor 1. Then in order to perform a proper reading by the channel 1 of the ADC, a delay of 10 ms is introduced. Afterwards, the ADC reads the value from the voltage divider and saves the value of the ADC in the global variable "ADCresult[0]". The output P3.2 of the Flex Sensor 2 is set to one, acting like the positive point for the voltage divider of the Flex Sensor 2. Then, as in the ADC channel 1, a delay of 10 ms is introduced in the ADC channel 2. After this period the ADC reads the value from the voltage divider and saves the ADC value in the global variable "ADCresult[1]". The output P3.1 of the Flex Sensor 3 is set to one, acting like the positive point for the voltage divider of the Flex Sensor 3. Then in order to do a proper reading by the ADC channel 3 a delay of 10 ms is done and after that, the ADC reads the value from the voltage divider and saves the ADC value in the global variable "ADCresult[2]". This procedure is repeated for all the outputs of the flex sensors belt with the assignments from Table 13.2.

When all ADC channels are read, the global variable "Ler_canal" is set to "0" in order to calculate the angles of each Flex Sensor. These formulas are introduced into the TIMER B function, Figure 13.17. In the TIMER B function, first, the procedure is to check if the global variable "Ler_canal" is equal to "0" or "1" in order to compute the angles from the flex sensors when the ADC is not reading the values from the voltage dividers.

When variable above is equal to "0" or "1" the angle of the Flex Sensor 1 is determined, then the algorithm checks if the patient button was pressed and calculates the angle of the Flex Sensor 2 and the procedure is repeated up to Flex Sensor 8, while always checking if the patient button was pressed. After the conclusion of this loop, a packet containing the angle values for each Flex Sensor is formed as well as, the patient counter value. Finally, the built packet sent to the computer by the serial port. By setting the ADC gains permission to read the voltage dividers from the Flex Sensors again the global variable "*Ler_canal*" is set to "2". The value of the angle sent to the computer from each Flex Sensor is the average value of the three measured values for the angle in order to reach a more stable measurement. The reason because the final value of the angle is the average of three measurements is because the three

Table 13.2   *Outputs of the flex sensors belt with the assignments*

Flex Sensor	Output pin	Channel	Storage variable
1	P3.3	1	ADCresult[0]
2	P3.2	2	ADCresult[1]
3	P3.1	3	ADCresult[2]
4	P3.0	4	ADCresult[3]
5	P1.7	5	ADCresult[4]
6	P1.6	6	ADCresult[5]
7	P1.5	7	ADCresult[6]
8	P1.4	8	ADCresult[7]

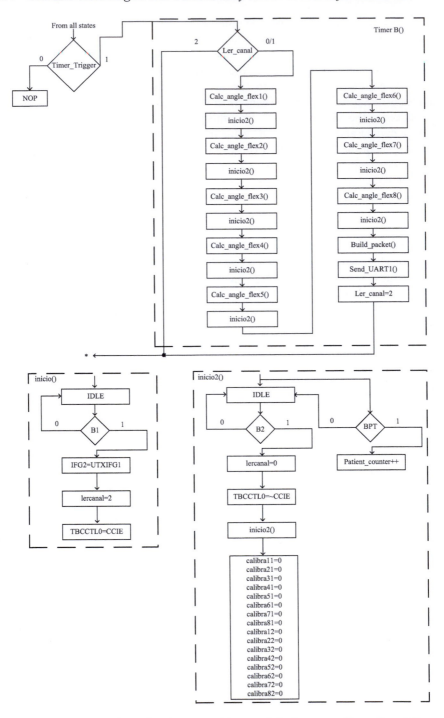

Figure 13.17 Algorithm for acquisition module firmware (Parts 2 and 3)

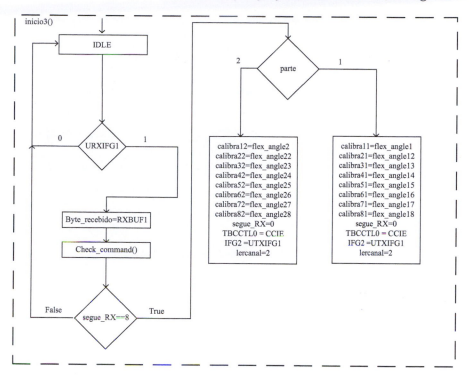

Figure 13.18   Algorithm for acquisition module firmware (Part 4)

readings were the minimum value of readings needed to achieve a stable result with respect to the trade-offs of time spent to compute the result.

The flowchart for the function "*inicio()*" shown is represented in Figure 13.17. It is responsible to check if the button B1 was pressed. If it is true the system interruptions are initiated. The purpose of the function "*inicio2()*" is 2-fold. The first purpose is to check if the button B2 was pressed. If it is true, it stops all the interruptions in the system and resets the calibration variables. This involves receiving the calibration values through the serial port. The second purpose is to check if the patient button (BPT) was pressed. If it is true then a unit is added to the patient counter.

The function "*inicio3()*" is used to receive the calibration data from the computer programme when the UARTRX flag indicates that it has data in the buffer, which needs to be read and processed, as shown in Figure 13.18.

13.6.4   First version of Flex Sensor view software

The first version of the Flex Sensor view software was conceived to acquire data from only one Flex Sensor, in order to check the accuracy of the conversions. The first version of this software is presented in Figure 13.19.

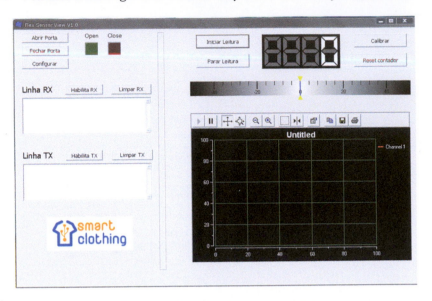

Figure 13.19   Flex Sensor view V1.0

The goal of the three buttons included in this software is to open, close and configure the serial port from the computer that is connected to the interface module. It shows two frames, "Line RX" and "Line TX", which show the data packets exchanged between the interface module and the computer. It includes one button to start the reading from the acquisition module and another to stop reading. Although an additional calibration button was added in this version it did not have any function. The calibration feature was only added in the next versions of the software. Finally, a button to reset the counter is included. It shows the number of times a movement was detected by the software.

13.6.5   Second version of Flex Sensor view software

The second version of the Flex Sensor view software, Figure 13.20, has some improvements. Examples are the capability to simultaneously show eight angles from the Flex Sensor, the patient counter and the option to save the data in a log file for later treatment. This version enables the communication between the computer and the acquisition interface by using a single packet for each Flex Sensor.

After pressing the button "Real-Time FLEX 1" a window is shown for each Flex Sensor with a chart where the evolution of the deformation angle of the respective Flex Sensor, can be monitored in real time, as shown in Figure 13.21.

A window like the one presented in Figure 13.22 is shown by pressing the button "All Real-Time FLEX Sensor". A chart is displayed with the representation of the deformation angle for each flex sensor along time, the patient counter, the threshold value for each Flex Sensor or a global threshold value for Flex Sensor (and their

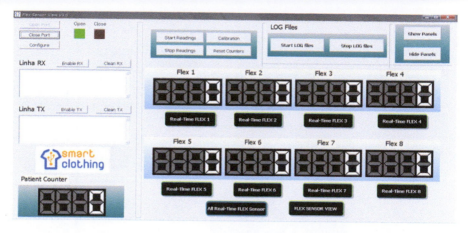

Figure 13.20   Flex Sensor view V2.0 main window

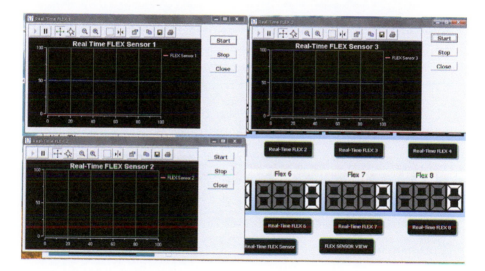

Figure 13.21   Flex Sensor view V2.0 individual plots of Flex Sensors

correspondent automatic counter for each Flex Sensor). These automatic counters compare the threshold (established before) with the instant value represented in the curves. If the instant value is larger than the threshold value a movement will be only counted if the instant value decreases until it is lower than that threshold after a few milliseconds.

Another button, which is presented in the first page of the programme, is the button "Flex Sensor View". When it is pressed instead of showing the evolution of

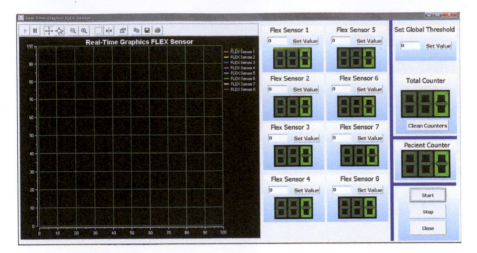

Figure 13.22   Flex Sensor view V2.0 of all Flex Sensors plots

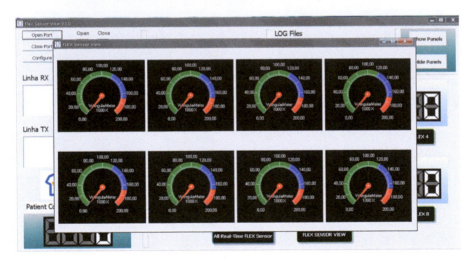

Figure 13.23   Flex Sensor view V2.0 Flex Sensor data view using gauges

the deformation angle as a curve it shows the instant angle deformation value in a gauge, as shown in Figure 13.23.

13.6.6   Third version of Flex Sensor view software

The third version of the Flex Sensor view software is basically the same as the Flex Sensor view software V2.0 but it has some improvements, mainly the following: all the data from each Flex Sensor is aggregated into one final packet jointly with the patient counter value. As such, only one packet is sent from the acquisition module

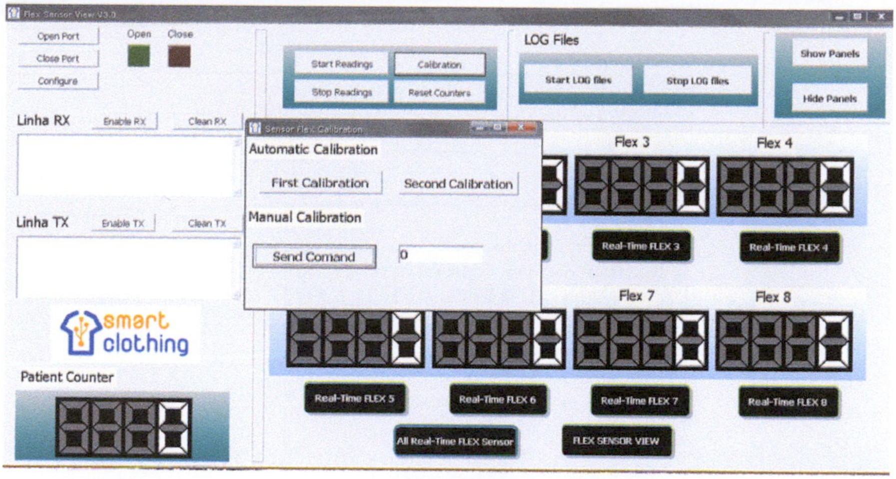

Figure 13.24 Flex Sensor view 3.0

to the computer. This feature avoids some delays in the communications between the acquisition module and the computer. In version 2.0 each packet sent from the acquisition module only included one value read from only one Flex Sensor. As the packets were sent one after other there were some delays in the reception by the software in the computer. Another feature added in this version is the possibility to calibrate the Flex Sensor in an automatic way, just by pressing a button, or manually, by transmitting the value and the desired Flex Sensor from the computer software, Figure 13.24.

To calibrate the Flex Sensor there are three steps that should be followed:

1. Stop the transmission of data from the acquisition module to the computer by pressing the button "B2" in the acquisition module.
2. Press the button "Calibration" in the main page of the programme located in the computer.
3. Choose either the automatic calibration or the manual calibration.
4. After the button is pressed the acquisition module calibrates the Flex Sensors and starts again sending the data from the Flex Sensors to the computer.

13.7 Packet protocol interface

13.7.1 Acquisition module to PC

It is important to establish a packet protocol between PCs in order to properly receive the data from the acquisition module. For the second version of the Smart-Clothing Flex Sensor belt, the packet protocol is presented in Figure 13.25.

The final version of the Smart-Clothing Flex Sensor belt (Figure 13.26) uses only one data packet to transmit the deformation angles of the eight Flex Sensors.

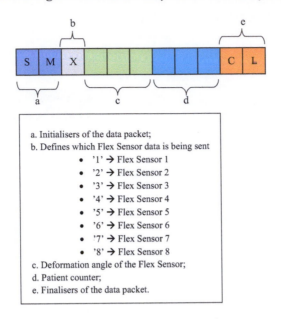

Figure 13.25   Data packet 1 for the first version

This data packet is sent only at the end of the routine controlled by a timer in the microcontroller, in order to maintain a constant flow of data between the acquisition module and the PC.

13.7.2   PC to acquisition module

A packet encapsulates the command that enables the acquisition module to properly receive the commands from the computer; e.g., the calibration routine. One of the commands sent from the computer to the acquisition module is the calibration. It has the following packet structure, as shown in Figure 13.27.

The purpose of the first part of the calibration command is the calibration of the flex sensors (if they are sensing values lower than 90° of deformation angle). Figure 13.28 shows the structure for the packet for automatic calibration.

The purpose of the second part calibration command is the calibration of the Flex Sensors (when they are sensing values higher than 90° of deformation angle).

The second command is the manual calibration of a single Flex Sensor which has the structure presented in Figure 13.29.

13.8   Results

Some initial tests were obtained for the first belt with a volunteer who was not a pregnant woman. The objective was to verify if the own respiratory movements or body

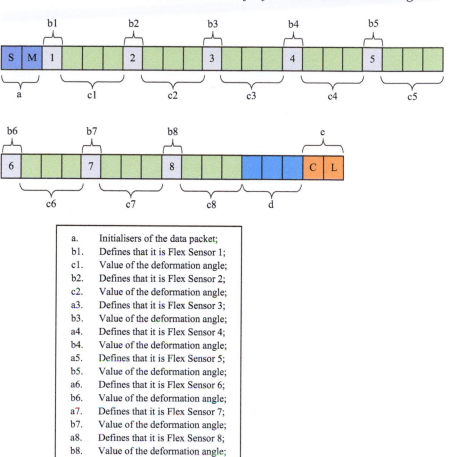

a. Initialisers of the data packet;
b1. Defines that it is Flex Sensor 1;
c1. Value of the deformation angle;
b2. Defines that it is Flex Sensor 2;
c2. Value of the deformation angle;
a3. Defines that it is Flex Sensor 3;
b3. Value of the deformation angle;
a4. Defines that it is Flex Sensor 4;
b4. Value of the deformation angle;
a5. Defines that it is Flex Sensor 5;
b5. Value of the deformation angle;
a6. Defines that it is Flex Sensor 6;
b6. Value of the deformation angle;
a7. Defines that it is Flex Sensor 7;
b7. Value of the deformation angle;
a8. Defines that it is Flex Sensor 8;
b8. Value of the deformation angle;
c. Patient counter;
d. Finalisers of the data packet.

Figure 13.26   Data packet 1 for the final version

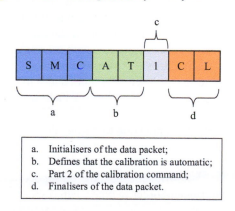

a. Initialisers of the data packet;
b. Defines that the calibration is automatic;
c. Part 2 of the calibration command;
d. Finalisers of the data packet.

Figure 13.27   Structure for the packet for automatic calibration (Part 1)

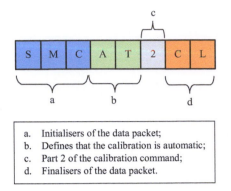

a. Initialisers of the data packet;
b. Defines that the calibration is automatic;
c. Part 2 of the calibration command;
d. Finalisers of the data packet.

Figure 13.28 Structure for the packet for automatic calibration (Part 1)

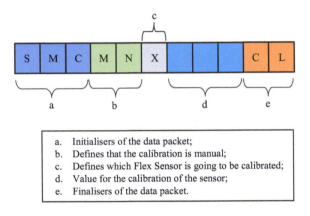

a. Initialisers of the data packet;
b. Defines that the calibration is manual;
c. Defines which Flex Sensor is going to be calibrated;
d. Value for the calibration of the sensor;
e. Finalisers of the data packet.

Figure 13.29 Command for manual calibration

motion influenced the angles of each Flex Sensor in the belt. We found that while the respiratory movements were only slightly perceived, when the woman moves quickly the sensors are able to detect the deforming effect from the belt. After these preliminary tests, the belts need to be tested in a pregnant woman, in order to detect the foetus movements and to compare these occurrences with the pregnant woman's subjective perception *presses the button* but the system did not detect foetal movements. By considering these initial tests, it seemed to be a good idea to implement a routine that automatically defines the value for a detection threshold-trigger whose values will be tuned.

Figure 13.30 presents the application to display the sensors deformation angles. For each sensor, an independent threshold trigger can be defined individually, or a unique threshold value can be defined as a whole for all the sensors. With the consideration of this threshold-trigger even if the sensor detects some motion artefact a boundary can be established in order to tune when it should count the deformation angle as a foetal movement.

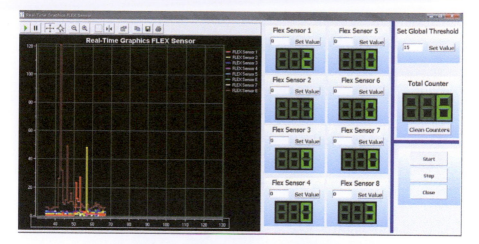

Figure 13.30   Application to display the deformation angles

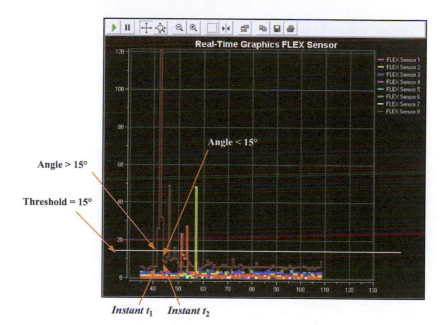

Figure 13.31   Real-time view chart for Flex Sensor view

Looking closely to Figure 13.3 we have obtained the image presented in Figure 13.31 where the peaks detected by the Flex Sensors are presented.

In this particular case for testing purposes instead of setting a threshold value for each Flex Sensor, a global threshold was set for all the Flex Sensors. The value used for the global threshold is equal to 15° (angle deformation). This means that

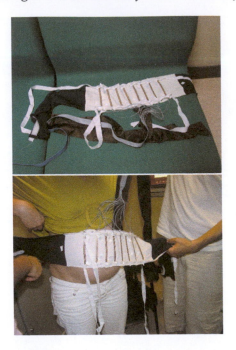

Figure 13.32   Belt for the practical tests of the Flex Sensor belt

the automatic counter from each Flex Sensor counts as a movement when the instant value of the Flex Sensor angle is bigger than the threshold value in one time instant and smaller in the next time instant. As an example of how the automatic counter works, the view chart from Figure 13.31, one can consider Flex Sensor 8 in the time instant $t1$. The value of the angle is equal to 28°, and the counter will count when this value decreases in the next time instant. In the time instant $t2$ value of the angle is equal to 8°, and the automatic counter will add one unit to the counter from Flex Sensor 8.

Some tests have been made with a pregnant woman by using the belt as shown in Figure 13.32. The volunteer during the tests most of the times was sitting in a chair, as shown in Figure 13.33.

A calibration of the flex sensors has been made in the belt before any test. We have done a test of how the system behaves if a sudden change of the pregnant woman position happens and the result is presented in the photo from Figure 13.34 and the chart from Figure 13.35. The change of position (from standing up to sitting) happened at the instant 190 s and there has been a lot of artefact movement detected during the change of position which causes to lose the calibration made at the beginning of the test. Another test has been made when the patient has been sitting in a chair. Two movements from the foetus have been detected. Nevertheless, this time, instead of presenting all the flex sensors in the same window, we enabled the option of the software to show separate windows for each flex sensor.

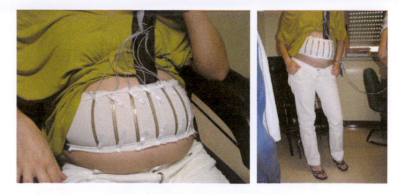

Figure 13.33   Pregnant woman seated on a chair or standing up during the tests

Figure 13.34   Portable computer to obtain the results (patient standing-up, and then patient seated)

The movements have been detected in the Flex Sensor 1 one at the instant time ≈420 s and the other at the instant time ≈460 s, as shown in Figure 13.36. In the chart, after the instant time 500 s motion artefact was detected and probably due to the mother respiration movements.

Simultaneously, the window that monitors the Flex Sensor 2 presented in Figure 13.37 shows two peaks, one at the instant time ≈420 s and the other at the instant time ≈460 s, higher than the other peaks of the chart. The movements detected in Flex Sensor 1 caused a deformation angle bigger than in the Flex Sensor 2 and if the Flex Sensor is placed side by side and separated by 8 cm we could conclude that the movements detected were more near the Flex Sensor 1.

Another test was made when the pregnant woman was sitting in a chair and is presented in Figure 13.38. Only one movement was detected by the system, while the volunteer claimed to have detected two movements. The instant time when the system and the woman detected the foetal movement simultaneously was at the instant time ≈910 s. The system detected the movement at the Flex Sensor 1, 2, 3, 4, 5 and 6, but with a stronger intensity in the Flex Sensor 3. The instant when the patient claim to

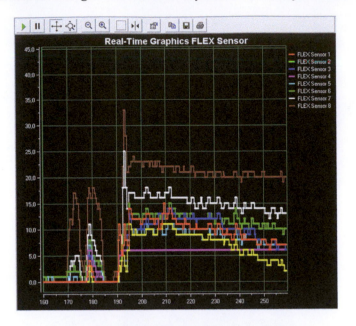

Figure 13.35 Results 1 (patient standing-up, and then patient seated)

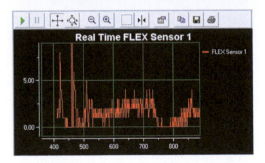

Figure 13.36 Results 2 (Patient seated)

felt a foetus movement was at the instant time ≈895, but the system has not detected any movement. At the instant ≈925 s the system claims to have detected a foetal movement. However, but it has been considered a false positive because the volunteer reported that she did not felt it.

13.9 Other solutions for the monitoring belt

13.9.1 Smart-Clothing on-off belt

Another solution to monitor the foetus in the last five weeks of pregnancy is the Smart-Clothing on-off belt. This belt is build based on pressure sensors built up with

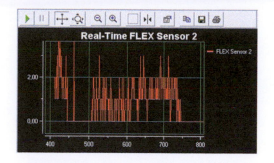

Figure 13.37 Results 3 (Patient seated)

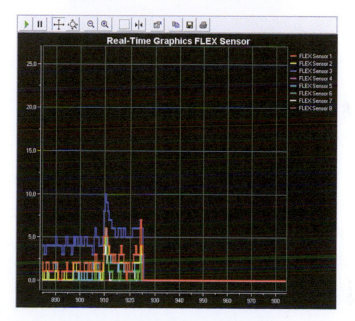

Figure 13.38 Results 4 (Patient seated)

conductive and semi-conductive tissues, according to the original idea proposed by the author from [11]. The production of this belt presented in Figure 13.39. The belt is made with two different types of conductive tissues: one of the types is conductive in all surface, while the other is only conductive in some areas. Tissues are deployed in layers so that the fabric (with conductive areas) stays in between the other layers. This will act as a switch when the outer layer of tissue is pressed letting the current to traverse from one to the other, according to the value of its resistance. This belt is made from conductive tissues that can be washed after disconnecting all the electronics associated to the belt.

The standalone Smart-Clothing on-off sensor belt is based on a matrix of 28 tissue squares, as shown in Figure 13.40. Each of these tissue squares acts like a switch that

Figure 13.39   On-off belt with a fabric pressure sensor

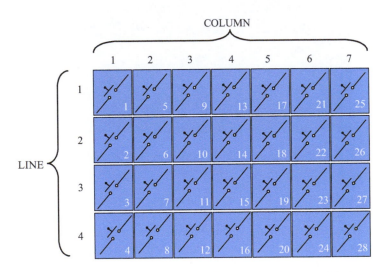

Figure 13.40   The matrix of the Smart-Clothing on-off belt

closes the circuit if it is pressed and leave the circuit open if it is not pressed. The main purpose of this belt is to detect which switch or switches are pressed, due to the movement of the foetus in the pregnant woman.

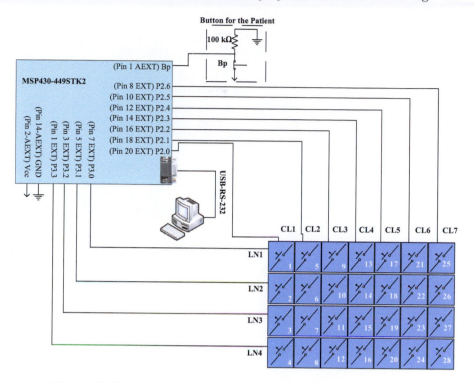

Figure 13.41 System diagram for the on-off belt acquisition module

The system diagram for the acquisition module is presented in Figure 13.41, where the lines and columns from the belt are connected to the MSP430-449STK2 module.

Besides the on-off belt, a button (or a pregnant woman counter) has been incorporated to be pressed by the pregnant woman when she feels or detects the foetus moving. These events will be very useful for comparison purposes, as they enable a comparison with the movements detected automatically by the belt. The initial idea is to read when the switches are pressed (or not). Lines are connected with different voltage values. They are read by using the ADC channels from the microcontroller. The column signals are searched to verify if there are any non-zero voltage values for the lines passing through. This idea has however been abandoned because the values of the voltage at the lines did not stay stable while passing through the conductive tissues (when pressed).

Another idea was to connect each individual tissue square (a switch) to the power supply and the other connector of the switch to a port in the acquisition module while detecting if there is any non-zero signal entering in the port of the microcontroller. This idea has also been abandoned due to hardware restrictions. To develop this idea 28 inputs of the acquisition module would be needed. As the chosen module does not have so many inputs available it has been impossible to explore this proposal.

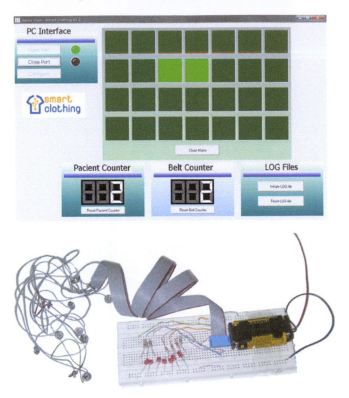

Figure 13.42   Visualisation software for the on-off belt system and breadboard with LEDs

The final and definitive idea consists of reading the switches by connecting the seven columns and the four lines to the input/output ports.

The acquisition module firmware has been produced as well as the visualisation software for the on-off belt system, as shown in Figure 13.42. Details on the packet for the communication between the on-off belt and the acquisition module are given in [12].

Tests have been made with this belt by experimenting its functionality in a pregnant woman. Results have shown that the belt is too sensitive to any casual movement of the pregnant woman. The belt detected some foetus movement but it was quite difficult to understand in the visualisation programme interface from the PC if the led blinked due to a foetus movement or due to the pregnant woman movement, as shown in Figure 13.42.

13.9.2   Belt with piezoelectric sensors

The team from the "Smart-Clothing" project [13] has developed another belt incorporating piezoelectric sensors. This type of sensors transduces the force to voltage

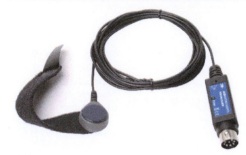

Figure 13.43   Piezoelectric sensor: former MLT1010, in 2009, nowadays Pulse Transducer, DIN, TN1012/ST (ADI Instruments)

(and vice-versa), present high sensitivity and proved to be an appropriate practical choice to detect mechanical movements, such as those originated from the foetus in a pregnant woman. Furthermore, the sensors could be very small, respond to a broad frequency range, do not react to static forces and are cheap.

In this belt, plastic pre-encapsulated piezoelectric sensors have been used with BNC connector to drive the electrical voltage signal to the signal processing circuit. This type of sensors (former Piezoelectric sensor MLT1010, in 2009, nowadays Pulse Transducer, DIN, TN1012/ST), as shown in Figure 13.43, is used by PowerLabs system from ADInstruments [14] at Faculdade de Ciências da Saúde from Universidade da Beira Interior.

Experimental tests of the piezoelectric sensor belt have been performed in Centro Hospitalar Cova da Beira. The volunteer was a woman at the 38th week of pregnancy with a belt that incorporates three piezoelectric sensors (one central sensor and two on both sides), adjusted to the abdomen by an elastic belt. The signals obtained from the sensors have been combined into one single signal, amplified, filtered and applied to an ADC converter, and then send via USB to a computer. At the computer, the signals have been processed and graphically presented to the medical team. The pregnant woman has a manual hand-held event marker (patient button) to mark the foetal movements when she feels it, for redundancy purposes, as shown in Figure 21 from [13]. There is another device, called Respisense [15], usually used to detect the baby breathing. The signals of event marker and Respisense are compared with those extracted from piezoelectric sensors. Figure 20 from [13] shows a block diagram of the circuit to acquire the signals from the sensors.

From these experimental results, presented in [13], it was possible to conclude that the system has a high capacity to detect foetal movements, isolating them from external interferences. However, the small foetal movements are easily hidden by signals with larger amplitude such as maternal walking, speaking and even breathing.

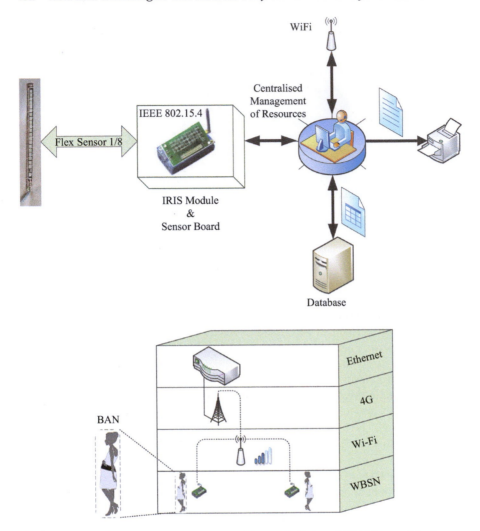

Figure 13.44   Patient monitoring using IEEE 802.15.4 and WiFi hierarchical
networking architecture

13.10   Wireless networking architecture

The Centralised Management of Resources (CMR) is formed by a base station, a PC, an application that is responsible to show the data and save all the records in a database, and a WiFi module to transmit data across a WLAN, as shown in Figure 13.44, within a multitechnology hierarchical network. As can be observed in Figure 13.44, the motes that we use to transmit the data across the network are the IRIS Motes, which is a 2.4 GHz IEEE 802.15.4 tiny wireless measurement system designed specifically for deeply embedded sensor networks [16].

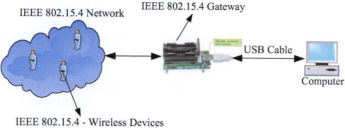

IEEE 802.15.4 Network

IEEE 802.15.4 Gateway

USB Cable

Computer

IEEE 802.15.4 - Wireless Devices
collect data from Flex Sensors

Figure 13.45   Pregnant woman monitoring using the IEEE 802.15.4 wireless monitoring

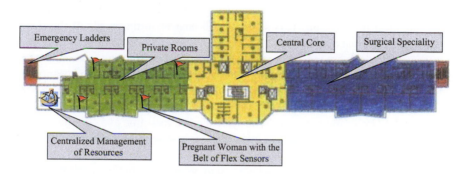

Emergency Ladders

Private Rooms

Central Core

Surgical Speciality

Centralized Management
of Resources

Pregnant Woman with the
Belt of Flex Sensors

Figure 13.46   Possible monitoring using IEEE 802.15.4 in a hospital

The IRIS Mote is one of the components of the MICA family from MEMSIC with the model XM2110CA [16]. Its primary characteristics are very similar to the MICAZ, the same rate of data transmission and radio frequency band, as the previous models the system supports TinyOS and is fully compatible with accessories of the family MICA. It is suitable for application that requires high speed in data transmissions, as for example real-time reading from a sensor in Indoor Building Monitoring. The main advantages are the following: the range up to 500 m between nodes without the need to amplify the signal, and 8 kB of RAM, almost the double of the existing modules in the past.

As described before an IEEE 802.15.4 network can be used to collect the different deformation angles caused by the foetus movements. Figure 13.45 presents a simple diagram of the remote monitoring system.

Figure 13.46 presents a possible scenario on the hospital floor. The CMR is located in a convenient area, the red spots represent the pregnant women with the belt. This represents the communication system installed in the pregnant woman and employed to deliver the data from the WBAN.

Note that we can easily move our CMR to another part of the hospital; hence, we are not limited by the geographical position of any components of the system.

13.11   Conclusions

Different versions of prototypes of sensor belts produced within the "Smart-Clothing" project, which aim at counting the movements of the foetus in a pregnant woman, have been discussed in this chapter.

The chapter starts by presenting a Flex Sensor belt produced within the "Smart-Clothing" project, which aims at counting the movements of the foetus from a pregnant woman. It detects the foetal movement based on the bending of the sensors, and an IEEE 802.15.4 network has been created so we can deliver all the data collected by the motes to our CMR.

An application is responsible to show the data to the clinical team (nurse/doctor) and save all the records in a database. If there is a need, it is also possible to transmit the data via WiFi, so the information can be shared or accessed by other authorised personnel. In the future, besides real-time data, the CMR will be able to provide long-term statistical information. This solution will answer to the needs of the market, especially in medical applications.

Nowadays there are increasing demands on remote monitoring, and onsite monitoring is becoming more and more important. It allows for easy access to instantaneous information; wireless devices can therefore help to increase overall efficiency and to enable better remote patient care. This standalone solution also can have the add-on of saving data into a memory card. The system gives real-time and continuous foetal monitoring while creating effective interfaces for querying sensor data and store all the medical record that can be later accessed by health professionals.

The belt was tested in a pregnant woman in order to verify their performance and tune the threshold-triggers. The results of these tests were that sometimes the system detects the movements from the foetus but there was still a lot of motion artefact mixed and some false positive movements detection happened during the tests.

Other type of belts that have been explored in the "Smart-Clothing" project, e.g., the on-off belt or the belt based on piezoelectric sensors. While the former is too much sensitive to any casual movement of the pregnant woman and has not shown potential to be explored in the current form, the belt that has piezoelectric sensors incorporated onto it has a high capacity to detect foetal movements, isolating them from external interferences.

As future work, a proposal is to implement other types of communication systems that could work together with the existing ones. For example, create a webpage where we can scroll through all the data produced in real time while sharing the information with other medical institutions.

Another proposal is to implement algorithms and filtering for noise and motion artefact signal suppression and to implement advanced algorithms for data treatment and aggregation.

13.12   Acknowledgements

This work was supported by UDR (Unidade de Detecção Remota), Department of Physics from University of Beira Interior, by IST-UNITE, by the PhD FCT (Fundação para a Ciência e Tecnologia) grant SFRH / BD / 38356 / 2007, by Fundação Calouste Gulbenkian, by "Projecto de Re-equipamento Científico" REEQ/1201/EEI/ 2005 (a Portuguese Foundation for Science and Technology project), and by the "Smart-Clothing" project. The authors would like to acknowledge the fruitful discussions and contributions from all the colleagues from the "Smart-Clothing" project. In special, we would like to thank Mrs. Andreia Rente and Prof. Rita Salvado for helping us in the production of the Flex Sensor and on-off belts.

References

[1] C. Gribbin and D. James, "Assessing fetal health", *Current Obstetrics & Gynaecology*, 2005;15:221–227.

[2] Cz. Evaggelos, E. Karvounis, C. Papaloukas, *et al.* "Remote maternal and fetal health monitoring during pregnancy", in *Proceedings of IEEE Information Technology Applications in Biomedicine-ITAB 2006*, Ioannina, Greece, 2006.

[3] *Maternal Fetal Monitoring System*, US 7,333,850 B2, Feb. 19, 2008.

[4] Fetal Wellbeing Monitoring Apparatus and Pad Therefor, WO 2008/010215 A2.R. Nicole, in press.

[5] A. Berger, A. Hazan and M. Rinott, "Fetal Wellbeing Monitoring Apparatus and Pad Therefor," U.S. Patent 8,075,500 B2 Dec. 13, 2011.

[6] L. M. Borges, A. Rente, F. J. Velez, *et al.* "Overview of progress in Smart-Clothing project for health monitoring and sport applications", in *Proceedings of ISABEL 2008 – First International Symposium on Applied Sciences in Biomedical and Communication Technologies*, Aalborg, Denmark, Oct. 2008.

[7] http://www.e-projects.ubi.pt/smart-clothing/.

[8] L. M. Borges, N. Barroca, F. J. Velez and A. S. Lebres, "Wireless Flex Sensor belt networks for foetal movement monitoring in low risk pregnancies", in *Proceedings of IMEKO XIX World Congress – Fundamental and Applied Metrology*, Lisbon, Portugal, Sep. 2009.

[9] Flexpoint, July 2018, http://www.flexpoint.com/wp-content/uploads/2015/10/Electronic-Design-Guide_150928.pdf.

[10] A. Rente, L. R. Salvado and P. Araújo, "Development of textile sensors and integration of sensors in belts for foetal movement monitoring", Report from the Smart-Clothing project, Universidade da Beira Interior, Covilhã. Portugal, April 2010. (in Portuguese).

[11] http://www.instructables.com/id/Flexible-Fabric-Pressure-Sensor/.

[12] L. M. Borges, N. Barroca, F. J. Velez and A. S. Lebres, "On–Off belt", Report from the Smart-Clothing project, Universidade da Beira Interior, Covilhã. Portugal, Sep. 2009.

[13] L. M. Borges, P. Araújo, A. S. Lebres, *et al.*, "Wearable sensors for foetal movement monitoring in low risk pregnancies", in A. Lay-Ekuakille and S. C. Mukhopadhyay (eds.), *Wearable and Autonomous Biomedical Devices and Systems for Smart Environment: Issues and Characterization*, Lecture

Notes in Electrical Engineering, Springer, Vol. 75, Norwell, MA, USA, 2010, pp. 115–136.

[14]   ADInstruments, July 2018: https://www.adinstruments.com/products/pulse-transducers.

[15]   Respisense, July 2018: http://www.respisense.com/en/index.php.

[16]   IRIS, July 2018, http://www.memsic.com/userfiles/files/Datasheets/WSN/IRIS_Datasheet.pdf.

Chapter 14

Radio frequency energy harvesting and storing in supercapacitors for wearable sensors

Luís M. Borges[1] and Fernando J. Velez[1]

The radio frequency energy harvesting (RF-EH) and storing solution proposed in this chapter allows for collecting only a small amount of energy. However, it is more stable than solar and wind power sources. In fact, RF-EH relies on ambient RF signals from communications systems; hence, it presents special characteristics not found in other types of energy sources. The amount of RF energy harvested depends on the schedules of the base stations from the telecommunication service, as well as on the fluctuations caused by multipath fading and shadowing. As such, this work introduces a novel energy-management algorithm whilst proposing a supercapacitor storing system that copes with these issues and allows for storing the energy harvested from the electromagnetic waves by means of N-stage Dickson voltage multiplier printed circuit board (PCB) boards (5 and 7 stages) optimized to guarantee the best conversion efficiency and output voltage. Since the objective is to harvest as much energy as possible from the electromagnetic spectrum, this work proposes an RF-EH prototype to harvest electromagnetic energy from the digital terrestrial television (DTT) frequency band (750–758 MHz). This band is chosen due to the potential arising from the wide/broad deployment of DTT in Portugal, which poses significant interest for EH. Regarding the supercapacitor storing system, the best electronic components as well as the most suitable values for the configuration parameters have been determined through an empirical approach for the RF-EH with supercapacitor storing system. As an ongoing work, we are addressing the possibility of using different RF-EH prototypes gathered in blocks that scavenge energy in the same or different frequency bands, where the sum of all the energy harvested from each prototype is stored in the supercapacitor-based storage system, is also being considered. Preliminary tests are very promising, as in a conference room, densely populated, with circa 300 people, the RF-EH system managed to build up and store energy even when mobile phones were not being used in the close proximity of the harvesting antennas.

[1] Instituto de Telecomunicações and Departamento de Engenharia Eletromecânica – Universidade da Beira Interior; Faculdade de Engenharia, Portugal

14.1   Introduction

Although in the past two decades there has been a strong research effort on legacy mobile communications and unlicensed wireless systems, several researchers consider that in the next 5 years, an inversion of the main trends will occur, and the focus will be in interdisciplinary research on wireless sensor networks (WSNs).

Nowadays, electromagnetic radio frequency (RF) energy is being transmitted from billions of radio transmitters around the world, including mobile cell phones, handheld radios, mobile base stations and television/radio broadcast stations. Therefore, there is a wide range of opportunities for harvesting the RF energy, enabling for the creation of energy self-sustainable WSNs without the need for dedicated transmitters [1]. In addition, one of the major limitations for the adoption of WSNs is the energy available for power supplying the wireless sensor nodes imposed by the finite battery capacity. Since nodes may be deployed in a hostile environment or may be used onto wearable devices, replacing batteries is often not feasible or convenient.

EH can dramatically extend the operating lifetime of nodes in WSNs. With this innovative technology, it enables to operate without the need for batteries and reduces the costly maintenance of nodes, which is mainly due to battery replacement.

Different energy sources have been considered for EH, namely solar, vibration, thermoelectric and RF electromagnetic energy.

In this work, we focus on the RF-EH, which allows for collecting only a small amount of energy, but it is more stable than solar and wind power sources. In RF-EH, we rely on ambient RF signals that are broadcasted by means of communications systems with special characteristics not found in other energy sources. The amount of RF energy harvested depends on the schedules of the base stations of the telecommunication service, as well as the fluctuations caused by multipath fading and shadowing.

This work introduces a novel energy-management algorithm that copes with these RF energy availability issues. In low-power embedded systems, the node reduces the power consumption by going to sleep mode during most of the time and only wakes-up periodically to execute the assigned procedures of acquiring sensor data and exchanging it with other node(s). The ratio between the time the node is awake and the total time is named as duty cycle. The implementation of a dynamic duty cycle that takes into account the amount of energy is being harvested and stored, as well as the amount of energy is going to be drained/consumed by the node each time it wakes up to execute the assigned tasks is of paramount importance. Additionally, we should keep in mind that if the duty cycle is too long the waste of energy will exceed the RF energy harvested and stored. Hence, the node consumes all the stored energy, leading to total shutdown of the node due to energy shortage. Typically, an embedded system with low-power microcontroller operates in one of the sleep, active or off modes. In terms of the order of magnitude from the current consumption, after the node is off, the power consumption is approximately five times more energy in the transition just to turn to the active mode, when compared with the transition from the sleep to the active mode. Thus, the dynamic duty cycle algorithm must ensure that the node does not operate with a duty cycle that will lead to complete shutdown of the embedded system due to lack of energy. Because of leakage current of storage capacitors, dynamic duty cycle algorithm should avoid low duty cycle. Our system supports green communication features that are environmentally friendly as it enables to create a self-sustainable system that harvests energy from electromagnetic waves.

Conventional rechargeable battery-based energy storage systems present some disadvantages because they do not allow a precise estimation of the remaining energy, require a higher periodic maintenance, have limitation on the number of recharge/discharge cycles while presenting a higher environmental impact when batteries are improperly disposed. Therefore, as mentioned above, in this work, initially we considered the use of capacitors to store the harvested energy. However, since capacitors present high leakage current characteristics, we decided to use supercapacitors, a more recent technology from the capacitors family. This efficient self-sustainable solution enables to power supply the embedded systems platforms, such as cognitive radio nodes (CRNs) through the use of supercapacitor-based RF-EH solution, that we call RF-SuperCap. Even though supercapacitors present a lower leakage problem, we had to deal with the supercapacitor leakage problem as it may result in a relatively large amount of energy loss in comparison to scavenged amount of energy obtained from an RF energy harvester. Nevertheless, supercapacitors yield more energy leakage than batteries, for example the leakage of a NiMH battery, which is the most leaky battery, is 30% per month, while the leakage of a supercapacitor is 5.9% per day [2]. Thus, we will present a more efficient self-sustainable solution to power supply the embedded systems platforms, such as cognitive radio nodes (CRNs) through the use of super-capacitor-based RF energy harvesting solution, that we call RF-SuperCap.

Implementing our RF-EH device based on a 5-stage Dickson voltage multiplier [3] paired with supercapacitors raises some interesting opportunities and challenges: rechargeable batteries reach their maximum voltage quickly, but it is difficult to find an analytical relationship between the stored energy and the output voltage. Besides, the energy stored in a supercapacitor is easily calculated as $E = 1/2CV^2$, where C and V are the capacitance and voltage of the supercapacitor, respectively. However, there is one drawback of knowing easily the stored energy in the supercapacitor. This drawback is related with the supercapacitor voltage ($V_{super-cap}$) that increases monotonically as it stores energy, until it eventually reaches a maximum value (2.5 V for the Vina Technology 10 F supercapacitors we have been using [4]). Thus, the voltage output of a supercapacitor block may be much lower or much higher than the operation voltage of the circuit it is supplying, which leads to the need of a circuit that allows to take advantage of all of the stored energy in the supercapacitor.

14.1.1 Motivation

The motivation to propose and develop a supercapacitor storing system is the need for storing the energy harvested from the electromagnetic waves by means of a 5-stage Dickson voltage multiplier prototype developed by our group [3]. Additionally, this system must be specified in order it is sufficient to power-supply a CRN.

The contributions of this work are summarized as follows:

- Implement a supercapacitor-based storage system for an RF-EH based on the 5-stage Dickson voltage multiplier.
- Propose a supercapacitor-based storage system that aims at being small, efficient and self-sustainable.

- Achieve the goal of power supplying a CRN just with the energy harvested from RF and stored by means of the proposed supercapacitor-based storage system.
- Propose and implement a dynamic duty cycle algorithm that ensure that the CRN does not operate with a duty cycle that will lead to complete shutdown of the embedded system, due to excess of energy consumption.
- Calculate the optimal stored energy level by the aggregate evaluation of supercapacitor leakage and the energy shortage risk.
- Evaluate the dynamic duty cycle algorithm.
- Assess the voltage acquired with the RF-EH prototype when inserted into (i) low traffic ambient and (ii) high traffic ambient.
- Study the possibility of using different RF-EH prototypes gathered in blocks that scavenge for energy in the same or different frequencies where the sum of all the energy harvested from each prototype is stored in the supercapacitor-based storage system propose by us.

14.2 Self-sustainable devices

The proposed supercapacitor-based storage system for RF-EH based on the 5-stage Dickson voltage multiplier, proposed in this work, must support computing and data-processing capacities of the CRNs from different hardware platforms. The authors from [5] present a survey regarding the comparison of key software device radios (SDR) platforms based on the published performance results in literature. In order to better understand the differences among the available SDR platforms, the main characteristics for key-platforms that use different approaches are summarized in Table 14.1. In addition, energy consumption is not defined for field programmable gate array (FPGA)-based platforms because it is heavily dependent on the configuration.

Table 14.1   Comparison of key SDR platforms based on the published performance results

Platform	Availability	Application	Language	Power consumption
USRP	Commercial	N/A	C++	≈PC
TI C64+	Commercial	Base station	C/ASM	6 W
MuSIC	Commercial	WCDMA	C/ASM	≤382 mW
Sandblaster	IP licence	WCDMA	C	171 mW
ARDBEG	Prototype	WCDMA	C	≤500 mW
BEAR	IP licence	MIMO OFDM	MATLAB®/C	231 mW
Magali	Prototype	MIMO OFDM	C/ASM	236 mW
ExpressMIMO	Prototype	MIMO OFDM	C	N/A
WARP	Commercial	MIMO OFDM	VHDL	N/A
Lyrtech	Commercial	N/A	MATLAB/VHDL	N/A
ASAP	Prototype	802.11a/g	N/A	198 mW
Genepy	Prototype	MIMO OFDM	C/ASM	192 mW

In this section, we describe the technical details from the RF-EH prototype and supercapacitors. The aim is to create a self-sustainable system that stores energy from the RF-EH prototype, whilst power supplying the system that monitors the supercapacitor voltage and controls the charge of the supercapacitor that allows for generating such power continuously.

14.2.1   RF energy harvesting using 5-stage Dickson voltage multiplier

In previous works [3,6], we have developed circuits by considering a dual-band wearable antenna responsible for harvesting the ambient energy, in the context of the PROENERGY-WSN project [7]. The aim of this work was to power supply a wireless sensor node. Besides the development of RF-EH prototypes, the work also describes the efficient dual-band antenna considered for collecting RF energy. Guidelines for the choice of textile materials to be used in future wearable antennas are also briefly presented. The design considerations for the dimensioning of a 5-stage Dickson voltage multiplier were also described along with the simulation and experimental results for the three prototypes (with match impedance circuit) of the 5-stage Dickson voltage multiplier. These concepts and experimental trials with the different prototypes are the key aspects to be considered during the design of the supercapacitor-based storage system.

14.2.1.1   RF energy-harvesting circuits

In this work [6], we have chosen the 5-stage Dickson voltage multiplier for RF-EH. According to [8], by adding more stages, the peak of the conversion efficiency curve shifts towards the higher received power region. In fact, more than 5 stages does not bring substantial improvement for the power levels considered, due to energy losses along the chain [9]. Moreover, since the wireless sensor nodes need at least 1.8 V for operation (i.e. approximately -10 dB m according to [8]), the 5-stage Dickson voltage multiplier is the one which presents the best performance in terms of conversion efficiency. As such, Prototypes 1, 2 and 3 presented in Figure 14.1 have been conceived in a PCB fabricated with two layers by using an FR-4 epoxy glass substrate.

In order to evaluate the performance of the 5-stage Dickson voltage multiplier with matched impedance, three different prototypes have been considered (1, 2 and 3). Prototypes 1 and 2 considers an open-circuit stub, whilst Prototype 3 considers a short-circuit stub, as shown in Figure 14.1.

14.2.1.2   Simulation and experimental results

In this section, we consider the three prototypes of the 5-stage Dickson voltage multipliers for RH-EH in the GSM bands that have already been considered in Chapter 2. Figure 14.2 presents the comparison between the simulated and experimental results, for Prototype 2, in terms of the conversion efficiency as a function of the RF received power. By observing Figure 14.2, we conclude that for a load impedance of 100 kΩ, Prototype 2 presents the best results in terms of conversion efficiency. Moreover, by

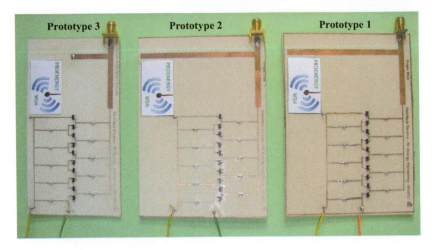

Figure 14.1   Prototypes of the 5-stage Dickson voltage multiplier with impedance matching

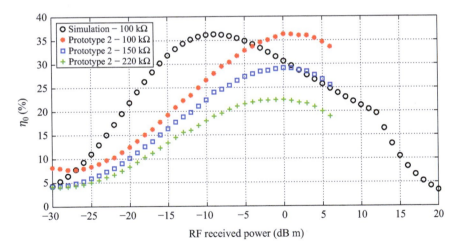

Figure 14.2   Impact of the RF received power on the conversion efficiency (η_0) for the Prototype 2 as a function of the load impedance

comparing the simulation and experimental results for a load impedance of 100 kΩ, it is observable that a deviation between the simulation and experimental results. This is explained by the fact that the developed prototype presents a mismatching between the antenna and the developed 5-stage Dickson voltage multiplier prototype. The worst case scenario occurs in the case we consider a load impedance of 220 kΩ.

Figure 14.3 presents simulation and experimental results for conversion efficiency as a function of the RF received power for different loads, for Prototype 3.

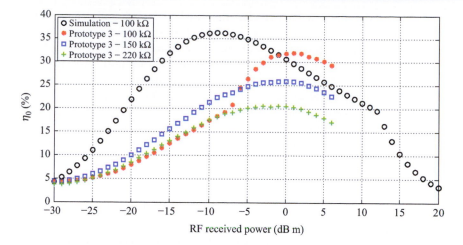

Figure 14.3   Impact of the RF received power on the conversion efficiency (η_0) for the Prototype 3 as a function of the load impedance

By analysing Figure 14.3, we conclude that for a load impedance of 150 kΩ, Prototype 3 presents better results in terms of conversion efficiency, when the RF received power is lower than −7 dB m. If the RF received power is higher than −7 dB m, Prototype 3 attains better results in terms of conversion efficiency by considering a load impedance of 100 kΩ.

In [3], the impact of the load impedance on the conversion efficiency of the 5-stage Dickson voltage multiplier is discussed by analysing its dependence on the RF-received power through simulations. Based on the simulation results from [3], the optimal conversion efficiency is achieved when the load impedance is 50 kΩ. If the resistive load value is too low or too high, the conversion efficiency significantly decreases. The deviations between the simulation and experimental results are explained by the PCB manual manufacturing techniques that were employed by us to develop each of the prototypes or we do not know the real values for the parameters from the diodes in the prototypes. Additionally, in the simulations we consider the parameter values supplied by the manufacturer which could induce some deviation in the simulation results presented. Different PCB fabrication techniques to produce the prototypes also resulted in different values for the conversion efficiency.

In Prototypes 2 and 3, the fabrication was achieved by means of a more rigorous control in the photopositive method and chemical bath procedure applied to the PCB, which resulted in enhanced values for the conversion efficiency. In Prototype 1, the fabrication of the PCB was less rigorous in terms of chemical bath. This fact could have resulted in a more abrasive procedure to the copper film of the PCB. By observing the results obtained, it is conclusive that the PCB fabrication allows for attaining better results if the fabrication is controlled in a more rigorous way.

Assuming the above-mentioned scenario, in which the advised voltage to power supply an IRIS mote is considered, i.e. 3 V, Prototype 1 can power supply the sensor node for an RF received power of -4 dB m, with a conversion efficiency $\eta_0 \approx 20\%$. With Prototype 2, the IRIS mote can be power supplied for an RF received power of -6 dB m, with $\eta_0 \approx 32\%$. Finally, Prototype 3 is capable of supplying an output voltage of 3 V for an RF received power of -5 dB m, with $\eta_0 \approx 26\%$.

Prototype 1 was able to harvest RF energy sufficient to supply a constant voltage (CV) of 1.41 V with a load of 100 kΩ, whereas in open-circuit, it was able to supply with a CV of 2.98 V. These results have been obtained by placing Prototype 1 at a distance of 25 m from the GSM 900 antenna transmitter. The maximum conversion efficiency, η_{0max}, attained for a load impedance of 100 kΩ, is presented in Table 14.2.

It is observable that Prototype 2 allows for achieving the highest value of conversion efficiency. Table 14.3 summarizes the results obtained whilst considering load impedance of 100 kΩ, for the maximum power collected, P_{colmax}. Similarly to the results for the maximum conversion efficiency, Prototype 2 is the one that allows for achieving the highest value of maximum power collected but at higher values of RF receiver power, i.e. $P_{RF} = 6$ dB m.

By combining the results and the spectral opportunities for EH identified in [3], it is possible that the levels of received power harvested from the environment are not enough to fulfil the goal of power supplying a WBAN. The data collected for the maximum levels of received power are around -27 dB m, which are insufficient to generate an output voltage of 1.8 V. Hence, the development of a supercapacitor storing system that allows for storing little harvested energy (and to use it when the output voltage from the supercapacitor is sufficient to power supply the IRIS mote) is of paramount importance.

Table 14.2   Maximum conversion efficiency for the three prototypes

Prototype	η_{0max}
1	22 (@ $P_{RF} = 0$ dB m)
2	36 (@ $P_{RF} = 0$ dB m)
3	32 (@ $P_{RF} = 1$ dB m)

Table 14.3   Maximum power collected for the three prototypes

Prototype	P_{colmax} (mW)
1	0.220 (@ $P_{RF} = 0$ dB m)
2	1.33 (@ $P_{RF} = 6$ dB m)
3	1.16 (@ $P_{RF} = 6$ dB m)

14.2.2   *Energy storage and supercapacitors*

Even though rechargeable batteries are the storage element more commonly used, we choose the supercapacitor due to the advantages that present against the rechargeable batteries in terms of offering low-maintenance, being self-sustainable and environmentally friendly. Authors from [10] present a performance comparison in terms of various energy-storing devices, as shown in Figure 14.4. On such a chart, the values of energy density (in W h/kg) are plotted versus power density (in W/kg). From Figure 14.4, it is observable that a supercapacitor has a considerable power density but a low energy density. This characteristic makes it suitable to provide energy during short power peaks, e.g. transmission of a data packet by a CRN. In addition, the supercapacitor also presents long lifetime that makes it suitable to smooth out the power that is sent out to embedded system or CRN.

Authors from [11] present a similar storing system that considers supercapacitors as the storing element. However, the power obtained to charge the supercapacitors is obtained from solar panels.

Since we consider to use supercapacitors as our main and unique storage element for the storage system, their advantages are as follows:

1. As mentioned above, the energy levels and charge/discharge cycles from the supercapacitors are easily predicated. The energy stored in the supercapacitor, $E = 1/2CV^2$, is determined by measuring its voltage. For now, we will only assume the use of one supercapacitor of 10 F with a maximum voltage of 2.5 V, but more ahead in this work, we will discuss the advantages and disadvantages of using blocks of supercapacitors in series or in parallel. By considering a single supercapacitor of 10 F with a maximum voltage of 2.5 V, we can operate a CRN

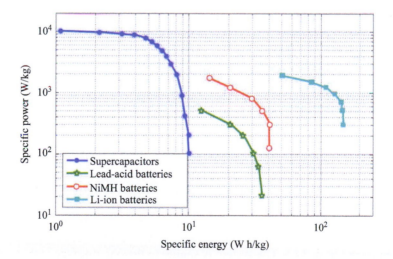

Figure 14.4   Ragone plot of electrochemical systems. © 2007. Extracted and adapted, with permission, from Reference [10]

that, for example, consumes 200 mW for about 2.60 min with a duty cycle of 100%. Therefore, since we intend to implement a dynamic duty cycle algorithm, this time of operation that we can achieve from a fully charged supercapacitor will also increase. In addition, if we consider to add blocks of supercapacitors in parallel, this will increase the capacity and therefore, increase the duration of operation of the CRN. With this procedure, we are able to precisely compute the remaining energy, allowing for a lifetime-aware manager.

2. Supercapacitors are manufactured from a porous material which is free of environmentally harmful acid and other corrosive chemicals and presents a near-infinite lifetime ($\approx$500,000 charge cycles for the supercapacitor we consider in this work [4]), implying lower maintenance costs and environmentally friendliness when compared to rechargeable batteries, which support, at most, $\approx$5,000 charge cycles.

3. Other special characteristic from the supercapacitors is the extremely low equivalent series resistance (ESR) of 120 mΩ, for the chosen supercapacitor. In addition, supercapacitors have a 10 times higher power density, as presented in Figure 14.4, allowing for repeating the charge and discharge pattern in order to absorb sudden peaks that may appear from our RF-EH prototype (e.g. in case there is a sudden increase of mobile traffic in the area of scavenging that leads to an increase of the output voltage from the prototype).

The **advantages** of considering supercapacitors may be summarized as

- unlimited cycle life, as compared to the electrochemical battery, since supercapacitors are not subject to the wear or age;
- on-hand charge methods which does not requires a full-charge circuit;
- quick charging times;
- low impedance and pulse current enhancements, by mounting them in parallel with batteries; and
- cost-effective storage, since a very high cycle count compensates the lower density the supercapacitor presents.

The **disadvantages** may be summarized as

- low energy density, usually in the order of magnitude of 1/5 to 1/10 of a battery;
- cannot use the full energy spectrum for some applications; and
- low voltage cells/blocks, since to get higher voltages, serial connection between supercapacitors is required.

14.3 Design of the RF energy harvesting/storage system

In order to develop a supercapacitor storing system for the energy harvested from the electromagnetic waves, we propose a system whose different building blocks are presented in Figure 14.5. The system is similar to the one from the authors in [12]. Instead of considering a solar panel as the energy source, we designed the circuit according to the requirements of the RF energy harvester system. It consists of four

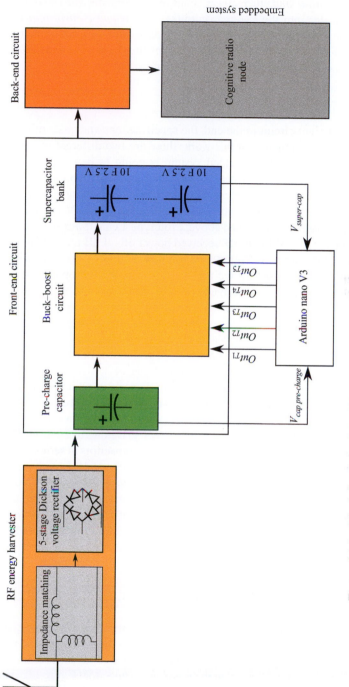

Figure 14.5   Block diagram of the RF energy harvesting with supercapacitor storing system

main blocks: the *front end*, to manage, control and transfer the energy from the RF energy harvester into the supercapacitor(s); a *front-end control unit* that manages/control the algorithm for the *front-end* block; a *back-end* circuit that converts the harvested RF energy stored inside the supercapacitor(s) to a CV and the *embedded system* blocks, i.e. the corresponding CRN to be powered-up.

14.3.1   Specifications

To achieve an efficient circuit in the end, the specifications and goals of our circuits must meet some requirements. In this work, there are two different sets of requirements, depending on which embedded system is going to be power-up. The first set of requirements consider a typical WSN node (e.g., IRIS mote) that is going to be powered-up by our RF-EH system with supercapacitor storing. As discussed in Section 14.2.1.2, to power supply an IRIS mote, a voltage of 3 V is required but at least 1.8 V is required. As mentioned above, Prototype 1 can power supply the sensor node with a voltage of 3 V for an RF received power of −4 dB m, with a conversion efficiency $\eta_0 \approx 20\%$, whereas Prototype 2 can power supply a voltage of 3 V for an RF received power of −6 dB m, with $\eta_0 \approx 32\%$. Finally, Prototype 3 supplies an output voltage of 3 V for an RF received power of −5 dB m, with $\eta_0 \approx 26\%$. In field tests, Prototype 1 was able to harvest RF energy sufficient to supply a CV of 1.41 V with a load of 100 kΩ, whereas in open-circuit it was able to supply a CV of 2.98 V. These results have been obtained by placing Prototype 1 at a distance of 25 m from the GSM 900 antenna transmitter. The maximum conversion efficiency, η_{0max}, attained for a load impedance of 100 kΩ, is presented in Table 14.2.

The second set of requirements consider a CR node in which Table 14.1 summarizes the main characteristics for key-platforms that use different approaches. As discussed in Section 14.2.2, in this work, it is only assumed the use of one supercapacitor of 10 F, with a maximum voltage of 2.5 V. Below, in Section 14.4.9, the advantages and disadvantages of using blocks of supercapacitors in series or in parallel will be discussed. By considering a single supercapacitor of 10 F, with a maximum voltage of 2.5 V, we can operate a CRN that, for example, consumes 200 mW for about 2.60 min with a duty cycle of 100%. Therefore, since we intend to implement a dynamic duty cycle algorithm, the time of operation that can be achieved from a fully charged supercapacitor will also increase. Additionally, by adding blocks of supercapacitors in parallel, the capacity will increase; hence, the duration of operation of the CRN will also increase. With this knowledge, we are able to precisely compute the remaining energy, allowing for a lifetime-aware manager (N.B.: It is more acceptable for this type of hardware that the maximum power supply available from the proposed supercapacitor storing system should be at least 5 V).

Since the output voltage from the RF energy harvester depends on the quantity of cellular communications traffic or on the DTT power in the scavenging area, there is the need to control the voltage that is delivered to pre-charge capacitor from the *front-end* circuit. The minimum voltage that the pre-charge capacitor ($V_{cap\,pre\text{-}charge}$) needs to achieve in order to initiate the procedure from the buck–boost circuit is 1.8 V, whereas the maximum voltage must be no more than the voltage the supercapacitor bank is designed for. In terms of voltage supply, the buck–boost circuit takes a maximum tolerance of ≈ 5%, while the maximum output ripple is at most 2%.

Figure 14.5 shows that the *front-end* circuit is controlled by an Arduino Nano V3 with the ATMEL 328P microcontroller [13]. The power source of the Arduino is not shown here, but there are two approaches that are going to be considered during this work. The first approach considers the use of an OLIMEX board [14] that makes use of a solar panel and a 1.5 V rechargeable battery that delivers an output voltage of 3.3 V. The Arduino is supplied with this system while the supercapacitor bank just delivers power supply to the chosen embedded system. This approach is the first one to be considered in this work. After optimizing the RF-EH system with supercapacitor storing the second approach will then be considered. The second approach considers the use of a pre-charged 1,000 μF capacitor connected to the microcontroller. The microcontroller makes use of the energy stored in the dedicated capacitor to power-up itself, while it controls the energy flow from the RF energy harvester to the super-capacitor bank. After a stable voltage is achieved in the supercapacitor bank and the dedicated capacitor drains all the energy, the microcontroller starts to use the energy stored in the supercapacitor bank in order to self-sustain its management activity. The main requirement to consider in this approach is that the RF energy harvester should be placed in a high traffic ambient that allows for obtaining high output voltages.

Figure 14.5 also presents the *back-end* circuit block. However, if the output voltage from the *front-end* circuit is sufficient to power supply the CR node directly, then this block is eliminated, which will increase the energy efficiency, and consequently decrease the energy consumption.

14.3.2 Characterization of traffic ambient scenarios

To assess the average voltage that each RF energy harvester prototype is able to deliver when placed in different traffic ambients, different field trials experiments can be considered, by taking into account the number of people/users that coexisted in each environment/ambient. Since a person in average carries at least one cellular phone, the rationale behind these field trials was to measure the rate of increase of the voltage in the pre-charge capacitor in each ambient. An ambient with high number of users corresponds to high traffic which, in turn, corresponds to a higher output voltage from the RF energy harvester. The case of low number of users corresponds to low traffic which in turn corresponds to lower output voltage from the RF energy harvester. Therefore, two mains scenarios have been defined: (i) *low traffic ambient* and (ii) *high traffic ambient*. For both scenarios, different values of capacitance of the pre-charge capacitor have been considered, in order to understand the time needed to attain the minimal voltage to deliver to the buck–boost circuit, i.e. 1.8 V. Nevertheless, in this work, the experiments performed only account for the worst case scenario considering the GSM 900/1800 bands, i.e. low traffic ambient. Therefore, since the RF-EH system with supercapacitor storing is going to be designed to operate with a minimum voltage output of 1.8 V, then if the RF energy harvester is placed on a high traffic ambient, the system will present enhanced values in terms of RF energy harvester voltage output, as well as of the voltage charged to the supercapacitor(s) bank.

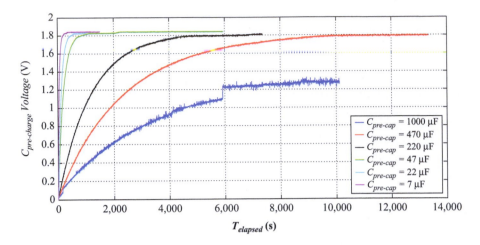

Figure 14.6   Voltage variation as function of the elapsed time for different values of the capacitance of the pre-charge capacitors within low ambient traffic

14.3.2.1   Low traffic ambient

For the *low traffic ambient*, the RF energy harvester prototype was placed in a zone for RF energy scavenging whose ambient contains a very low number of cellular device users (at most two users). Within these field trials, one assumes that the acquired voltage in the pre-charge capacitors is the minimum voltage value that we are able to deliver to the buck–boost circuit. Figure 14.6 presents the voltage variation as a function of the elapsed time for different values of pre-charge capacitors within *low ambient traffic*.

By analysing Figure 14.6 it is possible to conclude that the maximum achievable voltage within a *low ambient traffic* scenario is at most $\approx$1.8 V. Only for the case of $C_{pre-cap} = 1,000\ \mu\mathrm{F}$, $V_{cap\,pre-charge} = 1.8\,\mathrm{V}$ is not attained, even after 10,000 s scavenging RF energy. From Figure 14.6, it is also possible to conclude that as the capacitance of the pre-charge capacitor increases, the time elapsed to attain the maximum possible voltage in the pre-charge capacitor also increases, as expected. By comparing the elapsed time to attain the maximum voltage of 1.8 V for the different capacitor capacitance and the trade-off regarding the energy it can be stored and transferred by means of the buck–boost converter to the supercapacitor bank, a decision must be made to choose the best capacitance of the pre-charge capacitor, leading to the best values of capacitance of $C_{pre-cap} = 470$ and 220 μF.

14.3.2.2   High traffic ambient

For the *high traffic ambient*, the RF energy harvester prototype was placed in a zone with a high number of cellular devices users (more than two users) to scavenge RF energy. It is expected to observe that the acquired voltages from this experiment will be much higher than the ones acquired during the low traffic ambient trials from the previous section. This allows to charge at a higher rate the pre-charge capacitor and consequently charging the supercapacitor bank quicker by means of the buck–boost circuit.

14.3.3 Front end block

As shown in Figure 14.6, we should consider that the achievable voltage within a *low ambient traffic* scenario is at most ≈ 1.8 V in the pre-charge capacitor. This voltage can be assumed as the minimum acceptable to initiate the energy transferring to the supercapacitor(s). The goal of the *front-end* circuit is to keep the voltage of the pre-charge capacitor continuously controlled to remain within a minimum voltage of 1.8 V, whereas the maximum voltage must be no more than the voltage the supercapacitor bank is designed for. In terms of voltage supply, the buck–boost circuit will take a maximum tolerance of $\approx 5\%$, (V_{ripple}), while the maximum output ripple will be, at most, 2%.

Figure 14.5 presents the *front-end* circuit, which considers the CV method [15] which drives directly the power harvested from the RF energy harvester into the pre-charge capacitor. As mentioned above, the chosen capacitance of the pre-charge capacitor is the $C_{pre-cap} = 470$ or 220 μF. However, the most probable one is $C_{pre-cap} = 470$ μF as it a more appropriate trade-off between time to charge until the desired voltage and the amount of energy to be transferred to the supercapacitor. Some considerations regarding the choice of the pre-charge capacitor involve the overall efficiency which is increased when the pre-charge capacitor output is stored at a higher voltage, since step-down conversion is more efficient.

In terms of the response time for the variation of the voltage in the pre-charge capacitor when discharging the values are in the order of magnitude of milliseconds. This response time is slow (long) enough to facilitate the energy transfer with the Arduino platform containing a 104 μs-conversion-time SAR-ADC. The considered 1N4148 Schottky diode has a 720 mV voltage drop at 5 mA.

The use of multiple RF energy harvesters is a possibility to be considered since it allows for quickly obtaining the energy necessary to achieve the desired voltage in the pre-charge capacitor while initiating the energy transfer to the supercapacitor(s) bank by means of the buck–boost circuit.

The pre-charge capacitor voltage $V_{cap\,pre-charge}$ is monitored and controlled by the Arduino microcontroller and allows for transferring the accumulated energy into the supercapacitor(s) bank in small bursts. This energy transfer is accomplished by the front-end circuit, which is set to work in buck converter mode [16], while being controlled by the Arduino microcontroller, when the $V_{cap\,pre-charge}$ is higher than the voltage at the terminal of the supercapacitor(s) bank, denoted as $V_{super-cap}$. In turn, the front-end circuit switches to Boost mode when $V_{cap\,pre-charge} \leq V_{super-cap} + V_{loss}$, where V_{loss} is the voltage loss observed in the transistors and diodes.

After each time period of energy transfer, $V_{cap\,pre-charge}$ drops for an amount denoted as $\Delta V_{transfer}$ and transfers it to the 1.5 mH or 101 μH inductor. It is possible to theoretically calculate the amount $\Delta V_{transfer}$ that is transferred to the inductor. Here it is assumed that energy loss/gain in the pre-charge capacitor and in the inductor must be equal as shown below (in which the energy losses in the resistors, inductor, diodes and two transistors are neglected, for simplicity). At this point, we consider the possible use of different values from the capacitance from the pre-charge capacitor, in the range of $C_{pre\text{-}cap} = \{100; 200; 470; 1{,}000; 2{,}200\}$ μF. As mentioned above, we assume that a minimum voltage in pre-charge capacitor $V_{pre\text{-}cap\,min} = 1.8$ V is required and when the $V_{cap\,pre\text{-}charge} > V_{super\text{-}cap}$ to enable the energy transfer to the inductor. The maximum allowed voltage in the supercapacitor is denoted as $V_{super\text{-}cap\,max}$. In the following example, we assume $V_{ripple} = 0.6$ V.

$$V_{super\text{-}cap\,max} = 1.8 + 0.6 = 2.4 \text{ V} \tag{14.1}$$

$$\Delta V_{transfer} = V_{super\text{-}cap\,max} - V_{pre\text{-}cap\,min} \tag{14.2}$$

$$\Delta E(C) = \frac{1}{2} \times C_{pre\text{-}cap} \times (\Delta V_{transfer})^2 \tag{14.3}$$

$$\Delta E(100\ \mu F) = \frac{1}{2} \times 100 \times 10^{-6} \times (2.4^2 - 1.8^2) = 18\ \mu J$$

$$\Delta E(200\ \mu F) = \frac{1}{2} \times 200 \times 10^{-6} \times (2.4^2 - 1.8^2) = 36\ \mu J$$

$$\Delta E(470\ \mu F) = \frac{1}{2} \times 470 \times 10^{-6} \times (2.4^2 - 1.8^2) = 84.6\ \mu J$$

$$\Delta E(1{,}000\ \mu F) = \frac{1}{2} \times 1{,}000 \times 10^{-6} \times (2.4^2 - 1.8^2) = 180\ \mu J$$

$$\Delta E(2{,}200\ \mu F) = \frac{1}{2} \times 2{,}200 \times 10^{-6} \times (2.4^2 - 1.8^2) = 396\ \mu J \tag{14.4}$$

$$\Delta E(L) = \frac{1}{2} \times 101 \times 10^{-6} \times (I_2^2 - I_1^2) \tag{14.5}$$

$$\Delta E(C) \approx \Delta E(L) = 84.6\ \mu J \tag{14.6}$$

$$(I_2^2 - I_1^2) \approx \frac{84.6 \times 10^{-6}}{101 \times 10^{-6}} \times 2 \approx 1.675 \text{ A} \tag{14.7}$$

where 1.8 V is the selected minimum voltage in the pre-charge capacitor and is permitted to rise 0.6 V before the energy is transferred by enabling the NPN transistor. The NPN transistor enters into saturation by applying 5 V to the Out_{T2} input of the BC547C transistor. This causes the energy to be transferred to the inductor during a time interval Δt and is calculated as follows:

$$V_L = L \times \frac{dI}{dt} \approx L \times \frac{\Delta I}{\Delta t} \tag{14.8}$$

$$V_{pre\text{-}cap\,min} - V_{super\text{-}cap} - V_{CEsat} = 101 \times 10^{-6} \times \frac{\Delta I}{\Delta t} \tag{14.9}$$

We need to know the value of the currents I_1 and I_2 in order to calculate the Δt. By considering the relation $(I_2^2 - I_1^2) \approx 1.675$ A and the condition $I_1 < I_2$, the best solution to this relation is $I_1 = 0.3$ A and $I_2 = 1.3$ A. By replacing $\Delta I = I_2 - I_1 = 1.3 - I_1 = 1.0$ A in (14.9), Δt is determined as follows:

$$V_{pre\text{-}cap\,min} - V_{super\text{-}cap} - V_{CEsat} = 101 \times 10^{-6} \times \frac{1}{\Delta t}$$

$$1.8 - 0.2 - 0.7 = \frac{101 \times 10^{-6}}{\Delta t}$$

$$\Delta t = 112\,\mu s \tag{14.10}$$

where $112\,\mu s$ is the time needed to transfer $84.6\,\mu J$ from the pre-charge capacitor into the inductor when the difference between $V_{super\text{-}cap}$ and $V_{pre\text{-}cap\,min}$ is 1.6 V.

Now, if the voltage at the terminal of the supercapacitor(s) bank (VSC) is $V_{super\text{-}cap} = 20 \times 10^{-3}$, the time that it takes to transfer the same amount of energy is given by

$$V_{pre\text{-}cap\,min} - V_{super\text{-}cap} - V_{CEsat} = 101 \times 10^{-6} \times \frac{1}{\Delta t}$$

$$1.8 - 20 \times 10^{-3} - 0.7 = \frac{101 \times 10^{-6}}{\Delta t}$$

$$\Delta t = 94\,\mu s \tag{14.11}$$

This time interval, Δt, will get progressively longer when the voltage difference gets lower due to the rising VSC as it builds up energy in the supercapacitor. This observation can be observed in the calculations performed in (14.10) and (14.11) in which the example that presents the lowest voltage $V_{super\text{-}cap}$ at the supercapacitor(s) bank is the one that shows the shortest time to transfer the energy from the pre-charge capacitor to the inductor.

However, Figure 14.7 presents the variation of energy in the pre-charge capacitor for different values of capacitance ($C_{pre\text{-}charge}$) and V_{ripple}.

It is noticeable that as the value of the capacitance increases and V_{ripple} increases, the energy transferred to the inductor also increases, as expected. However, the time the pre-charge capacitor takes to charge until a $V_{pre\text{-}cap\,min} = 1.8$ V increases as the capacitance increases. Therefore, the choice of the best pre-charge capacitor capacitance for this work should rely on a trade-off between the time it takes to charge until the minimum voltage in the pre-charge capacitor and the amount of energy that is transferred to the inductor.

14.3.3.1 Specifications of the buck–boost converter

A boost converter (step-up converter) is a DC-to-DC power converter with an output voltage greater than its input voltage. The basic boost converter is presented in Figure 14.8 and 14.9 contains at least a diode and a transistor as well as one energy storage element, which in our work consists of a combination of a supercapacitor(s) bank and an inductor.

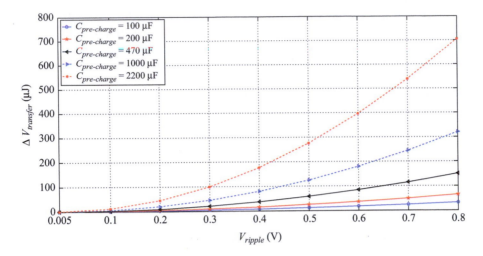

Figure 14.7   *Energy transferred from pre-charge capacitor to inductor as function of the capacitance and V_{ripple}*

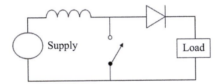

Figure 14.8   *Basic schematic of a boost converter*

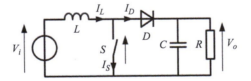

Figure 14.9   *Schematic of a boost converter with resistive and capacitive load*

The key principle of the boost converter is the ability of the inductor to oppose to changes in current by creating and destroying a magnetic field. In a boost converter, the output voltage is always higher than the input voltage. Figure 14.10 depicts the schematic for a boost power stage. By observing Figure 14.10, two boost converter configurations can be derived depending on if the switch is open or closed.

- In case the switch is closed, current flows through the inductor in clockwise direction and the inductor stores some energy by generating a magnetic field. On the left side of the inductor, the polarity is positive.

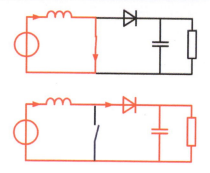

Figure 14.10   Two configurations of boost converter with resistive and capacitive load

- In case the switch is opened, current will be reduced as the impedance is higher. The magnetic field previously created will be destroyed to maintain the current flow towards the load. Therefore, the polarity will be reversed (i.e. the left side of the inductor the polarity is negative). As a result, two sources will be in series (V_i and V_L) causing a higher voltage to charge the capacitor through the diode D.

Other situation that can happen is when the switch is cycled fast enough, causing the inductor to not discharge fully in between charging stages which causes the load to see a voltage greater than the one from the input source alone when the switch is opened. In addition, while the switch is opened, the capacitor in parallel with the load is charged to this combined voltage. As soon as the switch is closed and the right hand side is shorted out from the left hand side, then the capacitor is able to provide the voltage and energy to the load. Here, the diode prevents, during this time, the capacitor from discharging through the switch. Another aspect that we should take into account is that the switch must be opened again fast enough to prevent the capacitor from discharging too much.

The basic principle of a boost converter consists of two distinct states (see Figure 14.11):

- In the on-state, the switch S is closed, as shown in Figure 14.10, resulting in an increase in the inductor current;
- In the off-state, the switch is open, and the only path offered to inductor current is through the flyback diode D, the capacitor C and the load R. Due to this change in the configuration, the energy accumulated during the on-state into the capacitor is transferred. The input current is the same as the inductor current. So, the current is not discontinuous as in the buck converter, and the requirements on the input filter are relaxed compared to a buck converter.

Every inductor has an inductive component and a resistive component. Since we are assuming only the DC component, the inductor acts like a resistance only. Theoretically, the inductor under DC conditions has a negligible resistance and a resistor needs to be added in series with the inductor in order to keep current under an acceptable value.

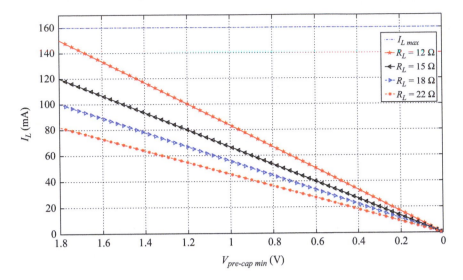

Figure 14.11   Current across the inductor of 101 μH as function of $V_{pre\text{-}cap\,min}$ for different values of the resistance for the inductor current limiting resistors

In order for the circuit to function properly, the external components need therefore to be calculated carefully. When the switch is in the on-state, the current across the inductor is given by

$$I_L(s) = \frac{V_{pre\text{-}cap\,min}(s)}{R_L} \tag{14.12}$$

where R_L is the resistance that we introduce in series with the inductor L.

For this example we consider $L = 101\,\mu H$, a maximum allowed current of $I_{Lmax} = 160\,mA$ and $V_{pre\text{-}cap\,min} = 1.8\,V$. In order to limit this current in the inductor, a resistor with resistance R_L needs to be connected in series with the inductor. By replacing the variables in (14.12), the minimum resistor value (R_L) is given by

$$R_L = \frac{V_{pre\text{-}cap\,min}(s)}{I_L}$$

$$R_L = \frac{1.8}{160 \times 10^{-3}}$$

$$R_L = 11.25\,\Omega \tag{14.13}$$

The value obtained for the minimum resistance of the resistor is 11.25 Ω. However, it should be rounded to an integer number since there is not this standard value of the resistance for commercially available resistors. In order to guarantee that the current across the inductor does not exceeds the maximum supported current, we should connect a resistor with resistance such as 12, 15, 18 or 22 Ω, from the 10%

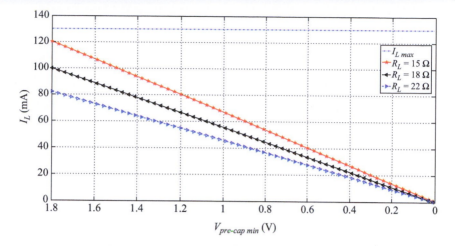

Figure 14.12   *Current across the inductor of 1.5 mH as function of $V_{pre\text{-}cap\,min}$ for different values of the resistance for the inductor current limiting resistors*

tolerance series. We opt to choose $R_L = 12\,\Omega$, in order to maximize the current across the circuit.

After finding the optimal value of the resistance to maximize the current across the circuit without damaging the inductor, we must calculate the time it takes to charge the inductor. When the switch is in the off-state, the time the inductor takes to transfer 63% of the stored energy to the capacitor is given by

$$\tau = \frac{L}{R_L} \tag{14.14}$$

By replacing the variables in (14.14) the time it takes the inductor to transfer the stored energy to the capacitor is given by

$$\tau = \frac{101 \times 10^{-6}}{12}$$
$$\tau = 0.000008\ \text{s} \tag{14.15}$$

These values should be kept as reference, since the power supply source to the boost converter is a capacitor that discharges (i.e. its voltage decreases) as soon as it starts to transfer the energy from the pre-charge capacitor to the inductor. In the presented example, the power supply source delivers a constant energy flow to the inductor.

Figures 14.11–14.13 present the behaviour that is expected to be observed for the current as the voltage in the pre-charge capacitor decreases, for different values of the inductance and associated current limiting resistors.

Figures 14.11, 14.12 and 14.13 show the evolution of the current across the 101 μH, 1.5 mH and 47 mH inductors as the voltage in the pre-charge capacitor

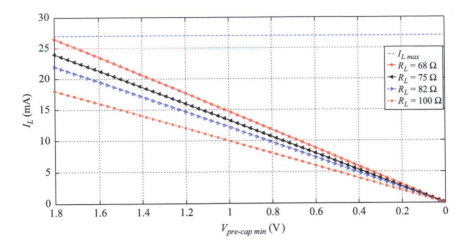

Figure 14.13   *Current across the inductor of 47 mH as function of $V_{pre\text{-}cap\,min}$ for different values of the resistance for the inductor current limiting resistors*

decreases due to the energy transfer to the inductor. In these cases, the maximum sup-ported currents across the inductors are $I_{L\,max} = 160, 130$ and 27 mA, respectively, and the minimum values for the resistance that avoid exceeding the supported current of the inductors are $R_L = 11.25, 13.84$ and $66.66\ \Omega$. By observing Figures 14.11, 14.12 and 14.13 it is possible to conclude that, in practice, respective values for the resis-tance $R_L = 12, 15$ and $68\ \Omega$ enable to achieve the maximum allowed current in the inductors when $V_{pre\text{-}cap\,min} = 1.8$ V. However, as the energy is transferred to the induc-tor, the current that is involved also decreases as expected. As the resistor value increases, the maximum allowed current across the inductor also decreases. Since the objective is to achieve the higher efficiency possible in this circuit, the choices rely on $R_L = 12, 15$ and $68\ \Omega$, respectively, as shown in Figures 14.11, 14.12 and 14.13.

Another aspect that we should take into account in the design of the algorithm implemented in the Arduino to control the Boost converter is the maximum voltage ripple that is allowed to be attained in the pre-charge capacitor. In order to have a quicker recharge of the pre-charge capacitor, the supercapacitor energy storing system algorithm must ensure that the voltage in the pre-charge capacitor is within a voltage interval that allows for a quick and efficient energy transfer to the inductor and consequently to the supercapacitor(s) bank.

By choosing $R_L = 12$ and $15\ \Omega$ to limit the current across the inductors, the algorithm should maintain the voltage in the pre-charge capacitor with a minimum discharge voltage of 1 and 1.2 V, as this capacitor is capable of delivering current in between 83 and 150 mA to the inductor (in series with $R_L = 12\ \Omega$), and in between 80 and 120 mA in the second case ($R_L = 15\ \Omega$).

It is a worth noting that, as the inductance increases, the maximum current allowed in the inductor decreases, which leads to lower current in the energy transfer between the pre-charge capacitor and the inductor. As a consequence, one may conclude that an inductor with an inductance higher than 1.5 mH is not suitable for our prototype. As such, in the experimental work, tests comprise the use of inductance with values within 101 μH and 1.5 mH.

The front-end schematic used to transfer the energy generated by the N-stage Dickson voltage multiplier into the supercapacitor(s) bank is presented in Figure 14.14. The front-end block consists of the boost converter with different transistors whose logic level is controlled by the Arduino microcontroller.

In our circuit, the process of measuring $V_{cap\,pre\text{-}charge}$ and $V_{super\text{-}cap}$ nodes take approximately 104 μs, i.e. the amount of time it takes the SAR-ADC of the Arduino to finish its conversion.

In Figure 14.14, the pre-charge capacitor may be associated with more capacitors, either in series or parallel. To achieve this modification, jumpers were used in the prototype in order to choose the connection to be used in between the capacitors. Jumpers $JP1$ and $JP2$ may be used to separate the pre-charge capacitor from the remaining circuit or connect oscilloscope probes and verify the voltage, as well as charge and discharge time duration. If the objective is to charge only one pre-charge capacitor (i.e. $C1$), by default, jumpers $JP1$ and $JP2$ are always attached while $JP3$ and $JP4$ are dis-attached. If the objective is to associate the capacitor $C2$ in series, the configuration of jumpers $JP1$, $JP2$, $JP3$ and JP4 should be the following: $JP1$ is attached, $JP2$ is dis-attached, $JP3$ is dis-attached and $JP4$ is attached. Otherwise, if the objective is to associate in parallel the capacitor $C2$, the configuration of jumpers $JP1$, $JP2$, $JP3$ and $JP4$ should be the following: $JP1$ is attached, $JP2$ is attached, $JP3$ is attached and $JP4$ is dis-attached.

Additionally, if the objective is to increase the capacitance or supported voltage in the supercapacitor(s) bank, it is also possible to associate the supercapacitors in series and parallel. Jumpers $JP6$ and $JP12$ can be used to separate the supercapacitor(s) bank from the remaining circuit or connect oscilloscope probes to verify the voltage and times of charge and discharge. To connect the supercapacitor(s) bank to the remaining circuit, jumpers $JP6$ and $JP12$ must be connected. Again, in the supercapacitor(s) bank, if the intention is to employ only the supercapacitor $C3$, jumpers $JP6$ and $JP12$ are attached, while jumpers $JP7$, $JP8$, $JP9$, $JP10$, $JP11$ are dis-attached. To associate the supercapacitor $C3$ in series with $C4$ then jumper $JP6$ is attached, $JP12$ is dis-attached, while jumper $JP7$ is dis-attached, $JP8$ and $JP9$ are attached, $JP10$ and $JP11$ are dis-attached. In case the configuration is to associate $C3$ in parallel with $C4$, then jumpers $JP6$ and $JP12$ are attached, $JP7$ is attached, $JP8$ and $JP9$ are dis-attached, and jumpers $JP10$ and $JP11$ are dis-attached. Other option available in this prototype is to associate one more supercapacitor, $C5$, in series or parallel with the supercapacitor $C4$. To associate the supercapacitor $C5$ in series with $C4$ then jumper $JP6$ is attached, $JP9$ and $JP10$ are dis-attached and $JP11$ is attached. To associate the supercapacitor $C5$ in parallel with $C4$ then jumper $JP6$ is attached, $JP9$ is attached, $JP10$ is attached and $JP11$ is dis-attached. Jumpers $JP12$, $JP8$ and $JP7$ are dis-attached or attached depending on the association in series or parallel of set $C4$ and $C5$.

To enable to transfer energy from the pre-charge capacitor to the inductor, the algorithm from the Arduino applies a logic level 1 to the OUT_{T2} node, while applying

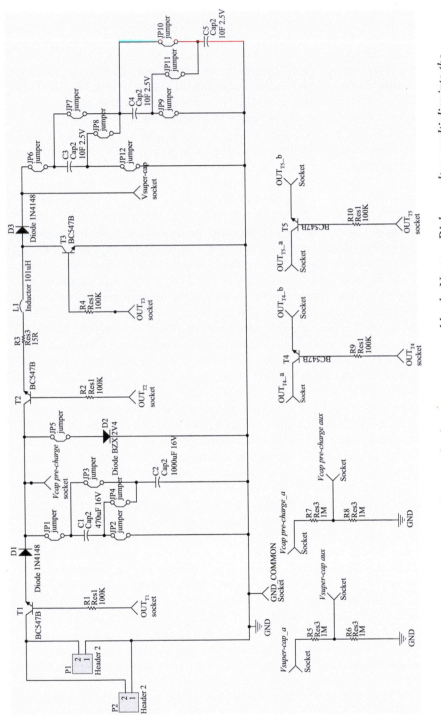

Figure 14.14   Front-end schematic used to transfer the energy generated by the N-stage Dickson voltage multiplier into the supercapacitor(s) bank

to node OUT_{T1} a logic level 0 to disconnect the front-end circuit from the N-stage EH prototype by a short duration. This stage corresponds to the on-state of the boost converter. The flyback diode $D1$ is necessary to avoid the discharge from the pre-charge capacitor when the energy transfer from the pre-capacitor to the inductor is occurring. As soon as the pre-charge capacitor attains a minimum discharge voltage defined in the algorithm of the Arduino microcontroller, it applies a logic level 0 to node OUT_{T2} and a logic 1 to node OUT_{T3}. This stage corresponds to the off-state of the boost converter. This allows to transfer the energy accumulated in the inductor to the supercapacitor(s) bank. The flyback diode D3 is required to avoid the discharge from the supercapacitor(s) bank when the energy transference from the inductor to the supercapacitor(s) bank is on course. After transferring the energy from the inductor to the supercapacitor(s) bank, the Arduino algorithm applies a logic 1 to node OUT_{T1}, a logic 1 to node OUT_{T2} and a logic 0 to node OUT_{T0}, in order to repeat all the process of energy transfer to the inductor, by means of the boost converter.

Another aspect from the front-end circuit is the option of using a Zener diode, $D2$, that can be useful to guarantee that the voltage across the circuit does not exceed the 2.4 V. In the case considered here, the Zener diode is a BZX2V4. This option is enabled by attaching jumper $JP5$.

Figure 14.14 depicts two headers ($P1$ and $P2$) that are used to attach the output of the N-stage Dickson voltage multiplier. It is possible to attach a unique N-stage Dickson voltage multiplier, but this supercapacitor-based storage system is able to cope with the option of attaching two N-stage Dickson voltage multipliers, each one harvesting RF energy from different frequency bands.

In addition, four auxiliary circuits are present in this PCB. Two circuits are resistor dividers that can be used for both $V_{cap\,pre\text{-}charge}$ and $V_{super\text{-}cap}$ nodes to match the Arduino power supply to these nodes' voltages (max: 10 V). The values of the resistors are set to 100 kΩ but can be modified according to the order of magnitude that the N-stage Dickson voltage multiplier is able to supply. The other two auxiliary circuits are two switches that can be used for any purpose that the user needs, for example the disconnection of supercapacitor(s) bank in the case of a sudden rise of the supported voltage of the supercapacitor. Figure 14.15 shows the PCB designed for the front-end from the circuit in Figure 14.14.

14.3.4 Storage system control software

The main function of the storage system control algorithm from the Arduino Nano V3 is to sample the $V_{cap\,pre\text{-}charge}$ and $V_{super\text{-}cap}$ voltages, applied to its input pins $ANALOG$ $IN\,A2$ and $A3$, respectively. It is responsible to monitor and control the energy flow from the 470 µF pre-charge capacitor to the inductor by enabling the NPN transistor $T2$, by means of applying a logic 1 to OUT_{T2}, and the NPN transistor $T3$, by means of applying a logic 1 to OUT_{T3}. The energy is then transferred from the inductor to the supercapacitor(s) bank.

The algorithm employed in the Arduino microcontroller is presented in Algorithm 1. This algorithm begins by defining the variables and constants that the system will rely on to control the energy flow. $V\_ccap\_min\_float = 1$ V is the

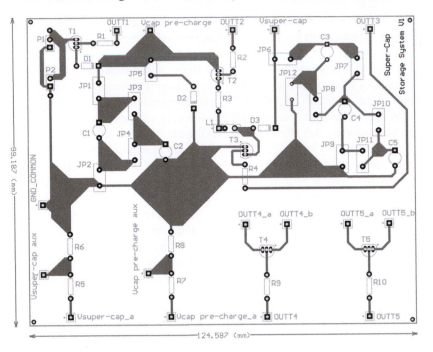

Figure 14.15   Front-end PCB prototype

minimum accepted voltage in the pre-charge capacitor after transferring energy to the inductor. $V\_scap\_max\_float = 2.4$ V is the maximum voltage supported by the supercapacitor(s) bank. $Vccap\_threshold\_float = 1.8$ V is the minimum voltage the pre-charge capacitor must attain to initiate the process. $num\_samples\_int = 10$ is the number of samples the ADC takes to read the voltages $V_{cap\,pre\text{-}charge}$ and $V_{super\text{-}cap}$. $step\_delay = 10$ ms is a variable used to adjust the duty cycle from the charge–discharge cycle. $delay\_samples_int = 10$ ms is the delay in between each sample from the ADC. $delay\_transistor2\_int = 500$ ms is the time duration of the energy flow from the pre-charge capacitor to the inductor. $delay\_transistor3\_int = 500$ ms is the time duration of the energy flow from the inductor to the supercapacitor(s) bank.

The algorithm initiates by setting the ADC registers. Then, it enters in an infinite loop, where the control source code is located.

The algorithm begins by verifying if the number of samples taken from the $V_{cap\,pre\text{-}charge}$ and $V_{super\text{-}cap}$ voltages has achieved the maximum allowed (i.e. $num\_samples\_int = 10$). While this condition is not true, the algorithm continues to acquire the ADC reading from the $V_{cap\,pre\text{-}charge}$ and $V_{super\text{-}cap}$ nodes with a delay of $delay\_samples_int = 10$ ms in between each sample reading. As soon as the max-imum number of samples is attained, the algorithm moves to the conversion of the ADC readings to V_ccap and V_scap that correspond to $V_{cap\,pre\text{-}charge}$ and $V_{super\text{-}cap}$ voltages, respectively.

After computing the voltages associated of each ADC reading for the $V_{cap\,pre\text{-}charge}$ and $V_{super\text{-}cap}$ nodes, the algorithm verifies if the voltage acquired from the

Algorithm 1: Arduino Nano V3 code to manage and control the supercapacitor storage system

---

Data:
```
#define TRANSISTOR1 2
#define TRANSISTOR2 3
#define TRANSISTOR3 4
float V_ccap = 0.0, V_scap = 0.0, V_scap_max_float=0.0 ;
float V_ccap_min_float=1.0, V_scap_max_float =2.4, V_ccap_min_float=1.8;
int samples_count = 0, sum = 0, sum2 = 0, step_delay=10, execute_cycle=0;
int num_samples_int=10, delay_samples_int=10;
int delay_transistor2_int=500, delay_transistor3_int=500;
while (1) do
    while (samples_count <num_samples_int) do
        sum += analogRead(A2);
        sum2 += analogRead(A3);
        samples_count++;
        delay(delay_samples_int);
    V_ccap= (((float)sum / (float)numero_amostras_int)* 5.015) / 1024.000;
    V_scap =(((float)sum2 / (float)numero_amostras_int)* 5.015) / 1024.000;
    if (V_scap <V_scap_max_float) then
        if (V_ccap≥ Vccap_threshold_float && execute_cycle==0) then
            execute_cicle=1;
        else
            if (V_ccap≤ V_ccap_min_float && execute_cycle==0) then
                digitalWrite(TRANSISTOR1,HIGH); //T1=On to charge Cpre-charge until it
                    reaches V_ccap_min_float
                digitalWrite(TRANSISTOR2,LOW); //T2=Off
                digitalWrite(TRANSISTOR3,HIGH); //T3=On

    if (execute_cycle==1) then
        /////////////// On-stage of boost converter – Buck Operation ///////////////
        digitalWrite(TRANSISTOR1,LOW); //T1=Off to stop charging Cpre-charge
        digitalWrite(TRANSISTOR2,HIGH); //T2=On to transfer energy from Cpre-charge to L
        digitalWrite(TRANSISTOR3,HIGH); //T3=On to transfer energy from Cpre-charge to L
        delayMicroseconds(delay_transistor2_int);
        /////////////// Off-stage of boost converter – Boost Operation ///////////////
        digitalWrite(TRANSISTOR1,LOW); //T1=Off to stop charging Cpre-charge
        digitalWrite(TRANSISTOR2,HIGH); //T2=On to transfer energy from Cpre-charge to L
        digitalWrite(TRANSISTOR3,LOW); //T3=Off to transfer energy from L to Csuper-cap
            bank
        delayMicroseconds(delay_transistor3_int);
        digitalWrite(TRANSISTOR1,HIGH);//T1=On to re-start charging Cpre-charge
        digitalWrite(TRANSISTOR2,LOW);
        digitalWrite(TRANSISTOR3,HIGH);
    execute_cycle=0; samples_count = 0; sum = 0; sum2 = 0;
    //// Duty Cycle Adjustment ////
    V_ccap= (((float)analogRead(A2)/ (float)numero_amostras_int)* 5.015) / 1024.000;
    if (V_ccap>V_ccap_min_float) && (delay_transistor3_int >0) then
        delay_transistor3_int=delay_transistor3_int+step_delay;
        delay_transistor2_int=delay_transistor2_int+step_delay;

    if (V_ccap<V_ccap_min_float) && (delay_transistor2_int >0) then
        delay_transistor3_int=delay_transistor3_int-step_delay;
        delay_transistor2_int=delay_transistor2_int-step_delay;
    sleep_8_seconds(); //Arduino sleep state until next verification
```

supercapacitor(s) bank is not exceeding the maximum voltage supported by the super-capacitor(s) bank, i.e. $V\_scap\_max\_float = 2.4$ V. If the condition is true then it moves to next condition that consists on comparing if the voltage of the pre-charge capacitor is greater or equal to the minimum voltage it must attain to initiate the process, i.e. $Vccap\_threshold\_float = 1.8$ V and the energy flow process is not being executed. If this condition is true, the energy flow process is initiated, i.e. $execute\_cycle = 1$.

Otherwise, the algorithm verifies if the instantaneous pre-charge capacitor voltage is less or equal to the minimum-accepted voltage in the pre-charge capacitor, i.e. $V\_ccap\_min\_float = 1$ V and the energy flow process is not being executed. If this condition is true, then the algorithm applies a logic 1 to OUT_{T1}, and a logic 0 to OUT_{T2} and OUT_{T3}, in order to charge the pre-charge capacitor.

When the energy flow process is initiated, i.e. $execute\_cycle = 1$, the algorithm initiates the buck operation in which the transistor $T1$ is turned off by applying a logic 0 to OUT_{T1}, while transistors $T2$ and $T3$ are turned on by applying a logic 1 to OUT_{T2} and OUT_{T2}, respectively. This procedure allows for the current to flow through to the inductor. The algorithm awaits for a delay of $delay\_transistor2\_int = 500$ ms to allow the energy flow from the pre-charge capacitor to the inductor. This allows the current to flow to the inductor, which increases the magnetically stored energy in the inductor, thereby reducing the voltage in the pre-charge capacitor.

After this step, the algorithm initiates the boost operation in which the transistor $T1$ is turned off by applying a logic 0 to OUT_{T1}, transistor $T2$ is turned on by applying a logic 1 to OUT_{T2}, while the transistor $T3$ is turned off by applying a logic 0 to OUT_{T3}. The algorithm awaits for a time $delay\_transistor3\_int = 500$ ms to allow the energy flow from the inductor to the supercapacitor(s) bank.

After completing this step, the algorithm restarts charging the pre-charge capacitor by applying a logic 1 to OUT_{T1}, while transistor $T2$ is turned off by applying a logic 0 to OUT_{T2} and transistor $T3$ is turned on by applying a logic 1 to OUT_{T3}.

In the end of the this three-stage process, the variable $execute\_cycle$ is reset. In addition, the variables related to the ADC samples reading are also reset.

After executing the energy-flow process to the supercapacitor(s) bank, the algorithm adjusts the duty cycle (of the process). To achieve this goal, the algorithm reads the $V_{cap\,pre\text{-}charge}$ voltage, which corresponds to the V_ccap variable in the algorithm. Then, if the voltage reading from the V_ccap variable is higher than $V\_ccap\_min\_float = 1$ V, which is the minimum accepted voltage in the pre-charge capacitor after transferring energy to the inductor, the algorithm will modify the time durations $delay\_transistor2\_int$ and $delay\_transistor3\_int$ by adding or subtracting a step delay, i.e. $step\_delay = 10$ ms. In this case, the time duration of the energy flow from the pre-charge capacitor to the inductor (i.e. $delay\_transistor2\_int$) and from the inductor to the supercapacitor(s) bank (i.e. $delay\_transistor3\_int$) is decreased by $step\_delay = 10$ ms.

If the previous condition is not valid, another condition is verified. If the voltage reading from the V_ccap variable is lower than the $V\_ccap\_min\_float = 1$ V, the algorithm will modify the time durations $delay\_transistor2\_int$ and $delay\_transistor3\_int$ by adding or subtracting a step delay, i.e. $step\_delay = 10$ ms. In this case,

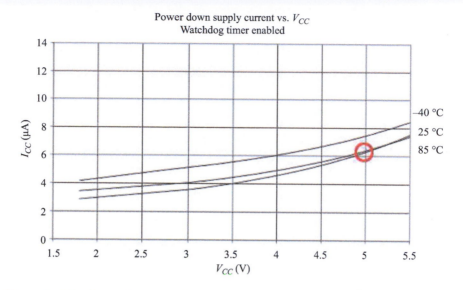

Figure 14.16   Power-down supply current versus power supply voltage (V_{cc}) for the Atmega328P processor. © 2014. Extracted, with permission, from Reference [17]

the time duration of the energy flow from the pre-charge capacitor to the inductor (i.e. *delay_transistor2_int*) and from the inductor to the supercapacitor(s) bank (i.e. *delay_transistor3_int*) is increased by *step_delay* = 10 ms.

The last instruction executed by the algorithm sets the Arduino to a deep sleep state that lasts for 8 s. This sleep state enables to save energy in the Arduino without sacrificing the control and management of the supercapacitor storage system. When the Arduino is set to sleep state, it consumes 6.66 μA, as shown in Figure 14.16 (with watchdog timer enabled) when the power supply of the Arduino is 5 V [17].

Another technique that we added into the algorithm of the Arduino aims at power saving. Aside the sleep state function already presented in Algorithm 1, we decided to employ a lower clock frequency in the Arduino microcontroller (an Atmega328P processor). By reducing the clock frequency we are also able to reduce the minimum supply voltage required for the Arduino to work properly. By default, clock frequency of the Atmega328P processor is 16 MHz. Since a higher clock frequency is not needed, we decided to reduce it to 4–8 MHz. Figure 14.17 presents the required supply voltage for each clock frequency from the Atmega328P processor. Below 1.8 V, the Arduino cannot operate. N.B.: Below 4 MHz, the minimum supply voltage is 1.8 V. Between 4 and 10 MHz, the required voltage can be calculated by $V[Volt] = 1.8 + ((M - 4) \times 0.15)$, where M is the desired clock frequency in MHz units, whereas between 10 and 20 MHz, the supply voltage is given by $V[Volt] = 2.7 + ((M - 10) \times 0.18)$ (where M is the desired clock frequency in MHz units). The maximum supply voltage supported by the Atmega328P processor is 5.5 V.

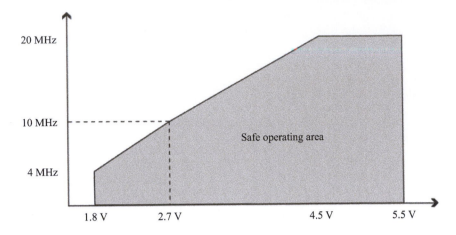

Figure 14.17   Required supply voltage for each clock frequency for the
           Atmega328P processor

Table 14.4   *Value of the required voltages*
           for each clock frequency

f (MHz)	*Voltage* (V)
4	1.80
5	1.95
6	2.10
7	2.25
8	2.40
9	2.55
10	2.70
11	2.88
12	3.06
13	3.24
14	3.42
15	3.60
16	3.78
17	3.96
18	4.14
19	4.32
20	4.50

Table 14.4 summarizes the voltages required for each value of the clock frequency. As such, we decided to implement the range of voltages in the code that correspond to the Atmega 328P processor working at a clock frequency in the range 4–8 MHz, i.e. a power supply in between 1.8 and 2.4 V, respectively. With these values for the power supply voltage, we are indeed able to use the voltage from the supercapacitor(s) bank to supply the Arduino.

From the Atmega328P processor datasheet [17] is possible to define different sleep modes as well as the associated power consumptions. These sleep modes and power consumptions are as follows:

- SLEEP_MODE_IDLE: 15 mA
- SLEEP_MODE_ADC: 6.5 mA
- SLEEP_MODE_PWR_SAVE: 1.62 mA
- SLEEP_MODE_EXT_STANDBY: 1.62 mA
- SLEEP_MODE_STANDBY: 0.84 mA
- SLEEP_MODE_PWR_DOWN : 0.36 mA

The purpose of this is simply to make the sensor node switch to its deepest sleep mode immediately after starting up. Once the Atmega328P processor begins to execute the code, it disables its watchdog timer. It then sets all unused pins as outputs and assigns them a logical value (in order to avoid floating input pins that can increase power consumption) while disabling the receiving circuitry of the node. Finally, it switches its processor and clock generators off. The Arduino remains in this deep sleep state during the time defined by us.

14.3.5 Back-end block

As shown in Figure 14.5, the back-end circuit is a typical buck converter. It converts the supercapacitor(s) voltage to a CV output to operate our embedded system, namely the CRN. To implement this circuit block, in this work, different ICs from Linear Technology have been considered for this harvesting applications. On the one hand, there are ICs whose purpose is EH. On the other, there are ICs to charge supercapacitors. The former one aggregates power-management products that convert energy from vibration (Piezo), photovoltaic (Solar) and thermal (thermoelectric cooling (TEC), thermoelectric generator (TEG), Thermopiles, Thermocouples) sources and provide high efficiency conversion to regulated voltages, or to charge batteries and supercapacitor storage elements. Note that one characteristic that makes these ICs suitable for this type of applications is the fact that boost converters operate with a minimum voltage of 20 mV. The latter includes ICs that allow for charging supercapacitors by means of buck–boost converter or by charge pump, depending on the type of IC. These ICs offer input or output current limiting, automatic cell balancing and a range of protection features that make them uniquely suited to supercapacitor charging.

For the choice of the ideal IC for this project, the features that have been accounted for by us are the minimum supply voltage, V_{in_min}, maximum supply voltage, V_{in_max}, maximum output voltage, V_{out_max}, maximum output current, I_{out}, and the quiescent supply current (I_{out}). Table 14.5 summarizes the different ICs considered in the design of the back-end circuit.

If the function of the back-end circuit is to convert the supercapacitor(s) voltage to a CV output, in order to operate our embedded system, the main features that should be taken into account is the minimum supply voltage (V_{in_min}) and the maximum output

Table 14.5   Energy harvester (#) and supercapacitors chargers (*) IC's

IC model	V_{in_min} (V)	V_{in_max} (V)	V_{out_max} (V)	I_{out} (A)	I_{supply} (mA)	Topology
LTC3105#	0.2	5	6	0.4	0.0024	Step-up DC/DC converter
LTC3108#	0.02	0.5	5	0.007	0.0002	Ultralow voltage step-up
LTC3109#	0.02	0.5	5	0.026	0.0002	Ultralow voltage step-up
LTC3588#	2.7	20	1.8, 2.5, 3.3, 3.6	0.100	0.00095	Piezoelectric full wave bridge rectifier
LTC3125*	1.6	5.5	5.25	1	0.015	Boost
LTC3127*	1.8	5.5	5.25	1	0.035	Buck–boost
LTC3128*	1.73	5.5	5.5	3	0.6	Buck–boost
LTC3226*	2.5	5.5	5.3	0.315	0.01	Charge pump
LTC4425*	2.7	5.5	5.5	3	0.02	Linear
MAX1686*	2.7	4.2	4.75	0.012	0.1	Charge pump
MAX1759*	1.6	5.5	3.3	0.1	0.05	Buck–boost
MAX1683*	2.0	5.5	$2*V_{in}$	0.00031	45	Voltage doubler

current, I_{out}, the IC can deliver. By observing Table 14.5, there are three types of ICs, as follows:

- The main advantage of the LTC3105, LTC3108 and LTC3109 energy harvesters ICs consists of very low minimum voltage needed to start working, compared to the other chips. However, output current it can deliver to the embedded system is also very low.
- The LTC3588 IC is a very good option since it can deliver a higher output current to the embedded system but at the cost of a higher minimum voltage supply, compared to the other energy harvester ICs.
- For the supercapacitors chargers, all ICs present a minimum supply voltage in between 1.6 and 2.7 V but have the advantage of being able to double the voltage that is being supplied to IC.

When the front-end block is going to deliver a voltage output in between 2.5 and 5 V, there is no justification to consider the back-end block to convert to a CV output. The back-end circuit is only justified in case the front-end circuit delivers, for example, a voltage of 16 V and the embedded system needs a voltage of 8 V, as it appropriately adjusts the voltage.

For now, in our proposed solution, the back-end circuit is going to be replaced by a Zener diode that maintains the voltage output of the front-end circuit in between the desired voltage for the embedded system (i.e. in the interval between 1.6 and 2.7 V).

14.4   Performance evaluation

Figure 14.18 shows the front-end prototype we have built, while Figure 14.19 shows the programme developed to evaluate the pre-charge capacitors and inductors, for different time durations *delay_transistor2_int* and *delay_transistor3_int*.

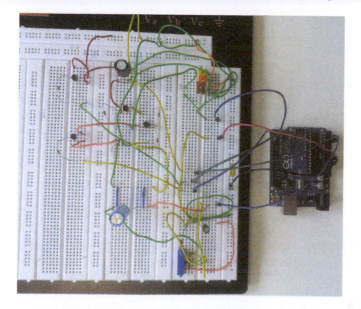

Figure 14.18   Prototype from the front-end circuit

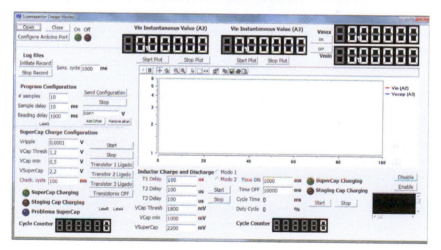

Figure 14.19   Programme to evaluate the pre-charge capacitors and inductors for different time durations delay_transistor2_int and delay_transistor3_int in the front-end circuit

It is observable in Figure 14.19 that the programme developed to control and charge the supercapacitor(s) bank by means of a buck–boost converter allows for tuning different parameters of the Arduino firmware whilst giving a real-time plot of the $V_{cap\,pre\text{-}charge}$ and $V_{super\text{-}cap}$ voltages, as well as an indication of the instantaneous voltage value of each one. Additionally, an algorithm is applied that analyses the

$V_{super-cap}$ voltage and shows the maximum and minimum voltages of the $V_{super-cap}$ as the test is ongoing. During the test trials, a programme configuration is sent to the Arduino where it is possible to define the number of samples, the delay in between the samples as well as the time in between the next cycle of samples. After defining these parameters, the user just has to click on the "send configuration" button which configures the Arduino. Apart from this ADC and sample configuration, the programme allows the user to enable and disable each one of the transistors manually.

For the inductor charge and discharge configuration, the parameters available to be tunable are the transistor enabling and disabling time durations (i.e. *delay_transis tor1_int*, *delay_transistor2_int*, *delay_transistor3_int*), the *VCapThresh* that corresponds to the minimum voltage the pre-charge capacitor must attain to initiate the process (i.e. *Vccap_threshold_float* = 1.8 V), the value of *VCap min* that corresponds to the minimum accepted voltage in the pre-charge capacitor, after transferring the energy to the inductor (i.e. *V_ccap_min_float* = 1 V) and the *VSuperCap*, which is the maximum voltage supported by the supercapacitor(s) bank (i.e. *V_scap_max_float* = 2.4 V). Besides the aforementioned configuration for the inductor charge and discharge, there are three LEDs that become active when the supercapacitor is receiving energy from the inductor (i.e. the supercapacitor is charging), when the pre-charge capacitor is charging, as well as when the supercapacitor(s) bank is reaching the maximum supported voltage. The supercapacitor overcharge warning causes the Arduino to disable all the transistors switches, in order to stop all the energy flows. Besides the information LEDs, there is an analogue counter that shows the number of charge and discharge cycles that have occurred in the pre-charge capacitor.

Before pursuing the solution of charging and discharging, the inductor by means of the transistor enabling and disabling time durations and different thresholds applied to the $V_{cap\,pre-charge}$ and $V_{super-cap}$ voltages readings, another strategy was pursued. This approach was based on the different thresholds applied to the $V_{cap\,pre-charge}$ and $V_{super-cap}$ voltage readings and the allowed voltage ripple in the pre-charge capacitor (identified in the programme screenshot as V_{ripple}). This approach allows for initiating the pre-charge capacitor energy to the inductor, and consequently to the supercapacitor. However, it was shown that the energy flow exchanges were not efficient.

To find the most suitable values for the electronic components of the buck–boost converter, different experiments have been conducted to find the best transition times for charging the pre-charge capacitor (*delay_transistor2_int*) as well as the duration of the energy transfer from the pre-charge capacitor to the inductor and from the inductor to the supercapacitor(s) bank (*delay_transistor3_int*). The programme has the option to log (with a editable recording interval) the $V_{cap\,pre-charge}$, $V_{super-cap}$ voltages as well as the corresponding cycle to each one of the voltages recorded. With the acquisition of this data, it was possible to plot different metrics that allowed to choose the best values for the capacitance of the pre-charge capacitor and inductance for the inductor, as well as the best time durations, for a more efficient charge of the supercapacitor(s) bank.

Table 14.6 presents the different trial cases tested with the front-end prototype by inserting the time durations *delay_transistor2_int* and *delay_transistor3_int* in the programme developed for this purpose. Besides the definition of the different

Table 14.6   Trial cases to evaluate the pre-charge capacitors
and inductors

Trial case	delay_transistor2_int (μs)	delay_transistor3_int (μs)
1	500	1,000
2	1,000	1,000
3	1,000	1,500
4	1,000	2,000
5	1,000	3,000
6	1,000	5,000
7	1,000	6,000
8	1,000	7,000
9	1,000	8,000
10	1,000	9,000
11	1,000	10,000

time durations, the inductance values for the inductor considered were 101 μH and 1.5 mH, whereas for the capacitance values of the pre-charge capacitor were 220, 470 and 1,000 μF. The minimum value for the resistance of the resistor value that avoids exceeding the supported current of the inductor is chosen according to Figures 14.11 and 14.12. i.e. $R_L = 12$ Ω for the 101 μH inductor while $R_L = 15$ Ω for the 1.5 mH inductor.

Initially, one 10 F (−2.5 V) supercapacitor unit is discharged. Figures 14.20 and 14.21 show snapshots of the oscilloscope output when the circuit is operating. Figure 14.21 shows the on and off states from the buck–boost converter. The time durations *delay_transistor2_int* and *delay_transistor3_int* presented in Figures 14.20 and 14.21 match the trial case 6, i.e. *delay_transistor2_int* = 1,000 μs and *delay_transi stor3_int* = 5,000 μs. In addition, for the remaining configuration, the ADC takes 10 samples with a sample delay of 10 ms and periodic loop of 100 ms.

Figure 14.21 shows that the curve with black line that corresponds to the voltage of the pre-charge capacitor (i.e. $V_{cap\,pre\text{-}charge}$) slowly increases along the time, until it reaches 1.8 V, while the Arduino control programme initiates the procedure of the buck–boost converter. This process is initiated by disabling the OUT_{T1} input (represented in dark blue), which disconnects the pre-charge capacitor from the harvesting device and allows the transfer of energy firstly to the inductor. Simultaneously, the OUT_{T2} input (represented in green) is enabled which allows to transfer the energy from the pre-charge capacitor to the inductor. In addition, the OUT_{T3} input (represented in cyan) is enabled in order the current/energy flow is not transferred from the inductor to the supercapacitor(s) bank. The aforementioned input configurations remain in these states during *delay_transistor2_int* = 1,000 μs. This phase is the so-called buck or on phase.

After 1,000 μs, the "boost" or "on" phase is initiated. During this phase, the OUT_{T1} input (represented in dark blue) and OUT_{T2} input (represented in green) continue enabled. However, the OUT_{T3} input (represented in cyan) is disabled in order to transfer the energy accumulated in the inductor to the supercapacitor(s) bank.

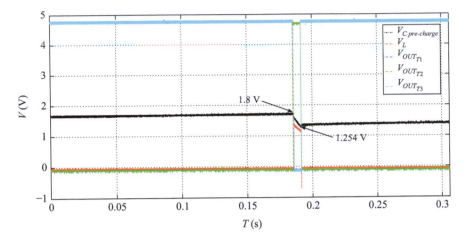

Figure 14.20   Outputs from the oscilloscope during a buck–boost operation

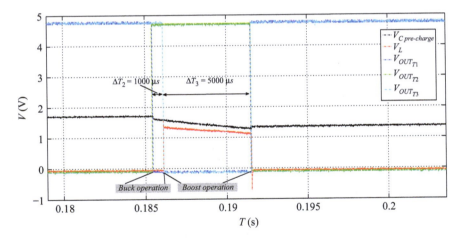

Figure 14.21   Close-up of the oscilloscope outputs during a buck–boost operation

The voltage measured at the inductor terminals is presented in Figure 14.21 by the red line. It can be observed that the voltage decreases as it transfers the accumulated energy to the supercapacitor(s) bank. The aforementioned input configurations remain in these states during *delay_transistor3_int* = 5,000 μs. After 5,000 μs, the inputs OUT_{T1}, OUT_{T2} and OUT_{T3} return to the initial setup, and the pre-charge capacitor is recharged. Figure 14.21 does not present the evolution of the voltage at the terminals of the supercapacitor(s) bank, since we had limited inputs at the Picoscope hardware that was used to obtain the presented traces.

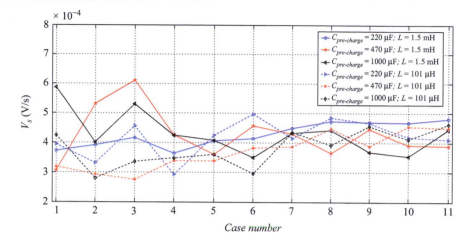

Figure 14.22 *Average voltage per unit of time in the $C_{super-cap}$, for different trial cases*

14.4.1 Average charging voltage in $C_{super-cap}$

After collecting all the data from the different combinations of values of the capacitance and inductance, for the different trials shown in Table 14.6, the first metric to be used is the average charging voltage in the $C_{super-cap}$, V_S. For all the trial cases, we registered at least 2,000 cycles of charging and discharging cycles for the pre-charge capacitor $C_{pre-charge}$ – but only considered the first 2,000 cycles.

This metric depends on the average time duration of each charge and discharge cycle of the $C_{pre-charge}$, i.e. $t_{cycle\_pre-charge\_cap}$, and the average voltage registered $C_{super-cap}$ in the same charge and discharge cycle of the $C_{pre-charge}$, i.e. $v_{cycle\_super-cap}$. Consequently, the average charging voltage in the $C_{super-cap}$, V_S, is given by (14.16), as follows:

$$V_S = \frac{v_{cycle\_super-cap}}{t_{cycle\_pre-charge\_cap}} \tag{14.16}$$

Figure 14.22 shows the comparison of the average voltage per unit of time in the supercapacitor for the different trial cases, shown in Table 14.6, and different values for the inductance and capacitance.

By observing Figure 14.22, it is possible to conclude that the maximum voltage per unit of time value is attained in the trial case number 3, with $C_{pre-charge} = 470 \ \mu F$ and $L = 1.5$ mH, and presents a value of $V_S \approx 6.09 \times 10^{-4}$ V/s, whereas the minimum voltage per unit of time value is attained for the same trial case number 3 with a $C_{pre-charge} = 470 \ \mu F$ and $L = 101 \ \mu H$ and presents a value of $V_S \approx 2.74 \times 10^{-4}$ V/s. In addition, for the trial cases number 1, 2 and 3, there are some substantial differences in the values of the V_S metric among the different cases and combinations for the inductance and capacitance values. This behaviour is explained by the short-time durations of the transistor-switch delays employed in these trials, as shown

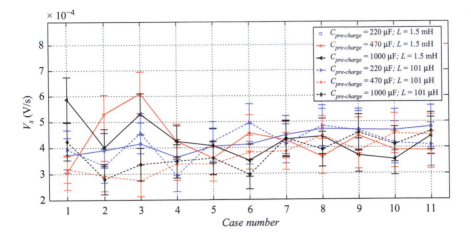

Figure 14.23   Average voltage per unit of time in the $C_{super\text{-}cap}$ for different trial cases with 95% confidence intervals

in Table 14.6. However, for the remaining trial cases, the differences in the V_S metric values are not so notorious. The values of V_S are in the range in between 3×10^{-4} and 5×10^{-4} V/s. From these observations, we can conclude that there are four component value combinations that stand out, which are the $C_{pre\text{-}charge} = 220\ \mu F$; $L = 1.5$ mH, $C_{pre\text{-}charge} = 220\ \mu F$; $L = 101\ \mu H$, $C_{pre\text{-}charge} = 470\ \mu F$; $L = 101\ \mu H$ and $C_{pre\text{-}charge} = 1,000\ \mu F$; $L = 101\ \mu H$.

Figure 14.23 shows the same evaluation for the metric from the analysis performed before in this section but with the representation of the 95% confidence intervals. The differences observed in the trial cases number 1, 2 and 3 (from Table 14.6) are due to the short-time durations of the transistor-switch delays and are characterized by the long range of confidence intervals in Figure 14.23. For the remaining trial case numbers, the range of confidence intervals are shorter, which indicates that there is more confidence in the average value of the metric V_S.

14.4.2   *Average charging voltage per cycle in $C_{super\text{-}cap}$*

Based on the data collected for the different trial cases and combinations of values for the capacitance and inductance, another metric that was computed was the average charging voltage per cycle in the $C_{super\text{-}cap}$, i.e. V_{cycle}. This metric allows for observing the increase of the voltage in the supercapacitor in average per cycle of charge and discharge. Again, we registered 2,000 cycles of charging and discharging cycles for the pre-charge capacitor $C_{pre\text{-}charge}$.

This metric relies on the number of the cycle of charge and discharge cycles of the $C_{pre\text{-}charge}$ (i.e. $n_{cycle\_pre\text{-}charge\_cap}$) and the average voltage registered $C_{super\text{-}cap}$, in the same charge and discharge cycle, $C_{pre\text{-}charge}$, $v_{cycle\_super\text{-}cap}$. For each cycle, this fraction is calculated and summed up and averaged by total number of cycles,

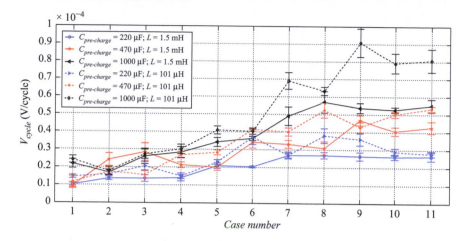

Figure 14.24   Average voltage per cycle in the $C_{super-cap}$ for different trial cases with 95% confidence intervals

i.e. 2,000. Consequently, the average charging voltage per cycle, V_{cycle}, in the $C_{super-cap}$ is calculated by (14.17), as follows:

$$V_{cycle} = \frac{1}{2,000} \times \sum_{k=1}^{2,000} \frac{v_{cycle\_super-cap}}{n_{cycle\_pre-charge\_cap}} \tag{14.17}$$

Figure 14.24 shows the results obtained for the average voltage per cycle in the supercapacitor for different trials and different values of the inductance and capacitance.

Figure 14.24 shows that, as trial case number increases, the V_{cycle} also increases for all the combinations of values of the capacitance and inductance. This increase is explained by the fact that the time duration *delay_transistor3_int* becomes longer as the trial case number increases which causes the inductor to spend more time transferring energy to the supercapacitor. However, for the first four trial cases (1, 2, 3 and 4 in Table 14.6), the increase is not so notorious. For the remaining trial cases, the increase rate occurs faster, except for the combination of components values $C_{pre-charge} = 220 \ \mu F$ and $L = 1.5 \ mH$. However, it remains almost constant after the trial case number 7. The maximum value for the voltage per cycle is attained in the trial case number 9 with $C_{pre-charge} = 1,000 \ \mu F; L = 101 \ \mu H$ and presents a value of $V_S \approx 0.908 \times 10^{-4}$ V/cycle, whereas the minimum voltage per cycle is attained for the trial case number 1, for all the combinations of values of the capacitance and inductance, except for the ones that use the capacitance value of $C_{pre-charge} = 1,000 \ \mu F$. Hence, the minimum value for V_{cycle} is in between $V_{cycle} \approx 0.11 \times 10^{-4}$ and $V_{cycle} \approx 0.14 \times 10^{-4}$ V/s.

From Figure 14.24, we conclude that to consider a pre-charge capacitor with larger capacity, such as $C_{pre-charge} = 1,000 \ \mu F$, allows for obtaining a higher increase of the voltage per cycle of charge and discharge. However, we should keep in mind that the time it takes to charge the pre-charge capacitor is longer as the capacitance increases. Therefore, we should opt for a pre-charger capacitor capacitance that does not compromise the time it takes to complete a full charge and discharge cycle between the pre-charge capacitor and the supercapacitor, by means of the energy flow that is passed to the inductor, and then from the inductor to the supercapacitor. An appropriate choice should be the combination of component values $C_{pre\text{-}charge} = 470\ \mu F;\ L = 101\ \mu H$, for a trial case number higher than or equal to 6. Here, one does not consider the characteristic of the supercapacitor which implies that, as the energy builds up in the supercapacitor, it takes longer time to add more energy into it.

14.4.3   *Normalized charging voltage in $C_{super\text{-}cap}$*

The normalized charging voltage in the $C_{super\text{-}cap}$ metric (i.e. $V_{super\text{-}cap\ norml}$) was derived by considering the data collected for the considered trial cases, as well as the combinations of values for the capacitance and inductance, by determining the ratio between the maximum voltage value attained in the supercapacitor, $max(V_{super\text{-}cap})$ and the total number of cycles of charge and discharge (i.e. *max_number_cycles*). These results for the average increase of the voltage per cycle in the supercapacitor. Here, again one does not consider the characteristic of the supercapacitor which implies that, as the energy builds up in the supercapacitor, it takes longer time to add more energy into it. As such, to obtain the intended metric, the calculated average increase of the voltage per cycle in the supercapacitor is multiplied by the total number of cycles considered in these tests, i.e. 2,000. Although in the previous sections we have just considered the first 2,000 cycles, in this section, we decided to use all the cycles registered for each combination of values for the components. Therefore, the calculated value for the voltage per cycle in this case could be divided by a higher value of registered cycles. The computation of the metric is presented in (14.18) and calculated as follows:

$$V_{super\text{-}cap\ norml} = 2{,}000 \times \frac{max[V_{super\text{-}cap}]}{max[number\_cycles]} \tag{14.18}$$

As shown in Figure 14.25, as the trial case number increases, $V_{super\text{-}cap\ norml}$ increases, for all the combination of values for the capacitance and inductance. This is because the time duration *delay_transistor3_int* becomes longer as the trial case number increases, which causes the inductor to spend more time transferring energy to the supercapacitor, as already observed in Figure 14.24. By comparing Figure 14.25 with Figure 14.24, one observes that the behaviour is similar. For the first four trial cases (1, 2, 3 and 4 from Table 14.6), the increase is not so notorious. Regarding the remaining trial cases, the increase rate occurs faster, except for the combination of values for the components $C_{pre\text{-}charge} = 220\ \mu F;\ L = 1.5\ mH$, which remain almost constant after trial case number 7.

By analysing Figure 14.25, we can conclude that by considering a larger pre-charge capacitor, with capacitance such as $C_{pre\text{-}charge} = 1{,}000\ \mu F$, allows for obtaining a higher increase of voltage per cycle of charge and discharge. We should keep in mind that the time it takes to charge the pre-charge capacitor is longer as the

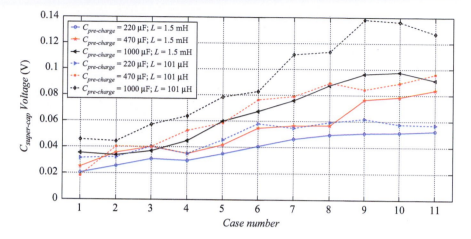

Figure 14.25   Normalized charging voltage in the $C_{super\text{-}cap}$ for different trial cases

capacitance increases. The best combinations of values for the components are $C_{pre\text{-}charge} = 1,000 \ \mu F$; $L = 101 \ \mu H$, and $C_{pre\text{-}charge} = 470 \ \mu F$; $L = 101 \ \mu H$, for a trial case number higher or equal than 6. Again, here one does not consider the characteristic of the supercapacitor which implies that, as the energy builds up in the supercapacitor, it takes longer time to add more energy into it.

14.4.4   Charging time of $C_{super\text{-}cap}$ with limited cycles

The charging time of $C_{super\text{-}cap}$ with limited cycles, $T_{super\text{-}capcharge}$,was derived by considering the data collected for the considered trials as well as the combinations of values for the capacitance and inductance.

To obtain the values presented in Figure 14.26, the MATLAB algorithm employed has analysed the data counts for each value acquired during an unit of time (i.e. 1 s). While not reaching the 2,000 cycles, the algorithm continues to sum the time elapsed. These charging times can be compared with the attained supercapacitor voltages, as shown in Figure 14.25.

Figure 14.26 presents the number of cycles the supercapacitor takes to charge and attain the voltages presented in Figure 14.25. It is observable that, for a pre-charge capacitor with a capacitance $C_{pre\text{-}charge} = 1,000 \ \mu F$, the number of cycles needed to charge and discharge is higher for $L = 101 \ \mu H$. However, by looking to Figure 14.25, this combination of values for the parameters also corresponds to the highest voltage in the supercapacitor. The higher number of cycles needed to attain the corresponding voltage in the supercapacitor is due to the time it takes to charge the pre-charge capacitor until it reaches the minimum voltage to initiate the process of transferring the energy to the inductor, and then to the supercapacitor(s) bank.

Additionally, as the trial case number increases, the number of cycles increases, for all the combinations of the values for the capacitance and inductance, as expected. This is due to the time duration *delay_transistor3_int*, which becomes longer as

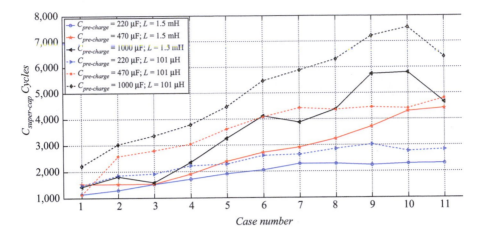

Figure 14.26   Supercapacitor charging time for 2,000 cycles

the trial case number increases and causes the inductor to spend more time transferring energy to the supercapacitor. The combination of values for the components $C_{pre\text{-}charge} = 470\,\mu F; L = 101\,\mu H$ and $C_{pre\text{-}charge} = 1{,}000\,\mu F; L = 1.5\,mH$ for the trial cases number 6 and 8 are the most suitable for our energy storing system, since there is a well-balanced trade-off between the achieved voltage that is build up in the supercapacitor(s) bank, and the time it takes to achieve that voltage in the supercapacitor(s) bank. Other combinations of values for the component and trial cases present no only longer or shorter times to charge but also higher or lower values of voltage achieved in the supercapacitor(s) bank, respectively, which may not lead to an appropriate trade-off between charging time and achieved voltage in the supercapacitor(s) bank.

14.4.5   Charging time per cycle of $C_{super\text{-}cap}$

Another metric that is useful to analyse for the considered trials is the average charging time spent per cycle ($T_{super\text{-}cap\ charge\ cycle}$) cases as well as the combination of values for the capacitance and inductance corresponding to the charging time of the $C_{super\text{-}cap}$.

To obtain the values presented in Figure 14.27, the MATLAB algorithm employed to analyse the data counts for each cycle considers the time spent to complete a cycle of charge and discharge.

Figure 14.27 presents the charging time spent to complete a cycle of charging and discharging the pre-charge capacitor (i.e. $C_{pre\text{-}charge}$). It is observable that for the component value combinations with the $C_{pre\text{-}charge} = 1{,}000\,\mu F$, the time it takes to complete the one cycle of charge and discharge is longer in the case of an inductance value of $L = 101\,\mu H$.

The combinations of values for the components $C_{pre\text{-}charge} = 470\,\mu F; L = 101\,\mu H$ and $C_{pre\text{-}charge} = 1{,}000\,\mu F; L = 1.5\,mH$ for the trial case number 6 and 8 are the ones that are more suitable for our energy storing system since the time spent is

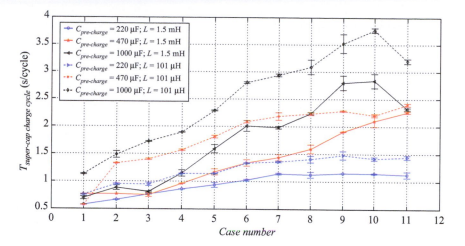

Figure 14.27   Charging time per cycle of $C_{super-cap}$ for different trial cases with 95% confidence intervals

acceptable (in the range in between 2 and 3 s/cycle). In addition, for the combination of component values $C_{pre-charge} = 470$ μF; $L = 101$ μH the time spent per cycle is approximately the same as the trial case number increases from 6 to 11. For trial cases lower than 6, the time spent to complete a cycle is shorter than the other trial cases but the attained variation of voltage acquired per cycle is also lower. The confidence intervals presented in Figure 14.27 have a short interval range, which corresponds to high confidence in the results and means that the averaged value presented in Figure 14.27 does not differ from the data collected during the different test.

14.4.6 $C_{super-cap}$ voltage variation per cycle

Following the analysis of the data collected for the charging time of the $C_{super-cap}$ for different combinations of values for the capacitance and inductance values, another useful metric is the voltage variation achieved per cycle in the supercapacitor, ΔV_{cycle}.

This metric was obtained by developing a data analysis algorithm in MATLAB that sums all the variations of voltage in the supercapacitor that occur during a cycle of charge and discharge of the pre-charge capacitor. In the end, it computes the average for each trial case and combination of capacitance and inductance values.

Figure 14.28 presents the voltage variation per cycle in the supercapacitor. It is observable that as the trial case number increases, the voltage variation for each combination of values for the components also increases, as expected. By comparing and associating the voltage variation per cycle from Figure 14.28 with the charging time per cycle of $C_{super-cap}$ from Figure 14.27, for the selected combinations of values from the components, one concludes that there is an adequate trade-off between the time spent to complete a cycle of charge and discharge and the voltage variation achieved during the cycle in the supercapacitor(s) bank.

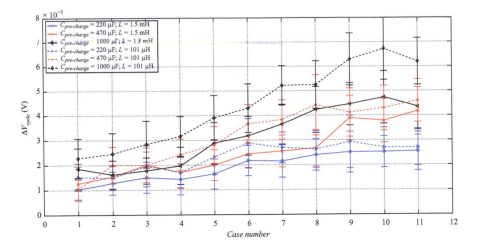

Figure 14.28   Voltage variation achieved per cycle in the $C_{super\text{-}cap}$ for different trial cases, with the representation of the 95% confidence intervals

For the trial case number 6 and combination of component values $C_{pre\text{-}charge} = 470\,\mu F$; $L = 101\,\mu H$, $T_{super\text{-}cap charge cycle} \approx 2.0771$ s, which corresponds to $\Delta V_{cycle} \approx 3.654 \times 10^{-5}$ V. In addition, for the trial case number 6 and combination of component values $C_{pre\text{-}charge} = 1{,}000\,\mu F$; $L = 1.5$ mH; $T_{super\text{-}cap charge cycle} \approx 2.077$ s corresponds to $\Delta V_{cycle} \approx 3.177 \times 10^{-5}$ V. In turn, for the trial case number 8 and combination of component values $C_{pre\text{-}charge} = 470\,\mu F$; $L = 101\,\mu H$ and $C_{pre\text{-}charge} = 1{,}000\,\mu F$; $L = 1.5$ mH, $T_{super\text{-}cap charge cycle} \approx 2.2239$ s for both combinations, but the voltage variation is $\Delta V_{cycle} \approx 4.4 \times 10^{-5}$ V and $\Delta V_{cycle} \approx 4.23 \times 10^{-5}$ V, respectively. In addition, the combination $C_{pre\text{-}charge} = 1{,}000\,\mu F$; $L = 101\,\mu F$, it may be an adequate choice if the intention is to transfer more energy each cycle, even though it takes more time to complete a cycle of charge and discharge of the pre-charge capacitor.

14.4.7   *Time and cycles estimation to attain different $C_{super\text{-}cap}$ voltages without harvesting device*

In order to estimate the time and number cycles needed to attain a certain voltage in the $C_{super\text{-}cap}$, we developed an algorithm that considers, as inputs, the voltage variation achieved per cycle, ΔV_{cycle}, presented in Figure 14.28, as well as the charging time per cycle, $T_{super\text{-}cap charge cycle}$, as shown in Figure 14.27. The algorithm finds the total time needed to achieve the different $C_{super\text{-}cap}$ voltage and the total number of cycles needed. The considered voltages for these time and cycles estimation are the following ones: $V_{super\text{-}cap max} = \{1.8, 2.0, 2.2\}$ V.

This estimation is useful to obtain insights regarding the estimated time needed to attain a certain voltage in supercapacitor(s) bank. Here, the characteristic of the supercapacitor is not considered, which implies that, as the energy builds up in

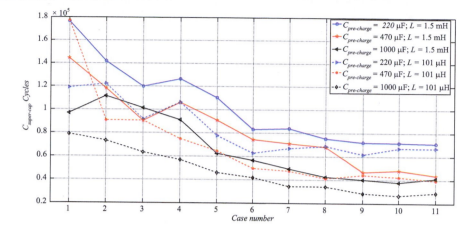

Figure 14.29 *Number of cycles estimation to attain* $V_{super\text{-}cap\,max} = 1.8$ V
in the $C_{super\text{-}cap}$

the supercapacitor, it takes longer time to add more energy into it. This analysis is performed for different combinations of values for the capacitance and inductance.

Figure 14.29 presents an estimation for the number of cycles, $E(C_{super\text{-}cap\,cycles})$, needed to achieve $V_{super\text{-}cap\,max} = 1.8$ V in the supercapacitor(s) bank, for different combinations of capacitance and inductance. It is observable that as the trial case number increases, the number of cycles needed to attain the desirable voltage in the supercapacitor(s) bank decreases. This decrease is explained by fact that the time duration *delay_transistor3_int* becomes longer as the trial case number increases which causes the inductor to spend more time transferring energy to the supercapacitor, leading to a higher voltage variation per cycle.

The maximum the number of cycles observed in Figure 14.29 corresponds to the combinations of component values $C_{pre\text{-}charge} = 470\,\mu\text{F}$; $L = 101\,\mu\text{H}$ and $C_{pre\text{-}charge} = 220\,\mu\text{F}$; $L = 1.5$ mH for the trial case number 1 with a $E(C_{super\text{-}cap\,cycles}) = 1.768 \times 10^5$ cycles which matches an estimated time $E(C_{super\text{-}cap\,ChargeTime}) = 9.925 \times 10^4$ s ($\approx$27.6 h), as shown in Figure 14.30.

Considering the combination of values already chosen in the previous metrics analysis, i.e. $C_{pre\text{-}charge} = 470\,\mu\text{F}$; $L = 101\,\mu\text{H}$ and $C_{pre\text{-}charge} = 1{,}000\,\mu\text{F}$; $L = 1.5$ mH, some lessons can be learned for the estimated number of cycles and time. For the combination of values for the components $C_{pre\text{-}charge} = 470\,\mu\text{F}$; $L = 101\,\mu\text{H}$ and trial case number 6, $E(C_{super\text{-}cap\,cycles}) = 4.925 \times 10^4$, which matches an estimated time of $E(C_{super\text{-}cap\,ChargeTime}) = 10.29 \times 10^4$ s ($\approx$28.6 h) from Figure 14.30, whereas for the combination of values for the component $C_{pre\text{-}charge} = 1{,}000\,\mu\text{F}$; $L = 1.5$ mH, $E(C_{super\text{-}cap\,cycles}) = 5.666 \times 10^4$ cycles, which corresponds to an estimated time of $E(C_{super\text{-}cap\,ChargeTime}) = 11.65 \times 10^4$ s ($\approx$32.5 h).

In turn, for the trial case number 8, the combinations of values for the component $C_{pre\text{-}charge} = 470\,\mu\text{F}$; $L = 101\,\mu\text{H}$ presents an estimated number of cycles $E(C_{super\text{-}cap\,cycles}) = 4.079 \times 10^4$, which matches an estimated time

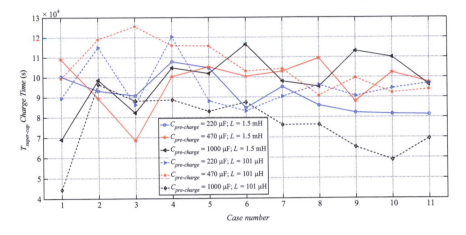

Figure 14.30 Time estimation to attain $V_{super\text{-}cap\,max} = 1.8\,V$ in $C_{super\text{-}cap}$

Table 14.7 Estimated charging time and number of cycles for $V_{super\text{-}cap\,max} = 1.8\,V$

$C_{pre\text{-}charge}$	L	$E[C_{super\text{-}cap\,ChargeTime}]$	$E[C_{super\text{-}cap\,cycles}]$
470 μF	101 μH (trial case 1)	9.925×10^4 s	1.768×10^4
220 μF	1.5 mH (trial case 1)	9.925×10^4 s	1.768×10^4
470 μF	101 μH (trial case 6)	10.29×10^4 s	4.925×10^4
1,000 μF	1.5 mH (trial case 6)	11.65×10^4 s	5.666×10^4
470 μF	101 μH (trial case 8)	9.073×10^4 s	4.079×10^4
1,000 μF	1.5 mH (trial case 8)	9.521×10^4 s	4.249×10^4

$E(C_{super\text{-}cap\,ChargeTime}) = 9.073 \times 10^4$ s ($\approx$25.2 h), as shown in Figure 14.30, whereas for $C_{pre\text{-}charge} = 1,000\ \mu F$; $L = 1.5$ mH, $E(C_{super\text{-}cap\,cycles}) = 4.249 \times 10^4$, which corresponds to an estimated time of $E(C_{super\text{-}cap\,ChargeTime}) = 9.521 \times 10^4$ s ($\approx$26.4 h). These results are shown in Table 14.7 ($V_{super\text{-}cap\,max} = 1.8$ V) (Figure 14.31).

Figure 14.32 shows an estimation for the number of cycles ($E(C_{super\text{-}cap\,cycles})$) needed to achieve $V_{super\text{-}cap\,max} = 2.0$ V in the supercapacitor(s) bank, for different combinations of the values for the capacitance and inductance.

The maximum number of cycles observed in Figure 14.32 corresponds to the combinations of values for the component $C_{pre\text{-}charge} = 470\ \mu F$; $L = 101\ \mu H$ and $C_{pre\text{-}charge} = 220\ \mu F$; $L = 1.5$ mH, for the trial case number 1, with $E(C_{super\text{-}cap\,cycles}) = 1.964 \times 10^5$, which matches an estimated time of $E(C_{super\text{-}cap\,ChargeTime}) = 1.103 \times 10^5$ s ($\approx$30.6 h), as shown in Figure 14.32.

For the chosen combinations of values for the components $C_{pre\text{-}charge} = 470\ \mu F$; $L = 101\ \mu H$ and $C_{pre\text{-}charge} = 1,000\ \mu F$; $L = 1.5$ mH, from the previous analysis of the metrics, the estimated number of cycles and time to attain $V_{super\text{-}cap\,max} = 2.0$ V in the $C_{super\text{-}cap}$ will increase, as expected, compared to the case of attaining $V_{super\text{-}cap\,max} = 1.8$ V in the supercapacitor(s) bank. For the combinations of values for the components

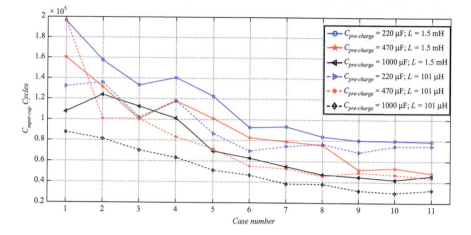

Figure 14.31   Number of cycles estimation to attain $V_{super\text{-}cap\,max} = 2.0\,V$ in $C_{super\text{-}cap}$

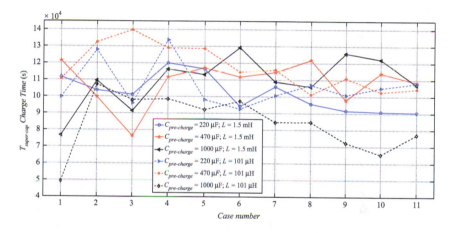

Figure 14.32   Time estimation to attain $V_{super\text{-}cap\,max} = 2.0\,V$ in $C_{super\text{-}cap}$

$C_{pre\text{-}charge} = 470$ μF; $L = 101$ μH and trial case number 6, $E(C_{super\text{-}cap\,cycles}) = 5.473 \times 10^4$, which matches an estimated time $E(C_{super\text{-}cap\,ChargeTime}) = 11.143 \times 10^4$ s ($\approx$31.8 h), as shown in Figure 14.32, whereas, for the combination of values for the components $C_{pre\text{-}charge} = 1{,}000$ μF; $L = 1.5$ mH, $E(C_{super\text{-}cap\,cycles}) = 6.295 \times 10^4$, which corresponds to an estimated time of $E(C_{super\text{-}cap\,ChargeTime}) = 12.93 \times 10^4$ s ($\approx$36.9 h).

In turn, for the case trial number 8, the combination of values for the components $C_{pre\text{-}charge} = 470$ μF; $L = 101$ μH presents an estimated number of cycles $E(C_{super\text{-}cap\,cycles}) = 4.533 \times 10^4$, which matches an estimated time $E(C_{super\text{-}cap\,ChargeTime}) = 10.09 \times 10^4$ s ($\approx$28 h), as shown in Figure 14.32, whereas, for the combinations of values for the components $C_{pre\text{-}charge} = 1{,}000$ μF;

Table 14.8   Estimated charging time and number of cycles for $V_{super\text{-}cap\,max} = 2.0\,V$

$C_{pre\text{-}charge}$	L	$E[C_{super\text{-}cap\,ChargeTime}]$	$E[C_{super\text{-}cap\,cycles}]$
470 μF	101 μH (trial case 1)	11.03×10^4 s	1.964×10^4
220 μF	1.5 mH (trial case 1)	11.03×10^4 s	1.964×10^4
470 μF	101 μH (trial case 6)	11.143×10^4 s	5.473×10^4
1,000 μF	1.5 mH (trial case 6)	12.93×10^4 s	6.295×10^4
470 μF	101 μH (trial case 8)	10.09×10^4 s	4.533×10^4
1,000 μF	1.5 mH (trial case 8)	10.58×10^4 s	4.721×10^4

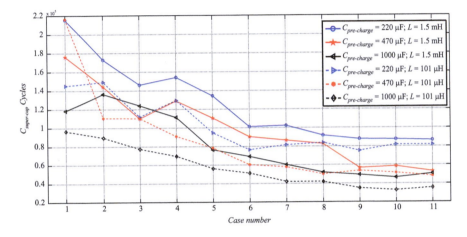

Figure 14.33   Number of cycles estimation to attain $V_{super\text{-}cap\,max} = 2.2\,V$ in $C_{super\text{-}cap}$

$L = 1.5$ mH, $E(C_{super\text{-}cap\,cycles}) = 4.721 \times 10^4$, which corresponds to an estimated time of $E(C_{super\text{-}cap\,ChargeTime}) = 10.58 \times 10^4$ s ($\approx$29.4 h). These results are shown in Table 14.8 ($V_{super\text{-}cap\,max} = 2.0$ V).

Figure 14.33 presents an estimation for the number of cycles ($E(C_{super\text{-}cap\,cycles})$) needed to achieve $V_{super\text{-}cap\,max} = 2.2$ V in the supercapacitor(s) bank for different combinations of the values for the capacitance and inductance.

The maximum the number of cycles observed in Figure 14.33 matches the component value combinations $C_{pre\text{-}charge} = 470$ μF; $L = 101$ μH and $C_{pre\text{-}charge} = 220$ μF; $L = 1.5$ mH for the trial case number 1 with a $E(C_{super\text{-}cap\,cycles}) = 2.16 \times 10^5$ which corresponds to an estimated time of $E(C_{super\text{-}cap\,ChargeTime}) = 1.213 \times 10^5$ s ($\approx$33.7 h) from Figure 14.34.

Considering the chosen combinations of values for the components $C_{pre\text{-}charge} = 470$ μF; $L = 101$ μH, $C_{pre\text{-}charge} = 1,000$ μF; $L = 1.5$ mH and $C_{pre\text{-}charge} = 1,000$ μF; $L = 101$ μF, from the previous choices for the metrics, the estimated number of cycles and time to attain $V_{super\text{-}cap\,max} = 2.2$ V will increase compared to the cases where $V_{super\text{-}cap\,max} = 1.8$ and 2.0 V were attained in the supercapacitor(s)

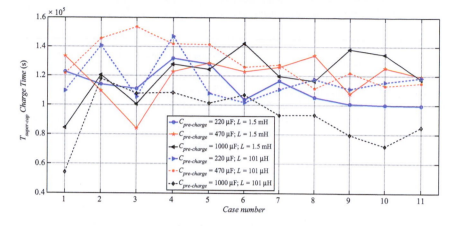

Figure 14.34   Time estimation to attain $V_{super\text{-}cap\,max} = 2.2$ V in $C_{super\text{-}cap}$

bank, as expected. For the combinations of values for the components $C_{pre\text{-}charge} = 470$ µF; $L = 101$ µH and trial case number 6, $E(C_{super\text{-}cap\,cycles}) = 6.02 \times 10^4$, which matches an estimated time $E(C_{super\text{-}cap\,ChargeTime}) = 12.57 \times 10^4$ s ($\approx$34.9 h), as shown in Figure 14.34. In turn, for the combinations of values for the components $C_{pre\text{-}charge} = 1{,}000$ µF; $L = 1.5$ mH, $E(C_{super\text{-}cap\,cycles}) = 6.925 \times 10^4$, which corresponds to an estimated time $E(C_{super\text{-}cap\,ChargeTime}) = 14.22 \times 10^4$ s ($\approx$39.5 h), whereas, for the combination of values for the components $C_{pre\text{-}charge} = 1{,}000$ µF; $L = 101$ µF, $E(C_{super\text{-}cap\,cycles}) = 5.118 \times 10^4$, which corresponds to an estimated time $E(C_{super\text{-}cap\,time}) = 10.69 \times 10^4$ s ($\approx$29.7 h), whereas, for the trial case number 8, the combination of values for the components $C_{pre\text{-}charge} = 470$ µF; $L = 101$ µH presents an estimated number of cycle $E(C_{super\text{-}cap\,cycles}) = 4.986 \times 10^4$, which matches an estimated time $E(C_{super\text{-}cap\,ChargeTime}) = 11.1 \times 10^4$ s ($\approx$30.8 h), as shown in Figure 14.34. In turn, for the combination of values for the component $C_{pre\text{-}charge} = 1{,}000$ µF; $L = 1.5$ mH, $E(C_{super\text{-}cap\,cycles}) = 5.193 \times 10^4$, which corresponds to an estimated time $E(C_{super\text{-}cap\,ChargeTime}) = 11.64 \times 10^4$ s ($\approx$32.3 h). Finally, for the combination of values for the components $C_{pre\text{-}charge} = 1{,}000$ µF; $L = 101$ µF, $E(C_{super\text{-}cap\,cycles}) = 4.191 \times 10^4$ which corresponds to an estimated time $E(C_{super\text{-}cap\,ChargeTime}) = 9.33 \times 10^4$ s ($\approx$25.9 h). These results are shown in Table 14.9 ($V_{super\text{-}cap\,max} = 2.2$ V).

One can learn from this analysis that $L = 101$ µH and $L = 1.5$ mH are the values for the inductance that present the most suitable trade-off between the time it takes and energy transfer to charge until the pre-charge capacitor attains the required voltage (1.8, 2.0 or 2.2 V).

From the two pre-chosen component value combinations, the one that presents best results in terms of time required to attain the desirable voltage in the supercapacitor(s) bank is the combination of component values $C_{pre\text{-}charge} = 470$ µF; $L = 101$ µH. However, the combination of component values $C_{pre\text{-}charge} = 1{,}000$ µF; $L = 101$ µH for trial cases higher than 6 presents adequate results for the time estimation to attain

Table 14.9   Estimated charging time and number of cycles for $V_{super\text{-}cap\,max} = 2.2\,V$

$C_{pre\text{-}charge}$	L	$E[C_{super\text{-}cap\,Charge\,Time}]$	$E[C_{super\text{-}cap\,cycles}]$
470 μF	101 μH (trial case 1)	12.13×10^4 s	2.16×10^4
220 μF	1.5 mH (trial case 1)	12.13×10^4 s	2.16×10^4
470 μF	101 μH (trial case 6)	12.57×10^4 s	6.02×10^4
1,000 μF	1.5 mH (trial case 6)	14.22×10^4 s	6.925×10^4
470 μF	101 μH (trial case 8)	11.1×10^4 s	4.986×10^4
1,000 μF	1.5 mH (trial case 8)	11.64×10^4 s	5.193×10^4
1,000 μF	101 μH (trial case 8)	9.33×10^4 s	4.191×10^4

$V_{super\text{-}cap\,max} = 1.8$ V in $C_{super\text{-}cap}$. Nevertheless, we should keep in mind that the time needed to charge the pre-charge capacitor is longer than for smaller capacitances.

14.4.8   Time estimation to attain different $C_{super\text{-}cap}$ voltages (with harvesting device)

In this section, in order to estimate the time and number cycles needed to attain a certain voltage in the $C_{super\text{-}cap}$, we present an algorithm that considers as inputs the voltage variation achieved per cycle (ΔV_{cycle}), as shown in Figure 14.28, and the charging time per cycle ($T_{super\text{-}cap\,charge\,cycle}$) shown in Figure 14.27, along with the time it takes to charge the pre-charge capacitor, as shown in Figure 14.6. The algorithm finds the total time needed to achieve the different values of the $C_{super\text{-}cap}$ voltage, and the total number of cycles needed. The considered voltages for these time and cycles estimation are $V_{super\text{-}cap\,max} = \{1.8, 2.0, 2.2\}$ V. For the estimation of cycles and time, we just considered the combination of values for the components $C_{pre\text{-}charge} = 220\,\mu$F; 470 μF and $L = 101\,\mu$H; 1.5 mH, since, in Figure 14.6, the considered pre-charge capacitances are the ones that present the most suitable trade-off between the time it takes and energy transfer to charge until it attain a voltage of 1.8 V and the energy it can transfer to the inductor.

This estimation is useful to obtain insights regarding the estimated time needed to attain a certain voltage in the supercapacitor(s) bank whilst considering the time it takes to charge the pre-charge capacitor until a minimum voltage of 1.8 V when the energy harvester is placed in a *Low traffic Ambient*. Here, the characteristic of the supercapacitor, whose energy build up takes longer time to be added into it, is not taken into account. This analysis is performed for different combination of values for the capacitance and inductance.

Figures 14.35 and 14.36 addresses the time and number of cycles estimation, respectively, to attain $V_{super\text{-}cap\,max}$ in the $C_{super\text{-}cap}$ with harvesting device when placed in a *Low Ambient Traffic* scenario.

The maximum charging time is achieved when considering the trial case number 1 for the combination of values for the components that consider $C_{pre\text{-}charge} = 470\,\mu$F; $L = 101\,\mu$H, which present the charging time estimation of $E(C_{super\text{-}cap\,time}) = 2.358 \times 10^9$ s, 2.62×10^9 s and 2.882×10^9 s for the $V_{super\text{-}cap\,max} = 1.8$, 2.0 and

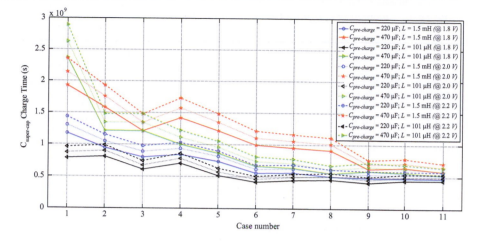

Figure 14.35 Time estimation to attain $V_{super\text{-}cap\,max}$ in the $C_{super\text{-}cap}$ with harvesting device

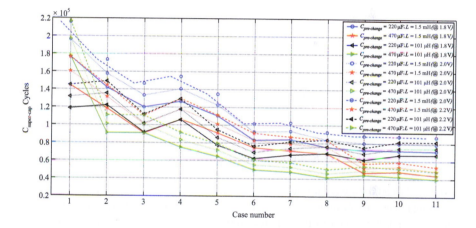

Figure 14.36 Number of cycles estimation to attain $V_{super\text{-}cap\,max}$ in the $C_{super\text{-}cap}$ with harvesting device

2.2 V, respectively. The maximum estimated number of cycles $E(C_{super\text{-}cap\,cycles})$ for the same combinations of values for the components observed in Figure 14.36 are $E(C_{super\text{-}cap\,cycles}) = 1.768 \times 10^5$, 1.964×10^5 and 2.16×10^5, for $V_{super\text{-}cap\,max} = 1.8$, 2.0 and 2.2 V, respectively. As the trial case number increases, the charging time estimation $E(C_{super\text{-}cap\,time})$ decreases.

By observing Figures 14.35 and 14.36, the combination of values for the components $C_{pre\text{-}charge} = 220\ \mu F$ and $470\ \mu F$; $L = 101\ \mu H$ are the ones that present the lowest values for the estimated charging time and number of cycles for each one of the voltages of $V_{super\text{-}cap\,max}$.

Table 14.10   Estimated charging time and number of cycles for trial case 6

		Trial case 6	
$C_{pre-charge}$	L	$E[C_{super-cap\ time}]$	$E[C_{super-cap\ cycles}]$
470 μF	1.5 mH (@ 1.8 V)	9.859×10^8 s (273,861 h)	4.925×10^4
470 μF	101 μH (@ 1.8 V)	6.569×10^8 s (182,472 h)	4.925×10^4
470 μF	1.5 mH (@ 2.0 V)	1.095×10^9 s (304,167 h)	5.473×10^4
470 μF	101 μH (@ 2.0 V)	7.299×10^8 s (202,750 h)	5.473×10^4
470 μF	1.5 mH (@ 2.2 V)	1.205×10^9 s (334,722 h)	6.02×10^4
470 μF	101 μH (@ 2.2 V)	8.029×10^8 s (223,028 h)	6.02×10^4

Table 14.11   Estimated charging time and number of cycles for trial case 8

		Trial case 8	
$C_{pre-charge}$	L	$E[C_{super-cap\ time}]$	$E[C_{super-cap\ cycles}]$
470 μF	1.5 mH (@ 1.8 V)	9.009×10^8 s (250,250 h)	6.754×10^4
470 μF	101 μH (@ 1.8 V)	5.441×10^8 s (151,139 h)	4.079×10^4
470 μF	1.5 mH (@ 2.0 V)	1.001×10^9 s (278,056 h)	7.622×10^4
470 μF	101 μH (@ 2.0 V)	6.087×10^8 s (169,083 h)	4.533×10^4
470 μF	1.5 mH (@ 2.2 V)	1.101×10^9 s (305,833 h)	8.255×10^4
470 μF	101 μH (@ 2.2 V)	6.65×10^8 s (184,722 h)	4.986×10^4

In Figure 14.36, the combination of values for the components that consider the $C_{pre-charge} = 470$ μF and $L = 101$ μH are the cases when the number of cycles estimated to attain the desirable $V_{super-cap\ max}$ voltage is lower.

Since trial cases number 6 and 8 are the ones that present best trade-off between time and energy transfer to the supercapacitor(s) bank, Tables 14.10 and 14.11 summarize the estimated charging time along with the estimated cycles for these values of the components.

By observing the values obtained for each combination of values for the components, for the trial cases number 6 and 8, when considering a harvesting device placed in a *Low Ambient Traffic* scenario, one concludes that it takes long time to charge a supercapacitor(s) bank for any $V_{super-cap\ max}$ voltage. The results obtained here can be considered as the worst case scenario in which the maximum estimated charging times that the energy storing system needs to attain the desirable $V_{super-cap\ max}$ voltage in the supercapacitor(s) bank is presented.

Therefore, an appropriate good option to follow, when *Low Ambient Traffic* scenario is the most common to happen, is to consider to charge the supercapacitor initially and then to use the harvested energy to maintain the voltage of the supercapacitor(s) bank more or less constant, while it supplies the CRN.

If the EH device is located in *High Ambient Traffic* scenario, the time it takes to attain the desirable $V_{super-cap\,max}$ voltage in the supercapacitor(s) bank is surely reduced drastically, due to the quick charge of the pre-charge capacitor, which is the most time-consuming fraction of the total time the energy storing system needs to attain the desirable voltage in the supercapacitor(s) bank. Approximately 80%–90% of the time estimated to attain the $V_{super-cap\,max}$ voltage in the supercapacitor(s) bank voltage depends on the time it takes to accumulate energy in the pre-charge capacitor by means of the EH device.

14.4.9 Implications of supercapacitors in series or parallel configurations

As stated before, supercapacitors are becoming more frequent in applications where electrical energy needs to be stored. Since these components can be charged and discharged many times, it can outlast a battery. In addition, in terms of lifetime, the supercapacitor presents longer lifetime than a battery. A supercapacitor is often chosen to supply power to low current load for many hours at a time or to supply high current loads but within shorter time intervals. Aside from the fact that the supercapacitor can be charged very quickly due to their low internal resistance, which is known as ESR, they can be discharged just as quickly as they are charged.

Supercapacitors do not pose hazard dangers like the lead acid batteries, but they cannot store as much power either. However, supercapacitors can be placed in series or in parallel to either increase the maximum charge voltage or the total capacitance of the supercapacitors bank.

In terms of voltage ratings, supercapacitors have very small voltage ratings, such as 2.5, 2.7 and 5.5 V. Due to this, some supercapacitors configurations should be made in order to make supercapacitors capable of charging up to high voltages, i.e. they should be placed in series. There are sections coming up on series/parallel configurations, as well as charging methods, and balancing methods.

Figure 14.37 presents the typical connection of supercapacitors in series, whereas Figure 14.38 shows the connection of supercapacitors in parallel.

Regarding the choice on the connections of the supercapacitors in series or parallel, it all depends on the need of either more capacitance or higher voltage at the terminals of the supercapacitor bank. If the option is to increase the capacitance of the supercapacitor bank while maintaining the maximum charge voltage of the supercapacitors bank, then the supercapacitors should be connected in parallel, as shown in Figure 14.38. In turn, if the option is to increase the maximum charge voltage of the supercapacitors bank, then the supercapacitors should be connected in series as shown in Figure 14.37.

The calculations of the energy in the supercapacitor is given by the following equation:

$$E = \frac{1}{2} \times CV^2$$

(14.19)

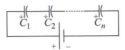

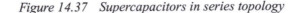

Figure 14.37   Supercapacitors in series topology

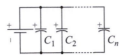

Figure 14.38   Supercapacitors in parallel topology

As an example, we consider a bank of 5 supercapacitors with capacitance $C = 10$ F each and a maximum charge voltage $V = 2.5$ V. Equation (14.20) enables to calculate the energy associated to one supercapacitor, as follows:

$$E = \frac{1}{2} \times 10 \times 2.5^2 = 31.25 \text{ J} \tag{14.20}$$

The energy can be converted to units by dividing it by 3.6. Therefore, the associated charge units of one 10 F supercapacitor is 31.25/3.6 = 8.6 mA h. Now, considering a bank of five supercapacitors connected in series topology, the total capacitance is calculated as follows:

$$C_{total\ series} = \frac{1}{(1/C_1) + (1/C_2) + (1/C_3) + (1/C_4) + (1/C_5)}$$

$$= \frac{1}{0.1 + 0.1 + 0.1 + 0.1 + 0.1} = 2 \text{ F} \tag{14.21}$$

The computation of the total maximum charge voltage of the supercapacitor bank is achieved by summing each supercapacitor maximum charge voltages, as follows:

$$V_{total\ series} = V_1 + V_2 + V_3 + V_4 + V_5 = 2.5 + 2.5 + 2.5 + 2.5 + 2.5 = 12.5 \text{ V} \tag{14.22}$$

Now, to know the energy that is stored in this supercapacitors bank, the computation is performed in (14.23), assuming the values obtained in (14.21) and (14.22), as follows:

$$E_{bank\ series} = \frac{1}{2} \times 2 \times 12.5^2 = 156.25 \text{ J} \tag{14.23}$$

In turn, the associated charge units of five 10 F supercapacitors in series is 156.25/3.6=43.40 mA h.

If we consider a bank of five supercapacitors connected in parallel topology, the total capacitance is calculated as follows:

$$C_{total\ parallel} = C_1 + C_2 + C_3 + C_4 + C_5 = 10 + 10 + 10 + 10 + 10 = 50 \text{ F} \tag{14.24}$$

The total maximum charge voltage of the supercapacitors bank is given by lowest maximum charge voltage value from the supercapacitors bank as follows, which in our case is $V_{total\,parallel} = 2.5$ V, since all the supercapacitors are equal. The energy that is stored in this supercapacitors bank is given by (14.25), assuming the values obtained in (14.24) and that $V_{total\,parallel} = 2.5$ V, as follows:

$$E_{bank\,parallel} = \frac{1}{2} \times 50 \times 2.5^2 = 156.25 \text{ J} \tag{14.25}$$

In turn, the associated charge units of five 10 F supercapacitors in parallel is 156.25/3.6=43.40 mA h.

If we assume that our embedded device that is going to be connected to the supercapacitors bank consumes $I_{supply} = 200$ mA, then the device holds for about 13 min with the supercapacitors bank. However, we should keep in mind that if the supercapacitors bank stops receiving energy from the buck–boost converter, the voltage at the terminals of the supercapacitors bank drops which leads to decreasing of the current supplying and therefore the calculated theoretical lifetime of 13 min will certainly decrease sharply. Therefore, it is important to keep transferring energy to the supercapacitors bank to keep approximately constant the voltage and current supplying to the embedded device.

Since supercapacitors can supply high current values to the embedded device, there is the need to limit the current that is supplied to the embedded device. To achieve this, a simple approach is done by dimensioning a resistor attached in series with the embedded device in order it limits the current being passed to the embedded device. Assuming the example that an embedded device consumes $I_{supply} = 200$ mA and the maximum supply voltage is $V_{supply} = 5$ V, then the resistor value is given by

$$R_{supply} = \frac{5}{200 \times 10^{-3}} = 25 \text{ }\Omega \tag{14.26}$$

Hence, we should compute the maximum power that should be supported by the resistor to limit the current to $I_{supply} = 200$ mA. This calculation is given as follows:

$$P_{supply} = R_{supply} \times 200 \times 10^{-3} = 0.25 \text{ W} \tag{14.27}$$

Some considerations that should be kept in mind is that if we have a 5.4 V supercapacitors bank and is being charged with a 5–5.4 V charge, then there is no problem in leaving the bank charging. Unlike batteries, the supercapacitors do not have the charge memory issue. When the supercapacitors bank is full, it will stop accepting energy. It will not damage the capacitors if charge is supplied to them, as long as the charging voltage is not higher than maximum charge voltage supported by the supercapacitors bank. Batteries, on the other hand, will lose integrity if the charger is left for too long. A good practice when designing a system that considers in the design a supercapacitor is that we should only charge the supercapacitor until 80%–90% of the referenced maximum charge voltage. This procedure prevents from destroying the supercapacitors.

Another consideration when connecting supercapacitors in series is that the voltages on each supercapacitor will vary mainly due to each individual leakage current.

It is highly recommended that the same capacitance values is used in supercapacitors banks in series topology. This is because if there simultaneously is a capacitor with higher capacitance and a capacitor with lower capacitance, they will discharge at different rates based on the load. This could lead to an effect called as voltage imbalance. If a voltage measuring is performed to each individual supercapacitor in a bank, then different voltages will be observed in each of them. Again, if we only charge to 80%–90% of the maximum charge voltage, different voltages on each of the supercapacitors will be observed, but they will be within the charge limit range.

Some issues are raised when associating supercapacitors with different maximum charge voltages. When charging the supercapacitors, this voltage should not be exceeded, or doing so can damage the component. In many applications, several supercapacitors are connected in series, to produce a supercapacitor bank with a higher voltage. But even if the proper charging voltage is used, the smallest supercapacitor in the string will charge up first. Hence, without a circuit to limit the voltage across each supercapacitor of the bank, the weakest supercapacitor in the series string will be overcharged as the rest of the supercapacitors in the string finish their charge. There are circuits in the literature that solve this overvoltage problem by balancing the supercapacitors bank with a voltage limiting circuit across each supercapacitor.

There are also circuits that use a battery and a supercapacitor along with a specific IC that allows to boost current from a small battery, by charging the supercapacitor with the battery. In such configuration, when charged, the load attached to the circuit can use a large amount of current in a short period of time.

14.4.10 Design of N-stage Dickson voltage multiplier for DTT band

As discussed in the Section 14.2.1, the development of a 5-stage Dickson voltage multiplier (with open and short circuit stub) was tested and considered as the EH device to supply energy to the supercapacitor storing system presented in Section 14.3. As mentioned in Chapter 2, from practical observation regarding the absence of improvement for $P_{RF} \leq -13$ dB m when more stages are considered, a 5-stage Dickson voltage multiplier and a 7-stage Dickson voltage multiplier are considered with two types of stubs (i.e. open and short circuit).

Figure 14.39 depicts the results from the simulations conducted to evaluate the impact of the RF-received power on the output DC voltage (V_{DC}) for a 5- and 7-stage Dickson voltage multipliers with a load impedance of 100 kΩ with impedance matching (with open and short circuit stub). It is observable that for $P_{RF} \leq 10$ dB m, the 5-stage Dickson voltage multiplier prototype presents higher output voltages than the 7-stage Dickson voltage multiplier prototype. However, for values of P_{RF} higher than 10 dB m, the output voltage values of the 7-stage Dickson voltage multiplier prototype are higher than the 5-stage Dickson voltage multiplier prototype as the RF received power increases.

The curves from Figure 14.40 represent the simulations results on the impact of the RF received power on the conversion efficiency (η_0) for a 5- and 7-stage Dickson voltage multipliers with a load impedance of 100 kΩ with impedance matching (with

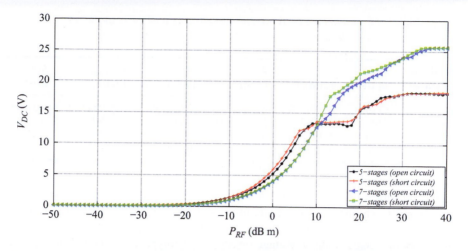

Figure 14.39 Simulation results on the impact of the RF received power on the output DC voltage (V_{DC}) for a 5- and 7-stage Dickson voltage multipliers with a load impedance of 100 kΩ with impedance matching (with open and short circuit stub)

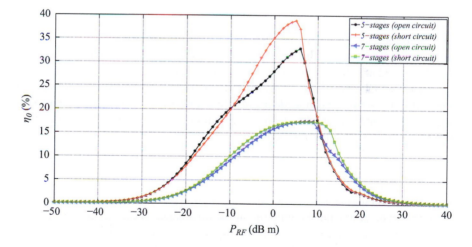

Figure 14.40 Simulation results on the impact of the RF received power on the conversion efficiency (η_0) for a 5- and 7-stage Dickson voltage multipliers with a load impedance of 100 kΩ with impedance matching (with open and short circuit stub)

Table 14.12   Open and short circuit stub dimensions and corresponding return losses

Number of stages	Type of stub	$W_{vertical}$ (mm)	$L_{vertical}$ (mm)	$W_{horizontal}$ (mm)	$L_{horizontal}$ (mm)	$S(1,1)$ (dB)
5	Open	3.359760	19.5534	3.359760	80.939100	−32.451
5	Short	3.359760	40.161200	3.359760	31.500600	−32.281
7	Open	3.359760	18.438	3.359760	113.247	−31.274
7	Short	3.359760	40.7037	3.359760	63.3221	−32.875

open and short circuit stub). It is shown that the 5-stage Dickson voltage multiplier prototypes (with open and short circuit stub) present higher conversion efficiency values than the 7-stage Dickson voltage multiplier prototypes. Additionally, by comparing both 5-stage Dickson voltage multiplier prototypes, the one that has a short circuit impedance matching circuit presents the highest value in terms of conversion efficiency, $\eta_0 \approx 38.8\%$ (@$P_{RF} = 5$ dB m). Hence, for lower values of RF received power (i.e. ≤ -10 dB m), the 5-stage Dickson voltage multiplier prototypes present significant conversion efficiencies compared to the 7-stage Dickson voltage multiplier prototypes. Only for higher values of RF received power (i.e. ≥ 10 dB m), the 7-stage Dickson voltage multiplier prototypes present conversion efficiency values higher than the ones from the 5-stage Dickson voltage multiplier prototypes.

Table 14.12 presents the dimensions of the stubs considered for each of the N-stage Dickson voltage multiplier prototypes, as well as the return losses associated to each prototype.

Figures 14.41 and 14.42 present the return losses of the 5-stage Dickson voltage multiplier with a load impedance of 100 kΩ (with open and short circuit stubs, respectively), and the obtained simulated values for the return losses, i.e. $S(1,1) = -32.451$ dB and $S(1,1) = -32.281$ dB, respectively.

Figures 14.43 and 14.44 present the return losses of the 7-stage Dickson voltage multiplier with a load impedance of 100 kΩ (with open and short circuit stubs, respectively), and the simulated values for return losses are $S(1,1) = -31.274$ dB and $S(1,1) = -32.875$ dB, respectively.

Figures 14.45 and 14.46 depict the ADS software circuit for the 5-stage Dickson voltage multiplier prototypes with a load impedance of 100 kΩ whilst considering an open and a short-circuit stub, respectively (as an example). Besides, these circuits already account for the different widths and lengths assumed in the PCB design of the prototype in order to compute with rigour the impedance of the circuit before adding the impedance matching circuit by means of the SmithChart and LineCalc utilities of the ADS software.

Figures 14.47 and 14.48 present the PCB layouts for the 5-stage Dickson voltage multiplier with open and short-circuit stub, respectively. It is observable in the figures the two layers considered for the prototype as well as the SMA connector that is going to be used to connect the antenna or the signal generator to evaluate the performance of the prototypes.

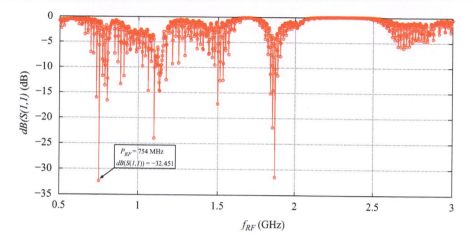

Figure 14.41   Frequency sweep procedure to determine the simulated return losses (S(1,1)) in a 5-stage Dickson voltage multiplier with a load impedance of 100 kΩ (with open circuit stub)

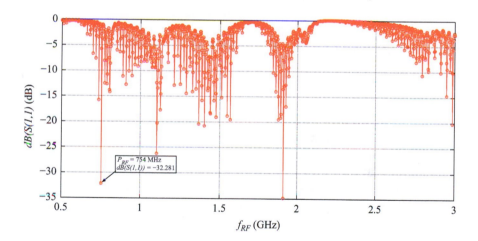

Figure 14.42   Frequency sweep procedure to determine the simulated return losses in a 5-stage Dickson voltage multiplier with a load impedance of 100 kΩ (with short circuit stub)

14.4.11   Experimental results for the 5- and 7-stage Dickson voltage multipliers

In order to evaluate the performance from the prototypes of the 5- and 7-stage Dickson voltage multipliers with match impedance within the 700 MHz DTT band, different experimental tests have been performed. These tests have comprised the use of a

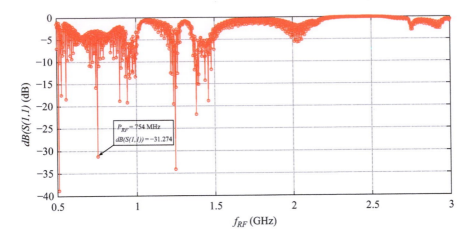

Figure 14.43   Frequency sweep procedure to determine the simulated return losses in a 7-stage Dickson voltage multiplier with a load impedance of 100 kΩ (with open circuit stub)

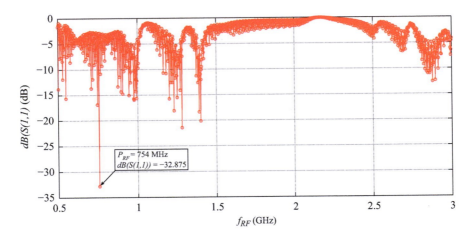

Figure 14.44   Frequency sweep procedure to determine the simulated return losses in a 7-stage Dickson voltage multiplier with a load impedance of 100 kΩ (with short circuit stub)

Rohde and Schwarz signal generator that enables to inject different values of RF power, in order to observe the DC voltage output.

Figures 14.49 and 14.50 show the simulation and experimental results for the output voltage and conversion efficiency, respectively, for the 5- and 7-stage Dickson voltage multipliers with match impedance (open and short circuit).

Figure 14.49 considers two levels for the output voltage that is going to be supplied to the pre-charge capacitor from the RF-EH with supercapacitor storing system,

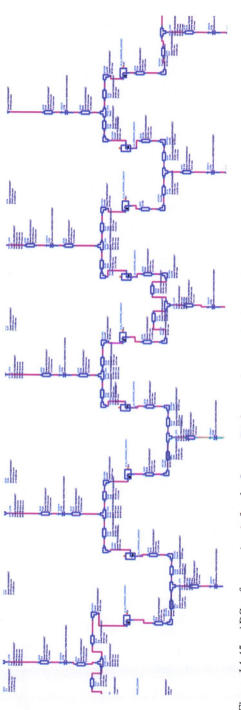

Figure 14.45   ADS software circuit for the 5-stage Dickson voltage multiplier with a load impedance of 100 kΩ with open circuit stub

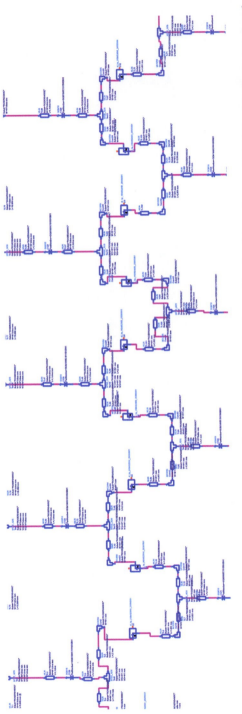

Figure 14.46 ADS software circuit for the 5-stage Dickson voltage multiplier with a load impedance of 100 kΩ with short circuit stub

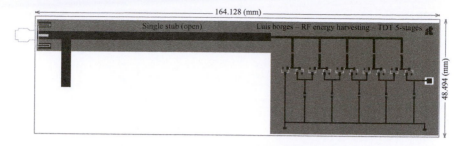

Figure 14.47   *PCB drawing for the 5-stage Dickson voltage multiplier with open circuit stub*

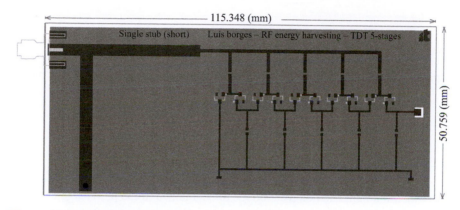

Figure 14.48   *PCB drawing for the 5-stage Dickson voltage multiplier with short circuit stub*

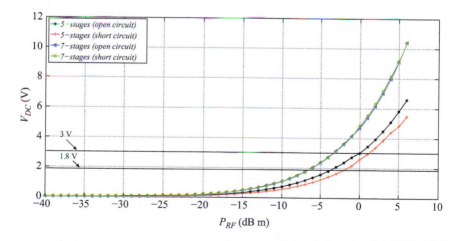

Figure 14.49   *Experimental results regarding the impact of the RF received power on output voltage for a 5- and 7-stage Dickson voltage multipliers with a load impedance of 100 kΩ with open and short circuit stubs*

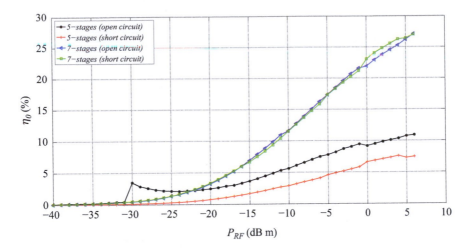

Figure 14.50   Experimental results for the impact of the RF received power on the conversion efficiency (η_0) for 5- and 7-stage Dickson voltage multipliers, with a load impedance of 100 kΩ, with open and short circuit stubs

i.e. 1.8 and 3 V, which are represented by solid black lines. These two levels represent the minimum and nominal voltage to be attained in the pre-charge capacitor, respectively.

By analysing Figure 14.49, we conclude that, by increasing the RF received power, we increase the output voltage. By considering the four prototypes, the maximum output voltage obtained experimentally is 6.59, 5.44, 10.39 and 10.37 V for the 5-stages (open circuit), 5-stages (short circuit), 7-stages (open circuit) and 7-stages (short circuit), respectively. These maximum voltages were attained at an RF received power of 6 dB m. By comparing the 5- and 7-stages prototypes, it is observable that the latter ones present higher output voltages than the former ones as the RF received power injected into the prototype increases. Regarding the two desirable levels to be attained by the prototypes, in order to supply the pre-charge capacitor from the RF-EH with supercapacitor storing system, results are summarized as follows. The 5-stages with open circuit stub reaches 1.8 V for $P_{RF} \geq -4$ dB m and reaches 3 V at $P_{RF} \geq 0$ dB m. The 5-stages with short circuit stub reaches 1.8 V for $P_{RF} \geq -2$ dB m and reaches 3 V at $P_{RF} \geq 1$ dB m. In turn, the 7-stages with open circuit stub reaches 1.8 V for $P_{RF} \geq -7$ dB m and reaches 3 V at $P_{RF} \geq -3$ dB m. Finally, the 7-stages with short circuit stub reaches 1.8 V for $P_{RF} \geq -4$ dB m and reaches 3 V at $P_{RF} \geq 0$ dB m. These results were expected, since the addition of more stages leads to the increase of the output voltage. Additionally, the prototypes with an open circuit stub present higher output voltages, compared to the prototypes with short circuit stubs.

Figure 14.50 depicts the conversion efficiency for the different 5- and 7-stages prototypes that were developed for the DTT band. In Figure 14.50, one observes

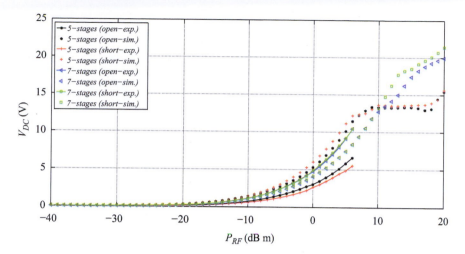

Figure 14.51 *Comparison of the experimental and simulation results for the impact of the RF received power on the output voltage for 5- and 7-stage Dickson voltage multipliers, with a load impedance of 100 kΩ, with open and short circuit stubs*

the maximum conversion efficiencies obtained experimentally are 10.9%, 7.6%, 27.1% and 27% for the 5-stages (open circuit), 5-stages (short circuit), 7-stages (open circuit) and 7-stages (short circuit), respectively. These maximum conversion efficiencies were attained at $P_{RF} = 6, 4, 6$ and 6 dB m, respectively. By comparing the 5- and 7-stage prototypes, it is observable that the latter ones present higher conversion efficiencies than the former ones, as the RF received power injected into the prototype increases, except for the 5-stages prototype (open circuit) that presents higher conversion efficiencies for the interval in between -30 and -23 dB m. The objective is to store the energy in our RF-EH with supercapacitor storing system (presented ahead in Section 4) and then supply the CRN. There are two levels of voltage (i.e. 1.8 and 3 V) that are desirable to be attained by the prototypes in order to fulfil the requirements. The results are summarized as follows: The 5-stage multiplier with open circuit stub reaches 1.8 V for $P_{RF} \geq -4$ dB m and reaches 3 V at $P_{RF} \geq 0$ dB m. The 5-stage multiplier with short circuit stub reaches 1.8 V for $P_{RF} \geq -2$ dB m and reaches 3 V at $P_{RF} \geq 1$ dB m. Regarding the 7-stage multiplier with open circuit stub reaches 1.8 V for $P_{RF} \geq -7$ dB m and reaches 3 V at $P_{RF} \geq -3$ dB m, whereas the 7-stages multiplier with short circuit stub reaches 1.8 V for $P_{RF} \geq -4$ dB m and reaches 3 V at $P_{RF} \geq 0$ dB m. These results were expected, since the addition of more stages leads to the increase of the conversion efficiency.

In order to compare the experimental results with the ones obtained by means of simulations, Figure 14.51 summarizes all these results.

To facilitate the analysis of the experimental results against the simulations ones, Figure 14.52 presents the comparison of the experimental and simulation results

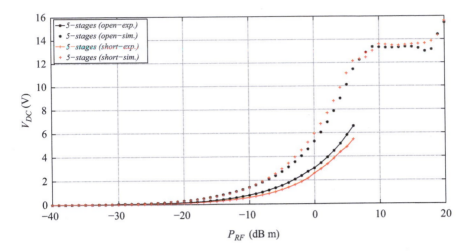

Figure 14.52   Comparison of the experimental and simulation results for the impact of the RF received power on output voltage for 5-stage Dickson voltage multipliers, with a load impedance of 100 kΩ, with open and short circuit stubs

regarding the impact of the RF received power on output voltage for a 5-stage Dickson voltage multiplier with a load impedance of 100 kΩ, with open and short circuit stubs.

From Figure 14.52, one can observe that the experimental results for output voltages from the 5-stage prototypes are lower than simulation results. Additionally, the saturation of the output voltage from the prototypes is not observable since to inject higher RF received power into the prototype could damage and burn the HSMS-2850 Schottky diodes from the amplifier due to excessive current. Therefore, in the experimental results, the maximum RF received power is 6 dB m. In turn, for the 7-stage Dickson voltage multiplier, the experimental results for the output voltage obtained from the prototypes are higher than the values computed by means of simulation.

Figure 14.53 presents the comparison between the experimental and simulations results for the 7-stage Dickson voltage multiplier. It is observable that the output voltage from the experimental results obtained from the prototypes are higher than the values computed by means of simulations. In addition, it is not observable the saturation of the output voltage from the prototypes since higher RF received power injected into the prototype could damage and burn the HSMS-2850 Schottky diodes due to excessive current. Also, since these prototypes present more stages, the saturation point of the voltage output will occur for higher values of RF received power that are not represented in Figure 14.53. As mentioned above, in the experimental results, the maximum RF received power is 6 dB m.

Figure 14.54 summarizes the conversion efficiencies for the different 5 and 7-stage prototypes in terms of experimental and simulation results.

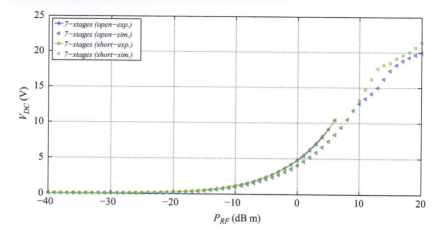

Figure 14.53   *Comparison of experimental and simulation results for the impact of the RF received power on output voltage for a 7-stage Dickson voltage multiplier with a load impedance of 100 kΩ, with open and short circuit stubs*

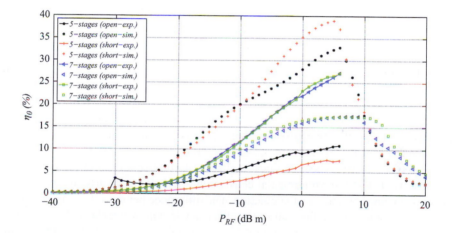

Figure 14.54   *Comparison of experimental and simulation results for the impact of the RF received power on the conversion efficiency (η_0) for 5 and 7-stage Dickson voltage multipliers, with a load impedance of 100 kΩ, with open and short circuit stubs*

Figure 14.55 depicts the experimental and simulation results comparison regarding the impact of the RF received power on the conversion efficiency for a 5-stage Dickson voltage multiplier with open and short circuit stubs.

In Figure 14.55, the difference between the simulation and experimental results from the conversion efficiency, $\Delta\eta_0$, is $\Delta\eta_0 \approx 21.48\%$ at $P_{RF} = 6$ dB m for the

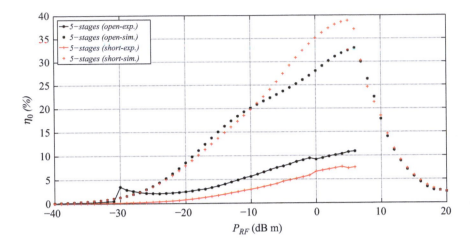

Figure 14.55   *Comparison of experimental and simulation results for the impact of*
the RF received power on the conversion efficiency (η_0) for a 5-stage
Dickson voltage multiplier, with a load impedance of 100 kΩ, with
open and short circuit stubs

5-stage Dickson voltage multiplier, with open circuit stub. In turn, the 5-stage Dickson voltage multiplier with short circuit stub presents a difference $\Delta \eta_0 \approx 31.23\%$ at $P_{RF} = 6$ dB m. Therefore, experimental results of the conversion efficiency for the 5-stage Dickson voltage multiplier prototypes present a loss compared to the simulation results.

Figure 14.56 presents a comparison of the experimental and simulation results for the impact of the RF received power on the conversion efficiency, for a 7-stage Dickson voltage multiplier, with open and short circuit stubs.

For the 7-stage Dickson voltage multiplier with open and short circuit stubs, it is observable that the experimental conversion efficiency is clearly higher than the results obtained by means of simulation. The difference between the simulation and experimental results for the conversion efficiency is $\Delta \eta_0 \approx 9.87\%$ at $P_{RF} = 6$ dB m for the 7-stage Dickson voltage multiplier with open circuit stub. In turn, the 7-stage Dickson voltage multiplier with short circuit stub presents a difference $\Delta \eta_0 \approx 9.772\%$ at $P_{RF} = 6$ dB m. Hence, the experimental results for the 7-stage Dickson voltage multiplier prototypes present a conversion efficiency gain compared to the simulation results.

As a final remark, and by observing the results obtained for the 5- and 7-stage Dickson voltage multiplier prototypes, the most suitable ones to harvest electromagnetic energy from the DTT frequency band are the 7-stage prototypes because of the higher efficiency that can be attained jointly with the output voltage that could be delivered. By comparing the results obtained for the 5-stage Dickson voltage multiplier prototypes at 945 MHz, as shown in Figures 2.5 and 2.6 from Chapter 2, with the results from Figures 14.51 and 14.55, it is noticeable that at 945 MHz, the output

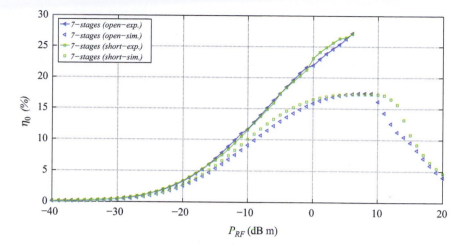

Figure 14.56 *Comparison of experimental and simulation results for the impact of the RF received power on the conversion efficiency (η_0), for a 7-stage Dickson voltage multiplier, with a load impedance of 100 kΩ with open and short circuit stubs*

voltage and conversion efficiency is higher than at 754 MHz. According to [18], the increase trend of the conversion efficiency as the RF received power increases is due to a decrease in the effect of losses due to the diode threshold voltage. At high power, larger efficiencies are possible, as the EH devices operation reaches a linear state, far above their diode turn-on voltage [18]. As the power level is reduced, the device efficiency decreases because the diode is "on" for a shorter fraction of the RF wave period. This is an expected result since the DTT frequency is lower than the GSM 900 frequency. This leads to a decrease of the output voltage and conversion efficiency in the DTT EH prototypes, due to a decrease of the devices efficiency caused by higher device parasitic losses encountered at higher frequencies. However, since the 7-stage Dickson voltage multiplier prototypes has achieved higher experimental output voltage and conversion efficiency, compared to the simulation results, and the DTT transmitters are dedicated only to transmit the TV signal at a considerably higher power than GSM base stations, it is expected that the 7-stages multiplier prototypes will harvest more electromagnetic energy than the 5-stages prototypes for GSM 900/1800 frequency.

14.4.12 Design of a half-wave dipole antenna for DTT band

The antenna is a device that converts the guided waves coming from a feeder cable, wave-guide or transmission line into radiating waves travelling in free space, and vice versa. Antennas are the key part of almost every wireless systems, since they allow for establishing the wireless connection. They gain a more prominent role when the devices to which they are attached to are supposed to completely rely on them, not only for communication, but also for RF-EH procedures, which is the case in this research work.

In this research, we consider the design of a traditional half-wave dipole antenna which provides an omnidirectional radiation pattern as well as an impedance of nearly 50 Ω. In addition, this dipole-abased antenna was considered since the simpler the matching circuits, the less power will the antenna dissipate and the Dickson voltage multiplier based circuit is intended to have a 50 Ω interface, in order the standard antennas, cables, connectors as well as measuring equipment could be utilized with it without considering an impedance matching circuits. Although textile antennas are not considered for the DTT band yet, this work constitutes a preliminary phase to facilitate future design of these antennas with textile conductive and dielectric materials.

Dipole antenna is a radio antenna that can be made of a simple wire, with a centre-fed-driven element. It consists of two metal conductors of rod or wire, in line with each other, with a small space between them. The RF voltage is applied to the antenna at the centre, between the two conductors. The geometry of this antenna, which is a cylindrical structure, is fully described by five parameters, including length, radius, feeding gap, frequency and wavelength [19].

For a half-wave dipole, the length of the dipole should be half of the wavelength but approximately 0.45 times of the wavelength. Typically, a dipole antenna is formed by two quarter wavelength conductors or elements placed back to back for a total length of $\lambda/2$. A standing wave on an element of a length $\lambda/4$ yields the greatest voltage differential, as one end of the element is at a node while the other is at an antinode of the wave.

In our case, instead of designing the two dipole on the same plane of RO4003 substrate, one of the two arms of the antenna was moved to the opposite side of the substrate, so that an edge-mount SMA connector could be easily soldered to them. By considering this design option, it was found that no effect on the characteristics of the antenna was detected, as expected, since the thickness of the substrate is actually very small compared to the length of the copper traces.

After designing the first version of the antenna (as shown in Figure 14.57), other versions were designed and produced. In the second version (shown in Figure 14.58), we sided one arm over the other, while in the third version (Figure 14.59), the width of the arms from the first version was enlarged. Finally, in the fourth version (Figure 14.60), we sided one arm over the other based on the third version of the antenna. By considering these changes, different effects were observed in the radiation pattern.

The width of the arms relies was enlarged because the dipole effectively establishes a trade-off between quality factor and bandwidth. Thinner lines lead to a more resonant antenna, potentially allowing a very effective impedance matching, but only if the antenna is designed and built with enough precision, so that it indeed resonates at the frequency it was designed to. Hence, the thinner the lines are, the more precise the manufacturing technique must be. However, some trade-offs should be taken into account since larger lines lead to a wider bandwidth antenna that is much less sensitive to frequency deviations, but at the cost of a lower quality factor [20].

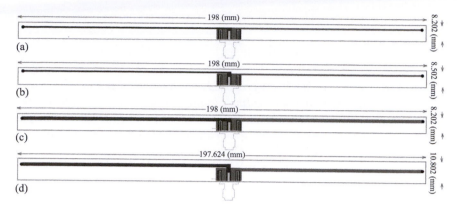

Figure 14.57 First versions of the proposed dipole-based PCB antenna to operate at 754 MHz with a) 8.202 mm, b) 8.502 mm, c) 8.202 mm and d) 10.802 mm substrate widths (the rectangular area represents the RO4003C substrate and the other two structures correspond to the arms of the dipole, with the upper arm being on the top side of the substrate and the lower arm being on the bottom side)

Figure 14.58 Second version of the proposed dipole-based PCB antenna to operate 754 MHz (the rectangular area represents the RO4003C substrate and the other two structures correspond to the arms of the dipole, with the upper arm being on the top side of the substrate and the lower arm being on the bottom side)

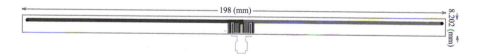

Figure 14.59 Third version of the proposed dipole-based PCB antenna to operate at a frequency of 754 MHz (the rectangular area represents the RO4003C substrate and the other two structures correspond to the arms of the dipole, with the upper arm being on the top side of the substrate and the lower arm being on the bottom side)

Figure 14.60 *Fourth version of the proposed dipole-based PCB antenna for a*
frequency of 754 MHz (the rectangular area represents the RO4003C
substrate and the other two structures correspond to the arms of the
dipole, with the upper arm being on the top side of the substrate and
the lower arm being on the bottom side)

Figure 14.61 *Half-wave dipole antenna*

For the computation of the dipole lengths of the antennas, we consider a resonating frequency of $f_c = 754$ MHz. In (14.28), the calculated the wavelength, λ, (SI unit: m) is associated to the resonating frequency.

$$\lambda = \frac{c}{f_c} = \frac{3 \times 10^8}{754 \times 10^6} = 0.397878 \text{ m} \tag{14.28}$$

where c is the speed of light (i.e. 3×10^8 m/s) and f_c is the resonating frequency. Equation (14.29) presents the computation of the radius (or width), R, (SI unit: m) of the each antenna dipole.

$$R = \frac{D}{2} = 0.001 \times \lambda = 0.001 \times 0.397878 = 0.000397878 \text{ m} = 0.397878 \text{ mm} \tag{14.29}$$

Equation (14.30) computes the total length dipole antenna, l, which is half of the calculated wavelength in (14.28).

$$l = \frac{\lambda}{2} = \frac{0.397878}{2} = 0.198939 \text{ m} = 198.939 \text{ mm} \tag{14.30}$$

Equation (14.31) computes the length of each dipole from the antenna, L, which is a quarter of the calculated wavelength in (14.28).

$$L = \frac{\lambda}{4} = \frac{0.397878}{4} = 0.099469 \text{ m} = 99.469 \text{ mm} \tag{14.31}$$

In Figure 14.61 the associated lengths to the half-wave dipole antenna is presented.

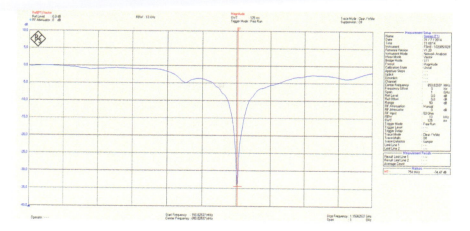

Figure 14.62   Experimental results for the return losses, S_{11}, from the proposed antenna for 754 MHz (marker centred at $f = 754$ MHz)

In order to calculate the theoretical gain of the half-wave dipole antenna, the following equation is considered, as in [21].

$$G_{\lambda/2} = \frac{60^2}{30 \times R_{\lambda/2}} \tag{14.32}$$

From the available literature, the gain of a dipole antennas can be $G_{\lambda/2} = 2.15$ dB i if $L = 0.5$ m and $G_{\lambda/2} = 1.76$ dB i if $L \ll 0.1$ m. In our case, since the obtained length of the antenna dipole is in between these values, the gain of our dipole antenna is in between 1.76 and 2.15 dB i.

Compared to an isotropic antenna (a mathematically perfect antenna that radiates in all directions equally), a full-wave antenna has 3 dB (two times) more gain, a half-wave, 2.15 dB (1.7 times) and a quarter-wave, 0.15 dB. Therefore, a full-wave antenna provides more gain, but at the expense of size.

In Figures 14.62, 14.63 and 14.64, the experimental results for the return losses, S_{11}, obtained from the spectrum analyser for the frequency markers at 754, 750 and 758 MHz, respectively, are presented.

At the frequency we considered to design the antenna, the return loss is of $S_{11} = -34.37$ dB at $f_c = 754$ MHz. To determine the associated frequency bandwidth to the antenna, values for the frequency in the range that is in between a set of frequencies that fulfil the condition $S_{11} < -10$ dB are considered. It is observable that the proposed antenna presents a range of frequencies that fulfil this condition between 725 and 787 MHz ($BW = 62$ MHz). This bandwidth value is suitable for the range of frequencies where we intend to use the antenna for electromagnetic EH.

Figure 14.63 shows that the antenna has return loss of $S_{11} = -28.77$ dB, at $f_c = 750$ MHz.

Finally, for the frequency marker at $f_c = 758$ MHz, the antenna presents return losses of the order of $S_{11} = -26.45$ dB.

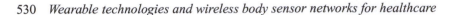

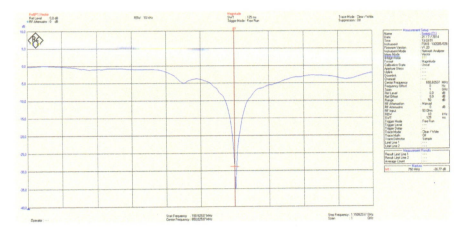

Figure 14.63   Experimental results for the return losses, S_{11}, from the proposed antenna for 754 MHz (marker centred at $f = 750$ MHz)

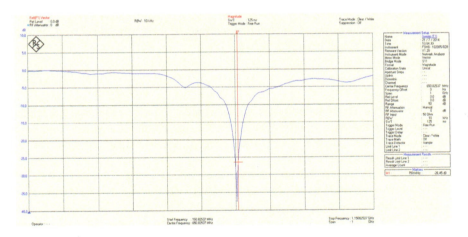

Figure 14.64   Experimental results for the return loss, S_{11}, from the proposed antenna for 754 MHz (marker centred at $f = 758$ MHz)

By considering the aforementioned values for the return losses at different values of frequencies, one concludes that the gains that will be achieved with this type antenna are promising.

14.4.13   Experimental results for the complete solution of the RF energy harvesting and storing system

In order to test the RF-EH with supercapacitor storing system, the 7-stage Dickson voltage multiplier (open stub) was considered as the EH prototype to be evaluated at different input power levels whilst considering a frequency of 754 MHz in the signal

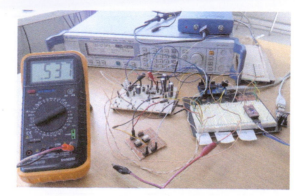

Figure 14.65   DTT RF energy harvester prototype experimental apparatus

generator, as shown in Figure 14.65. The output power from the signal generator was adjusted to three power levels, i.e. -10, -5 and -2 dB m, staying in each power levels for about 22,550 s ($\approx$6.5 h) and then taking $V_{super\text{-}cap}$ average voltage at a rate of 50 samples/s.

Within the supercapacitor storing system configuration, different values of the capacitance for the pre-charge capacitor can be chosen, as well as the capacitance and maximum charge voltage ($V_{s\text{-}cap\ charge\_max}$) of the supercapacitor bank. For this experiment, the following configurations were assumed: $C_{pre\text{-}cap} = 1,000\ \mu F$ by configuring their jumpers $JP1$, $JP3$ (attached) and $JP2$, $JP4$ (dis-attached); $C_{super\text{-}cap} = 5\ F$ and $V_{s\text{-}cap\ charge\_max} = 5$ V in which the jumpers should be configured as $JP6$, $JP8$, $JP9$ are attached and $JP7$, $JP10$, $JP11$, $JP12$ are dis-attached.

For this experimental test, the programme loaded onto the Arduino platform to control the energy transference is the one presented in Algorithm 1, with the exception that it does not consider the deep sleep mode.

Figure 14.66 shows the voltage building up in the supercapacitor(s) bank as a function of time for different received powers generated by the signal generator. One observes that the voltage in the supercapacitor(s) at the beginning of the experiment is 0 V and increases with time, as expected. One also observes that for a $P_{RF} = -10$ dB m, the curve presents a slight growing compared to the other curves. The curve that achieves the highest voltage after elapsing the experiment time is the one corresponding to the $P\_RF = -2$ dB m. This result is expectable, because of the behaviour of the voltage output from the RF energy harvester for different received powers. The explanation is as follows: $V_{DC} = 3.61$, 2.34 and 1.436 V for $P_{RF} = -2$, -5 and -10 dB m, respectively. The maximum voltages attained in the supercapacitor(s) are $V_{super\text{-}cap} = 0.5199$, 0.3758 and 0.1075 V at $P_{RF} = -2$, -5 and -10 dB m, respectively.

In order to evaluate how the DTT RF energy harvester and storing systems behave when assuming different voltages that are pre-charged in the supercapacitor, we conducted experimental tests in which the $P_{RF} = -2$ dB m and it lasts for about 19,300 s

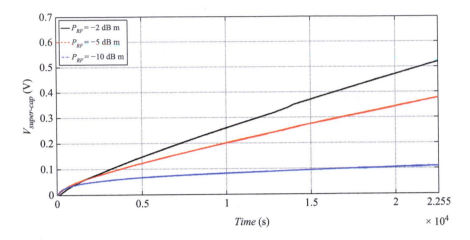

Figure 14.66   Voltage in the supercapacitor(s) bank from the DTT RF energy harvester for different values of the received power as function of time

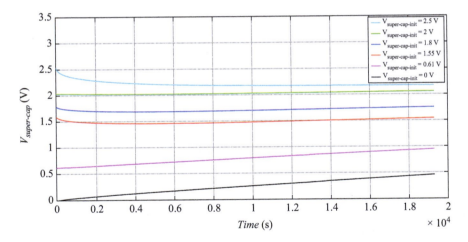

Figure 14.67   Voltage in the supercapacitor(s) bank from the DTT RF energy harvester for different pre-charged voltages in the supercapacitor ($V_{super-cap-init}$), for $P_{RF} = -2$ dB m

(≈ 5.3 h). During this experiment, the average voltage $V_{super-cap}$ was taken at a rate of 50 samples/s, as in the previous test.

Figure 14.67 presents the different curves that represent the voltage in the supercapacitor(s) bank for different pre-charged voltages, $V_{super-cap-init}$, for a constant value of $P_{RF} = -2$ dB m. One observes that only for the values of $V_{super-cap-init} = 0$, 1.8

and 2 V, the voltages attained in the supercapacitor(s) increase as a function of the time, whereas the remaining ones present a voltage decay. The curve corresponding to the case in which the pre-charged voltage is $V_{super-cap-init} = 2.5$ V is the one that shows a larger decay of the voltage as a function of the time. Only after 1.2×10^4 s, the voltage remains constant at nearly 2.2 V. The other curve that matches the case of a pre-charged voltage of $V_{super-cap-init} = 1.55$ V also presents a decay in the voltage. However, in comparison with the curve for $V_{super-cap-init} = 2.5$ V, it shows a lower voltage decay rate. Additionally, the curve for a voltage in the supercapacitor(s) of $V_{super-cap-init} = 1.55$ V shows an increase in the voltage after 0.8×10^4 s. In the end of the test, it attained almost the initial pre-charged voltage in the supercapacitor, i.e. 1.55 V. From Figure 14.67, it is conclusive that at a received power of $P_{RF} = -2$ dB m and with the supercapacitor pre-charged at a voltage of $V_{super-cap-init} = 2$ V, the system is capable of maintaining a stable voltage in the supercapacitor(s) bank. Regarding the curve for a voltage in the supercapacitor(s) of $V_{super-cap-init} = 1.8$ V, it shows a lower voltage decay as a function of time compared to the curves where a voltage decay is also present. In addition, after 1.8×10^4 s, the voltage reaches 1.7 V in the supercapacitor(s) bank, and then it starts to increase. By observing Figure 14.67, one can conclude that for voltages pre-charged in the supercapacitor(s) bank higher than 2 V present a high voltage decay as a function of time, which implies that there is the need of a higher received power in the RF energy harvester, to solve the high voltage decay observed in the curves. Therefore, the requirement of a minimum voltage of 1.8 V in the supercapacitor(s) bank is doable for a received power of $P_{RF} = -2$ dB m within the DTT band. Also, it is conclusive that here it is observable, the characteristic of the supercapacitor that as the energy builds up in the supercapacitor it takes more time to add more energy to the supercapacitor.

To verify the implications of modifying the transition times parameters in the firmware of the microcontroller that accounts for the energy transfer from the pre-charge capacitor to the inductor, *delay_transistor2_int*, and from the inductor to the supercapacitor(s) bank, *delay_transistor3_int*, an experiment was conducted for about 23,171 s ($\approx$6.4 h). The results from this experiment are shown in Figure 14.68. In this experiment, the supercapacitor(s) bank was pre-charged with 1.8 V and the power level injected in the RF energy harvester is $P_{RF} = -2$ dB m.

First of all, Figure 14.68 depicts three curves that maintain the value for the delay parameter *delay_transistor2_int* $= 1,000$ µs and vary the value for the delay parameter *delay_transistor3_int* within the interval of {2,500; 5,000; 10,000} µs. In addition, there is one curve that maintains the value for the delay parameter *delay_transistor3_int* $= 5,000$ µs (which is the default value in the microcontroller firmware) but varies the value for the delay parameter *delay_transistor2_int* $=$ 3,000 µs. The latter delay configuration allows for comparing the results obtained with the curve with the delays configuration *delay_transistor2_int* $= 1,000$ µs and *delay_transistor3_int* $= 5,000$ µs. From Figure 14.68, it is observable that for the curves that vary the value for the delay parameter *delay_transistor3_int*, the one configured with the shorter delay (i.e. 2,500 µs) presents a faster recovering behaviour from the voltage decay observed in the beginning of the experiment, compared with the curve that has the configured default delay of

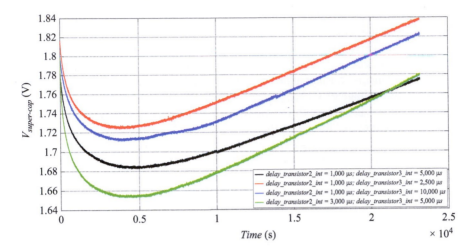

Figure 14.68   *Voltage in the supercapacitor(s) bank from the DTT RF energy harvester for different transition times with a voltage $V_{super-cap-init} = 1.8\ V$ in the supercapacitor, for $P_{RF} = -2\ dB\ m$*

delay_transistor3_int $= 5,000\,\mu s$. Also in the figure, the curve that corresponds to the delays configuration *delay_transistor2_int* $= 3,000\,\mu s$ and *delay_transistor3_int* $= 5,000\,\mu s$ shows a higher voltage decay between 0 and 0.5×10^4 s, compared to curve that presents a delay configuration of *delay_transistor2_int* $= 1,000\,\mu s$ and *delay_transistor3_int* $= 5,000\,\mu s$. Therefore, on the one hand, by increasing the delay value *delay_transistor2_int* leads to a less efficient trade-off of the voltage that is building up in the supercapacitor(s) bank. On the other, if there is a decrease of the value specified for the delay parameter *delay_transistor2_int*, it shows a better efficiency in terms of the voltage that is build up in the supercapacitor(s) bank as a function of time and the voltage decay is not so sharp as in the curve with the delay configuration *delay_transistor2_int* $= 1,000\,\mu s$ and *delay_transistor3_int* $= 5,000\,\mu s$.

14.4.14   *Simultaneous multiband GSM and DTT RF energy harvester*

Before proposing a multiband GSM and DTT RF energy harvester, there is the need to assess the output voltages that the DTT RF energy harvesters are capable of supplying in a real scenario. Therefore, an experiment was conducted, in which the DTT RF energy harvester prototypes with the antenna attached were placed in line-of-sight (LoS) to the nearest DTT antenna. This experiment lasts for an hour and consisted in placing each DTT RF energy harvester in LoS with the DTT antenna while taking the voltage reading, as presented in Figure 14.69. The distance in LoS from the RF energy harvester to the DTT tower is approximately 600 m.

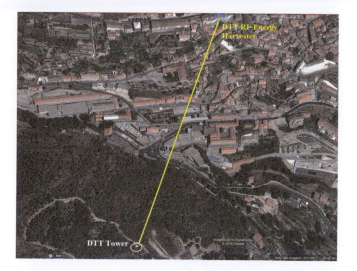

Figure 14.69   Location of the DTT antenna and DTT RF energy harvester

Table 14.13   DTT RF energy harvester prototypes voltage output

Number of stages	Type of stub	V_{DC} (V)
5	Open	0.334
5	Short	0.750
7	Open	1.035
7	Short	1.020

The results obtained for the average output voltages, V_{DC}, are summarized in Table 14.13. In addition, the results presented in Table 14.13 were measured during a rainy day.

Figure 14.70 presents the multiband GSM and DTT RF energy harvester block diagram that combines the energy harvested from an RF energy harvester for the GSM band jointly with the energy harvested from an RF energy harvester for the DTT band. For each RF energy harvester, there is an impedance matching circuit that filters the frequency band intended for the prototype. The powers received in the GSM and DTT bands are denoted by P_{RF_IN1} and P_{RF_IN2}, respectively. After filtering the signal by means of the impedance matching circuits, the received power at the GSM and DTT RF energy harvester is attenuated by factors α_1 and α_2, respectively. In the literature [22], it is mentioned that this loss could be at most 3%. The received powers, $\alpha_1 P_{RF\_IN1}$ and $\alpha_2 P_{RF\_IN2}$, are then rectified by means of 5- and 7-stage Dickson voltage rectifiers, which results in a DC voltage in the output of RF energy harvester. In order to take advantage of all the RF energy that is scavenged from each

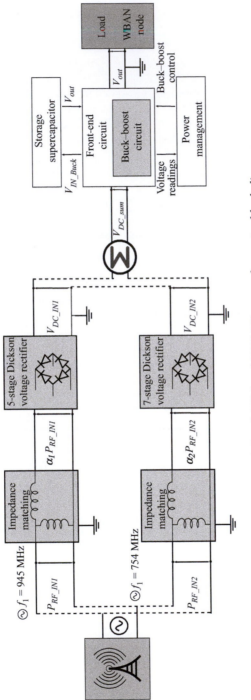

Figure 14.70 Multiband GSM and DTT RF energy harvester block diagram

of the frequency bands, both output voltages, V_{DC_IN1} and V_{DC_IN2}, are combined into the total output voltage from the multiband RF energy harvester, i.e. V_{DC_sum}. The total output voltage is delivered to the front-end circuit (i.e. buck–boost circuit) which is controlled by a power management unit. This unit acquires the different voltages at different stages of the buck–boost converter and manages the transfer of the energy received from the multiband RF energy harvester to the storage unit (composed by supercapacitors). The voltage variation associated to the energy transfer between the buck–boost converter and storage unit is denoted by V_{IN_Buck} in Figure 14.70. Finally, the energy stored is transferred to the load, which in this case is represented by a CRN. The voltage variation associated to the discharge from the storage unit is denoted as V_{OUT}, as shown in Figure 14.70.

14.5   RF harvesting/storage system design upgrade

After testing the first version of the RF harvesting and storage system, some modifications were needed in order to improve the energy transference from the RF-EH to the supercapacitor(s) bank. One of the modifications that was performed is the replacement of the transistor that switches on and off the connection between the RF-EH board and the supercapacitor storing system by an optocoupler. This modification was necessary, since the switching performed by means of a transistor was interfering with the voltage that was being charged in the pre-charge capacitor.

By replacing the transistor with an optocoupler TLP521-1, the switching is galvanic since the on and off switching is made by means of optic beams encapsulated in a chip.

Another required modification was the addition of an header to connect the embedded system to the supercapacitor(s) bank.

Finally, since the objective is to self-sustain the microcontroller that controls the energy flow from the pre-charge capacitor to the inductor and from the inductor to the supercapacitor(s) bank, then to enable the deep sleep state presented in the Algorithm 1, some modifications in the energy storing system board must be performed. It must be guaranteed that the on/off switching control, performed by the microcontroller, between RF energy harvester and the pre-charge capacitor, does not interrupt the voltage from flowing to the pre-charge capacitor from the RF energy harvester when the microcontroller is in deep sleep state.

As such, there are two possible solutions to solve this issue. First, instead of using an NPN BC547B transistor or the NPN TLP521-1 optocoupler, which is normally not conductive, the solution is to use a PNP transistor or optocoupler. The NPN transistors are normally closed. That is, if nothing is flowing into the base, then the transistor is not allowing any current to flow from the collector to the emitter. But when current is allowed to flow into the base on an NPN transistor, then current will be allowed to flow from collector to the emitter. In turn, the PNP transistors are the opposite (normally open). These type of transistors will allow current to pass from the collector to the emitter, until you apply current to the base (then it turns off). Second, the other solution is to use another type of transistors, the so called junction gate field-effect transistor (JFET). This type of transistors present a channel that conducts

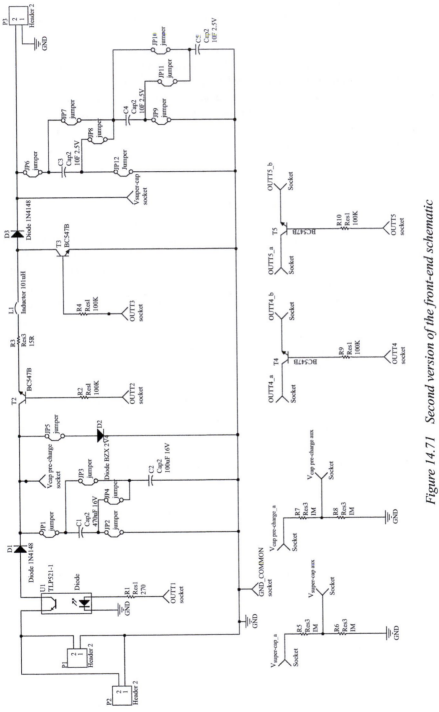

Figure 14.71 Second version of the front-end schematic

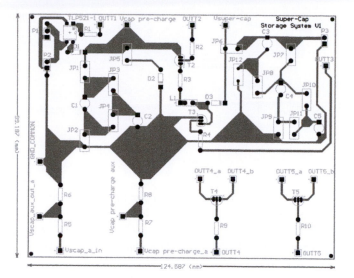

Figure 14.72   Second version of the front-end PCB

current even when no gate drive is provided, i.e. a JFET is normally conductive. Available JFETs that can be used in this work include the ND2410L, ND2020L, BST80, BS107 or BF245B.

Figure 14.71 presents the second version of the front-end circuit (i.e. supercapacitor storing system) with the aforementioned modifications. It is observable in Figure 14.71 that the transistor *T1* from Figure 14.14 was replaced by the TLP521-1 optocoupler. In case we want to change for an NPN optocoupler, the solution relies on replacing the TLP521-1 optocoupler by an NPN optocoupler, e.g. the Texas Instruments TIL920. Additionally, the header *P3* was also added in order to connect the embedded system to the supercapacitor(s) bank.

Figure 14.72 shows the second version of the front-end PCB. The necessary modifications were performed, and some components were repositioned in order to accommodate the new electronics components that were added in the version of supercapacitor storing system.

Figure 14.73 shows the prototype for F-EH with textile antenna for GSM 900–1,800 MHz frequency band, PCB antenna for 754 MHz DTT band, buck–boost converter, Dickson amplifiers, supercapacitor energy storing system and switching circuits, together with the computer that runs the software to visualize how the energy builds-up.

14.6 Conclusions

RF-EH is becoming a mature technology with the enormous potential to realize the vision for uninterrupted self-powered networks that include the cognitive radio networks. In the last years, there have been an effort to evolve the RF-EH devices

*Figure 14.73   Real operation of the prototype for RF-EH with textile antenna for
GSM 900–1,800 MHz frequency band, PCB antenna for 754 MHz
DTT band, buck–boost converter, Dickson amplifiers, supercapacitor
energy storing system and switching circuits*

in terms of scavenging RF energy from various commercial wireless communication
systems. In this work, different and important observations have been reported concerning the different RF-EH devices as well as the energy storing systems that can
be used to store the energy for posterior use by a cognitive radio. Regarding the RF-
EH device, the GSM 900/1800 bands provide the highest power density for RF-EH
in urban and suburban environments. Improvement in the conversion efficiency and
sensitivity of rectennas for these frequency bands is one of the biggest challenges. By
comparing the obtained results with the ones for the DTT RF-EH, it is noticeable that
the RF-EH device for the GSM band presents an output voltage and conversion efficiency higher than the DTT RF-EH device. This is an expected result since the DTT
frequency is lower than the GSM 900/1800 frequency, which leads to a diminishing
of the output voltage and conversion efficiency in the DTT EH prototypes. However,
since the DTT RF-EH prototypes achieved better results in terms of output voltage
and conversion efficiency (compared to the simulation results) and the DTT transmitters are only dedicated to transmit TV signal at higher power levels than GSM base
stations, it is expected that the DTT RF-EH prototypes will be more capable to harvest
electromagnetic energy than the GSM RF-EH prototypes. We also present guidelines
for the choice of the number of stages for the RF energy harvester, depending on the
requirements from the embedded system we want to power supply, which is useful for
other researchers that work in the same area. In addition, regarding the component
values choice for the inductor and pre-charge capacitor of the supercapacitor energy
storing system proposed within the framework of this work, we present detailed experimental results that help to decide which component values are more suitable to attain
the best overall efficiency of the RF-EH with supercapacitor storing system. From
the two pre-chosen component value combinations, the one that presents best results
in terms of time required to attain the desirable voltage in the supercapacitor(s) bank
is the combination of component values $C_{pre\text{-}charge} = 470\,\mu\text{F}$; $L = 101\,\mu\text{H}$. However,
the combination of component values $C_{pre\text{-}charge} = 1,000\,\mu\text{F}$; $L = 101\,\mu\text{H}$ for trial

cases higher than 6 presents adequate results in terms of time estimation to attain $V_{super\text{-}cap\,max} = 1.8\,V$ in the $C_{super\text{-}cap}$, but we should keep in mind that the time needed to charge the pre-charge capacitor takes more time than smaller capacitances.

By observing the values obtained for each combination of component values, when considering a harvesting device placed in a *Low Ambient Traffic* scenario, for trial cases number 6 and 8, one concludes that it takes too long time to charge a supercapacitor(s) bank for any $V_{super\text{-}cap\,max}$ voltage. The results obtained here can be considered as the worst case scenario in which the maximum estimated charging times that the energy storing system needs to attain the desirable $V_{super\text{-}cap\,max}$ voltage in the supercapacitor(s) bank is presented. Therefore, a viable option to follow if *Low Ambient Traffic* scenario is the most common to happen, is to consider to charge the supercapacitor initially and then using the harvested energy to maintain the voltage of the supercapacitor(s) bank more or less constant while it supplies the CRN.

If the EH device is located in *High Ambient Traffic* scenario, the time it takes to attain the desirable $V_{super\text{-}cap\,max}$ voltage in the supercapacitor(s) bank surely be reduced drastically due to the quick charge of the pre-charge capacitor, which is the most time-consuming fraction of the total time the energy storing system needs to attain the desirable voltage in the supercapacitor(s) bank. Approximately 80%–90% of the time estimated to attain the $V_{super\text{-}cap\,max}$ voltage in the supercapacitor(s) bank voltage depends on the time it takes to accumulate energy in the pre-charge capacitor by means of the EH device.

Additionally, in this work, we also propose a multiband GSM and DTT RF energy harvester block concept that combines the energy harvested from an RF Energy harvester for the GSM band jointly with the energy harvested from an RF Energy harvester for the DTT band.

As a future work, the increase of test duration will be considered, in order to observe how the energy builds up in the supercapacitor(s) bank and when the supercapacitor(s) bank is pre-charged with a certain voltage.

In addition, tests will be performed with the multiband GSM and DTT RF energy harvester, scavenging together RF energy with the respective antenna to the supercapacitor storing system in a real-environment scenario to observe how the energy builds up. Preliminary tests, in a conference room densely populated, with circa 300 people, showed that the RF-EH system managed to build up and store energy, even when we were not using mobile phones in the close proximity of the harvesting antennas (the system for such trial is shown in the picture from Figure 14.73).

Another suggestion for future work involves the use of an optimization framework design and implementation for RF-EH that allows for finding the best transition values.

References

[1] Jabbar H., Song Y., Jeong T. 'RF energy harvesting system and circuits for charging of mobile devices'. *IEEE Transactions on Consumer Electronics*. 2010;56(1):247–253.

[2]   Taneja J., Jeong J., Culler D. 'Design, modeling, and capacity planning for micro-solar power sensor networks'. *in Proc. of International Conference on Information Processing in Sensor Networks (IPSN '08)*; 2008. pp. 407–418.

[3]   Barroca N., Saraiva H.M., Gouveia P.T., *et al.* 'Antennas and circuits for ambient RF energy harvesting in wireless body area networks'. *in Proc. of IEEE 24th International Symposium on Personal Indoor and Mobile Radio Communications (PIMRC)*; 2013. pp. 532–537.

[4]   Vina 10F 2.5V Super-capacitor Datasheet; 2014. Datasheet. Available from: http://www.supercapacitorvina.com/product/edlc.html.

[5]   Dardaillon M., Marquet K., Risset T., Scherrer A. 'Software defined radio architecture survey for cognitive testbeds'. *in Proc. of the 8th International Wireless Communications and Mobile Computing Conference (IWCMC 2012)*; Limassol, Cyprus; 2012. pp. 189–194.

[6]   Borges L.M., Barroca N., Saraiva H.M., *et al.* 'Design and evaluation of multi-band RF energy harvesting circuits and antennas for WSNs'. *in Proc. of International Conference on Telecommunications (ICT)*; Lisbon, Portugal; 2014. pp. 308–312.

[7]   PROENERGY-WSN; 2012. PROENERGY-WSN Project: Prototypes for Efficient Energy Self-sustainable Wireless Sensor Networks. Available from: http://www.e-projects.ubi.pt/proenergy-wsn/.

[8]   Nintanavongsa P., Muncuk U., Lewis D., Chowdhury K. 'Design optimization and implementation for RF energy harvesting circuits'. *IEEE Journal on Emerging and Selected Topics in Circuits and Systems*. 2012;2(1):24–33.

[9]   Yan H., Montero J.G.M., Akhnoukh A., de Vreede L.C.N., Burghartz J. 'An integration scheme for RF power harvesting'. *in Proc. of The 8th Annual Workshop on Semiconductor Advances for Future Electronics and Sensors*; Veldhoven, Netherlands; 2005. pp. 64–66.

[10] Mastragostino M., Soavi F. 'Strategies for high-performance supercapacitors for HEV'. *Journal of Power Sources*. 2007;174(1):89–93. Hybrid Electric Vehicles. Available from: http://www.sciencedirect.com/science/article/pii/S0378775307012177.

[11] Fahad A., Soyata T., Wang T., Sharma G., Heinzelman W., Shen K. 'SOLARCAP: Super capacitor buffering of solar energy for self-sustainable field systems'. *in Proc. of the 2012 IEEE International SOC Conference (SOCC)*; 2012. pp. 236–241.

[12] Fahad A., Soyata T., Wang T., Sharma G., Heinzelman W.B., Shen K. 'SOLARCAP: Super capacitor buffering of solar energy for self-sustainable field systems'. *in SoCC*; IEEE; 2012. pp. 236–241. Available from: http://dblp.uni-trier.de/db/conf/socc/socc2012.html#FahadSWSHS12.

[13] Arduino Nano V3; 2014. Datasheet. Available from: http://arduino.cc/en/Main/arduinoBoardNano.

[14] MSP430-SOLAR Board; 2014. Datasheet. Available from: https://www.olimex.com/Products/MSP430/Power/MSP430-SOLAR/resources/MSP430-SOLAR.pdf.

[15] Rufer A., Barrade P. 'A supercapacitor-based energy-storage system for elevators with soft commutated interface'. *IEEE Transactions on Industry Applications*. 2002;38(5):1151–1159.

[16] Pressman A. *Switching Power Supply Design*. 2nd ed. New York, NY: McGraw-Hill, Inc.; 1998.

[17] ATmega 3280; 2014. Datasheet. Available from: http://ww1.microchip.com/downloads/en/DeviceDoc/ATmega48A-PA-88A-PA-168A-PA-328-P-DS-DS40002061A.pdf

[18] Valenta C.R., Durgin G.D. 'Harvesting wireless power: Survey of energy-harvester conversion efficiency in far-field, wireless power transfer systems'. *IEEE Microwave Magazine*. 2014;15(4):108–120.

[19] Singh P., Sharma A., Uniyal N., Kala R. 'Half-wave dipole antenna for GSM applications'. *International Journal of Advanced Computer Research*. 2012;2(4):354–357.

[20] Dobkin D.M. *The RF in RFID: Passive UHF RFID in Practice*. Newton, MA: Newnes; 2007.

[21] Stutzman W.L., Thiele G.A. *Antenna Theory and Design*. 2nd ed. New York, NY: J. Wiley; 1997. Includes bibliographical references (p. 636–641) and index.

[22] Pavone D., Buonanno A., D'Urso M., Corte F.D. 'Design considerations for radio frequency energy harvesting devices'. *Progress In Electromagnetics Research B*. 2012;45:19–35.

Chapter 15
Conclusion

Fernando J. Velez[1] and Fardin Derogarian[1]

This chapter not only gives a final overview of the main topics covered in the book but also presents a taxonomy for the classification of wearable devices that was proposed, and the literature review identified papers whose authors applied SWOT analysis to this market by examining their strengths, weaknesses, opportunities and some of the major threats they face. Such work gives insights on the primary areas of innovation in wearable healthcare, which include infant safety and care, elderly care, chronic disease management, military support, sports medicine and preventive medicine. Among the several identified obstacles, one particular concern is the interoperability with other wireless systems, not only interference constraints but also aspects of data collection and tracking in healthcare systems. According to the authors of the study, not only Health Information Technology for Economic and Clinical Health (HITECH) act has helped drive the standardization of healthcare data terminology and encourage system interoperability but also ITU-T, ISO/IEC JTC 1, W3C (World Wide Web Consortium) and Open Interconnect Consortium/Open Connectivity Foundation (OIC/OCF) have ongoing works in relation to wearable devices and their standardization. This is certainly vital to the widespread implementation and success of the wearable healthcare technology, as it is an important step towards enhancing the interaction of wireless body-area networks (BANs) (WBANs) with traditional healthcare databases and electronic health record (EHR) systems.

15.1 Classification taxonomy and primary areas of innovation in wearable healthcare

The fields of wearable systems and wireless body sensor networks (WBSNs) for biomedical application rapidly change the way people live. Recent advances in the electronics industry have made the size of the devices and sensors small enough to embed thousands of them in textiles. Wearable technologies benefit from advances in other fields, including new telecommunications techniques and protocols, signal processing, and various types of power supplying such as radio frequency (RF) energy harvesting (RF-EH). However, there are many technological challenges in

[1]Instituto de Telecomunicações and Departamento de Engenharia Eletromecânica – Universidade da Beira Interior; Faculdade de Engenharia, Portugal

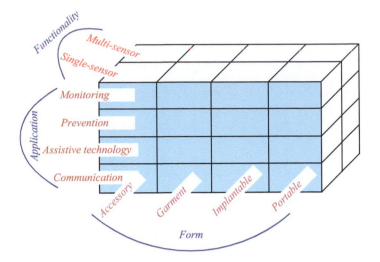

Figure 15.1   Dimensional taxonomy from wearable technology in healthcare.
© 2015. Adapted, with permission, from Reference [1]

each of these specific fields. This book has addressed and discussed technologies and methodologies for overcoming challenges in the context of wearable technologies and WBSNs for healthcare applications.

As the concepts evolve, the identification of a need for the classification of wearable devices for healthcare has been identified by the authors from [1] in terms of application, form and functionality dimensions, as shown in Figure 15.1.

As mentioned in [1], the value of this taxonomy includes the following:

- An easy to understand classification which can assist young scholar to acquire a deeper understanding of the emerging phenomenon.
- Clarify the underlying principles behind wearable technologies and clear the ambiguity around all forms of wearable technologies, whether implantable, such as smart adhesives or tattoos or edible such as endoscopy capsules.
- Assist designers and researchers to either add novel products to the existing base or extend the possibilities of wearable computing by applying the taxonomy.

As described in [1], the authors allow application overlap in the proposed taxonomy, e.g. Google glasses are designed to aid both clinical applications targeting healthcare professionals and non-clinical applications targeting consumers. The authors found that the four forms (accessory, garment, implantable and portable) are independent in their own existence but there are some systems where a garment may be interoperating with a portable device such as a smartphone. The application dimension addresses the main purpose of the wearable technology and is divided into four levels: monitoring and assistive technology (mainly used by patients), as well as prevention and communication, which are typically used by consumers. The functionality distinguishes between devices that measure one parameter and does a single function from the ones that measure more and do multiple functions. In the book, the authors have shown how the taxonomy applies to various wearable products, e.g. Muse or eSight.

Given the rising interest of the wearable healthcare industry, the authors from [2] applied SWOT analysis to this market by examining their strengths, weaknesses, opportunities, and some of the major threats they face. Their work gives insights on the primary areas of innovation in wearable healthcare, which include infant safety and care, elderly care, chronic disease management, military support, sports medicine, and preventive medicine. The authors also found that several obstacles still stand in the way of the wearable market's success; these include threats to data security and privacy, regulatory requirements, cost of system operation and management, and adoption rates below the average. According to the authors from [2], significant effort is needed to address the identified technological, societal, and governmental barriers that are preventing the wearable healthcare market from reaching its full potential. Incentives by governments that fuel innovation and encourage adoption by healthcare professionals and patients are deemed necessary for the wearable market's continued growth.

One particular concern is the interoperability with other wireless systems, not only interference constraints but also aspects of data collection and tracking in healthcare systems. Not only HITECH act has helped drive the standardization of healthcare data terminology and encourage system interoperability [3] but also ITU-T, ISO/IEC JTC 1, W3C and OIC have ongoing works in relation to wearable devices and their standardization [4]. This is certainly vital to the widespread implementation and success of the wearable healthcare technology and enhances the interaction of WBANs with traditional healthcare databases and EHR systems.

According to [4], on June 2015, ITU-T has established a new SG (SG20: IoT and its applications including smart cities and communities) to focus their work on IoT whose work will carry on writing a draft recommendation, which was originally proposed at a different SG (SG13: future networks including cloud computing, mobile and next-generation networks) for wearable devices with respect to their specific requirements and capabilities. The outcome SG20 will facilitate works on not only Internet of Things (IoT) but also their applications such as wearable devices.

In ISO/IEC Joint Technical Committee (JTC), the SC 29/WG 11, known as the Moving Picture Experts Group (MPEG), has been writing standards for media codecs since 1988. They have standardized numerous compression formats such as MPEG-4, MPEG-DASH, and so on. MPEG has recently established the Wearable MPEG Ad-hoc group for standardization of wearable devices. The ad-hoc group is currently discussing related use cases and their corresponding requirements and interfaces. JTC 1 has established WG 10 in 2014 that specialize IoT and its future applications. WG 10 is developing their IoT reference architecture and has initiated research on standardization gaps, network level technologies, identification and so forth. The WG is receiving use cases on IoT applications and carry on further studies on respective standardization requirements and capabilities of wearable devices.

The W3C has engaged in Web standardization since 1994, and they have recently published HTML5, the latest web markup language that allows developers to create and implement Web applications running on browsers. Since a number of HTML5 features were designed especially for low-powered portable devices, HTML5 is considered as the potential candidate for a universal application platform of IoT. In this sense, according to [4], in January 2015, W3C launched the "Web of Things Interest Group" to bridge different IoT platforms through the Web. Meanwhile, W3C established the "Web Bluetooth Community Group (CG)" to discuss and develop JavaScript API for Bluetooth. W3C also formed the "Wearable Web CG" to discuss technical specification issues for Web technology on wearable devices and Internet of wearable things environment. According to [4], both CGs are open to the public in order to attract more participation from experts in wearable devices.

The OIC was formed in 2014 by a number of global enterprises in order to develop specifications, certification and branding for interoperability between a plethora of devices. Later, it became OCF, and it has been maintained by The Linux Foundation. Unlike other standards development organizations, which are mentioned above, OIC/OCF solely focuses on IoT and its applications. In addition, to foster wide deployment of OIC standards, the consortium delivers open-source implementations that incorporate OIC specifications. The OCF aims to expand the IoT ecosystem by deploying the OIC/OCF standards and source codes that allows any products, which support certain wireless technologies, to communicate with any OIC/OCF-certified products. The OIC/OCF's Standards Work Group (SWG) handles approving specifications developed by their Task Groups (TG).

In June 2015, the Healthcare TG (HCTG) was formed under SWG to evaluate use cases, derive interoperability, legal, regional requirements and develop profiles and technical specifications where necessary for the healthcare vertical within the general framework of OCF, by specifying technical basis of the core framework specification which had built on the scheduled to be published in late 2015. The core framework is a vertical agnostic core architecture; standards for healthcare domain has developed resource model for various healthcare sensor or devices until now. Examples of IoT open-source framework implementing of all OCF applications including not only healthcare but also smart home, industrial, automotive and so forth. Since wearable devices are fundamental to healthcare applications, it is expected that use cases, requirements, capabilities and resources associated with wearable devices will be considered as an essential part of the healthcare specification.

According to [5], the essential element of OCF standard which should be implemented to add new vertical domain is "resource model" for that domain. HCTG which develops OCF standards for healthcare domain has developed resource model for various healthcare sensor or devices until now. The authors from [5] mention that OCF standards are implemented by IoTivity, an IoT open-source framework. The main goal of IoTivity is to provide seamless device-to-device connectivity regardless of kinds of operating system or communication protocol to satisfy various requirements of IoT. IoTivity is distributed with Apache license 2.0, thus anyone can use it but revealing source codes based on it is not mandatory. IoTivity supports various connectivity technologies: Wi-Fi, Ethernet, bluetooth low energy (BLE), near field communications (NFC) and so on. It works on various operating systems: Linux, Android, Arduino, Tizen and so on. Details are provided in [5], where extensive lists of healthcare devices and resources are presented.

15.2 Final overview

While conceiving the book structure, the goal of the editors and co-authors was to put together the efforts, research and developments attempts, and results of more than one decade of research on wearable technologies, IoT and WBSNs for healthcare within their own groups. The aim is sharing efforts, approaches, failures and achievements in this interdisciplinary field of investigation and inspire new generations of students, researchers and practitioners, who will make use of these outcomes as their starting point while extending this very promising field towards innovation.

In this book, besides Chapter 1, the following main parts (and chapters) have been presented:

- Overview of WBSN communication applications and scenarios (Chapters 2 and 3)
- Devices and systems (Chapter 4)
- Textile materials for wearable applications (Chapter 5)
- Propagation aspects and cognitive radio (CR) (Chapters 6 and 7)
- Link layer, media access control (MAC) sub-layer and synchronization aspects (Chapters 8 and 9)
- Applications of wearable technologies and WBSN (Chapters 10–15)

Chapter 2 provided an overview of sets of WBSN applications, started by presenting taxonomies to the classification of applications while introducing the main characterization parameters and highlighting aspects of their standardization within IEEE 802.15.6. RF-EH is becoming a mature technology with the enormous potential to realize the vision for uninterrupted self-powered networks of biomedical sensors, capable of perpetual monitoring of patients vital signs. Along this way, several important observations have been reported, which they are summarized and considered. An IC has also been introduced which is designed to use in low-power wearable systems. The aforementioned IC integrates most of network functionalities and reduces significantly power consumption and circuit size.

A wearable BAN system for both medical and non-medical applications, especially those including a large number of sensors at BAN scale (<250), embedded in textile and with high data rate ($<9+9$ MHz) communication demands was presented in Chapter 3. The overall system includes an on-body central processing module connected to a computer via a wireless link and a wearable sensor network. A wired network, composed of sensors, base node and conductive yarns, has been considered for the wearable components. From the standpoint of the network, each sensor node is a four-port router capable of handling packets from destination nodes to the base station. The communication module has been implemented in a low power field programmable gate arrays and a microcontroller. The measurement results obtained from the prototype show that the proposed circuits can work up to 18 Mbps which is high enough to satisfy most of BAN application requirements. The ability of the circuit to detect low-quality signals in the communication lines, low power operation, high number of sensors, high data rate and reliable structure make it a suitable system for use in wearable applications with conductive yarns. The proposed BAN system uses the SRMCF routing protocol, which has also been described in this chapter. SRMCF is energy-efficient, reactive and does not require routing tables at every sensor node. Experimental and simulation results show that minimum cost forwarding, failure recovery mechanism, absence of equal cost paths problem and improved performance make SRMCF a good candidate for energy-efficient, practical BAN applications.

Chapter 4 analysed the state-of-the-art in wearable human activity recognition (HAR) systems with an application-oriented approach. Several real-world domains are identified that benefit from activity recognition; monitoring daily activity to support medical diagnosis, for rehabilitation, or to assist patients with chronic impairments was shown to provide key enhancements to traditional medical methods such as multiple sclerosis, diabetes, emotion and mood, heart-related treatments, osteoarthritis, pulmonary, obesity, autism and fall detection, as well as the sports and entertainment, social, military, child, game sectors that attract a great deal of interest. Application requirements including sensory information, activity types, and performance requirements (such as energy consumption, processing, accuracy, obtrusiveness and user flexibility) provide a baseline for the discussed solutions. A general architecture is first presented along with a description of the main components of any HAR system. A comprehensive background to the HAR procedures, designing, implementing and evaluating HAR systems have been provided. Sequence of techniques of the HAR chain, i.e. signal processing, pattern recognition, and machine-learning techniques have been described and some open problems and ideas have been highlighted.

Chapter 5 presented an overview of the textile materials commonly applied for wearable antennas and systems. Because off-the-shelf fabrics represent a common choice for antenna substrates, the characterization of their electromagnetic properties is essential to carry out an optimal antenna design and achieve excellent antenna performance. Therefore, a multitude of methods described in the literature to characterize textile materials have been discussed. Furthermore, a new resonator-based experimental technique was introduced to characterise several textile fabrics of interest. This method is based on the theory of resonance-perturbation, which consists in computing the electromagnetic parameters of the material under test, at a single frequency (2.25 GHz), by measuring the shift in the resonance frequency and in the value of the Q-factor of a microstrip patch antenna. During the dielectric characterization of textile and other related materials using the resonator-based experimental technique, it was observed that when positioning the rougher face of the material under test (MUT) in contact with the resonator board, the extracted ϵ_r values were lower than the ones extracted with the material upside-down. This effect is attributed to the amount of air that is confined between the textile and the resonator board because of the roughness of the fabric. Therefore, the experimental results presented in this chapter have mainly focussed on the roughness of the MUT and the consequent existence of pores on its surface. The resonator-based experimental method discussed in this chapter proved to be an efficient, simple and fast technique for the characterization of the electromagnetic properties of textile materials for wearable antennas and bodycentric communication.

Chapter 6 provided an intensive study of the human-movement identification using the radio signal strength in WBAN. Developing the human-movement identification system mainly consisted of three parts, which were data collection, data pre-processing and classifier training. The WBAN measurements were conducted to collect the WBAN channels for static and dynamic movements. After that, the WBAN channel data underwent preprocessing procedures, which involved mean normalization, data segmentation, feature extraction and feature scaling. Various conditions for the human movement identification system have been examined. It was found that a window size used in the data segmentation and a receiver location on human body had strong impact on the classification results. The classification results of different Tx–Rx pairs showed slightly different accuracy rate. The P(Y)–P(Y) channel provided slightly lower accuracy rate than the other pairs due to the large number of walking instances classified as running instances and vice versa, and the large number of sitting instances classified as walking or running instances. The investigation results presented in this chapter support that the accurate human-movement identification using only the radio-signal strength is feasible. This approach enables many context-aware applications in WBAN without additional installation of special hardware. However, it is worth noting that this human-movement-identification approach is based on the WBAN channel characteristics, which are highly influenced by a sensor location and an antenna orientation; hence, the classifier is limited to use in the specific location and the orientation for which it is trained.

Currently, MBANs are being introduced in unlicensed frequency bands, where the risk of mutual interference with other electronic devices can be high. Techniques developed during the evolution of cognitive radio can potentially alleviate these problems in medical communication environments. In addition, these techniques can help increase the efficiency of spectrum usage to accommodate the rapidly growing demand for wireless medical body area network (MBAN) solutions and enhance coexistence with other collocated wireless systems. A viable architecture of an MBAN with practical CR features based on ultra wideband radio technology is proposed in Chapter 7. Ultra-wide-band (UWB) signals offer many advantages to MBANs, and some features of this technology can be exploited for effective implementation of CR while considering aspects of RF-EH in the choices for protocol efficiency optimization. The physical and MAC layer aspects of the proposal were proposed in addition to the implementation challenges. Dynamic spectrum access can be used to mitigate spectrum scarcity. This is accomplished by enabling an unlicensed user (i.e. secondary user) to adaptively adjust its operating parameters and exploit the spectrum which is unused by licensed users (i.e. primary users), in an opportunistic manner.

An IEEE 802.15.4 MAC layer performance enhancement by employing RTS/CTS combined with packet concatenation has been proposed in Chapter 8. The use of the RTS/CTS mechanism improves channel efficiency by decreasing the deferral time before transmitting a data packet. The proposed solution has shown that, even for the case with retransmissions, if the number of TX packets is lower than 5 (i.e. the number of aggregated packets), IEEE 802.15.4 employing RTS/CTS combined with concatenation achieves higher values for the throughput in comparison to IEEE 802.15.4 with no RTS/CTS even for shorter packet sizes. The advantage comes from not including the backoff phase into the retransmission process like the IEEE 802.15.4 basic access mode (i.e. BE = 0). By comparing the analytical and simulation results, it can be concluded that there is a perfect match. This actually verifies the accuracy of the proposed retransmission model. Performance results for the minimum average delay, D_{min}, as a function of the number of Tx packets and by assuming a fixed payload size of 3 bytes (i.e. L_{DATA} = 3 bytes) show that, by using IEEE 802.15.4 with RTS/CTS with packet concatenation, for 5, 7 and 10 aggregated packets, D_{min}, decreases by 8%, 14% and 18%, respectively. For more than 28 aggregated packets, D_{min} decreases by approximately 30%.

Chapter 9 discussed a one-way method for synchronization at the MAC layer of nodes and a circuit based on that in a wearable sensor network. The proposed approach minimizes the time skew with an accuracy of half of clock cycle on average. The synchronization is based on one-way master-to-slave message exchange implemented in the MAC layer, in order to avoid the nondeterministic delays caused by data processing and buffering in the higher levels of the protocol stack. By directly sending and processing the timing information without buffering, the proposed approach leads to an average clock skew of a few nanoseconds. The circuit generates two synchronized values: a programmable clock signal and a real-time counter for time-stamping purposes. Experimental evaluation with IC implementation obtained an average one-hop clock skew of 4:6 ns which is the time required for signal propagation from sender output to the receiver input. Based on theoretical calculations, in a multi-hop network, the global average time skew grows linearly with hop count; this is supported by the experimental results. The low skew values provided by this approach satisfy the requirements of many BAN applications. Even for networks whose nodes are 10 hops away from the time reference node, the average global skew will typically be under 50 ns. A value of ten hops exceeds the largest inter-node distance of many, if not all, existing wearable systems. The proposed circuit achieves the maximum synchronization performance that could be achieved by PTP, but with fewer timing messages and calculations, less complexity and better energy efficiency.

Chapter 10 described the development, operation and characteristics of a wearable body sensor network for human locomotion data capture for gait analysis. Gait analysis and fall monitoring are two fields where this combination of technologies has been explored. That requires acquiring knowledge on the biomechanics of gait and human movement and on the characteristics and acquisition of electromyographic (EMG) signals, which are required to develop the kinematics models that govern the analysis and detection tools. The system explores technical fabrics to build a more comfortable and easier to manipulate e-leggings, being conductive yarns used to wire-up sensor nodes, as well as surface EMG signals. The architecture and main characteristics of the sensor node are described. Experimental results obtained in a biomechanics laboratory, for a real-time human locomotion data acquisition, are presented which show that the proposed body sensor network performance is equivalent to that of a commercial system with the advantage of consuming lower wireless transmission power and providing also the measurement of an additional quantity – angular velocity.

Chapter 11 described the case of eWALL, an e-Health environment designed to cater for patients with mild cognitive impairments or chronic obstructive pulmonary disease. It details the required software and hardware technological innovations to design an affordable, easy-to-install smart "caring home" cognitive environment, which "senses" intuitively the wishes and "learns" the needs of the person living in the home. As a result, eWALL provides unobtrusive daily support, notifying informal and formal caregivers when necessary and serving as a bridge to supportive services offered by the outside world. The eWALL platform integrates sensors, actuators, reasoners and the cloud to enable the patient-relevant personalisation.

Chapter 12 presented the VitalResponder (VR) platform which is wearable technology wireless devices that are connected to a body-area aggregator and are capable of acquiring medical quality electrocardiogram, body temperature, actigraphy and several environmental measures such as luminosity, temperature, humidity, pressure, altitude, toxic gases and Global Positioning System coordinates. The VR platform presented in different perspectives of its current implementation status has been an evolving system that is capable of fulfilling requirements of different first response scenarios to improve the security and well-being of professionals such as fire fighters (FFs) and police officers (POs). It relies on the synergies between innovative wearable sensing, geolocation, WBANs and ICT to be able to respond to those scenarios. The wearable sensing component has been evolving in two directions: (1) to accommodate ambient and physiological measures and (2) from a T-shirt form factor to a patch for the physiological variables collection. This component has already resulted in two patents that may be explored in the coming years. Team data communication has been a major topic of R&D within the project and a MANET built on IEEE 802.11 technology approach has been developed to provide the needed connectivity in the target scenarios. This solution is disconnection-tolerant and has shown excellent results. The ICT integration has been evolving to be more and more complete both at the data collection and at the different data visualization components and has shown to be highly adaptable to the different requirements our first responder partners have been imposing to the VR team. The platform has been intensively used for many hours of data collection in drills, testing protocols and in real operations (e.g. fire-fighting in the summer season). These data have been used for both stress and fatigue and exposure level metric studies in order to obtain new requirements to iteratively develop the different components of the system so that it reaches its current state. This project is continuously evolving and currently a business concept model is being designed in order to bring to the market the results achieved with this project, taking into account a previous study on supply chain strategy for the specific product targets of the project.

In Chapter 13, an easy-to-wear belt with a telemedicine system for continuous monitoring of the foetal health in the last weeks of pregnancy was proposed. Solutions applying different types of sensors have been discussed.

An RF-EH and storing was presented in Chapter 14, which allows for collecting a small amount of energy. An energy-management algorithm and a supercapacitor storing system have been introduced that allow storing the energy harvested from the electromagnetic waves by means of N-stage Dickson voltage multiplier PCB boards (5 and 7 stages). Regarding the RF-EH device, the GSM 900/1800 bands provide the highest power density for RF-EH in urban and suburban environments. Improvement in the conversion efficiency and sensitivity of rectennas for these frequency bands is one of the biggest challenges. By comparing the obtained results with the ones for the DTT RF-EH, it is noticeable that the RF-EH device for the GSM band presents an output voltage and conversion efficiency higher than the DTT RF-EH device. A multiband GSM and DTT RF energy harvester block concept also has proposed that combines the energy harvested from an RF Energy harvester for the GSM band jointly with the energy harvested from an RF Energy harvester for the DTT band.

Chapter 15 not only gives a final overview of the main topics covered in the book but also presents a taxonomy for the classification of wearable devices that was proposed, and the literature review identified papers whose authors applied SWOT analysis to this market by examining their strengths, weaknesses, opportunities and some of the major threats they face [2]. Their work gives insights on the primary areas of innovation in wearable healthcare, which include infant safety and care, elderly care, chronic disease management, military support, sports medicine and preventive medicine. Among the several identified obstacles, one particular concern is the interoperability with other wireless systems, not only interference constraints but also aspects of data collection and tracking in healthcare systems. Not only HITECH act has helped drive the standardization of healthcare data terminology and encourage system interoperability [3] but also ITU-T, ISO/IEC JTC 1, W3C and OIC/OCF have ongoing works in relation to wearable devices and their standardization [4]. This is certainly vital to the widespread implementation and success of the wearable healthcare technology, as it is an important step towards enhancing the interaction of WBANs with traditional healthcare databases and EHR systems.

References

[1] Alrige M., Chatterjee S. 'Toward a taxonomy of wearable technologies in healthcare'. *International Conference on Design Science Research in Information Systems*; Springer; 2015. pp. 496–504.

[2] Casselman J., Onopa N., Khansa L. 'Wearable healthcare: Lessons from the past and a peek into the future'. *Telematics and Informatics*. 2017;34(7): 1011–1023.

[3] Davis Z., Khansa L. 'Evaluating the epic electronic medical record system: A dichotomy in perspectives and solution recommendations'. *Health Policy and Technology*. 2016;5(1):65–73.

[4] Cha H., Lee W., Jeon J. 'Standardization strategy for the Internet of wearable things'. *2015 International Conference on Information and Communication Technology Convergence (ICTC)*; 2015. pp. 1138–1142.

[5] Lee J.C., Jeon J.H., Kim S.H. 'Design and implementation of healthcare resource model on IoTivity platform'. *2016 International Conference on Information and Communication Technology Convergence (ICTC)*; IEEE; 2016. pp. 887–891.

Index

abacus ring 4

access points (APs) 45, 122

accuracy rate 187–8, 199–202, 204, 206, 551

ACK/NACK packet 253

acknowledgement (ACK) control frames 230

activities of daily living (ADL) 89, 361–2, 418

Activity Trackers 7

adaptive and QoS routing 227

adaptive frequency hopping (AFH) algorithm 56, 221, 332

ADDBA (Add Block Acknowledgement) 231

additional "emergency" channel (AEC) 216

ad hoc-based architecture 122–4

AdHocDroid MANET 397

advanced encryption standard (AES) encryption algorithm 121, 124

aggregator 399

Agilent 85070E dielectric measurement probe kit 169–70

agile spectrum access 220

ambient assisted living (AAL) systems 362, 364

AmbiUnit 394–5, 399

AMQP (Advanced Message Queuing Protocol) 398

Analogue-to-Digital Converter (ADC) 420

antenna, defined 525

ANT™ protocol 332

Apple watch 7

applications for wearable technologies and WBSNs with energy harvesting 31

classification of applications and characterization parameters 33–7

integrated circuit for low power wearable system 46–8

medium access control (MAC) 44

double stage MAC for radio cognitive networks 44–5

opportunistic RF-EH for WBAN 45–6

multi-band RF-EH solutions in the GSM 900/1800 bands 40–2

RF energy harvesting (RF-EH), spectrum opportunities for 37–8

single-band RF-EH solutions 39

supercapacitor-based energy storing system 42–4

application-specific IC (ASIC) 39

Arduino 369–70, 469, 481, 485–7, 489–90

Artificial Neural Networks (ANN) 113

automatic gain controller (AGC) 56

Automatic Goal Setting 378

average backoff window 247

BACK mechanism 231, 265, 267, 282

with BACK Request 231–2

backoff exponent (BE) 245

base station (BS) 53, 62, 103, 294, 457–8

Bayesian classifier 98, 113

Bayesian Network (BN) classifier 113

beacon-enabled networks 246, 254